MOLLUSCS AS CROP PESTS

Molluscs as Crop Pests

Edited by

G.M. Barker

Landcare Research
Hamilton
New Zealand

CABI *Publishing*

CABI *Publishing* is a division of CAB *International*

CABI Publishing
CAB International
Wallingford
Oxon OX10 8DE
UK

Tel: +44 (0)1491 832111
Fax: +44 (0)1491 833508
Email: cabi@cabi.org
Web site: www.cabi-publishing.org

CABI Publishing
10 E 40th Street
Suite 3203
New York, NY 10016
USA

Tel: +1 212 481 7018
Fax: +1 212 686 7993
Email: cabi-nao@cabi.org

A catalogue record for this book is available from the British Library, London, UK.

Library of Congress Cataloging-in-Publication Data
Molluscs as crop pests / G.M. Barker.
p. cm.
Includes bibliographical references.
ISBN 0-85199-320-6 (alk. paper)
1. Mollusks. 2. Agricultural pests. I. Barker, G.M.

SB998.M64 M65 2002
632′.643--dc21 2001037839

ISBN 0 85199 320 6

Typeset by AMA DataSet Ltd, UK.
Printed and bound in the UK by Biddles Ltd, Guildford and King's Lynn

Contents

The colour plate section can be found following p. 244

Contributors

Keith L. Andrews, Entomology Department, Forbes 410, Box 210036, University of Arizona, Tucson, AZ 85721-0036, USA

Stuart E.R. Bailey, School of Biological Sciences, University of Manchester, Oxford Road, Manchester M13 9PT, UK

Geoff H. Baker, CSIRO Entomology, GPO Box 1700, Canberra, ACT 2601, Australia

Gary M. Barker, Landcare Research, Private Bag 3127, Hamilton, New Zealand

Robert A. Byers, *Current address*: Department of Entomology, 501 ASI Building, Pennsylvania State University, University Park, PA 16802, USA

Rafael Caballero, Entomology Department, Forbes 410, Box 210036, University of Arizona, Tucson, AZ 85721-0036, USA

Chia-Pao Chang, Taiwan Miaoli District Agricultural Improvement Station, 42 Min-Chu Road, Ta-Hu, Miaoli, Taiwan

Robert H. Cowie, Center for Conservation Research and Training, University of Hawaii, 3050 Maile Way, Silmore 409, Honolulu, HI 96822, USA

Mieke Daneel, Institute for Tropical and Subtropical Crops, Private Bag X11208, Nelspruit 1200, South Africa

Karen de Jager, Institute for Tropical and Subtropical Crops, Private Bag X11208, Nelspruit 1200, South Africa

Chen De-niu, Institute of Zoology, Academia Sinica, Beijing 100080, China

Albert Ester, Applied Plant Research, PO Box 430, 8200 AK Lelystad, The Netherlands

David M. Glen, IACR – Long Ashton Research Station, Department of Agricultural Sciences, University of Bristol, Long Ashton, Bristol BS41 9AF, UK

Zhang Guo-qing, Institute of Zoology, Academia Sinica, Beijing 100080, China

Ronald B. Hammond, Department of Entomology, Ohio Agricultural Research and Development Center, Ohio State University, Wooster, OH 44691, USA

Ian Henderson, 30 Barlings Road, Harpenden AL5 2AP, UK

Gérard Hommay, UR Biologie des Interactions Virus Vecteur, INRA, 28 rue de Herrlisheim, 68021 Colmar, Cedex, France

Wu Jiqing, Shaanxi Plant Protection Institute, Yanglin, Shaanxi Province, China

Rina Kaminsky, Dirección de Investigación Científica, Universidad Nacional Autónoma de Honduras y Laboratorio de Parasitología, Hospital-Escuela, Apartado Postal 1587, Tegucigalpa, Honduras, Central America

Wang Man-sheng, Shaanxi Plant Protection Institute, Yanglin, Shaanxi Province, China

Robert Moens, Département Lutte biologique et Resources phytogénétiques, CRA Chemin de Liroux, 2, B-5030 Gembloux, Belgium

Gordon Port, Department of Agricultural and Environmental Science, University of Newcastle, Newcastle upon Tyne NE1 7RU, UK

S.K. Raut, Department of Zoology, University of Calcutta, 35 Ballygunge Circular Road, Calcutta 700019, India

Alfredo Rueda, Escuela Agrícola Panamericana, Zamorano, PO Box 93, Tegucigalpa, Honduras, Central America

Nick J. Sakovich, University of California Cooperative Extension, 669 County Square Drive, Suite 100, Ventura, CA 39003-5401, USA

Graeme Sanderson, Agricultural Research and Advisory Station, NSW Agriculture, PO Box 62, Dareton, NSW 2717, Australia

Willem Sirgel, Zoology Department, University of Stellenbosch, Private Bag X1, Matieland 7602, South Africa

Mark M. Stevens, NSW Agriculture and Cooperative Research Centre for Sustainable Rice Production, Yanco Agricultural Institute, PMB Yanco, NSW 2703, Australia

Rita Triebskorn, Animal Physiological Ecology, University Tübingen, Tübingen, Germany, and Steinbeis Transfer Center Ecotoxicology and Ecophysiology, Kreuzlingerstr. 1, 72108 Rottenburg, Germany

Xu Wenxian, Shaanxi Plant Protection Station-in-Chief, Xian, Shaanxi Province, China

Cheng Xingmin, Shaanxi Plant Protection Institute, Yanglin, Shaanxi Province, China

Liu Yanhong, Shaanxi Plant Protection Station-in-Chief, Xian, Shaanxi Province, China

Preface

Molluscs have been largely neglected in the pest-control literature, and yet gastropod mollusc species currently constitute some of the most significant and intractable threats to sustainable agriculture. Instances of crop losses from herbivorous gastropods have been reported throughout recorded history. However, the 20th century witnessed the emergence of gastropods as important crop pests in temperate and tropical regions. The increased pest status has been associated with cultivation of new crops, intensification of agricultural production systems, and the spread through human trade and travel of species adapted to these modified environments. Furthermore, in some crops, the significance of gastropods is only now becoming apparent with the decline in the importance of other pest groups, such as insects, for which effective control strategies have been developed.

Two chapters review progress towards the development of chemical control strategies, one addressing the toxicology of chemicals, the other on the deployment of molluscicides in baits. These chapters highlight the substantive progress made in identification of molluscicidal chemicals and the development of formulations for field use against gastropod pests of crops. These chapters also serve to emphasize that statistically – and biologically – robust procedures for the screening and evaluation of pesticides are not yet well entrenched as standard practice in applied malacology.

A series of chapters focus on specific crop situations, providing a synopsis of the current pest status of gastropod species or species groups and progress towards the development of solutions. These pest-oriented chapters highlight the emergence of gastropod species as pests when the natural environment is disturbed for the establishment of crops or associated with invasive species establishing in new areas. There has, however, been little consideration of the disturbance or loss of the natural regulatory agents that has led to the eruptive population biology of the pest species. Indeed, only in a limited number of cases has there been any attempt to incorporate an understanding of population dynamic processes into the development of pest-management strategies.

The use of the generic terms 'snails' and 'slugs' is widespread in the applied malacology and popular literatures, but has been discouraged in the preparation of the chapters in this book. 'Snail' refers to a gastropod possessing a fully developed shell, capable of housing the retracted animal. 'Slug' refers to the gastropod body form where the shell is reduced to the extent that it is no longer capable of housing the animal. The slug form has evolved many times in gastropods, and many taxa of widely divergent origins have, by parallel evolution, assumed a remarkable similarity in body form. Slugs are simply snails with a reduced shell, and therefore not a natural group of closely related animals. While there are some behavioural-ecology differences between animals towards the extremities of the snail–slug body form continuum, the use of these terms greatly de-emphasizes the tremendous diversity in biologies, ecologies and pestiferousness among species. An applied medical entomologist would not contemplate addressing Anophelinae mosquitoes (Culicidae) and Phlebotominae sandflies (Psychodidae) under the generalized term 'flies', and yet we repeatedly see the complex of gastropod species in the phylogenetically disparate families Agriolimacidae, Arionidae and Milacidae, infesting cereal crops in Europe, referred to as slugs. The arbitrary subdivision of gastropods into snails and slugs and the failure to recognize the individuality of species in terms of physiologies, behaviours and ecologies have been major barriers to understanding the ecology of pest infestations and in the development of effective management strategies.

The nomenclature of taxa has been standardized throughout this book, to reflect recent taxonomic revisions. While some names remain contentious – for example, the adoption of *Cantareus aspersus* (Müller) for the species long referred to as *Helix aspersa* Müller – the need for a consistent, modern usage across chapters outweighed the preferences of individual authors. The most important of these nomenclatural updates are cross-referenced in the index.

1 Chemical Control of Terrestrial Gastropods

Ian Henderson[1] and Rita Triebskorn[2, 3]

[1]*30 Barlings Road, Harpenden, UK;* [2]*Animal Physiological Ecology, University Tübingen, Tübingen, Germany;* [3]*Steinbeis Transfer Center Ecotoxicology and Ecophysiology, Kreuzlingerstr. 1, 72108 Rottenburg, Germany*

History of Chemical Control of Terrestrial Gastropods

'Snails, earwigs and all other creatures, hurt not the vines, nor the land nor the fruit of the trees, nor the vegetables . . . but depart into the wild mountains . . .'. So prayed the martyr Trypho in the 10th century AD (Taylor, 1894), confirming that terrestrial gastropods have had pest status since antiquity. Psychic control methods then appear to fall into abeyance, although the depredations continue, and, while the Reverend Gilbert White noted, in England in 1777, that slugs 'much injure the green wheat' (White, 1777), he did not invoke divine intervention.

The improved farming methods of the agricultural revolution produced better crops, and an increasing awareness of the need for pest control to secure these increases in output. In 1821, while riding near Newbury, William Cobbett observed 'a piece of wheat with cabbage leaves laid all over it at eight to ten feet from each other. It was to catch the slugs' (Cobbett, 1930). A shrewd commentator on the agricultural scene, he dismissed this early attempt at 'mechanical' control of pestiferous gastropods and stated that 'the only effectual way to destroy them is to sow lime, in dust and not slaked . . . at dusk . . . the slug is wet . . . the smallest dust of hot lime kills him, and a few bushels to the acre are sufficient'.

This early endorsement of chemical control is borne out in the subsequent development of control methods, with chemicals of one sort or another being deployed either as repellent coverings placed around plants, as dusts or sprays broadcast over the soil surface or, more latterly, as poisons incorporated into attractive foodstuffs as baits.

The chemicals thus used have been chosen largely empirically, the better ones being discovered by trial and error with whatever materials were to hand. Writers at the end of the 19th century and beginning of the 20th recommended readily available materials, such as lime, salt, soot and ashes (Theobald, 1895), carbolic acid and sawdust (French, 1906) and tobacco dust (Gahan, 1907). Other simple chemicals promoted as molluscicidal dusts and sprays included Bordeaux mixture (Lovett and Black, 1920), essentially a suspension of cupric hydroxide originally developed as a viticulturalists' fungicide, and a range of inorganic metal salts, such as aluminium sulphate (Durham, 1920), copper sulphate (Anderson and Taylor, 1926), potassium aluminium sulphate (Anon., 1930) and iron sulphate (Palombi, 1948). At about this time a variety of simple chemicals were also recommended as repellent barriers around plants, ranging from naphthalene (Massee, 1928) to corrosive sublimate (mercuric chloride) (Miles *et al.*, 1931). None were very effective and all were severely limited in their use by their phytotoxic effects: today's legislators would look askance at their widespread use on environmental grounds.

The use of poison baits seems to have first been suggested by Tyron (1899), who recommended baits containing the early fungicide Paris green (copper acetoarsenate), and many authors subsequently advocated baits containing copper and arsenical compounds. Other inorganic salts, such as barium, calcium and sodium fluorosilicate, were also advanced as stomach poisons (Shropshire and Compton, 1939). Baits incorporating derris root (*Lonchocarpus* Kunth spp.; Fabaceae), which contains the naturally occurring insecticide rotenone, were first suggested by Thompson (1928), marking a move toward baits deploying more complex organic poisons.

The first major advance in chemical control was made with the serendipitous discovery, *c.* 1934 in South Africa, of the molluscicidal properties of metaldehyde, a solid polymer of acetaldehyde, then on sale as a solid fuel for picnic stoves (Gimingham, 1940). In the UK it was first mentioned in the amateur gardening press (Hadden, 1936) and 4 years later was the most popular and generally recommended bait poison for use against terrestrial gastropod pests (Gimingham, 1940). Metaldehyde baits then remained the mainstay of terrestrial gastropod chemical control until the advent of carbamate-based baits in the 1950s.

Although undoubtedly more efficient than the chemicals used up to that time, metaldehyde-containing baits did not always protect crops successfully. The short field life of the baits and the need to synchronize applications with periods of pest activity reduced effectiveness, a problem only partly solved by later developments in carrier formulation. Consequently, the search for contact-acting molluscicides that could be applied as dusts or sprays, as were insecticides, fungicides and herbicides, still continued. Indeed, many of the chemicals so tested were themselves taken from these sources. The herbicide dinitroorthocresol (DNOC) successfully killed *Deroceras reticulatum* (Müller) (Agriolimacidae) in field trials in Belgium, but at an impracticably high rate (van den Bruel and Moens, 1958). Broadcast applications of the nitrogenous fertilizer calcium cyanamide were also shown to reduce damage by gastropods on a field scale, but again at 'fertilizer' rather than pesticide rates (300 kg ha^{-1})

(van den Bruel and Moens, 1956). Its use was further restricted by phytotoxic effects on young plants, and the material was never widely adopted.

The discovery, in Switzerland in the early 1940s, of the insecticidal properties of dichlorodiphenyltrichloroethane (DDT) (Lauger *et al.*, 1944) began the era of synthetic organic pesticides, and many were subsequently evaluated for molluscicidal activity. Often the results of tests for molluscicidal activity were conflicting. Buckhurst (1947) found that DDT gave excellent results in small plot trials, but Frömming (1950) found it ineffective in laboratory tests. Thomas (1949) thought benzene hexachloride (BHC) 'very poisonous' to *Deroceras agrestis* (Linnaeus), but Cleland (1952) found both DDT and BHC ineffective as seed dressings. In the USA, Schread (1958) obtained some control of the agriolimacid *D. reticulatum* and the limacid *Limax maximus* Linnaeus using dieldrin, while Stephenson (1959) claimed a reduction in damage to potatoes (*Solanum tuberosum* Linnaeus; Solanaceae) using aldrin in the UK. Getzin and Cole (1964) tested a number of pesticides against gastropods in laboratory tests and concluded that chlorinated hydrocarbons and organophosphates were poor molluscicides, with the exception of zinophos (*O,O*-diethyl-*O*-pyrazinyl phosphorothioate) and phorate (*O,O*-diethyl-*S*-ethylthiomethyl phosphorodithioate). The activity of phorate in granular formulation was then confirmed in small plot tests by Judge (1969) and as an emulsion in field trials on maize (*Zea mays* Linnaeus; Gramineae) by Barry (1969). However, despite the number of biologically active new molecules coming forward, no commercially successful molluscicides emerged from either the chlorinated hydrocarbon or the organophosphate classes of pesticides.

The next significant event was the detection of molluscicidal activity in another group of synthetic pesticides, the carbamates. In a screening programme started in 1954 to find a substitute for the arsenical baits used to control the introduced helicid snail *Cantareus aspersus* (Müller) in Californian citrus (*Citrus* Linnaeus spp.; Rutaceae) groves, Pappas and Carman (1955) obtained good results with baits containing isolan (1-isopropyl-3- methyl-5-pyrazolyl dimethylcarbamate). Other carbamates were subsequently shown to be effective against terrestrial gastropods. Ruppel (1959) demonstrated the activity of sevin (1-naphthyl-*N*-methylcarbamate) baits, although it appeared to be less effective against *D. reticulatum* than against the milacid *Milax gagates* (Draparnaud). In comparisons between metaldehyde baits and those containing various organochlorine, organophosphorus and carbamate insecticides, Getzin and Cole (1964) concluded that, while most were poor molluscicides, the carbamates were generally active. They reported good control with baits containing Bayer 37344 (4-methylthio-3,5-xylyl-*N*-methylcarbamate), later 'methiocarb', a synthetic molecule originally of interest because of its insecticidal and acaricidal properties (Unterstenhöfer, 1962). This compound was developed as a 4% a.i. bait formulation (Martin and Forrest, 1969) and under the trade names Draza and Mesurol achieved equal or greater commercial success then the metaldehyde-based baits already available. Other carbamate baits were subsequently developed, such as those incorporating cloethocarb (2-(2-chloro-1-methyloxy-ethoxy)phenyl methylcarbamate) (Harries *et al.*, 1980) and thiodicarb (3,7,9,13-tetramethyl-5,11-dioxa- 2,8,14-2,8,14-trithia-4,7,9,12-tetra-aza-pentadeca, 12-diene-6,10-dione) (Yang and Thurman, 1981). A recent variation on the theme of redeploying insecticides as active ingredients in baits for gastropod control is the product Malice, which contains bensultap (*S,S′*-2-dimethylaminotrimethylene di(benzenethiosulphonate)), an insecticide originally derived from a programme of chemical synthesis based on nereistoxin, a poison found in a species of marine annelid worm (Sakai, 1969). A related insecticide, cartap hydrochloride (*S,S*-(2-dimethylaminotrimethylene)bis(thiocarbamate)), applied to seeds reduced damage by *D. reticulatum* in wheat (*Triticum aestivum* Linnaeus; Gramineae) fields in southern England (Scott *et al.*, 1984) but was considered too toxic to birds to warrant commercial development.

The development of new, specifically molluscicidal chemicals was favoured by the World Health Organization, interested from the 1960s in controlling freshwater gastropods that act as intermediate hosts for trematodes causing human schistosomiasis. Trifenmorph (*N*-trityl-morpholine) was developed for this market and sold as Frescon (Boyce *et al.*, 1966), but also proved effective against the amphibious species, *Lymnaea truncatula* (Müller) (Lymnaeidae), in pasture and was employed to reduce transmission of the liver fluke, *Fasciola hepatica* Linnaeus (Fasciolidae) to domesticated ungulates (Crossland *et al.*, 1969). Its use, however, was never extended to terrestrial gastropod pests. Another synthetic molecule developed around this time for schistosomiasis control was niclosamide (2,5-dichloro-4-nitrosalicylanilide), subsequently marketed as Bayluscide (Gönnert and Schraufstätter, 1958). In recent years this molluscicide has been applied to the control of species of *Pomacea* Perry (Ampullariidae) (Schnorbach, 1995), which are increasingly important pests of rice (*Oryza sativa*) Linnaeus; Gramineae in the Asian region (Halwart, 1994; Cowie, Chapter 5, this volume).

More than 1000 plant species have been evaluated since 1933 as sources of naturally occurring chemicals to control aquatic gastropods involved in trematode transmission (Marston and Hostettman, 1985), and some 70 natural products with molluscicidal activity have been isolated. In some instances the active components are saponins and, interestingly, a synthetic surfactant, sodium dodecyl sulphate, has recently been found effective against *Pomacea* in rice (Tzeng *et al.*, 1994). While the search for naturally occurring molluscicides has tended to focus on aquatic gastropods as targets, some plant-derived compounds active against terrestrial pest species have also been discovered, although their commercial viability is yet to be established. Hussein *et al.* (1994) isolated a cardenolide, usharin, from an Egyptian desert shrub and found it highly toxic by contact to the helicid *Theba pisana* (Müller). Likewise, Hagin and Bobnick (1991) isolated 6-hydroxy-1,2,3,4-tetrahydro-*b*-

carboline-3-carboxylic acid from the couch grass *Agropyron repens* (Linnaeus) Beauvois (Gramineae). While inactive against three species of freshwater gastropods, it was toxic to *D. reticulatum*, *Deroceras laeve* (Müller) and in the arionid *Arion subfuscus* (Draparnaud) on contact and by ingestion in baits.

The early use of simple metal compounds to poison gastropod pests was unsuccessful, mainly because these materials were quickly dispersed when broadcast applied, or were too distasteful when incorporated in baits. Recently, it has been found that, by combining a metal with an appropriate organic ligand, molluscicides can be produced which are effective in bait formulations. Baits containing iron and aluminium chelates have been shown to be as effective as metaldehyde- and carbamate-based baits under field conditions (Henderson *et al.*, 1989, 1990). While the optimist might detect progress on this aspect of chemical control, the pessimist might feel that, after 170 years or so, it has come full circle.

Historically, therefore, chemical control of terrestrial gastropods began with the ad hoc use of whatever unsophisticated materials were to hand, progressed to the use of chemicals developed for other pest or parasitological problems and has only recently turned to the discovery of materials specifically for the purpose. The main constraint on progress has always been the difficulty of delivering the toxin to the target. To this is now added the increasing cost of developing new molluscicides, particularly environmental safety testing, which must be recovered from a relatively small market. In the immediate future it seems likely that chemical control will continue to depend on pesticidal molecules and formulations whose development and registration costs have already been covered by more extensive markets, or on chemicals of such limited toxicity that rigorous environmental safety testing is not required.

Modes of Action

Since molluscicides are applied either as baits or broadcast in solid or emulsified form, terrestrial gastropods encounter them via food uptake or by dermal contact, and the chemicals act either as stomach or contact poisons. Generally, stomach-poison activity increases with increasing lipophilicity in the molecule because of the lipophilic character of the plasma membranes that have to be penetrated. Another factor influencing the effectiveness of molluscicides delivered by the oral route is their stability under different pH conditions. In the digestive tract conditions range from pH 5.7 in the oesophageal crop to pH 7.9 in the distal intestine (Kelly *et al.*, 1996). For contact-action poisons, the relationship between lipophilicity, acidity and molluscicidal activity is less simple, since the molecules have first to penetrate the thick hydrophilic mucus layer and then to dissolve in the underlying lipophilic plasma membrane of the skin (Briggs and Henderson, 1987).

Ingested molluscicides are only partly transported down the digestive tract in the food mass, since a large proportion is quickly resorbed by the cells of the oesophagus (Triebskorn *et al.*, 1990). After being released from these cells into the haemolymph, molluscicidal molecules are distributed in the body by the haemolymph and are quickly transported to peripheral tissues and organs. Using radiolabelled metaldehyde and cloethocarb, it has been shown that, only a few minutes after ingestion by *D. reticulatum*, molluscicides have left the oesophageal-crop cells and entered the cells of the stomach and the digestive gland via the cell bases that abut the haemolymph space (Triebskorn *et al.*, 1990). Soon afterwards, metaldehyde could also be detected in the skin, the reproductive tract and the nervous system (R. Triebskorn, unpublished data). The rapid resorption of a metaldehyde metabolite, paraldehyde, by the cells of the digestive tract has also been demonstrated by Booze and Oehme (1985). Clark *et al.* (1995) and Triebskorn *et al.* (1999) have shown that iron chelate molluscicides also pass out of the digestive tract and into the haemolymph soon after ingestion, before finally becoming concentrated in the digestive gland. Unlike individuals treated with a non-toxic chelate, which accumulated large amounts of iron in the digestive gland, animals killed by a toxic chelate characteristically contained less iron in the digestive gland and more in the body wall and the reproductive tract.

Cells in which the molluscicidal chemicals or their metabolites can be localized after ingestion or dermal contact have to be considered as primary targets for such chemicals. Methods used to detect such primary targets include X-ray analysis, autoradiography, energy-filtering transmission electron microscopy (EFTEM), atomic absorption spectroscopy (AAS) (summarized by Triebskorn, 1995; Triebskorn *et al.*, 1996, 1999) and inductively coupled plasma atomic emission spectrophotometry analysis (ICPAES) (Bullock *et al.*, 1992). However, even in parts of the body that do not come into direct contact with the toxic molecule, pathological effects can be found which are related to the reaction of the animals' metabolism to the toxic conditions. These reactions may be considered as secondary reactions to the toxic input.

After dermal application, the primary targets for molluscicides are the epithelial cells of the skin, including the mucus cells (Triebskorn *et al.*, 1998). Henderson (1970) showed that reactions in the skin follow not only after direct contact with metaldehyde itself, but also after exposure to the vapour of its main metabolite, acetaldehyde. The gastropod skin is known to be involved in the acquisition and resorption of various molecules and free ions from the environment (Henderson, 1970; Zylstra, 1972; Machin, 1977; Ryder and Bowen, 1977; Bullock *et al.*, 1992). Chemicals passing through the skin reach the haemolymph and are thus transported throughout the body.

The mode of action of molluscicidal chemicals depends on their properties and is best understood for carbamates, such as methiocarb and cloethocarb. These act primarily as cholinesterase inhibitors (e.g. Casida,

1963). Perhaps the simplest observation on the mode of action is the desiccation effect caused by metaldehyde. Two other aspects of the mode of action of molluscicidal chemicals are their influence on metabolic enzymes and energy supply, and the detoxification processes that may limit the effectiveness.

Neurotoxic effects

Terrestrial gastropods poisoned by carbamates, such as methiocarb, soon become immobilized as the muscle tonus is lost (Mallet and Bourgaran, 1971; Godan, 1983). The carbamates behave as reversible, competitive inhibitors of acetylcholinesterase, which allow local accumulation of acetylcholine at organs innervated by cholinergic nerves (Metcalf, 1955; Casida, 1963), usually acting on the anionic and esteric binding sites of cholinesterase (Metcalf and Fukuto, 1965; Eldefrawi and Eldefrawi, 1990).

This interaction of carbamates with cholinesterases is not specific to gastropods as the same mode of action occurs in earthworms (Young and Wilkins, 1989b) and carabid beetles (Buchs *et al.*, 1989). Inhibition of cholinesterases in gastropods can also be caused by phosphoric acid esters (Pessah and Sokolove, 1983; Bakhtawar and Mahendru, 1987) and by bensultap, a sulphonate (Atger *et al.*, 1990).

Symptoms of metaldehyde poisoning in gastropods include increased mucus secretion, convulsions and paralysis (Booze and Oehme, 1986). Details of the mechanism by which metaldehyde or acetaldehyde causes convulsions are still unknown, but Booze and Oehme (1986) have shown that acetaldehyde acts as a releasing agent for noradrenaline and 5-hydroxytryptamine (serotonin) in gastropods. They suggest that metaldehyde may act as a releasing agent for the neurotransmitter GABA, thus acting directly on the central nervous system. This thesis is supported by the work of Homeida and Cooke (1982). Recent electron-microscope studies have revealed that mucus cells in the skin of *D. reticulatum* are innervated by neurons that can be stained with antibodies against serotonin and those binding to amino acid decarboxylase or dopamine-*b*-hydroxylase, two enzymes involved in the synthesis of serotonin and catecholamines. In animals treated with metaldehyde, this immuno-staining is much more intense than in controls, suggesting that metaldehyde may not only increase the release of serotonin but also its rate of synthesis and recycling (Triebskorn *et al.*, 1998). Neurotoxic effects may result in alterations of locomotion (Wedgwood and Bailey, 1988) and of feeding behaviour (Wright and Williams, 1980; Wedgwood and Bailey, 1986; Bourne *et al.*, 1988; Bailey, 1989; Bailey *et al.*, 1989), which are of particular significance when attempting to deliver the chemicals in baits. Wright and Williams (1980) suggested that the observed reduction in feeding caused by molluscicidal chemicals is due to a paralysing effect on the digestive-tract wall immediately after ingestion, terminating the feeding response. A direct effect on the nerves controlling the digestive tract may also be involved. Roach (1968) has shown that digestive-tract motility in *Arion ater* (Linnaeus) is largely independent of central control but is 'coordinated' by an extensive nerve plexus in the gut wall. Bailey *et al.* (1989) claimed that both metaldehyde and methiocarb shorten meals by interfering with the neural control of feeding, and that metaldehyde may also disable the muscles involved in feeding. The effects observed by Bailey *et al.* (1989) could also be due to the impact of metaldehyde on neuronal transmission, as demonstrated with *Lymnaea stagnalis* Linnaeus by Mills *et al.* (1990, 1991a,b,c), as metaldehyde induced bursting activity and paroxysmal depolarizing shifts in the motor neurons of the feeding system.

Desiccation effects

Increased mucus production followed by increased mucus secretion is one of the first reactions of gastropods to many kinds of stressors, including mechanical stimuli or chemical irritation caused by molluscicidal chemicals (Godan, 1983; Triebskorn and Ebert, 1989; Triebskorn *et al.*, 1998). One effect of the extruded mucus is to form a protective barrier preventing direct contact between the toxin and the epithelia of the skin or digestive tract, so reducing the toxicity of the chemicals (Port and Port, 1986; Triebskorn and Ebert, 1989). At the same time the mucus may also dilute the chemical. In some cases the mucus may even detoxify the chemical, provided that it is pH-sensitive or unstable under certain (usually acid) pH conditions, as are both metaldehyde and cloethocarb. On the other hand, the animals risk desiccation due to the high water content of the extruded mucus, and lose large amounts of ions entailing high energy costs for resynthesis (Triebskorn *et al.*, 1996, 1998). Port and Port (1986) claim that dehydration is the most frequent cause of death following metaldehyde poisoning in gastropods, and desiccating effects have also been noted in cases of molluscicide poisoning in vertebrates, such as dogs (*Canis domesticus* Linnaeus) (Canidae) and cats (*Felis catus* Linnaeus) (Felidae) (Maddy, 1975; Booze and Oehme, 1986). Increased mucus secretion following metaldehyde ingestion may also induce additional toxic effects, as the acidic mucus hydrolyses the metaldehyde to acetaldehyde, particularly in the anterior part of the digestive tract (Booze and Oehme, 1985).

Effects on metabolic enzymes and energy metabolism

Carbamates are not only cholinesterase inhibitors but in many cases are general inhibitors of esterases (Casida, 1963). They have been used to differentiate esterases and to elucidate the nature of the enzymatically active sites (Augustinsson, 1960). Tegelstrom and Wahren (1972) have shown that, while certain isoforms of esterases can be inhibited both by eserine and by methiocarb plus eserine, others are inhibited by methiocarb and not by eserine. It has also been shown that both cloethocarb and methiocarb can induce activity in non-specific esterases in the digestive gland of *D. reticulatum* within a few hours of treatment, although in the cells of the digestive tract these enzymes are inhibited (Triebskorn, 1991a). These apparently contradictory results may be due to the fact that non-specific esterases include carboxylesterases, arylesterases and acetylesterases, which may differ in sensitivity to carbamates and may be unevenly distributed in the digestive tract. The high esterase activity in the digestive gland, in molluscicide-treated individuals, may be due to the observed destruction of the (secondary) lysosomal system in digestive cells (Triebskorn and Künast, 1990). It may also arise through the release of enzymes into the cytoplasm and autolysis of the cells, as has been reported by Banna (1980) for aquatic species treated with trifenmorph. However, after metaldehyde ingestion, a strong increase in the activity of non-specific esterases has been found in the resorptive cells of the digestive gland (Triebskorn, 1991a), even though the lysosomal system showed less damage than after carbamate poisoning (Triebskorn, 1989).

A second intracellular digestive enzyme system, the acid phosphatases, which are found in primary lysosomes, has been shown to be influenced by molluscicide treatment in *D. reticulatum*. Enzymes of this group can catalyse the breakdown of ester bonds in orthophosphate esters under acid conditions and are involved in the attack on pyrophosphate bonds. They can also act as transphosphorylases. The activity of these enzymes has been shown to be completely inhibited after cloethocarb treatment, and strongly reduced after methiocarb and metaldehyde treatment (Triebskorn, 1991a). Inhibition of acid phosphatase by trifenmorph has also been demonstrated in *Bulinus truncatus* (Andouin) (Planorbidae) by Banna (1980). Kela and Bowen (1995) describe a reduction, but not a total inhibition, of acid phosphatases in the digestive gland of *L. stagnalis* after treatment with a naturally occurring molluscicide.

Reduction of an enzyme's activity may not be due to direct inhibition of the enzyme by a chemical, as described, for example, by Dauterman and Hodgson (1990). It might also be related to the cessation of protein synthesis, due to the effect of the toxin on the general metabolism of the animal, or be caused by the destruction of the membrane compartment in which the enzyme is located. On the other hand, activation of an enzyme may not necessarily be related to a toxin-specific induction of the respective enzyme synthesis or to a specifically induced increase in its activity. In many cases changes in enzyme activity levels are related to general metabolic responses accompanying increased mucus production and secretion or to other detoxification mechanisms (see below), which need large amounts of energy (Triebskorn *et al.*, 1996). In achieving acute toxicity, which is the goal for an effective molluscicide, the early phase of metabolic activation is usually followed by a decrease and finally a cessation of enzymatic activity coincident with cell and animal death. In this context, the observed molluscicide-induced alteration of metabolic enzyme activity in terrestrial gastropods can be interpreted, at least in part, as metabolic reactions reflecting different stages of intoxication–detoxication.

Since detoxification processes, including mucus production and extrusion, are energetically expensive, the required energy must be provided by increasing intracellular digestion and activating various enzyme systems. These include the above-mentioned intracellular digestive enzymes, and also the enzymes involved in transport processes, such as the alkaline phosphates or the ATPases. Alkaline phosphatases break down ester compounds of orthophosphate acids under alkaline conditions. A special type of alkaline phosphatase with a pH optimum of 7.5 is capable of catalysing the hydrolytic breakdown of ATP. This is the cell-membrane ATPase, the activity of which depends on the presence of sodium and potassium ions. After treatment with both carbamates and metaldehyde these transport enzymes are initially activated, but are then totally inhibited (Triebskorn, 1991a; Triebskorn *et al.*, 1996). Correlated in many cases with this activation of transport enzymes after exposure to molluscicidal chemicals is the reduction in intracellular energy stores and increased biotransformation (Triebskorn *et al.*, 1996). Kela and Bowen (1995) observed a shift of alkaline phosphatase from the brush border of the digestive gland to the lumen and a marked inhibition of ATPase activity in animals treated with a naturally occurring plant molluscicide.

Effects of detoxification

Detoxification processes are reactions to limit damage by toxins or to excrete them. With regard to the efficacy of molluscicides, detoxification reactions are obstacles to be overcome. Detoxification includes: (i) enzymatic reactions leading to the biotransformation of chemicals by oxidation (phase I reactions); (ii) hydrolysis or conjugation (phase II reactions); and (iii) several other biochemical processes, such as the induction of stress proteins or metallothioneins involved in direct interactions with the toxins and maintenance of intracellular homoeostasis. The general aim of all these processes is either to facilitate excretion of the toxin or its transformation into a non-toxic storage product, which is especially important for metal poisons.

On hydrolysis, the main breakdown product of metaldehyde is acetaldehyde, which in turn may be reduced to ethanol with nicotinamide adenine dinucleotide (NADH), the reaction being catalysed by alcohol dehydrogenase. Although oxidative processes are not involved in this degradation, a significant increase of several mixed-function oxygenases (phase I enzymes, MFO enzymes) has been observed in metaldehyde-treated *D. reticulatum* (Triebskorn, 1991a). This is possibly due to an indirect impact of the molluscicide on oxygenases, which are normally involved in the metabolism of steroid hormones.

Carbamates are normally detoxified or biotransformed by oxidation of the phenol ring and *N*-methyl group by microsomal oxidases, by hydrolysis and by several types of conjugation (formation of glucuronides, sulphates and glucosides) (Wegler, 1970; Wilkinson, 1976). One of the most important metabolic pathways of the *N*-methyl- and the *N,N*-dimethylcarbamates is the *N*-methyl hydroxylation of the carbamyl moiety (Wilkinson, 1976). Metcalf and Fukuto (1965) claim that aliesterases may be involved in detoxification of carbamates, while Gordon and Eldefrawi (1960) speak of 'carbamate esterases'. However, most of our knowledge of carbamate biotransformation is restricted either to vertebrates or to insects, and there is little information on terrestrial gastropods. Photometric measurements of enzyme activities in the digestive gland of *D. reticulatum* treated with methiocarb have revealed the induction of two enzymes involved in the oxidative degradation of carbamates: arylhydrocarbon hydroxylase and NAD phosphate (NADPH)-neotetrazolium reductase. No such induction was found when the animals were treated with cloethocarb, but activation of the enzyme glucose-6-phosphate dehydrogenase, which catalyses the formation of the $NADPH_2$ needed for the above reaction, has been detected (Triebskorn, 1990).

The role of the digestive gland in the biotransformation of molluscicides is very important. In experiments with *D. reticulatum* using radiolabelled cloethocarb and metaldehyde, radioactivity accumulated in the basophilic cells of the digestive gland a short time after application of the chemicals (Triebskorn *et al.*, 1996; R. Triebskorn, unpublished data), and was accompanied by increased activity of the phase I detoxification enzymes in the same cells. Ultrastructural changes in the basophilic cells have also been shown to follow exposure to metaldehyde and carbamate (Triebskorn, 1989; Triebskorn and Künast, 1990), and similar changes have been related to an induction of phase I detoxification enzymes in vertebrates by Klaunig *et al.* (1979). The metal residue deriving from isotopically labelled iron chelate molluscicides applied to *D. reticulatum* was found concentrated in the digestive gland irrespective of the chelate used (Clark *et al.*, 1995), and its localization in the digestive cells was confirmed using EFTEM (Triebskorn *et al.*, 1996, 1999). The observations fit well with the known role of the digestive gland of gastropods in metal detoxification (e.g. Dallinger *et al.*, 1989; Marigomez and Dussart, 1996).

Cytological Effects

Cytological effects of chemicals of known molluscicidal activity have been investigated in various tissues by light- and electron-microscope studies (Ryder and Bowen, 1977; Triebskorn, 1989, 1991b, 1995; Kela and Bowen, 1995; Rondelaud and Dreyfuss, 1996; Triebskorn and Köhler,

1996; Triebskorn *et al.*, 1998, 1999). Light-microscope investigations are useful in detecting gross structural effects on target tissues, such as the lesions induced in the oesophageal crop of *D. reticulatum* by cloethocarb (Triebskorn and Florschütz, 1993) or necrosis in the skin after oral or dermal application of metaldehyde (Triebskorn *et al.*, 1998). Other changes demonstrated include alterations in glycogen and lipid content in the oesophageal crop and digestive gland, and qualitative and quantitative differences in mucus induced by molluscicides (Triebskorn *et al.*, 1998).

Transmission electron-microscope studies have shown that cellular damage caused by molluscicidal chemicals is most pronounced in the following cell types: (i) mucous cells in the skin, digestive tract and salivary gland; (ii) digestive cells of the digestive gland; (iii) basophilic cells of the digestive gland; and (iv) epithelial cells of the oesophageal crop (Triebskorn, 1989, 1991b; Triebskorn and Ebert, 1989; Triebskorn and Künast, 1990: Triebskorn *et al.*, 1998). These effects are discussed in more detail below.

At the subcellular level, different organelles exhibit different spectra of reaction (Köhler and Triebskorn, 1998). The endoplasmic reticulum (ER) emerges as a very sensitive organelle in that it reacts to low dose levels of chemicals soon after exposure. Alterations such as dilatation, degranulation and vesiculation of the cisternae and formation of concentrically arranged cisternae occur, probably due to the induction of detoxification enzymes rather than to pathological effects of the chemical. Mitochondria are also sensitive to many stressors and rapidly exhibit gross pathological symptoms, such as swelling, shrinkage, disruption of cristae or the formation of intramitochondrial cristae (Ghadially, 1988; Köhler and Triebskorn, 1998). The observed changes in the mitochondria may be due to interactions with the lipophilic components of the mitochondrial membrane disrupting ion transport.

In gastropods treated with molluscicides the most noticeable effects on the Golgi bodies are seen in the mucocytes: these include dilatation of the cisternae and an increase in their number, increased vesicle formation and the breakdown of the Golgi-system membranes (Triebskorn *et al.*, 1998).

Mucocytes of the digestive tract and skin

One type of mucous cell predominates in the oesophagus (including the crop), intestine and salivary gland of *D. reticulatum*. It is pear-shaped and characterized by a basally located nucleus, large cisternae in a wide, laminar, granular ER, large Golgi complexes and mucous vacuoles, which coalesce in the apical region of the cells. Three types of mucous cells can be distinguished in the skin:

1. Pedal gland cells which are located in the sole. The ultrastructure of this cell type is comparable to that of the mucous cells of the digestive tract and the secretory product varies in appearance from spotted and electron-dense to fluffy and less electron-dense.
2. Mantle gland cells are the most numerous type of mucous cell found in the skin of *D. reticulatum*, occurring in the mantle, dorsal epidermis and foot. They are characterized by their prominent Golgi apparatus, consisting of ten to 20 stacked lamellae, and large mucous vacuoles containing slightly electron-dense material. In mature cells they displace the nucleus and cytoplasm to the periphery.
3. Peripodal gland cells, which are club-shaped cells located in the foot fringes. This cell type is characterized by a basally located nucleus and by the very long cisternae of the rough ER.

In all types of mucous cells in the skin and digestive tract, chemical-specific and dose- and time-dependent responses to molluscicides occur, which reflect either the increased demand for mucus (phase 1) or pathological effects (phase 2) (Triebskorn *et al.*, 1998). Dilatation, vesiculation or degranulation of the ER, dilatation of the Golgi membranes and swelling of mucous vacuoles are all phase 1 reactions to molluscicide exposure. Phase 2 is characterized by the general breakdown of the ER membranes, Golgi systems and vacuoles, and by swelling or bursting of the mitochondria, karyolysis and, in some instances, autolysis of complete cells (Triebskorn *et al.*, 1998).

When cells are compared after similar exposure times, cellular damage is generally less severe after exposure to carbamates than to metaldehyde. Histological preparations of *D. reticulatum* tissues made after carbamate ingestion show that mucocytes can alter the type of mucus produced, as their immediate reaction is to secrete large amounts of acidic mucus into the digestive-tract lumen and then to resynthesize large amounts of neutral mucus within the cells (Triebskorn *et al.*, 1998). Increased secretion into the digestive tract can also be seen immediately following metaldehyde ingestion. Subsequently, however, the destruction of the Golgi apparatus and ER brings mucus production to a halt.

Cells of the digestive gland

Three cell types can be distinguished in the digestive gland of *D. reticulatum*:

1. Columnar digestive cells, which dominate the epithelium in untreated animals.
2. Cone-shaped crypt cells, which have secretory functions and are characterized by having large amounts of granular ER, numerous Golgi bodies and secretory vesicles. These cells increase in number under toxic conditions (Cajaraville *et al.*, 1990).

3. Goblet-like excretory cells, characterized by large vacuoles containing electron-dense material.

Cytological effects induced by carbamates and by metaldehyde have been recorded, particularly in the first two types of cell (Triebskorn, 1989; Triebskorn and Künast, 1990; Triebskorn *et al.*, 1996). In the digestive cells of molluscicide-treated animals, pinocytosis ceased and the microvilli were shortened. This was followed by increased coalescence of vacuoles and their fusion with lysosomes, indicating the activation of intracellular digestive processes. Finally, the cells became dominated by large vacuoles and large secondary lysosomes. Increased vacuolization of digestive cells in *L. stagnalis* following treatment with a naturally occurring molluscicide has been reported by Kela and Bowen (1995), and in *D. reticulatum* exposed to various metals (Triebskorn and Köhler, 1996). Generally, after molluscicide treatment, lysosomes containing acid phosphatases are not found near the apices of the digestive cells, as they are in the cells of untreated animals, and glycogen storage is drastically reduced.

In the basophilic digestive-gland cells of molluscicide-treated *D. reticulatum*, structural modifications of the ER have been observed which appear to be correlated with the induction of enzymes involved in biotransformation and cellular transport. All chemicals showing molluscicidal activity examined so far have significantly reduced the storage products in the basophilic cells. Epithelial necrosis in the freshwater gastropods *Lymnaea glabra* (Müller) and *Bellamya dissimilis* Müller (Viviparidae) has been reported after the application of the molluscicide niclosamide and the pesticides endosulfan, methylparathion, quinalphos and 2,2-dichlorovinyldimethylphosphate (DDVP) (Jonnalagadda and Rao, 1996; Rondelaud and Dreyfuss, 1996).

Epithelial cells of the oesophageal crop

Apart from a few cells that are translucent under electron microscopy, the epithelium of the oesophageal crop is dominated by columnar storage cells, which are characterized by large amounts of lipid and carbohydrate. Shortly after exposure to metaldehyde, to pesticides such as carbamates, pentachlorophenol and lindane or to environmental pollutants such as heavy metals, these storage products are greatly reduced. When the toxic load exceeds a certain limit, epithelial cell damage occurs and pathological symptoms, such as the bursting of mitochondria and the disruption of the ER membranes and the microvillous border, appear (Triebskorn *et al.*, 1996). Morphological damage to the epithelial cells of the oesophageal crop has also been described by Bourne *et al.* (1991), following ingestion of sublethal doses of metaldehyde and methiocarb. These authors demonstrated that endocytosis in the oesophageal crop was also impaired. Manna and Ghose (1972) found that, in

Achatina fulica Bowich (Achatinidae) treated with endrin, cell damage within the digestive tract was most severe in the intestine and rectum, while pronounced shrinkage of muscles occurred in the oesophageal crop.

Techniques for Evaluating Molluscicidal Activity

Many comparative studies of molluscicidal activity have been published. The diversity of methods used indicates the perceived difficulty of evaluating chemicals in a way relevant to their ultimate deployment in the field. Evaluation methods can be conveniently thought of as falling into three categories, namely laboratory experiments, terrarium trials and field trials. Laboratory experiments under controlled conditions establish whether a candidate compound has any useful activity and indicate its likely mode of action so that an appropriate formulation for delivery can be chosen. This is often followed by tests of the chosen formulation in enclosed terraria designed to resemble field conditions, but still confining the animals to allow accurate assessments of mortality and detailed observation of behaviour. The definitive assessment of a molluscicide is its performance in the relevant cropping situation. Gastropod numbers and associated crop damage are compared on treated and untreated areas in a field trial. The size of plots is determined by the post-treatment assessments planned and the design is replicated to facilitate statistical analysis (Anon., 1986).

Laboratory experiments

The first methodical examination of a molluscicidal chemical was probably that of Cragg and Vincent (1952), in which the effect of metaldehyde on *D. reticulatum* was investigated. The animals were confined with metaldehyde, but not in contact with it, to check for fumigant activity. The animals were also held on filter-paper soaked in an aqueous solution and dusted with the powdered compound to test for contact activity. Stomach-poison activity was assayed by injecting the compound into the lumen of the anterior oesophageal crop in solid form and in aqueous solution. These tests showed that metaldehyde was toxic by dermal contact and by ingestion but did not allow for accurate determination of the lethal dose. Methods for measuring the median lethal dose (LD_{50}) were subsequently developed (Henderson, 1968, 1969, 1970) and have been used to compare the molluscicidal activity of various compounds as stomach poisons, including several carbamate pesticides (Hunter and Johnston, 1970).

Forcibly injecting chemicals into the digestive tract allows accurate assessment of the lethal dose but obviously gives no indication of its acceptability when incorporated into bait. Various voluntary ingestion methods have been used to discover whether a candidate molluscicide is suitable for use in bait and to determine the optimum concentration to use. Quantified measurements of bait ingestion by *D. reticulatum* and *Tandonia budapestensis* (Hazay) (Milacidae) were obtained by Wright and Williams (1980) when these species were confined with wheat-flour baits containing an inert tracer, chromic oxide. Wedgwood and Bailey (1986) devised a method of analysing the feeding responses of *Arion hortensis* de Férussac by attaching molluscicidal baits to a record-player stylus, thus picking up and recording the bite frequency electronically. The results obtained by both methods unfortunately support the general conclusion that the addition of molluscicidal chemicals reduces ingestion of bait by the gastropods and compromises the ingestion of a lethal dose. With a view to improving bait performance, voluntary ingestion methods have also been used to identify different carriers, with which the molluscicidal chemical is mixed, that increase ingestion (e.g. Frain and Newell, 1982; Henderson *et al.*, 1992; Clark *et al.*, 1997). This topic is reviewed in depth by Bailey (Chapter 2, this volume).

Getzin and Cole (1964) screened proprietary pesticides against *Prophysaon andersoni* (Cooper) (Arionidae) by spraying animals directly in a Potter spray tower. However, Getzin and Cole (1964) observed that most pesticides applied in this way were ineffective because they were quickly removed by mucus secretion. The increased mucus secretion in response to noxious compounds made it uncertain how much of the topically applied dose penetrated the test animals. Subsequently, Young and Wilkins (1989a) devised a technique in which mucus secretion was delayed by anaesthetizing agriolimacids and applying measured amounts of test formulations to an area of the dorsal epidermis that had been wiped clear of mucus. This method was used to compare the toxicity of six compounds, including three metal salts, and methiocarb as the molluscicidal standard. Another method, appropriate only to shelled species, is that of Hussein *et al.* (1994), in which known amounts of test compounds are applied in solution or emulsion to the body surface of individual animals with a micropipette. On animal contraction, the applied materials are contained within the shell and in intimate contact with the animal's tissues.

Since terrestrial gastropods are most likely to acquire contact poisons by crawling on treated surfaces, many laboratory evaluation methods involve exposure of animals to compounds deposited on inert surfaces, rather than treating the animals directly. When *D. reticulatum* were held on glass plates coated with a range of phenolic and other chemicals, the relative toxicities differed from those obtained when the same chemicals were injected into the oesophageal-crop lumen (Briggs and Henderson, 1987). When comparing the activity of a range of iron and aluminium compounds as contact poisons against this species, substitution of a wet-soil substrate for a dry-glass one reduced activity generally, although the attenuation was less with chelated compounds than with simple metal salts (Henderson and Martin, 1990).

More recently, techniques have been developed for assessing the antifeedant or deterrent properties of secondary plant metabolites and other compounds for use as topically applied plant protectants. Barratt *et al.* (1993), for example, dipped cabbage (*Brassica oleracea* Linnaeus var. *capitata* Linnaeus) (Brassicaceae) leaf discs in extracts of over 50 plants and fungi and compared the fresh-weight loss after exposure to *D. reticulatum*: reductions in feeding of up to 87.5% were recorded. Dawson *et al.* (1996) applied test materials to glass plates and to leaf surfaces by electrostatic spraying and monitored the proportion of time spent on the treated and untreated surfaces using time-lapse video recording. At the standard test concentration of 50 $\mu g\ cm^{-2}$, common salt was only slightly repellent, while several proprietary surfactants were effective repellents. However, these surfactants were ineffective in the field because they were rapidly lost from plant surfaces under wet conditions.

Terrarium trials

Trials in enclosed areas of soil, with or without plants, are often used to predict how test formulations will perform under field conditions. Judge (1969) followed laboratory experiments with 74 candidate molluscicides by evaluating 23 of the most active ones under such simulated field conditions. Groups of *D. laeve* and *D. reticulatum* were confined in wooden trays containing pea (*Pisum sativum* Linnaeus) (Fabaceae) seedlings growing in potting compost. Trays were treated with test formulations, either sprayed on to the plants or applied as granules, and kept in a glasshouse. After 7 days, mortality in the gastropods and damage to the plants were recorded. Of the 23 compounds found active in laboratory tests, only four were effective in the terrarium trials, including the organophosphate phorate and the oxime carbamate aldicarb, which were 'spectacularly' molluscicidal under these conditions. In subsequent field trials with granular formulations, 'none of these treatments achieved conspicuous success in slug control' (Judge and Kuhr, 1972).

Field trials

While molluscicide efficacy can be expressed in terms of its effect on gastropod numbers, the inherent difficulty of accurately assessing field populations in certain cropping situations means that this approach is less reliable than quantifying the effect on the crop itself. Population estimates based on numbers found in soil-surface refuge traps vary with the level of activity, which is weather-dependent. Estimates based on numbers found using soil-sampling methods, such as that of South (1964), are more accurate but are generally regarded as too laborious for extensive use. A third method, which appears to offer accuracy comparable to soil sampling but which is less laborious, has been described by Ferguson *et al.* (1989). In the 'defined area trap' all gastropods at the surface within a metal cylinder driven into the soil are collected over a period until the area is exhausted, thus providing an estimate of gastropod numbers per unit area.

Post-treatment counts of the number of gastropods killed by a test material have long been used as an evaluation procedure, most recently by Meredith (1996). While dead gastropods are reassuring proof of toxicity, the real measure of efficacy is the size of the residual population. Ultimately, because of the difficulty of getting accurate population estimates, evaluations based on numbers at best give an indication of the effectiveness of new treatments in comparison with existing products. Field tests of baits against *C. aspersus* in citrus orchards were less difficult for Pappas and Carman (1955): test baits were broadcast into individual trees and dead and live *C. aspersus* in and beneath each tree counted 10 days later. Unfortunately, few gastropod pest situations are as experimenter-friendly.

Reduction of crop damage is the most unequivocal measurement of a molluscicide's efficacy. When present in large numbers, terrestrial gastropods can reduce the gross yield of crops by damaging the parent plants. Even in small numbers they may greatly reduce the market value of the harvested produce, either by causing blemishes or merely by being present in the end-product. The type of damage inflicted therefore dictates the field assessment method and how accurate it must be.

When assessing treatments applied to wheat seed to control damage by *D. reticulatum*, Scott *et al.* (1984) counted damaged seeds in soil samples taken after germination and also compared grain yields from whole plots at harvest. Results of the two methods were in general agreement: the most effective teatment reduced the proportion of seeds initially damaged from 44% to 7% and an increase in grain yield of 50% was subsequently recorded. Glen and Orsman (1986) used an indirect method to compare the effectiveness of two bait formulations with a broadcast application of aluminium sulphate applied to bare-soil plots. Glasshouse-grown seedlings of Chinese cabbage (*Brassica chinensis* Linnaeus) were transplanted into the plots 2 days after treatments were applied and damage to the leaves of the indicator plants was visually assessed over the following 12 days. Treatments prevented damage to the indicator plants for 4 days, implying that protection was due to a brief depression of feeding activity rather than a reduction in the pest population. In Brussels sprouts (*Brassica oleracea* Linnaeus convar. *oleracea* var. *gemmifera*), where mild surface grazing of the outer leaves of the sprouts renders the crop unsaleable, Dawson *et al.* (1996) evaluated sprays repellent to *D. reticulatum* in field plots by removing individual sprouts from sample plants and visually scoring for damage.

Field-trial evaluation techniques are therefore dictated by the cropping situation involved and with such polyphagous animals a range of methods is required. In a group of pests whose activity levels are notoriously unpredictable, reliable field assessment demands a degree of replication that is often difficult to satisfy.

Application Technologies

Most pesticides are applied in spray, dust or granular formulations but only occasionally as baits. In contrast, molluscicides directed against terrestrial gastropods are only occasionally delivered as sprays or dusts but are more usually deployed in baits. For this reason, application technology is largely concerned with the composition of baits and how, where and when to apply them.

The first effective use of a molluscicide in a bait was probably with metaldehyde, in 1936, when recipes for making formulations using loose bran were published in the UK gardening press (Hadden, 1936). Thereafter, in efforts to extend the post-application life of such baits, various additives were tested. Thomas (1948) increased the effectiveness of metaldehyde/bran baits with a casein glue binder and the majority of commercial baits are now produced in compressed pellet form to delay disintegration. Such pellets are usually, though not invariably, grain-offal-based with the toxicant dispersed throughout the carrier medium. One departure from this generality is a formulation recently developed in New Zealand in which the active ingredient (metaldehyde) is incorporated into an edible matrix which is then applied as a coating over an inert granular core (Barker *et al.*, 1991). This formulation is claimed to be more easily dispensed from application machinery and to give a higher density of bait particles per unit area.

Because most commercial baits are similar in size to fertilizer granules, they can be broadcast using existing tractor-mounted machinery. They can also be applied by purpose-built distributors, usually involving a gravity-fed hopper discharging on to a horizontally spinning disc. These distributors can also be mounted on low-ground-pressure vehicles, such as 'quad' bikes, which can operate in the wet soil conditions that favour pest activity, or on vehicles conducting other operations, such as fungicide spraying, to reduce crop disturbance.

In crops susceptible to gastropod damage at or before germination, such as cereals, baits can be mixed with the seed and sown with it. This may give better control in very coarse seed-beds but in general soil-surface applications made prior to sowing are more effective (Green *et al.*, 1992).

Deployment of molluscicides against terrestrial gastropods in formulations other than baits is less common. Aqueous sprays of copper sulphate were included in UK recommendations for controlling gastropods in arable crops as recently as 1984 (Anon., 1984), but in practice the use of copper and other metal salts as dusts or sprays has been largely superseded. Metaldehyde has also been applied in emulsified form as a spray, if its introduction into an overhead irrigation system can be so regarded. Howitt (1961) reduced damage to cocksfoot (*Dactylis glomerata* Linnaeus; Gramineae)/clover (*Trifolium repens* Linnaeus; Fabaceae) pasture in the western USA in this way, although it is not clear whether the molluscicide was toxic by contact or after ingestion of treated foliage. Electrostatic spraying has been used experimentally to give high, localized deposition on growing plants (Dawson *et al.*, 1996), although the materials tested were repellents rather than molluscicidal.

Molluscicides are routinely sprayed into watercourses for the control of aquatic species that act as hosts for trematodes causing human and animal schistosomaisis, but since there are few aquatic molluscan crop pests this method of application is rarely applicable. A notable exception is the so-called golden apple snail (*Pomacea* sp.), which attacks paddy rice in Asia: molluscicides for the control of this pest can be delivered as aqueous emulsions in sprays (Schnorbach, 1995; Palis *et al.*, 1996).

Molluscicides are also deployed as seed dressings: Scott *et al.* (1984) successfully treated wheat seeds to prevent grain hollowing of wheat in the UK. At treatment rates high enough to be effective in laboratory tests many of the candidate compounds were too phytotoxic, but in field trials good protection culminating in increases in grain yield were given by a methiocarb seed dressing. This was applied to the seed as a slurry in a Rotostat seed-treatment machine using a methylcellulose sticker. This technique has been further developed in the Netherlands by Ester *et al.* (1996) using polymer 'film-coating' techniques, which allow higher seed loadings without phytotoxic effects.

Transfer of pesticides to their targets is an inefficient process. The proportion of the amount of insecticide applied that actually effects control has been estimated at between 0.02% and 0.03% (Graham-Bryce, 1975). Delivery of molluscicides to aquatic species is helped by continuing accumulation from the surrounding water. Using radiolabelled 4′-chloronicotinanilide Duncan *et al.* (1977) found that, after 2 h , the rate of uptake by *Biomphalaria glabrata* (Say) (Planorbidae) from flowing water remained constant. With terrestrial species the situation is more difficult. They are relatively large in comparison with most insect pests and they are only intermittently active in the crop or on the soil surface. Exposed surfaces are protected by a disposable layer of aqueous mucus and shelled species have additional protection. These factors severely constrain the choice of biocides suitable for chemical control, which have to be delivered in a relatively large amount, and reduce the options open for their delivery.

Although application of molluscicides in bait formulation is relatively inefficient it remains the best option available under many circumstances. Thus far, application techniques have developed to optimize the usage of relatively ineffective chemicals and are only likely to change significantly following the emergence of new molluscicides and better formulations.

Conclusions

The importance of terrestrial gastropods as crop pests has greatly increased over the past 30 years, and in the process demands for effective controls have outstripped the development of chemical control measures.

The increased pest status has been brought about by a number of contributory factors. In temperate regions, changes in husbandry practices, such as autumn rather than spring sowing of cereal crops, the expansion of dense-canopy crops, such as oilseed rape (*Brassica napus* Linnaeus var. *oleifera* Linnaeus), and the adoption of minimum-tillage regimes, which conserve soil moisture, have all favoured the survival and reproduction of these moisture-dependent animals. In tropical and subtropical areas the most noticeable developments have been the introduction and spread of exotic species, which have become serious pests, the most dramatic examples being the giant African snail, *A. fulica* (Mead, 1961; Raut and Barker, Chapter 3, this volume), and more recently the golden apple snails, *Pomacea* spp. (Halwart, 1994; Cowie, Chapter 5, this volume), and the bean slug, *Sarasinula plebeia* (Fischer) (Vaginulidae) (Rueda *et al.*, Chapter 4, this volume). Such changes, occurring against a background of increasing demand in food quantity and quality, have combined to make terrestrial gastropods more important crop pests than ever before.

The development of chemical control measures has not kept pace for a number of reasons – some technical, others economic. Perhaps crucial has been the fragmented nature of the market for molluscicidal products, which made commercial research investment relatively unattractive. Primary screens for new molecules did not normally contain a test gastropod, although compounds that demonstrated activity against major pest groups, such as insects, might subsequently be assessed for molluscicidal activity. It is noticeable that even today terrestrial gastropod control still relies heavily on metaldehyde, a chance amateur discovery, and on carbamate pesticides originally developed as acaricides and insecticides.

The technical problems surrounding effective delivery to terrestrial gastropods stem largely from the conflicting requirements for adequate water solubility to penetrate a mucous barrier and adequate persistence in wet environments to match the intermittent activity of the pests. Failure to solve this conundrum has resulted in the near-universal adoption of bait carriers. While these offer some advantages in terms of low rates of active ingredient and a degree of target specificity, they still compare poorly with control measures for other pests, notably insects, themselves a fairly inefficient process. The proportion of the applied dose of insecticide that is transferred to the target has been estimated at 0.02–0.03% (Graham-Bryce, 1975). When the comparative toxicity of the materials used is low the process becomes even less efficient: the LD_{50} of the synthetic pyrethroid deltamethrin to *Musca domestica* Linnaeus (Muscidae) is equivalent to 0.02 mg kg^{-1}, while the equivalent value for metaldehyde to *D. reticulatum* is 1.0 g kg^{-1} (Henderson and Parker, 1986).

Improved formulation of existing molluscicides has probably achieved as much as it can and innovation is required to make significant advances in chemical control of terrestrial gastropods. The recent research interest in modes of action of molluscicides provides a more rational approach to the search for better materials, while the extension of screening to naturally occurring compounds has produced a chemically diverse range of active materials (e.g. Adewunmi and Monache, 1989; Airey *et al.*, 1989; Abdel-Aziz *et al.*, 1990; Watkins *et al.*, 1996). Two compounds, one isolated from a grass (Hagin and Bobnick, 1991) and one from a desert shrub (Hussein *et al.*, 1994), have been found to be more active than some synthetic pesticides. Fundamental studies on the biosynthesis of gastropod mucus (Cotterell *et al.*, 1993, 1994) raise the prospect of novel, target-specific control agents, while investigations into the chemosensory mechanisms of *D. reticulatum* hold out the promise of plant protection based on behaviour-modifying chemicals that do not present a toxic hazard (Dodds *et al.*, 1996).

Whether acting as poisons, as feeding deterrents or in other behaviour-modifying ways and whether deployed as baits, sprays or seed dressings, it is hard to escape the conclusion that what is required is a number of new, better, active ingredients. If the impetus of current research can be maintained they may well be found.

References

Abdel-Aziz, A., Brain, K. and Bashir, A.K. (1990) Screening of Sudanese plants for molluscicidal activity and identification of leaves of *Tacea leontopetaloides* (L.) Oktye (*Taccaceae*) as potential new exploitable resource. *Phytotherapy Research* 4, 62–65.

Adewunmi, C.O. and Monache, F.D. (1989) Molluscicidal activity of some coumarins. *Fitoterapia* 60, 1.

Airey, W.J., Henderson, I.F., Pickett, J.A., Scott, G.C., Stephenson, J.W. and Woodcock, C.M. (1989) Novel chemical approaches to mollusc control. In: Henderson, I.F. (ed.) *Slugs and Snails in World Agriculture.* Monograph No. 41, British Crop Protection Council, Thornton Heath, pp. 301–307.

Anderson, A.W. and Taylor, T.H. (1926) *The Slug Pest.* Bulletin of University of Leeds Department of Agriculture No. 143, University of leeds and the Yorkshire Council for Agricultural Education, Leeds, 14 pp.

Anon. (1930) Slugs and snails. *Queensland Agricultural Journal* 34, 115.

Anon. (1984) *Slugs and Snails.* Leaflet 115 (amended 1984), Ministry of Agriculture, Fisheries and Food, 12 pp.

Anon. (1986) Guideline for the biological evaluation of molluscicides. *OEPP/EPPO Bulletin* 16, 189–196.

Atger, J.C., Delpuech, I. and Maurin, G. (1990) Un nouveau produit anti-limaces à base de bensultap. In: *ANPP – Deuxieme Conference Internationale sur les*

Ravageurs en Agriculture, Versailles. Association National pour la Protection des Plantes, Paris, pp. 305–312.

Augustinsson, K.-B. (1960) Butyryl- and propionylcholinesterases and related types of eserine-sensitive esterases. In: Boyer, P.D., Lardy, H. and Hyrback, K. (eds) *The Enzymes.* Academic Press, New York, pp. 521–540.

Bailey, S.E.R. (1989) Foraging behaviour of terrestrial gastropods: integrating field and laboratory studies. *Journal of Molluscan Studies* 55, 263–272.

Bailey, S.E.R., Cordon, S. and Hutchinson, S. (1989) Why don't slugs eat more bait? A behavioural study of early meal termination produced by methiocarb and metaldehyde baits in *Deroceras caruanae.* In: Henderson, I.F. (ed.) *Slugs and Snails in World Agriculture.* Monograph No. 41, British Crop Protection Council, Thornton Heath, pp. 385–390.

Bakhtawar, P.M. and Mahendru, V.K. (1987) Toxicity of organo-phosphorus molluscicide on the snail pest *Helix vittata* (Hornell). *Journal of Advanced Zoology* 8, 84–87.

Banna, H.B.M. (1980) Histochemical studies of some enzymes in the tissues of the schistosome vector snail *Bulinus truncatus* (Andouin) with special reference to the effects of a molluscicide. II. Hydrolases. *Histochemical Journal* 12, 145–152.

Barker, G.M., Pottinger, R.P., Lloyd, J.M., Addison, P.J., Firth, A.C. and Stewart, A.P. (1991) A novel bait formulation for slug and snail control. *Proceedings of the New Zealand Weed and Pest Control Conference* 44, 195–200.

Barratt, B.I.P., Lorimer, S.D., Perry, N.B., Bashi-Hammer, M., Foster, L.M. and Ferguson, C.M. (1993) A bioassay to evaluate native plant extracts as slug feeding deterrents. *Proceedings of the New Zealand Plant Protection Conference* 46, 194–196.

Barry, B.D. (1969) Evaluation of chemicals for control of slugs on field corn in Ohio. *Journal of Economic Entomology* 62, 1277–1279.

Booze, F. and Oehme, F.W. (1985) Metaldehyde toxicity: a review. *Veterinary and Human Toxicology* 27, 11–20.

Booze, F. and Oehme, F.W. (1986) An investigation of metaldehyde and acetaldehyde toxicities in dogs. *Fundamental and Applied Toxicology* 6, 440–446.

Bourne, N.B., Jones, G.W. and Bowen, I.D. (1988) Slug feeding behaviour in relation to control with molluscicidal baits. *Journal of Molluscan Studies* 54, 327–338.

Bourne, N.B., Jones, G.W. and Bowen, I.D. (1991) Endocytosis in the crop of the slug, *Deroceras reticulatum* (Müller) and the effects of the ingested molluscicides, metaldehyde and methiocarb. *Journal of Molluscan Studies* 57, 71–80.

Boyce, C.B.C., Crossland, N.O. and Shiff, C.J. (1966) A new molluscicide, *N*-trithomorpholine. *Nature* 210, 1140.

Briggs, G.G. and Henderson, I.F. (1987) Some factors affecting the toxicity of poisons to the slug *Deroceras reticulatum* (Müller) (Pulmonata: Limacidae). *Crop Protection* 6, 341–346.

Buchs, W., Heimbach, U. and Czarnecki, E. (1989) Effects of snail baits on non-target carabid beetles. In: Henderson, I.F. (ed.) *Slugs and Snails in World Agriculture.* Monograph No. 41, British Crop Protection Council, Thornton Heath, pp. 245–252.

Buckhurst, A.S. (1947) Slug control with DDT. *Fruit Grower* 103, 146.

Bullock, J.I., Coward, N.P., Dawson, G.W., Henderson, I.F., Larkworthy, L.F., Martin, A.P. and McGrath, S.P. (1992) Contact uptake of metal compounds and their molluscicidal effect on the field slug, *Deroceras reticulatum* (Müller) (Pulmonata: Limacidae). *Crop Protection* 11, 329–334.

Cajaraville, M.P., Diez, G., Marigomez, J.A. and Angulo, E. (1990) Responses of basophilic cells of the digestive gland of mussels to petroleum hydrocarbon exposure. *Diseases of Aquatic Organisms* 9, 221–228.

Casida, J.E. (1963) Mode of action of carbamates. *Annual Review of Entomology* 8, 39–58.

Clark, S.J., Coward, N.P., Glenn, W.D., Henderson, I.F. and Martin, A.P. (1995) Metal chelate molluscicides: the redistribution of iron diazaalkanolates from the gut lumen of the slug, *Deroceras reticulatum* (Müller) (Pulmonata: Limacidae). *Pesticide Science* 44, 381–388.

Clark, S.J., Dodds, C.J., Henderson, I.F. and Martin, A.P. (1997) A bioassay for screening materials influencing feeding behaviour in the field slug *Deroceras reticulatum* (Müller). *Annals of Applied Biology* 130, 379–385.

Cleland, J.W. (1952) Slug control. *New Zealand Entomologist* 1, 6.

Cobbett, W. (1930) *Rural Rides in the Southern, Western and Eastern Counties of England, Together with Tours in Scotland and in the Northern and Midland Counties of England, and Letters from Ireland.* Cole, G.D.H. and Cole, M. (eds), Peter Davies, London.

Cotterell, J.M., Henderson, I.F., Pickett, J.A. and Wright, D.J. (1993) Evidence for glycosaminoglycans as a major component of trail mucus from the terrestrial slug, *Arion ater* L. *Comparative Physiology and Biochemistry* 104B, 455–468.

Cotterell, J.M., Henderson, I.F. and Wright, D.J. (1994) Studies on the glycosaminoglycan component of trail mucus from the terrestrial slug, *Arion ater* L. *Comparative Biochemistry and Physiology* 107B, 285–296.

Cragg, J.B. and Vincent, M.H. (1952) The action of Metaldehyde on the slug *Agriolimax reticulatus* (Müller). *Annals of Applied Biology* 39, 392–406.

Crossland, N.O., Bennett, M.S. and Hope-Cawdery, M.J.H. (1969) Preliminary observations on the control of *Fasciola hepatica* with the molluscicide *N*-tritylmorpholine. *Veterinary Record* 84, 182–184.

Dallinger, R., Janssen, H.H., Bauer-Hilty, A. and Berger, B. (1989) Characterization of an inducible cadmium-binding protein from hepatopancreas of metal-exposed slugs (Arionidae, Mollusca). *Comparative Biochemical Physiology* 92C, 355–360.

Dauterman, W.C. and Hodgson, E. (1990) Metabolism of xenobiotics. In: Hodgson, E. and Kuhr, R.J. (eds) *Safer Insecticides: Development and Use.* Marcel Dekker, New York, pp. 19–55.

Dawson, G.W., Henderson, I.F., Martin, A.P. and Pye, B.J. (1996) Physiochemical barriers as plant protectants against slugs (Pulmonata: Gastropoda). In: Henderson, I.F. (ed.) *Slug and Snail Pests in Agriculture.* Symposium Proceedings No. 66, British Crop Protection Council, Farnham, pp. 439–444.

Dodds, C.J., Ford, M.G., Henderson, I.F., Leake, L.D., Martin, A.P., Pickett, J.A., Wadhams, L.J. and Watson, P. (1996) Slug chemical ecology: electrophysiological and behavioural studies. In: Henderson, I.F. (ed.) *Slug and Snail Pests in Agriculture.* Symposium Proceedings No. 66, British Crop Protection Council, Farnham, pp. 73–81.

Duncan, J., Brown, N. and Dunlop, R.W. (1977) The uptake of the molluscicide, 4′-chloronicotinanilide into *Biomphalaria globrata* (Say) in a flowing water system. *Pesticide Science* 8, 345–353.

Durham, H.E. (1920) Of slugs. *Gardeners Chronicle* 68, 46, 85–86, 110.

Eldefrawi, M.E. and Eldefrawi, A.T. (1990) Nervous-system-based insecticides. In: Hodgson, E. and Kuhr, R.J. (eds) *Safer Insecticides: Development and Use.* Marcel Dekker, New York, pp. 155–209.

Ester, A., Darwinkel, A., Floot, H.W. and Nijenstein, J.H. (1996) Control of field slugs (*Deroceras reticulatum*) in winter wheat using seeds treated with pesticides. In: Henderson, I.F. (ed.) *Slug and Snail Pests in Agriculture.* Symposium Proceedings No 66, British Crop Protection Council, Farnham, pp. 165–172.

Ferguson, C.M., Barratt, B.I.P. and Jones, P.A. (1989) A new technique for estimating density of the field slug *Deroceras reticulatum* (Müller). In: Henderson, I.F. (ed.) *Slugs and Snails in World Agriculture.* Monograph No. 41, British Crop Protection Council, Thornton Heath, pp. 331–336.

Frain, J.M. and Newell, P.F. (1982) Meal size and feeding assay for *Deroceras reticulatum* (Müll.). *Journal of Molluscan Studies* 48, 98–99.

French, C. (1906) Slugs. *Journal of the Department of Agriculture of Victoria* 8, 445–447.

Frömming, E. (1950) Untersuchungen uber die Wirksamkeit des DDT -proparates Duolit auf Landlungsschnecken. *Anzeiger Schadlingsbekampfung* 23, 57–60.

Gahan, A.B. (1907) *Greenhouse Pests of Maryland.* Bulletin of the Maryland Agricultural Experiment Station No. 119, 36 pp.

Getzin, L.W. and Cole, S.G. (1964) *The Evaluation of Potential Molluscicides for Slug Control.* Technical Bulletin of the Washington Agricultural Experiment Station 653, Pullman, 5 pp.

Ghadially, F.N. (1988) *Ultrastructural Pathology of the Cell and Matrix*, Vols I and II, 3rd edn. Butterworths, London, 1340 pp.

Gimingham, C.T. (1940) Some recent contributions by English workers to the methods of insect control. *Annals of Applied Biology* 27, 161–175.

Glen, D.A. and Orsman, I.A. (1986) Comparison of molluscicides based on metaldehyde, methiocarb and aluminium sulphate. *Crop Protection* 5, 371–375.

Godan, D. (1983) *Pest Slugs and Snails: Biology and Control.* Springer Verlag, Berlin, 445 pp.

Gönnert, R. and Schraufstätter, E. (1958) A new molluscicide: molluscicide Bayer 73. In: *Proceedings of the 6th International Conference of Tropical Medicine and Malaria, Lisbon*, Vol. 2, pp. 197–202.

Gordon, H.T. and Eldefrawi, M.E. (1960) Analog-synergism of several carbamate insecticides. *Journal of Economic Entomology* 53, 1004–1009.

Graham-Bryce, I.J. (1975) The future of pesticide technology: opportunities for research. In: *Proceedings of the 8th British Insecticide and Fungicide Conference, Brighton.* British Crop Protection Council, Thornton Heath, pp. 901–914.

Green, D.B., Corbett, S.J., Jackson, A.W. and Nowak, K.J. (1992) Surface versus admixed applications of slug pellets to winter wheat. In: *Proceedings of the Brighton Crop Protection Conference – Pests and Diseases.* British Crop Protection Council, Thornton Heath, pp. 587–592.

Hadden, N.G. (1936) A remedy for slugs. *Gardening Illustrated*, 2 May, 255.

Hagin, R.D. and Bobnick, S.J. (1991) Isolation and identification of a slug-specific molluscicide from quackgrass (*Agropyron repens* L. Beauv.). *Journal of Agricultural and Food Chemistry* 39, 192–196.

Halwart, M. (1994) The golden apple snail *Pomacea canaliculata* in Asian rice farming systems: present impact and future threat. *International Journal of Pest Management* 40, 199–206.
Harries, V., Aldolphi, H., Kiehs, K. and Neumann, U. (1980) BAS 263. I: Insektizide und nematizide Wirkung eines neuen Carbamates fur den Einzatz in Mais und anderen landwirstschaftlichen Kulturen. *Medelingen von de Faculteit Landbouwetenschappen Rijksuniversiteit Gent* 45, 739–748.
Henderson, I.F. (1968) Laboratory methods for assessing the toxicity of contact poisons to slugs. *Annals of Applied Biology* 62, 363–369.
Henderson, I.F. (1969) A laboratory method for assessing the toxicity of stomach poisons to slugs. *Annals of Applied Biology* 63, 167–171.
Henderson, I.F. (1970) The fumigant effect of metaldehyde on slugs. *Annals of Applied Biology* 65, 507–510.
Henderson, I.F. and Martin, A.P. (1990) Control of slugs with contact-action molluscicides. *Annals of Applied Biology* 116, 273–278.
Henderson, I.F. and Parker, K.A. (1986) Problems in developing chemical control of slugs. *Aspects of Applied Biology* 13, 341–347.
Henderson, I.F., Briggs, G.G., Coward, N.P., Dawson, G.W. and Pickett, J.A. (1989) A new group of molluscicidal compounds. In: Henderson, I.F. (ed.) *Slugs and Snails in World Agriculture*. Monograph No. 41, British Crop Protection Council, Thornton Heath, pp. 289–294.
Henderson, I.F., Martin, A.P. and Parker, K.A. (1990) Laboratory and field assessment of a new aluminium chelate slug poison. *Crop Protection* 9, 131–134.
Henderson, I.F., Martin, A.P. and Perry, J.N. (1992) Improving slug baits: the effects of some phagostimulants and molluscicides on ingestion by the slug *Deroceras reticulatum* (Müller) (Pulmonata: Limacidae). *Annals of Applied Biology* 121, 423–430.
Homeida, A.M. and Cooke, R.G. (1982) Pharmacological aspects of metaldehyde poisoning in mice. *Journal of Veterinary Pharmacological Therapy* 5, 77–81.
Howitt, A.J. (1961) Chemical control of slugs in orchard grass–ladino white clover pastures in the Pacific Northwest. *Journal of Economic Entomology* 54, 778–781.
Hunter, P.J. and Johnston, D.L. (1970) Screening carbamates for toxicity against slugs. *Journal of Economic Entomology* 63, 305–306.
Hussein, H.I., Kamel, A., Abou-Zeid, M., El-Sebae, A.-K.H. and Sulch, M.A. (1994) Uscharin, the most potent molluscicidal compound tested against land snails. *Journal of Chemical Ecology* 20, 135–140.
Jonnalagadda, P.R. and Rao, B.P. (1996) Histopathological changes induced by specific pesticides on some tissues of the fresh water snail, *Bellamya dissimilar* Müller. *Bulletin of Environmental Contamination and Toxicology* 57, 648–654.
Judge, F.D. (1969) Preliminary screening of candidate molluscicides. *Journal of Economic Entomology* 62, 1393–1397.
Judge, F.D. and Kuhr, R.J. (1972) Laboratory and field screening of granular formulations of candidate molluscicides. *Journal of Economic Entomology* 65, 242–245.
Kela, S.L. and Bowen, I.D. (1995) The histopathological effect of *Detarium microcarpum* extract, a naturally occurring plant molluscicide, on the mid-gut and the digestive gland of *Lymnaea stagnalis*. *Cell Biology International* 19, 175–181.

Kelly, C.R., Greenwood, S. and Bailey, S.E.R. (1996) Can different pH environments in slug digestive tracts be exploited to improve the efficacy of molluscicide baits? In: Henderson, I.F. (ed.) *Slug and Snail Pests in Agriculture.* Symposium Proceedings No. 66, British Crop Protection Council, Farnham, pp. 83–90.

Klaunig, J.E., Lipsky, M.M., Trump, B.F. and Hinton, D.E. (1979) Biochemical and ultrastructural changes in teleost liver following subacute exposure to PCB. *Journal of Environmental Toxicology* 2, 953–963.

Köhler, H.-R. and Triebskorn, R. (1998) Assessment of the cytotoxic impact of heavy metals on soil invertebrates using a protocol integrating qualitative and quantitative components. *Biomarkers* 3, 109–127.

Lauger, P., Martin, H. and Muller, P. (1944) Uber Konstitution und toxische Wirkung von naturlichen und neuen synthetischen insekttotenden Stoffen. *Helevetic Chimica Acta* 27, 892.

Lovett, R.L. and Black, A.B. (1920) *The Grey Garden Slug with Notes on Allied Forms.* Bulletin of the Oregon Agricultural Experiment Station 170, Corvallis, 143 pp.

Machin, J. (1977) Role of integument in molluscs. In: Gupta, B.L., Moreton, R.B., Oschman, J.L. and Wall, B.J. (eds) *Transport of Ions and Water in Animals.* Academic Press, London, pp. 735–762.

Maddy, K.T. (1975) Poisoning of dogs with metaldehyde in snail and slug poison bait. *Californian Veterinarian* 29, 24–25.

Mallet, C. and Bougaran, H. (1971) Action molluscicides de divers carbamates. In: *XXième Symposium International de Phytopharmacie et de Phytiatrie,* Vol. 12, pp. 207–215.

Manna, B. and Ghose, K.C. (1972) Histo-pathological changes in the gut of *Achatina fulica* caused by endrin, a molluscicide. *Indian Journal of Experimental Biology* 10, 461–463.

Marigomez, I. and Dussart, G.B.J. (1996) Cellular basis of the adaptation to metal pollution in sentinel slugs: eco(toxico)logical implications. In: *ESCPB 17th Annual Meeting, 'Adaptation to Stress in Aquatic and Terrestrial Ecosystems' Antwerp,* p. 81.

Marston, A. and Hostettman, K. (1985) Plant molluscicides. *Phytochemistry* 24, 639–652.

Martin, T.J. and Forrest, J.D. (1969) Development of Draza in Great Britain. *Pflanzenschutz-Nachrichten Bayer* 22, 205–243.

Massee, A.M. (1928) Notes on insect pests for the years 1926–1927. *Report of East Malling Research Station* 14/15, (Suppl. 2), 157–162.

Mead, A.R. (1961) *The Giant African Snail: a Problem in Economic Malacology.* University of Chicago Press, Illinois, 257 pp.

Meredith, R.H. (1996) Testing bait treatments for slug control. In: Henderson, I.F. (ed.) *Slug and Snail Pests in Agriculture.* Symposium Proceedings No. 66, British Crop Protection Council, Farnham, pp. 157–164.

Metcalf, R. (1955) Carbamates. In: *Organic Insecticides, Their Chemistry and Mode of Action.* Interscience Publishers, New York, pp. 317–329.

Metcalf, R. and Fukuto, T.R. (1965) Carbamate insecticides: effects of chemical structure on intoxication and detoxication of phenyl-*N*-methylcarbamates in insects. *Journal of Agricultural Food Chemistry* 13, 220–231.

Miles, H.W., Wood, J. and Thomas, I. (1931) On the ecology and control of slugs. *Annals of Applied Biology* 18, 370–400.

Mills, J.D., McCrohan, C.R., Bailey, S.E.R. and Wedgwood, M.A. (1990) Effects of metaldehyde and acetaldehyde on feeding responses and neuronal activity in the snail *Lymnaea stagnalis. Pesticide Science* 28, 89–99.

Mills, J.D., McCrohan, C.R. and Bailey, S.E.R. (1991a) Electro-physiological response to metaldehyde in neurones of the feeding circuitry of the snail *Lymnaea stagnalis. Pesticide Biochemistry and Physiology* 42, 35–42.

Mills, J.D., McCrohan, C.R. and Bailey, S.E.R. (1991b) Effects of metaldehyde and acetaldehyde on specific membrane currents in neurons of the pond snail *Lymnaea stagnalis. Pesticide Science* 34, 243–247.

Mills, J.D., McCrohan, C.R. and Bailey, S.E.R. (1991c) Effects of the molluscicide metaldehyde on neuronal activity of *Lymnaea stagnalis.* In: Kits, K.S., Boer, H.H. and Joosse, J. (eds) *Molluscan Neurobiology.* North Holland, Amsterdam, Oxford and New York, pp. 209–213.

Palis, F.V., Macatula, R.F. and Browning, L. (1996) Niclosamide, an effective molluscicide for the golden apple snail (*Pomacea canaliculata* Lamarck) control in Philippine rice production systems. In: Henderson, I.F. (ed.) *Slug and Snail Pests in Agriculture.* Symposium Proceedings No. 66, British Crop Protection Council, Farnham, pp. 213–218.

Palombi, A. (1948) I molluschi dannozi all'agricoltura. *Rivista fitosanitaria* 2, 7–13.

Pappas, J.L. and Carman, G.E. (1955) Field screening tests with various materials against the European brown snail in citrus in California. *Journal of Economic Entomology* 48, 698–700.

Pessah, J.N. and Sokolove, P.G. (1983) The interaction of organophosphate and of carbamate insecticides with choline esterases in the terrestrial pulmonate *Limax maximus. Comparative Biochemistry and Physiology* 74C, 291–297.

Port, C.M. and Port, G.R. (1986) The biology and behaviour of slugs in relation to crop damage and control. *Agricultural Zoology Review* 1, 255–299.

Roach, D.K. (1968) Rhythmic muscular activity in the alimentary tract of *Arion ater* (L.) (Gastropoda: Pulmonata). *Comparative Biochemical Physiology* 24, 865–878.

Rondelaud, D. and Dreyfuss, G. (1996) The development of tissue lesions in the snail *Lymnaea glabra* exposed to a sublethal dose of molluscicide. *Veterinary Research* 27, 79–86.

Ruppel, R.F. (1959) Effectiveness of sevin against the gray garden slug. *Journal of Economic Entomology* 52, 360.

Ryder, T.A. and Bowen, J.D. (1977) The slug foot as a site of uptake of copper molluscicide. *Journal of Invertebrate Pathology* 30, 381–386.

Sakai, M. (1969) Nereistoxin and Cartap: their mode of action as insecticides. *Review of Plant Protection Research* 2, 17–28.

Schnorbach, H.-J. (1995) Die golden apple snail (*Pomacea canaliculata* Lamarck), ein Schadling in Reis mit zunehmender Bedeutung und Bekampfungsmoglichkeit mit Bayluscid. *Pflanzenschutz-Nachrichten Bayer* 48, 327–362.

Schread, J.C. (1958) *Control of Slugs, Sowbugs, Centipedes and Millipedes in the Greenhouse and Garden.* Connecticut Agricultural Experiment Station Circular 203, New Haven, Connecticut, 7 pp.

Scott, G.C., Pickett, J.A., Smith, M.C. and Woodcock, C.M. (1984) Seed treatments for controlling slugs in winter wheat. In: *Proceedings British Crop Protection*

Conference – Pests and Diseases. British Crop Protection Council, Thornton Heath, pp. 133–138.
Shropshire, L.H. and Compton, C.C. (1939) *Saving Garden Crops from Insect Pests*. Illinois Agricultural Experiment Station Circular 437, Urbana, 53 pp.
South, A. (1964) Estimation of slug populations. *Annals of Applied Biology* 53, 251–258.
Stephenson, J.W. (1959) Aldrin controlling slug and wireworm damage to potatoes. *Plant Pathology* 8, 53–54.
Taylor, J.W. (1894) *Monograph of the Land and Freshwater Molluscs of the British Isles*. Taylor Brothers, Leeds, 454 pp.
Tegelstrom, J. and Wahren, H. (1972) The effect of an *N*-methyl carbamate on esterases from snail, mouse and man studied by starch gel electrophoresis. *Comparative Biochemical Physiology* 43B, 339–343.
Theobald, F.V. (1895) Mollusca injurious to farmers and gardeners. *Zoologist* 19, 201–211.
Thomas, C.A. (1949) *The Symphylid or Garden Centipede and Other Greenhouse Pests*. Bulletin of the Pennsylvania Agricultural Experiment Station 508, 25 pp.
Thomas, D.C. (1948) The use of metaldehyde against slugs. *Annals of Applied Biology* 35, 207–227.
Thompson, H.W. (1928) Further tests of poison baits in South Wales. *Welsh Journal of Agriculture* 4, 342–347.
Triebskorn, R. (1989) Ultrastructural changes in the digestive tract of *Deroceras reticulatum* (Müller) induced by a carbamate molluscicide and by metaldehyde. *Malacologia* 31, 131–156.
Triebskorn, R. (1990) Die Wirkung von Molluskiziden auf den Verdauungstrakt einheimischer Schadschnecken. PhD thesis, Heidelberg University, Heidelberg.
Triebskorn, R. (1991a) The impact of molluscicides on enzyme activities in the hepatopancreas of *Deroceras reticulatum* (Müller). *Malacologia* 33, 255–272.
Triebskorn, R. (1991b) Cytological changes in the digestive system of slugs induced by molluscicides. *Journal of Medical and Applied Malacology* 3, 113–123.
Triebskorn, R. (1995) Tracing molluscicides and cellular reactions induced by them in slug tissues. In: Cajaraville, M.P. (ed.) *Cell Biology in Environmental Toxicology*, pp. 193–220.
Triebskorn, R. and Ebert, D. (1989) The importance of mucus production in slugs' reaction to molluscicides and the impact of molluscicides on the mucus producing system. In: Henderson, I.F. (ed.) *Slugs and Snails in World Agriculture*. Monograph No. 41, British Crop Protection Council, Thornton Heath, pp. 373–378.
Triebskorn, R. and Florschütz, A. (1993) Transport of uncontaminated and molluscicide-containing food in the digestive tract of the slug *Deroceras reticulatum* (Müller). *Journal of Molluscan Studies* 59, 35–42.
Triebskorn, R. and Köhler, H.-R. (1992) Plasticity of the endoplasmic reticulum in three cell types of slugs poisoned by molluscicides. *Protoplasma* 169, 120–129.
Triebskorn, R. and Köhler, H.-R. (1996) The impact of heavy metals on the grey garden slug *Deroceras reticulatum* (Müller): metal storage, cellular effects and semi-quantitative evaluation of metal toxicity. *Environmental Pollution* 93, 327–343.

Triebskorn, R. and Künast, C. (1990) Ultrastructural changes in the digestive system of *Deroceras reticulatum* (Mollusca, Gastropoda) induced by lethal and sublethal concentrations of the carbamate molluscicide cloethocarb. *Malacologia* 32, 87–104.

Triebskorn, R., Künast, C., Huber, R. and Brem, G. (1990) Tracing a ^{14}C-labelled molluscicide through the digestive system of *Deroceras reticulatum* (Müller). *Pesticide Science* 28, 31–330.

Triebskorn, R., Henderson, I.F., Martin, A.P. and Köhler, H.-R. (1996) Slugs as target and non-target organisms for environmental chemicals. In: Henderson, I.F. (ed.) *Slug and Snail Pests in Agriculture*. Symposium Proceedings No. 66, British Crop Protection Council, Farnham, pp. 65–72.

Triebskorn, R., Christensen, K. and Heim, I. (1998) Effect of orally and dermally applied metaldehyde on mucus cells of slugs (*Deroceras reticulatum*) depending on temperature and duration of exposure. *Journal of Molluscan Studies* 64, 467–487.

Triebskorn, R., Henderson, I.F. and Martin, A.P. (1999) Detection of iron in slugs (*Deroceras reticulatum*) contaminated with iron-chelates by means of energy filtering transmission electron microscopy (EFTEM). *Pesticide Science* 55, 55–61.

Tyron, H. (1899) Plant pests: *Vaginula* slugs (*Vaginula hedleyi* and *V. leydigi*). *Queensland Agricultural Journal* 5, 63–70.

Tzeng, D.D.S., Tzeng, H.C., Lee, M.H. and Yeh, Y. (1994) Sodium dodecyl sulfate as an alternative agent for the control of golden apple snail *Pomacea canaliculata* (Lamark) in rice fields. *Proceedings of the National Science Council, ROC, Part B, Life Sciences* 18, 3.

Unterstenhöfer, G. (1962) Mesurol, a polyvalent insecticide and acaracide. *Pflanzenschutz Nachrichten, Bayer* 15, 177–189.

van den Bruel, W.E. and Moens, R. (1956) Une méthode de lutte efficace, utilisable en plein champ contre les limaces. *Parasitica* 12, 8–15.

van den Bruel, W.E. and Moens, R. (1958) Nouvelles observations sur les propriétés des hélicides. *Bulletin de l'Institut Agronomique et des Stations de Recherche de Gembloux* 26, 281–304.

Watkins, R.W., Mosson, H.J., Gurney, J.E., Cowan, D.P. and Edwards, J.P. (1996) Cinnamic acid derivatives: novel repellent seed dressings for the protection of wheat against damage by the field slug, *Deroceras reticulatum*. *Crop Protection* 15, 77–83.

Wedgwood, M.A. and Bailey, S.E.R. (1986) The analysis of single meals in slug feeding on molluscicidal baits. *Journal of Molluscan Studies* 52, 259–260.

Wedgwood, M.A. and Bailey, S.E.R. (1988) The inhibitory effects of the molluscicide metaldehyde on feeding, locomotion and faecal elimination of three pest species of terrestrial slugs. *Annals of Applied Biology* 112, 439–457.

Wegler, R. (1970) *Chemie der Pflanzenschutz- und Schadlings-bekampfungsmittel*, Vol. 6. Springer Verlag, Heidelberg and New York, 460 pp.

White, G. (1777) *The Natural History and Antiquities of Selbourne, in the County of Southampton*, Vol. 1. Bell, T. (ed.), van Voorst, London, 507 pp.

Wilkinson, C.F. (1976) *Insecticide Biochemistry and Physiology*. Heyden Verlag, London.

Wright, A.A. and Williams, R. (1980) The effect of molluscicides on the consumption of bait by slugs. *Journal of Molluscan Studies* 46, 265–281.

Yang, H.S. and Thurman, D.E. (1981) Thiodicarb – a new insecticide for integrated pest management. In: *Proceedings of the British Crop Protection*

Conference – Pests and Diseases. British Crop Protection Council, Thornton Heath, pp. 687–697.

Young, A.G. and Wilkins, R.M. (1989a) A new technique for assessing the contact toxicity of molluscicides to slugs. *Journal of Molluscan Studies* 53, 533–536.

Young, A.G. and Wilkins, R.M. (1989b) The response of invertebrate acetylcholinesterase to molluscicides. In: Henderson, I.F. (ed.) *Slugs and Snails in World Agriculture.* Monograph No. 41, British Crop Protection Council, Thornton Heath, pp. 121–128.

Zylstra, U. (1972) Uptake of particular matter by the epidermis of the freshwater snail *Lymnaea stagnalis. Netherlands Journal of Zoology* 22, 299–306.

2 Molluscicidal Baits for Control of Terrestrial Gastropods

STUART E.R. BAILEY

School of Biological Sciences, University of Manchester, Oxford Road, Manchester M13 9PT, UK

Formulation

Edible baits containing a toxicant are the principal means of delivery of molluscicidal chemicals in terrestrial-gastropod control programmes. Typically, bait formulations contain 2–8% of toxicant or active ingredient. The bulk is a foodstuff, often wheat (*Triticum aestivum* Linnaeus; Gramineae) bran or barley (*Hordeum vulgare* Linnaeus; Gramineae) flour, which is intended to act as both an attractant and a feeding stimulant. These attractant and feeding-stimulant properties may be enhanced by the addition of specific materials, such as proteins, dextrose and casein (e.g. Schnorbach and Matthaei, 1990). Various adjuvants are often added, particularly in commercial formulations: a stabilizer may be added to increase the life of the active ingredient, a fungicide and binder to increase the life of the bait in the field and repellents and colouring agents to reduce poisoning of non-target mammal and bird species.

Active ingredients

A variety of molluscicidal compounds have been evaluated as active ingredients in baits for terrestrial-gastropod control (Henderson and Triebskorn, Chapter 1, this volume). Metaldehyde-based baits became widely used after the accidental discovery of the molluscicidal properties of metaldehyde fuel tablets. Today metaldehyde-based products predominate in the molluscicide bait market in most regions of the world. In Britain, for example, bait products containing metaldehyde are used on 55% of the crop area treated with chemicals for gastropod control, compared with 40% for methiocarb- and 5% for thiodicarb-containing baits (Garthwaite and Thomas, 1996). Metaldehyde, the cyclic tetramer of acetaldehyde, is used as a pesticide only against gastropods, unlike other organic chemicals used in molluscicidal baits. It is, however, toxic to mammals (oral medium lethal dose (LD_{50}) in *Rattus* Fischer sp., Muridae = 630 mg kg^{-1}). Metaldehyde tends to depolymerize readily, especially under acid conditions, and many commercially available bait products incorporate a stabilizer to offset the acidity of the edible carriers. One company, Lonza, produces nearly all metaldehyde, but formulations are produced by many independent companies (Heim *et al.*, 1996).

The carbamate methiocarb (4-methylthio-3,5-xylyl-*N*-methyl-carbamate; mercapturdimethyl) has been available since the late 1960s. It is produced by Bayer and marketed worldwide as an active ingredient in baits for gastropod control as Draza or Mesurol. Methiocarb is relatively short-lived in contact with soil, with a half-life of 4–56 days, depending on soil type (Kuhr and Dorough, 1976): it tends to denature in alkaline conditions and thus bait integrity can be important in extending their field life.

Among carbamate compounds that have gained prominence in recent years are thiodicarb (3,7,9,13-tetramethyl-5,1-dioxa-2,8,14-2,8,14-trithia-4,7,9,12-tetra-azapentadeca,12-diene-6,10-dione) and cloethocarb (2-(2-chloro-1-methyloxyethoxy)phenyl methylcarbamate). Thiodicarb baits (Skipper) were introduced commercially for gastropod control in the late 1980s by Rhône-Poulenc. Field evaluations show thiodicarb to be as effective as methiocarb and to have a long period of activity (Gaulliard and Laverriere, 1989; Ferguson *et al.*, 1995), and laboratory trials suggest that thiodicarb may give better control in wet conditions and methiocarb better in dry conditions (Ferguson *et al.*, 1995)

Mixes of active ingredient have been extensively evaluated in the past in attempts to overcome perceived disadvantages of individual compounds, such as temperature-dependent mortality from metaldehyde (Mallet and Bougaran, 1971) and differential species susceptibility to metaldehyde and carbamates (Crawford-Sidebotham, 1970), and to reduce the total concentration of active ingredient without reducing the efficacy. Prior to the recognition of the environmental hazards of arsenates, calcium arsenate was extensively used in combination with metaldehyde (see Godan, 1983). Mallet and Bougaran (1971) combined carbamate PE-9683 with metaldehyde to produce baits that not only were effective against gastropods at low temperature, but also produced rapid and irreversible toxicological effects. Although both Crowell (1977) and Bourne *et al.* (1990) found high recovery rates from a 2% methiocarb–2% metaldehyde mix, Bourne *et al.* (1990) found that 2% methiocarb combined with 1.0–0.25% metaldehyde caused higher mortality in the agriolimacid *Deroceras reticulatum* (Müller) than 4% metaldehyde alone. None the less, mixtures of active ingredients are not a feature of the contemporary molluscicide-bait market.

Bowen and Jones (1985) suggested adding surfactants to increase the efficacy and simultaneously reduce the concentration of molluscicide required, and Bowen *et al.* (1996) identified a number of useful additives, some based on beehive products (propolis), to methiocarb-based baits.

Feeding-depressant action of active ingredients

Although first administered as an attractant, metaldehyde was shown by Godan (1983) to be a feeding inhibitor at concentrations required to cause mortality in gastropods. The carbamates are also feeding inhibitors. Ingested molluscicides have been shown to cause immediate inhibition of feeding in *D. reticulatum*, the arionid *Arion hortensis* de Férussac, the milacid *Tandonia budapestensis* (Hazay) (Wedgwood and Bailey, 1988) and various helicid species (Coupland, 1996). Molluscicides also cause locomotor immobilization in the helicids *Cantareus aspersus* (Müller), *Cepaea nemoralis* (Linnaeus) and *Cepaea hortensis* (Müller) (Stringer and Morgan, 1969). Slug forms usually die 1–2 days after ingestion of metaldehyde or 3–4 days after ingestion of methiocarb (Martin and Forrest, 1969), while snails die 4 days after ingesting metaldehyde (Stringer and Morgan, 1969).

Working with methiocarb, Crowell (1967) emphasized the need to avoid a concentration of active ingredient that was too low and thus sublethal, or too high, deterring the gastropod from eating it and thus also sublethal. Wright and Williams (1980) showed that both metaldehyde and methiocarb are feeding depressants, to the extent that quantities of bait ingested often result in sublethal doses of these toxins. Henderson and Parker (1986) summarized the dilemma: meal size falls with increasing concentration of metaldehyde or methiocarb and, consequently, although less bait has to be ingested to acquire a lethal dose at higher concentrations, the interaction of toxicity with repellence and speed of action of a particular molluscicide determines the optimum effective concentration. Methiocarb and metaldehyde show similar patterns, but more active poisons, such as phenol and ioxynil, deter feeding at such low concentrations that a lethal dose is never ingested. Wedgwood and Bailey (1988) showed that metaldehyde reduced the probability of a meal continuing once started: meals on non-toxic baits lasted 1800–2300 s (700–900 bites) compared with 200–400 s on a bait containing 0.5% metaldehyde (200–300 bites) and only about 100 bites on baits containing 4–8% metaldehyde. However, there was little evidence that the presence of metaldehyde altered the likelihood of starting a meal. Bailey *et al.* (1989) showed that incorporation of methiocarb into baits similarly shortened the duration of meals by *D. reticulatum* – from 640 bites at 0.5%, to 400 bites at 4% and 200 bites at 8% methiocarb.

The toxic effects that Triebskorn and Ebert (1989) have demonstrated in the digestive-tract epithelia of *D. reticulatum* fed metaldehyde or carbamate are probably not those which cause the early termination of feeding on the toxic bait. Because short meals often result in a sublethal quantity of toxin being ingested, it is of interest to investigate the cause of the early termination of the meal. Wedgwood and Bailey (1988) and Bailey *et al.* (1989) showed that the reduction in amount ingested arises mostly from the reduction in number of bites, partly from the reduction in bite size as the meal progresses and partly from the increased amount of time pausing before the end of a meal. Immediately after a metaldehyde meal, *Deroceras caruanae* (Pollonera) refused a non-toxic bait (Bailey *et al.*, 1989). In contrast, after a methiocarb meal, these animals usually accepted a non-toxic bait. Differences in the rate of loss of locomotor coordination, as measured by the animal's speed of righting the body when turned on to its left side, largely accounted for these differences: immediately after a meal on bait containing 4% metaldehyde, *D. caruanae* had reduced locomotor coordination compared with animals that had fed on a non-toxic bait. Animals that had the typical short meals on baits containing 4–8% methiocarb exhibited only slight loss of locomotor coordination immediately after the meal, but after 30 min locomotor loss was well developed.

Wedgwood and Bailey (1988) suggested that the antifeedant effect of metaldehyde is twofold – both a toxic effect and an aversive effect. McCrohan *et al.* (1995) showed that metaldehyde applied to the lips of the freshwater lymnaeid species *Lymnaea stagnalis* Linnaeus caused a significant and immediate reduction in number of bites, suggesting that metaldehyde acts as an aversive taste stimulus: the isolated muscles of the buccal mass, oesophageal crop and gizzard (but not the anterior oesophagus) showed stronger and sometimes more frequent contractions, indicating a toxic action. Wright and Williams (1980) suggested that reduced intake is caused by paralysis of the digestive tract. This may be so in the case of metaldehyde, as meals on non-toxic baits are shortened if the animal first takes 30 bites from a 4% metaldehyde bait (Bailey *et al.*, 1989). Animals were shown to engage in a substantial meal (868 bites) after a full meal (434 bites) on a methiocarb-containing bait. Since bait reacceptance rate for animals that had fed a short time on baits containing 4–8% methiocarb was higher than for those that had fed longer on 0.5–2% methiocarb baits, the feeding depression may not be due to distastefulness (deterrence). Bailey *et al.* (1989) suggested that methiocarb feeding depression may be similar to the processes of declining food arousal – since size, frequency and regularity of bites, which all fall throughout a normal meal, are all affected by methiocarb.

Although the importance of particle size of biologically active chemicals has long been realized (Maas, 1978), scant consideration is given in the published literature to the effect of particle size in overcoming the deterrency of molluscicides. Metaldehyde is produced as elongate crystals (up to 2 mm), which are milled to produce a fined material for incorporation into baits. S.E.R. Bailey (unpublished) found that baits containing finely milled metaldehyde were eaten for much shorter periods than those containing unmilled metaldehyde at the same concentration. It is apparent, therefore, that there is merit for an investigation of meal sizes on baits containing different-size particles of this molluscicide, together with a study of the action of the radular teeth and jaw on the ingested molluscicide particles.

Microencapsulation might possibly overcome the feeding-depressant action of the molluscicide and thus avoid the early termination of meals on baits. Microencapsulation is used to delay the action of orally administered drugs on humans and other mammals, but often relies on the marked differences in pH of different regions of the digestive tract. Investigations of digestive tract pH in gastropods indicate much smaller changes in pH than those evident in mammals (Kelly *et al.*, 1996; Walker *et al.*, 1996). A bait product incorporating encapsulated particles of metaldehyde is available in the UK. An alternative to microencapsulation might be to bind the active ingredient within some protective and yet digestible material, such as starch (Schroder, 1985).

Attractants and feeding stimulants

Howling (1991) has usefully drawn the distinctions between attractiveness (including any effect of a bait on gastropod behaviour before contact is made), acceptability (the likelihood of commencing to feed once contact is made) and palatability (effects of the formulation, including the active ingredient, on the length of a meal once started). These distinctions have seldom been made, especially in field trials. Although Pickett and Stephenson (1980) showed that *D. reticulatum* responds to plant volatiles, it is not clear to what extent these animals use odour or any other sensation to locate food items or to what extent food items are encountered in the field by chance. None the less, Stephenson (1972) was able to show that bran is attractive to *D. reticulatum* and that the attractive component was readily extracted in water. Stephenson and Pickett (1977) showed that, in *D. reticulatum*, sucrose accounted for almost all the phagostimulant activity of aqueous extracts of 11 common plants, with glucose, fructose and some amino acids also, but weakly, active. Consistent with these results, Henderson *et al.* (1992) showed that addition of sucrose to agar gel markedly increased consumption by *D. reticulatum*, while the incorporation of other sugars was significantly less phagostimulatory. These animals did not respond to the incorporation in the gel of the artificial sweeteners saccharin and aspartame. Senseman (1977) showed that starch, while odourless and thus not attractive, was a powerful feeding stimulant in *Ariolimax californicus* Cooper (Arionidae).

Although metaldehyde by itself 'caught' some gastropods when applied in gardens (most of which were probably *D. reticulatum*), Barnes and Weil (1942) found that baits comprising metaldehyde and various foodstuffs (meals, powdered breadcrumbs) effected higher mortality. A mixture of metaldehyde and bran, and metaldehyde with non-foodstuffs such as sand and soil, caught few gastropods. Barnes and Weil (1942) concluded that the foodstuffs were the main attractant, while the metaldehyde was an arrestant. In a comparison of species responses, Daxl (cited by Godan, 1983) showed that *D. reticulatum* were attracted more to a vegetable diet than to bran or blood meal, but the limacid *Limacus flavus* (Linnaeus) preferred a carcass meal. Thomas (1948) found casein to be attractive to *T. budapestensis* and *D. reticulatum*, but not to *A. hortensis*. Howling (1991) found that *D. reticulatum* searched randomly in empty arenas and those with commercial metaldehyde baits, but were attracted from a distance to commercial methiocarb bait: the difference in attractiveness may be attributed to the bait formulation rather than the active ingredients.

Beer has long been known to be attractive to *D. reticulatum* and other gastropods (e.g. Taylor, 1902–1907), and Smith and Boswell (1970) found that commercial and laboratory-prepared methiocarb and metaldehyde baits were more attractive if moistened with beer. Frain (1981) suggested that the attractiveness of beer lay in the sediment and that the fermentation product (a mixture of yeast cells and wort pressed from the vats after fermentation) was more attractive than the beer. Fermenting barley malt is very attractive to *D. reticulatum* (Moens and Fraselle, 1980). However, field trials with *D. reticulatum* were inconclusive as to the benefit of adding yeast to metaldehyde-based baits as a means of increasing bait attractiveness (Frain, 1981).

Henderson *et al.* (1992) demonstrated that addition of four molluscicidal compounds at concentrations of 0.001% and above progressively reduced the volume of sucrose–agar gel ingested by *D. reticulatum*. Metaldehyde reduced ingestion more than methiocarb, ferric acetylacetonate was intermediate and ioxynil completely deterred feeding even at 0.001%. They concluded that there was little scope for improving the performance of baits with additives that attempt to increase palatability: of overriding importance is the effect of the molluscicide on feeding, and a major advance in bait effectiveness awaits the discovery of poisons that are more readily ingested.

Adjuvants

Various adjuvants are used in commercial molluscicide-bait products to resist weathering, reduce the growth of moulds and reduce the risks of inadvertent poisoning of mammals and birds, but the exact nature of these are often commercial secrets. Chlorosilanes as waterproofing agents were tried by Webley (1966), and Stephenson (1972) showed that gelatin bait (with an aqueous bran extract and an experimental molluscicide) hardened by formaldehyde did not disintegrate and continued to kill gastropods over a long period of wet weather. Fungicides may be used to delay microbial attack (Martin and Forrest, 1969), but can act as a repellent to gastropods (Webley, 1965). Materials such as nonanoic acid vanillylamide may be added to baits to discourage mammals from consuming baits (Schnorbach and Matthaei, 1990) and blue dye added to discourage birds (Martin and Forrest, 1969).

Size and density of pellets

For broadcast applications, Mallet (1968) recommended 28 baits m^{-2}, based on 40 bait particles g^{-1}. Working with bait pellets 2 mm × 5 mm to 5 mm × 10 mm in size, Hunter and Symonds (1970) reckoned that 25 baits m^{-2} (10–20 cm apart) was optimal for effecting mortality in *D. reticulatum* in the field. Their empirical assessment of optimal bait density was in agreement with the optimal bait distribution estimated theoretically from the model $P = 1 - e^{-2xy/A}$, where P is the probability of encounter, x the attraction area of a bait, y the distance moved by *D. reticulatum* while foraging and A the area within which each bait is placed (e.g. $A = 100$ if one bait occupies 100 units of area). As parameters in the model, Hunter and Symonds (1970) chose a foraging distance of 85 cm (based on measures of nightly distances travelled by individuals and Newell's (1966) observations that 40–50% of the movements of *D. reticulatum* were non-foraging), and an attraction area of 4 cm (estimated from the detection distance exhibited by *D. reticulatum* when offered the proprietary baits used in their experiments). Thus, if bait pellets are 5–10 cm apart, the probability of encounter, P, was estimated to closely approximate 1, if 20 cm apart, P was 0.82, but, if baits were 40 cm apart, P fell to 0.42. This model indicated that no gain in encounter frequency could be expected from reducing mean interbait distances below 10 cm. However, Hunter and Symonds (1970) recognized that the surface activity may be less in some gastropod species and may be reduced by weather and microtopographic conditions. This led Hunter and Symonds (1970) to recommend increasing the number of baits per unit area for field use (25–100 bait particles m^{-2}), either by higher quantities of bait per hectare or by smaller bait-particle size. Products with around 60–90 bait pellets g^{-1} (corresponding to approximately 2 mm × 4 mm) have since become common commercial practice.

Barker *et al.* (1991) developed baits with an inert core, which, at 130 bait particles g^{-1}, allows application rates exceeding those recommended by Hunter and Symonds (1970), without increasing the environmental load of the active ingredient. Barker *et al.* (1991) have shown that a commercial product of this inert-core formulation, with 1.8% metaldehyde by weight concentrated to 9% in the edible outer coating and applied at 130–260 baits m^{-2} (10–20 kg ha^{-1}), is effective against a range of gastropod species in various crop situations.

Early work using clumps of bait pellets is of limited relevance, except in indicating differential attractiveness between species – Thomas (1948) reckoned that the distance of attraction was 915 mm for *T. budapestensis* and 457 mm for *A. hortensis*. Mayer (1957) presented results for baits spread as a band around strawberries (*Fragaria* × *ananassa* Duchesne; Rosaceae) and lettuces (*Lactuca sativa* Linnaeus; Asteraceae). There is renewed interest in banded treatments, as this could provide necessary protection for crops from species that invade from the field edge: for example, Frank (1996) showed that most of the damage to oilseed rape (*Brassica napus* Linnaeus var. *oleifera* Linnaeus; Brassicaceae) in Switzerland arose from *Arion lusitanicus* Mabille invading from wild-flower strips. Although Frank (1996) found *D. reticulatum* far into fields, mark–release studies by Pinder (1969) and Hogan (1985) revealed no net movement of this species after an initial dispersal of 2–3 m from their release point. Furthermore, Fleming (1989) found considerable enzyme polymorphism in *D. reticulatum* within a field, suggesting little movement of animals.

The introduced Mediterranean helicid *Theba pisana* (Müller) and hygromiids *Cernuella virgata* (da Costa), *Cochlicella acuta* (Müller) and *Prietocella barbara* (Linnaeus) are pestiferous in cereal–pasture rotations in South and Western Australia (Baker, 1989; see also Baker, Chapter 6, this volume). Due to the fact that these species – particularly *C. virgata* – tend to migrate from areas of high density in pastures into the cereal crops indicates that strategic banding of baits along fence lines may offer an effective control strategy. Various helicid species are also pests in vines, shrubs and trees – *C. aspersus* and *Cepaea* spp. in blackcurrants (*Ribes nigrum* Linnaeus; Grossulariaceae) (Stringer and Morgan, 1956); and *C. aspersus* in citrus (*Citrus* Linnaeus spp.; Rutaceae) (Sakovich *et al.*, 1996, see also Sakovich, Chapter 17, this volume) – and baiting is generally restricted to periods when the animals are on the ground.

Factors Affecting Field Efficacy

Weather

It is well recognized that the prevailing weather strongly influences gastropod activity and consequently the efficacy of molluscicide treatments. Gastropods are for the most part active in crops at night or following rain on overcast days. In addition to weather conditions, other field parameters affect the efficacy of treatments: recovery is enhanced in cloddy soils, since sublethally poisoned species may shelter in crevices, and the competing attractiveness of the crop may prevent the pests from feeding sufficiently on the baits. Thirdly, weather also affects persistence of the bait (Port and Port, 1986).

Gastropods that have been poisoned by feeding on molluscicide-containing baits may recover under certain climatic conditions. This recovery from poisoning occurs where the molluscicide dose ingested is sublethal, either because repellency or early onset of toxicity reduces feeding at the bait, as discussed above, or because under the physiological and environmental conditions prevailing the ingested molluscicide that is normally lethal is reduced in cytological effect (Bieri and Schweizer, 1990; Triebskorn and Schweizer, 1990; Henderson and Triebskorn, Chapter 1, this volume).

Early studies of the action of metaldehyde attribute excessive mucus secretion as the most common cause of death in gastropods ingesting the material. While recording that metaldehyde stimulates water loss via mucus secretion, Cragg and Vincent (1952) noted that death is not simply due to desiccation. Webley (1965) counted gastropods (predominantly *D. reticulatum*) at sites where metaldehyde baits had been applied and documented that, of the animals present and exhibiting symptoms of poisoning in the morning, 15% had recovered and moved away by the afternoon of the same day. Martin and Forrest (1969) removed gastropods 'caught' under tiles baited with 10 g of methiocarb or metaldehyde baits to moist recovery chambers. They found that mortality of *D. reticulatum* in the recovery chambers had risen to 94% by the tenth day in the case of animals exposed to methiocarb, but only to 58% in animals exposed to metaldehyde. Bailey and Wedgwood (1991), similarly, found higher mortalities in both *D. reticulatum* and *Arion distinctus* Mabille resulting from single meals on 4% methiocarb baits than on metaldehyde baits when placed in moist recovery chambers. Godan (1983) summarized the reversal of the early optimistic views on metaldehyde: 'disadvantages became apparent . . . such as its great weather dependency and the resulting rapid recovery of slugs which at first appeared lethally poisoned'. There is a common perception that the problem of recovery is largely confined to gastropods feeding on metaldehyde baits, although Triebskorn and Schweizer (1990) have shown that irreversible cellular damage occurs in *D. reticulatum* following ingestion of metaldehyde, even under wet conditions. Moreover, Kemp and Newell (1985) suggest that metaldehyde-poisoned *D. reticulatum* which recover show no aversion to a second metaldehyde meal and are less likely to survive the further ingestion.

Higher temperatures enhance the toxicity of metaldehyde (Thomas, 1945; Cragg and Vincent, 1952; Webley, 1965; Wright and Williams, 1980), although it is not possible to determine whether this is due to higher amounts being ingested at higher temperatures. Mallet and Bougaran (1971) described laboratory experiments that showed that metaldehyde effected higher mortality in *D. reticulatum* above 20°C than did methiocarb, while the converse applied at temperatures of 16°C and below. More recent work by Bieri and Schweizer (1990) demonstrated that metaldehyde effected mortality in *D. reticulatum* at temperatures as low as 2°C, although the mortality took longer to occur. At 5°C, for example, mortality in metaldehyde-dosed animals was 55% by day 10, rising to 90% by day 20, while at 15°C 100% occurred by day 10.

Species and age differences in susceptibility

In many field situations, a mixture of gastropod species and ages are present and these differ in their susceptibility to molluscicides and to different bait formulations. These differences are sufficiently large to make it necessary, in situations where a particular species is a problem, for the choice of formulation to be considered. Godan (1983) compiled a series of decreasing resistance to molluscicides based upon published reports, with *T. budapestensis* being least sensitive and *D. reticulatum* most sensitive among European slug forms. Thomas (1948) reported that snails are more sensitive than slugs, and Coupland (1996) recorded large differences in the efficacy of metaldehyde formulations against different helicids and hygromiids: *C. acuta* was much more resistant than *T. pisana* and *C. virgata*. Moreover, species differ in their response to different formulations, and Crawford-Sidebotham (1970) concluded that methiocarb and metaldehyde baits give biased catches of different species. Airey (1986) also reported differences in susceptibility of *A. hortensis* and *D. reticulatum* to methiocarb bait, although no difference was found with metaldehyde bait. The lower temperatures used by Airey (1986) and the different alternative food sources could account for the differences between the findings of Airey (1986) and Crawford-Sidebotham (1970).

Juveniles of limacid species are reported as less susceptible to the carbamate isolan than are hatchlings or reproductive adults (Godan, 1983, Table 20). In contrast, the agriolimacid *D. reticulatum* was equally susceptible as juveniles and adults. Frain and Newell (1982) suggested that juvenile *D. reticulatum* were less likely to eat molluscicidal baits and, indeed, Kemp and Newell (1985) found that *D. reticulatum* less than 0.8 g were less susceptible to poisoning by methiocarb or metaldehyde bait formulations.

Effect of competing food resources

The formula used by Hunter and Symonds (1970) to estimate optimum bait application rates (see above) assumes that any individual gastropod will consume any bait to which it is attracted or which it randomly encounters in a treated field. This may not be so if that animal has recently fed on the crop or weeds: baits compete with the crop plant and other food resources for the gastropods' attention. Doucette (1984) found that metaldehyde bait was effective in causing mortality in *Arion ater* (Linnaeus) on fallow ground, but, when applied in vegetable or flower gardens, it was only moderately effective, apparently because bait attraction to the animals was reduced in the presence of plant foliage. Airey (1986) found that, while methiocarb baits were more effective at killing *A. hortensis* than metaldehyde baits when no alternative food was present, the difference in effectiveness disappeared if potato (*Solanum* tuberosum Linnaeus; Solanaceae) leaf discs were provided as alternative food. However, in experiments with *D. reticulatum*, the presence of leaf discs did not reduce methiocarb's superiority.

Bourne *et al.* (1988) also found that *D. reticulatum* suffered lower mortality when baits were given in the presence of alternative food, even though, when introduced into terraria, they invariably fed on molluscicide baits (4 and 6% metaldehyde and 4% methiocarb) in preference to crop seedlings. However, although most animals were initially paralysed by the ingested molluscicide and damage was slight, by day 14 the majority of the *D. reticulatum* had recovered and the crop was severely damaged. Crop protection lasted for only 4–5 days. Glen and Orsman (1986), similarly, state that protection is greatest in the first 4 days.

Frain and Newell (1982) found that the 24 h food consumption by *D. reticulatum* declines over successive days when the animals had access to only one food type. They suggested that, since this decline in consumption can be overcome by the presentation of novel foods, an effective control strategy in the field may be to apply a mixture of baits, incorporating different attractants and phagostimulants. Foraging gastropods would then find a variety of baits, each likely to stimulate an enhanced meal size. Inherent in this idea is that gastropods exhibit 'neophilia' in their food preferences and that gastropods recover from sublethal ingestion of molluscicides.

Learning and habituation

Symonds (1975) suggested that gastropods feeding on baits and consuming sublethal doses of metaldehyde are more susceptible to subsequent poisoning. Both Gelperin (1975), working on *Limax maximus* Linnaeus (Limacidae), and Kemp and Newell (1985), working on *D. reticulatum*, showed that animals that had fed on a metaldehyde-containing bait and survived were not deterred from feeding on another. Further, Kemp and Newell (1985) showed that the mortality rate among individuals that survived and subsequently fed on a second metaldehyde bait was higher than that expected in naïve animals, suggesting a cumulative effect. Although methiocarb meals also produced no aversive conditioning, except perhaps in juveniles, Kemp and Newell (1985) could find no evidence of cumulative vulnerability to methiocarb.

Bait persistence

Testing of persistence is fundamental to commercial product development, but generally the results of such tests are not published. In a field-cage experiment (with known numbers of *D. reticulatum* added to

1 m^2 plots at 6-day intervals), Chabert (1996) found that a commercial 5% metaldehyde bait formulation caused 90% mortality during the first 6-day period and 60% mortality during the next 6 days, but caused only 13% mortality during the third 6-day period. In comparison, a commercial 4% methiocarb bait effected low mortalities for the first and second 6-day periods (68 and 34%, respectively). In a second experiment with the same commercial formulations, the metaldehyde-based bait efficacy was above 80% during the first 6-day period in both wet and dry conditions, but declined more in wet (to 13%) than in dry (to 57%) conditions in the next 6-day period. Methiocarb-based baits again effected lower mortality than did those containing metaldehyde, and also declined in efficacy faster in the wet conditions.

Resistance

Carbamate compounds are acetylcholinesterase (AChE) inhibitors, and resistance to carbamate insecticides based on reduced sensitivity to AChE is known in many insects. Pessah and Sokolove (1983) explored the interaction of AChE-like enzymes from *L. maximus* with several carbamates. Low-toxicity carbamates, such as dioxacarb, strongly inhibit the enzyme in the haemolymph but weakly inhibit AChE in the foot tissues, while methiocarb and its sulphoxide strongly inhibit both. Young and Wilkins (1989) investigated the activity of AChEs from *D. reticulatum*, *A. distinctus*, *T. budapestensis* and the earthworm *Lumbricus terrestris* Linnaeus (Lumbricidae) 5 days after a dermal application of methiocarb. The five isoenzymes of AChE from *D. reticulatum* varied in their sensitivity to inhibition by methiocarb, although the activity (K_m) and concentration of inhibitor required to reduce activity by 50% (I_{50}) did not differ significantly between the survivors from the methiocarb-treated group and untreated controls. AChEs from *D. reticulatum, A. distinctus* and *T. budapestensis* exhibited low sensitivity (high I_{50}s) to methiocarb, compared with insects, and AChE from *D. reticulatum* was 30 times less sensitive than that from *L. terrestris*. AChE activity in *D. reticulatum* was not inhibited by metaldehyde.

If gastropod AChE isoenzymes are genetically based, AChE sensitivity might potentially limit the effectiveness of control measures. However, Young and Wilkins (1989) considered that resistance is unlikely to develop with present control measures because the selection pressure exerted by baits is low and affects only a proportion of the gastropod population. However, if control measures regularly reduce populations to low numbers or a high percentage of the population ingests sublethal doses, then selection pressure may increase to the point where resistance becomes important. Fisher and Orth (1975) reported the possible resistance of *C. aspersus* in Oregon to metaldehyde.

Field Assessment

While laboratory tests generally support the view that methiocarb-based bait products are more effective against gastropod populations than those based on metaldehyde, the results of many field trials indicate similar levels of efficacy (Port and Port, 1986). Where differences in field efficacy have been reported, methiocarb was more often rated better. However, efficacy comparisons in the field have often been based on counts of dead animals, and this section details some inadequacies of that approach.

Symonds (1975) suggested that, where slug forms were the target pests, this method of assessment would underestimate the performance of metaldehyde, because some of the animals sublethally poisoned by this material and consequently not included in the dead-animal counts may subsequently ingest a lethal dose.

Frain and Newell (1983) pointed out that the method of estimating bait effectiveness by comparing the numbers of dead animals lying on the surface of treated plots assumes that molluscicidal materials act in comparable ways, such that the relationships between numbers found and numbers killed apply equally across the treatments. Although their study of slug populations (predominantly *D. reticulatum*) on a pasture of white clover (*Trifolium repens* Linnaeus; Fabaceae) and ryegrass (*Lolium perenne* Linnaeus; Gramineae) revealed significantly more dead gastropods on methiocarb-treated plots than on metaldehyde-treated ones, estimates of the residual populations by trapping showed no differences in the abundance of adult *D. reticulatum* on methiocarb- and metaldehyde-treated plots (although abundances were lower than on untreated control plots). Frain and Newell (1983) suggested that the different modes of action of metaldehyde and methiocarb left more methiocarb-poisoned *D. reticulatum* on the surface, leading to a serious underestimation of the effectiveness of metaldehyde. Frain and Newell (1983) also pointed out that the pale, bloated bodies characteristic of carbamate-poisoned *D. reticulatum* are more visible and thus more likely to be recorded than the dark, shrivelled bodies of metaldehyde-poisoned animals, possibly leading to a further underestimation of the relative effectiveness of metaldehyde. However, the video recordings of Bailey and Wedgwood (1991) tend to refute Frain and Newell's (1983) explanation for the discrepancy between numbers of dead animals on the surface and the residual populations of live animals. The video recordings showed that, after feeding on methiocarb baits, *D. reticulatum* moved over 2 m (further than animals fed baits free of molluscicide) and fed again and 62% returned to shelters by dawn (compared with 90% of unpoisoned animals), while metaldehyde-fed animals were rapidly immobilized (moving less than 1 m) and seldom fed again and only 8% regained shelters; these observations suggest that it is the metaldehyde-poisoned animals that would be more likely to be found on the surface after dawn.

While the explanation offerred by Frain and Newell (1983) may be disputed, they admirably focus attention on the need to assess the effectiveness of a molluscicide treatment in reducing gastropod populations, which is based on a comparison of populations estimated by trapping before and after molluscicide treatment. In their comparison of residual live populations and counts of dead animals as approaches to estimating efficacy, treatments with methiocarb baits produced significantly higher numbers of dead animals than those of metaldehyde, but the numbers of animals subsequently active on the plots (caught in traps) were similar for methiocarb- and metaldehyde-treated plots. Glen *et al.* (1991) found that methiocarb-treated plots initially showed a 70% reduction in gastropod numbers (principally *D. reticulatum* and five *Arion* species) compared with untreated plots, but abundance in untreated plots declined in subsequent months to similar levels. This highlights the dynamic nature of gastropod populations and the inappropriateness of reliance on the count of dead animals as the criterion for assessing molluscicide efficacy.

In most crop situations, prevention of yield loss is the primary objective of molluscicide applications. It follows, therefore, that crop yield should be the primary measure of molluscicide-treatment effectiveness. There are several reasons why other methods are used instead, which include the longer duration of trials, during which time several compounding factors may operate on the crop to obscure the performance of the bait. Therefore a number of partial parameters of performance are used. G.M. Barker (personal communication, 2000) comments on the combination of measures used to estimate the efficacy of a product by Barker *et al.* (1991). These measures range from estimates of mortality in replicated field cages seeded with known standard numbers of live gastropods in each cage at the start of the trial, through estimation of the residual population of gastropods to sampling plant loss or damage, to estimates of crop yield. All these methods have good statistical power. However, estimates of mortality of the pest or of the pest's residual population are in themselves poor indicators of the efficacy if the relationship between gastropod numbers and yield reduction is not well established. Barker notes that in some cases a high mortality may not be necessary if all that is required is temporary relief from gastropod grazing pressure: the inhibition of feeding of sufficient of the gastropod population over the critical phase of crop development is the important concern.

Effects on Non-target Organisms

Gimingham (1940) found that metaldehyde was ineffective against soil arthropod pests, such as elatrid and tipulid larvae, and isopods. Metaldehyde-containing baits have no measurable effect on earthworms (Bieri *et al.*, 1989a), and adverse effects have been recorded for only one species of carabid (*Carabus granulatus* Linnaeus) (Buchs *et al.*, 1989).

Methiocarb has insecticidal activity and Martin and Forrest (1969) reported that certain fauna other than gastropods were killed in small numbers by bait formulations, including earthworms, carabid and staphylinid beetles, phalangids, tipulid larvae and isopods. Effects on earthworms were studied by Symonds (1975) and Bieri *et al.* (1989a) and on carabids by Buchs *et al.* (1989). However, at the recommended application rates, methiocarb baits are not expected to have ecologically significant adverse effects on earthworm populations, based on the results of laboratory tests by Wellmann and Heimbach (1996). Populations of soil-dwelling collembolans, acarines and proturans are not significantly affected by methiocarb baiting (Martin *et al.*, 1969). Kelly and Curry (1985) found that a single application of methiocarb bait did not reduce abundance in members of the predator and decomposer fauna of winter wheat, but Bieri *et al.* (1989b) documented a reduced abundance of staphylinid and carabid beetles following application of methiocarb bait in grassland. Purvis (1996) concluded that, although methiocarb-based baits are toxic to carabids that are active at the time of the application, the longer-term impact on carabid populations is negligible for most species, as indicated by population recoveries between annual applications to arable crops in Britain. However, populations of one carabid species monitored by Purvis (1996), namely *Bembidium obtusum* Serv., which lacks a distinct dispersal phase, showed no such recovery. In the Boxworth project in Britain (Burn, 1992), a broadcasting treatment of methiocarb baits in cereal crops reduced the level of predation on sentinel prey during winter. In the same experiment, methiocarb baits incorporated into the soil at drilling had no effect on predation rate. Ester and Geelen (1996) demonstrated that methiocarb pellets can be combined with the biocontrol nematode *Phasmarhabditis hermaphrodita* (Schneider) (Rhabditidae) to achieve effective integrated control of *D. reticulatum* in sugar beet (*Beta vulgaris* Linnaeus; Chenopodiaceae) grown in a rye (*Secale cereale* Linnaeus; Gramineae) cover crop.

Metaldehyde is toxic to vertebrates, producing reproductive impairment and posterior paralysis (see review by Booze and Oehme, 1985). The pharmacology of metaldehyde poisoning is very incompletely known. In the mammalian digestive tract, metaldehyde is depolymerized to acetaldehyde, which accumulates in the bloodstream to toxic levels. If metaldehyde exerts some of its actions in gastropods by the intermediary of acetaldehyde, then it may be possible to enhance its action by delaying the oxidation of acetaldehyde by, for example, disulphuram, as used in alcohol-aversion therapy in humans (Brien and Loomis, 1983). Early formulations of molluscicidal baits were acceptable to domesticated animals, and Homeida and Cooke (1982) listed several examples of poisoning. The contemporary use of baits of small pellet size and containing a mammalian repellent have reduced this danger (Kitchell *et al.*, 1978). Fletcher and Hardy (1983) considered that there were surprisingly few reported incidents of mammal poisoning from molluscicides in Britain, given the quantities applied annually to garden and agricultural crops. However, such poisoning cases do occur – usually involving dogs (*Canis domesticus* Linnaeus; Canidae) feeding on metaldehyde-bait products. Many of the deaths in dogs were the result of animals gaining access to the product in insecure storage. In Australia, 82% of poisoning cases seen by veterinary practices were due to molluscicidal baits (Studdert, 1985).

Martin and Forrest (1969) found no evidence of adverse effects on wildlife from field trials involving the application of a commercial methiocarb-bait product. However, Tarrant and Westlake (1988) showed that methiocarb baits were toxic to the woodmouse *Apodemus sylvaticus* Linnaeus (Muridae) and, with consideration given to the animal's home range and the quantities of baits applied, they considered the practice of methiocarb-bait application to cereal crops to present a risk to local populations of *A. sylvaticus* in Britain. Johnson *et al.* (1992), in the Boxworth project, found that surface application of methiocarb baits killed most *A. sylvaticus* in the fields within 2–3 days, but the population recovered quickly due to immigration of juvenile mice into the fields. The effects were greatly reduced when the baits were drilled rather than broadcast. However, Johnson *et al.* (1992) could find no long-term effect on *A. sylvaticus* numbers and noted that autumn applications of methiocarb baits to winter wheat coincided with the period of maximum abundance and dispersal in mice populations. It is recognized, however, that methiocarb-bait applications in early summer or to fields isolated from woods and copses from which animals could immigrate may have more serious effects on *A. sylvaticus* populations.

Johnson *et al.* (1992) produced evidence from a laboratory trial that secondary poisoning of *A. sylvaticus* by feeding on gastropods killed by methiocarb was unlikely. However, other wildlife may be at risk from the consumption of poisoned gastropods, although much of the information is anecdotal (South, 1992). Keymer *et al.* (1991) found concentrations of up to 80 mg kg^{-1} acetaldehyde in dead hedgehogs, *Erinaceus europaeus* Linnaeus (Erinaceidae). Gemmeke (1997) exposed *E. europaeus* to *D. reticulatum* that had been poisoned by metaldehyde or methiocarb baits. None of the six animals given access to metaldehyde-poisoned *D. reticulatum* exhibited symptoms, despite four of these animals each consuming 196–200 of these gastropods in one night (4.1–5.1% body weight). Of 11 *E. europaeus* exposed to methiocarb-poisoned *D. reticulatum*, one died after eating 25 poisoned gastropods, another showed locomotor disturbances for some hours after consuming 98 and nine showed no symptoms, despite some individuals consuming more than 100 poisoned *D. reticulatum*.

Conclusions

Molluscicidal baits containing metaldehyde or some carbamates, such as methiocarb or thiodicarb, can give good control of gastropods, but they often fail to do so. The principal problem, restated by many different authors, is that of persuading the gastropod to consume a lethal dose: if the concentration is low, then the animal suffers the effects of poisoning and stops eating too soon and, if the concentration is too high, the animal will reject the bait. Chemical masking does not work, but physical masking in the form of microencapsulation of the active ingredients might repay more attention, as might mixtures of active ingredients. Consideration of synergists has probably been delayed by our lack of knowledge of the mode of action of toxins in gastropods.

The area where advances have been made is in getting the gastropod in contact with the bait, by the use of attractants or by the adoption of smaller pellets, which give more 'killing points' without increasing the application rate. Perhaps the future will bring greater attention to focal applications, rather than whole-field broadcasting. There have also been advances in our methods of study, with the realization that the number of gastropods killed is less important than the number left unaffected, and that loss of crop yield or market quality is the crucial parameter. Useful distinctions have been made between bait attractiveness and palatability. We are also now less likely to ignore subtle factors such as cumulative action of metaldehyde over several meals, the presence of the crop as a competing food resource and changing individual and different species' food preferences. Perhaps too much attention in the past has been paid to just one species – *D. reticulatum*?

In the future, baits will be subjected to tighter environmental protection constraints, and chemical control of gastropod pests may develop as repellents and seed dressings. However, baits appear to be compatible with other elements of integrated pest management, such as the use of biocontrol nematodes.

References

Airey, W.J. (1986) The influence of an alternative food on the effectiveness of prorietary molluscicidal pellets against two species of slugs. *Journal of Molluscan Studies* 52, 206–213.

Bailey, S.E.R. and Wedgwood, M.A. (1991) Complementary video and acoustic recordings of foraging by two pest species of slugs on non-toxic and molluscicidal baits. *Annals of Applied Biology* 119, 163–176.

Bailey, S.E.R., Cordon, S. and Hutchinson, S. (1989) Why don't slugs eat more bait? A behavioural study of early meal termination produced by methiocarb and metaldehydebaits in *Deroceras reticulatum*. In: Henderson, I.F. (ed.) *Slugs and Snails in World Agriculture*. Monograph 41, British Crop Protection Council, Thornton Heath, pp. 385–390.

Baker G.H. (1989) Damage, population dynamics, movement and control of pest helicid snails in southern Australia. In: Henderson, I.F. (ed.) *Slugs and Snails in World Agriculture*. Monograph 41, British Crop Protection Council, Thornton Heath, pp. 175–185.

Barker, G.M., Pottinger, R.P., Lloyd, J.M., Addison, P.J., Firth, A.C. and Stewart, A.P. (1991) A novel bait formulation for slug and snail control. *Proceedings of the New Zealand Weed and Pest Control Conference* 44, 195–200.

Barnes, H.F. and Weil, J.W. (1942) Baiting slugs using metaldehyde mixed with various substances. *Annals of Applied Biology* 29, 56–58.
Bieri, M. and Schweizer, H. (1990) Effet létal du metaldehyde sur la petite Limace grise (*Deroceras reticulatum*) maintenue sur un surface de contact détrempée en relation avec la dose de matière active et la température. In: *ANPP – Deuxième Conférence Internationale sur les Ravageurs en Agriculture, Versailles*, Vol. 1, pp. 151–158.
Bieri, M., Schweizer, H., Christensen, K. and Daniel, O. (1989a) The effect of metaldehyde and methiocarb pellets on *Lumbricus terrestris* Linne. In: Henderson, I.F. (ed.) *Slugs and Snails in World Agriculture.* Monograph 41, British Crop Protection Council, Thornton Heath, pp. 237–244.
Bieri, M., Schweizer, H., Christensen, K. and Daniel, O. (1989b) The effect of metaldehyde and methiocarb pellets on surface welling organisms in grassland. In: Henderson, I.F. (ed.) *Slugs and Snails in World Agriculture.* Monograph 41, British Crop Protection Council, Thornton Heath, pp. 391–392.
Booze, T.F. and Oehme, F.W. (1985) Metaldehyde toxicity: a review. *Veterinary and Human Toxicology* 27, 11–19.
Bourne, N.B., Jones, G.W. and Bowen, I.D. (1988) Slug feeding behaviour in relation to control with molluscicidal baits. *Journal of Molluscan Studies* 54, 327–338.
Bourne, N.B., Jones, G.W. and Bowen, I.D. (1990) Feeding behaviour and mortality of the slug *Deroceras reticulatum* in relation to control with molluscicidal baits containing various combinations of metaldehyde with methiocarb. *Annals of Applied Biology* 117, 455–468.
Bowen, I.D. and Jones, G.W. (1985) Getting pesticides into cells. *Industrial Biotechnology* 5, 29–32.
Bowen, I.D., Antoine, S. and Martin, T.J. (1996) Reformulation studies with methiocarb. In: Henderson, I.F. (ed.) *Slug and Snail Pests in Agriculture.* Symposium Proceedings 66, British Crop Protection Council, Farnham, pp. 397–404.
Brien, J.F. and Loomis, C.W. (1983) Pharmacology of acetaldehyde. *Canadian Journal of Physiology and Pharmacology* 61, 1–22.
Buchs, W., Heimbach, U. and Czarnecki, E. (1989) Effects of snail baits on non-target carabid beetles. In: Henderson, I.F. (ed.) *Slugs and Snails in World Agriculture.* Monograph 42, British Crop Protection Council, Thornton Heath, pp. 245–252.
Burn, A.J. (1992) Interactions between cereal pests and their predators and parasites. In: Greig-Smith, P., Frampton, G. and Hardy, A. (eds) *Pesticides, Cereal Farming and the Environment: the Boxworth Project.* HMSO, London.
Chabert, A. (1996) Active duration of molluscicides. In: Henderson, I.F. (ed.) *Slug and Snail Pests in Agriculture.* Symposium Proceedings 66, British Crop Protection Council, Farnham, pp. 173–180.
Coupland, J.B. (1996) The efficacy of metaldehyde formulations against helicid snails: the effect of concentration, formulation and species. In: Henderson, I.F. (ed.) *Slug and Snail Pests in Agriculture.* Symposium Proceedings 66, British Crop Protection Council, Farnham, pp. 151–156.
Cragg, J.B. and Vincent, M.H. (1952) The action of metaldehyde on the slug *Agriolimax reticulatus* (Müller). *Annals of Applied Biology* 39, 392–406.
Crawford-Sidebotham, T.J. (1970) Differential susceptibility of species of slugs to metaldehyde/bran and to methiocarb baits. *Oecologia* 5, 303–324.

Crowell, H.H. (1967) Slug and snail control with experimental poison baits. *Journal of Economic Entomology* 60, 1048–1050.
Crowell, H.H. (1977) *Chemical Control of Terrestrial Slugs and Snails.* Bulletin 628, Oregon Agricultural Experimental Station, 70 pp.
Doucette, C.F. (1984) An introduced slug, *Arion ater* (L.), and its control with metaldehyde. *Journal of Economic Entomology* 47, 370.
Ester, A. and Geelen, P.M.T. (1996) Integrated control of slugs in a sugar beet crop growing in a rye cover crop. In: Henderson, I.F. (ed.) *Slug and Snail Pests in Agriculture.* Symposium Proceedings 66, British Crop Protection Council, Farnham, pp. 445–450.
Ferguson, C.M., Firth, A.C. and Wyatt, R.T. (1995) Evaluation of a new slug bait incorporating thiodicarb. *Proceedings of the New Zealand Plant Protection Conference* 48, 314–317.
Fisher, T. and Orth, R.E. (1975) Differential mortality of brown garden snail to metaldehyde. *California Agriculture* 29(6), 7–8.
Fleming, C.C. (1989) Population structure of *Deroceras reticulatum.* In: Henderson, I.F. (ed.) *Slugs and Snails in World Agriculture.* Monograph 41, British Crop Protection Council, Thornton Heath, pp. 413–415.
Fletcher, M.R. and Hardy, A.R. (1983) *Research and Development Report: Pesticide Science 1982.* Reference Book of the Ministry of Agriculture, Fisheries and Food 252, HMSO, London.
Frain, M.J. (1981) Chemoreception and feeding in the grey field slug *Deroceras reticulatum* (Müller) with reference to molluscicide formulation. PhD thesis, University of London, London.
Frain, J.M. and Newell, P.F. (1982) Meal size and a feeding assay for *Deroceras reticulatum* (Mull). *Journal of Molluscan Studies* 48, 98–99.
Frain, J.M. and Newell, P.F. (1983) Testing molluscicides against slugs: the importance of assessing the residual population. *Journal of Molluscan Studies* 49, 164–173.
Frank, T. (1996) Sown wildflower strips in arable land in relation to slug density and slug damage in rape and wheat. In: Henderson, I.F. (ed.) *Slug and Snail Pests in Agriculture.* Symposium Proceedings 66, British Crop Protecton Council, Farnham, pp. 289–296.
Garthwaite, D.G. and Thomas, M.R. (1996) The usage of molluscicides in agriculture and horticulture in Great Britain over the last 30 years. In: Henderson, I.F. (ed.) *Slug and Snail Pests in Agriculture.* Symposium Proceedings 66, British Crop Protection Council, Farnham, pp. 39–46.
Gauliard, J. and Laverriere, P. (1989) Skipper: nouvel antilimace. *Phytoma* 406, 55–56.
Gelperin, A. (1975) Rapid food aversion learning by a terrestrial mollusc. *Science* 189, 567–570.
Gemmeke, H. (1997) Investigations on the threat to hedgehogs posed by poisoned slugs. *Meta News* 5, 7pp.
Gimingham, C.T. (1940) Pests of vegetable crops. *Annals of Applied Biology* 27, 167–168.
Glen, D.M. and Orsman, I.A. (1986) Comparison of molluscicides based on metaldehyde, methiocarb or aluminium sulphate. *Crop Protection* 5, 371–375.
Glen, D.M., Wiltshire, C.W. and Butler, R.C. (1991) Slug population changes following molluscicide treatment in relation to distance from edge of treated area. *Crop Protection* 10, 408–412.

Godan, D. (1983) *Pest Slugs and Snails: Biology and Control.* Springer Verlag, Berlin, 445pp.

Heim, I.U., Blum, R.A. and Loliger, A.W. (1996) Producing and marketing a molluscicide. In: Henderson, I.F. (ed.) *Slug and Snail Pests in Agriculture.* Symposium Proceedings 66, British Crop Protection Council, Farnham, pp. 47–52.

Henderson, I.F. and Parker, K.A. (1986) Problems in developing chemical control of slugs. *Aspects of Applied Biology* 13, 341–347.

Henderson I.F., Martin, A.P. and Perry, J.N. (1992) Improving slug baits: the effects of some phagostimulants and molluscicides on ingestion by the slug, *Deroceras reticulatum* (Müller) (Pulmonata: Limacidae). *Annals of Applied Biology* 121, 423–430.

Hogan, J.M. (1985) The behaviour of the grey field slug *Deroceras reticulatum* (Müller), with particular reference to control in winter wheat. Unpublished PhD thesis, University of Newcastle-upon-Tyne.

Homeida, A.M. and Cooke, R.G. (1982) Pharmacological aspects of metaldehyde poisoning in mice. *Journal of Veterinary Pharmacology and Therapeutics* 5, 77–82.

Howling, G.G. (1991) Slug foraging behaviour: attraction to food items from a distance. *Annals of Applied Biology* 119, 147–153.

Hunter, P.J. and Symonds, B.V. (1970) The distribution of bait pellets for slug control. *Annals of Applied Biology* 65, 1–7.

Johnson, I.P., Flowerdew, J.R. and Hare, R. (1992) Populations and diet of small rodents and shrews in relation to pesticide usage. In: Greig-Smith, P., Frampton, G. and Hardy, A. (eds) *Pesticides, Cereal Farming and the Environment: the Boxworth Project.* HMSO, London.

Kelly, C.R., Greenwood, S. and Bailey, S.E.R. (1996) Can different pH environments in slug digestive tracts be exploited to improve the efficacy of molluscicide baits? In: Henderson, I.F. (ed.) *Slug and Snail Pests in Agriculture.* Symposium Proceedings 66, British Crop Protection Council, Farnham, pp. 83–90.

Kelly, M.T. and Curry, J.P. (1985) Studies on the arthropod fauna of a winter wheat crop and its response to the pesticide methiocarb. *Pedobiologia* 28, 413–421.

Kemp, N.J. and Newell, P.F. (1985) Laboratory observations on the effectiveness of methiocarb and metaldehyde baits against the slug *Deroceras reticulatum* (Müll.). *Journal of Molluscan Studies* 51, 228–229.

Keymer, I.F., Gibson, E.A. and Reynolds, D.J. (1991) Zoonoses and other findings in hedgehogs (*Erinaceus europaeus*): a survey of mortality and review of the literature. *Veterinary Record* 128, 245–249.

Kitchell, R.L., Schubert, T.A., Mull, R.L. and Knaak, J.B. (1978) Palatability studies of snail and slug poisons using dogs. *Journal of the American Veterinary Medicine Association* 173, 85–90.

Kuhr, R.J. and Dorough, H.W. (1976) *Carbamate Insecticides: Chemistry, Biochemistry and Toxicology.* CRC Press, Cleveland, Ohio.

Maas, W. (1978) Influence of particle size on pesticidal activity. In: Geissbuhler, G., Brooks, G.T. and Kearney, P.C. (eds) *Advances in Pesticide Science (Part 3): 4th International Congress of Pesticide Chemistry, Zurich*, pp. 772–779.

McCrohan, C.R., Mills, J.D., Cheng, S.C. and Bailey, S.E.R. (1995) Inhibition of feeding responses by the molluscicide, metaldehyde. *Acta Biologica Hungarica* 46, 241–245.

Mallet, C. and Bougaran, H. (1971) Action molluscicide de divers carbamates. *Mededelingen Landbouwhogeschool en de Opzoekingsstations van de Staat de Gent*, 207–215.

Mallet, M. (1968) Expérimentation sur les appats dans la lutte contre les limaces. *La Défense des Végétaux* 22, 253–264.

Martin, T.J. and Forrest, J.D. (1969) Development of ^{R}Draza in Great Britain. *Pflanzenschutz-Nachrichten Bayer* 22, 205–243.

Martin, T.J., Davis, M.E. and Morris, D.B. (1969) Development work with methiocarb in Great Britain. In: *Proceedings of the 5th British Insecticide and Fungicide Conference*, Vol. 2, pp. 434–441.

Mayer, K. (1957) Die Schneckenbekampfung mit Metaldehydpraparaten. *Nachrichtenblatt der Deutschen Pflanzenschutzd, Braunschweig* 9, 36–41.

Moens, R. and Fraselle, J. (1980) La lutte contre les limaces en cultures d'orchidées. *Mededelingen Faculteit Landbouw Rijksuniversitet Gent* 45, 613–625.

Newell, P. (1966) The nocturnal activity of slugs. *Medical and Biological Illustration* 16, 146–156.

Pessah, I.N. and Sokolove, P.G. (1983) The interaction of organophosphate and carbamate insecticides with cholinesterases in the terrestrial pulmonate *Limax maximus. Comparative Biochemistry and Physiology* 74C, 291–297.

Pickett, J.A. and Stephenson, J.W. (1980) Plant volatiles and components influencing behaviour of the field slug *Deroceras reticulatum* (Müll.). *Journal of Chemical Ecology* 6, 435–444.

Pinder, L.C.V. (1969) The biology and behaviour of some slugs of economic importance, *Agriolimax reticulatus, Arion hortensis* and *Milax budapestensis.* Unpublished PhD thesis, University of Newcastle-upon-Tyne.

Port, C.M. and Port, G.R. (1986) The biology and behaviour of slugs in relation to crop damage and control. *Agricultural Zoology Reviews* 1, 255–299.

Purvis, G. (1996) The hazard posed by methiocarb slug pellets to carabid beetles: understanding population effects in the field. In: Henderson, I.F. (ed.) *Slug and Snail Pests in Agriculture.* Symposium Proceedings 66, British Crop Protection Council, Farnham, pp. 189–196.

Sakovich, N.J. (1996) An integrated pest management (IPM) approach to the control of the brown garden snail, (*Helix aspersa*) in California citrus orchards. In: Henderson, I.F. (ed.) *Slug and Snail Pests in Agriculture.* Symposium Proceedings 66, British Crop Protection Council, Farnham, pp. 283–287.

Schnorbach, H.-J. and Matthaei, H.-D. (1990) Molluscicides. In: *Ullman's Encyclopaedia of Industrial Chemistry*, Vol. 16, pp. 649–653.

Schroder, U. (1985) Crystallized carbohydrate spheres for slow release and targetting. *Methods in Enzymology* 112, 116–128.

Senseman, D.M. (1977) Starch: a potent feeding stimulant for the terrestrial slug *Ariolimax californicus. Journal of Chemical Ecology* 3, 707–715.

Smith, F. and Boswell, A.L. (1970) New baits and attractants for slugs. *Journal of Economic Entomology* 63, 1919–1922.

South, A. (1992) *Terrestrial Slugs: Biology, Ecology and Control.* Chapman & Hall, London, 428pp.

Stephenson, J.W. (1972) Gelatin as a carrier for *S*2-cyanoethyl *N*-[(methylcarbamoyl)oxy]thioacetimidate, an experimental molluscicide. *Pesticide Science* 3, 81–87.

Stephenson, J.W. and Pickett, J.A. (1977) Responses of slugs to plant constituents. In: *Rothamsted Experimental Station Report for 1977*, Part 1, p. 100.

Stringer, A. and Morgan, N.G. (1969) Population and control of snails in blackcurrant plantations. In: *Proceedings of the 5th Insecticide and Fungicide Conference*, pp. 453–457.

Studdert, V.P. (1985) Epidemiological features of snail and slug bait poisoning in dogs and cats. *Australian Veterinary Journal* 62, 269–271.

Symonds, B.V. (1975) Evaluation of potential molluscicides for the control of the field slug, *Agriolimax reticulatus* (Müll.). *Plant Pathology* 24, 1–9.

Tarrant and Westlake (1988) Laboratory evaluation of the hazard to woodmice *Apodemus sylvaticus* from the agricultural use of methiocarb molluscicide pellets. *Bulletin of Environmental Contamination and Toxicology* 40, 147–152.

Taylor, J.W. (1902–1907) *Monograph of the Land and Freshwater Mollusca of the British Isles (Testacellidae, Limacidae, Arionidae)*, Parts 8–13. Taylor Brothers, Leeds.

Thomas D.C. (1948) The use of metaldehyde against slugs. *Annals of Applied Biology* 35, 207–227.

Triebskorn, R. and Ebert, D. (1989) The importance of mucus production in slugs' reaction to molluscicides and the impact of molluscicides on the mucus producing system. In: Henderson, I.F. (ed.) *Slugs and Snails in World Agriculture.* Monograph 41, British Crop Protection Council, Thornton Heath, pp. 373–378.

Triebskorn, R. and Schweizer, H. (1990) Influence du molluscicide métaldehyde sur les mucocytes du tract digestif de la petite Limace grise (*Deroceras reticulatum* Müller). In: *ANPP – Deuxième Conférence Internationale sur les Ravageurs en Agriculture, Versailles*, Vol. 1, pp. 183–190.

Walker, A.J., Miller, A.J., Glen, D.M. and Shewry, P.R. (1996) Determination of pH in the digestive system of the slug *Deroceras reticulatum* (Müller) using ion-selective microelectrodes. *Journal of Molluscan Studies* 62, 387–390.

Webley, D. (1965) Aspects of trapping slugs with metaldehyde and bran. *Annals of Applied Biology* 56, 37–54.

Webley, D. (1966) Waterproofing of metaldehyde on bran baits for slug control. *Nature* 212, 320–321.

Wedgwood, M.A. and Bailey, S.E.R. (1988) The inhibitory effects of the molluscicide metaldehyde on feeding, locomotion and faecal elimination of three pest species of terrestrial slug. *Annals of Applied Biology* 112, 439–457.

Wellmann, P. and Heimbach, F. (1996) The effects of methiocarb slug pellets on the earthworm *Lumbricus terrestris* in a laboratory test. In: Henderson, I.F. (ed.) *Slug and Snail Pests in Agriculture.* Symposium Proceedings 66, British Crop Protection Council, Farnham, pp. 181–188.

Wright, A.A. and Williams, R. (1980) Effect of molluscicides on the consumption of bait by slugs. *Journal of Molluscan Studies* 46, 265–281.

Young, A.G. and Wilkins, R.M. (1989) The response of invertebrate acetylcholinesterase to molluscicides. In: Henderson, I.F. (ed.) *Slugs and Snails in World Agriculture.* Monograph 41, British Crop Protection Council, Thornton Heath, pp. 263–269.

3 *Achatina fulica* Bowdich and Other Achatinidae as Pests in Tropical Agriculture

S.K. Raut[1] and G.M. Barker[2]

[1]*Department of Zoology, University of Calcutta, 35 Ballygunge Circular Road, Calcutta 700019, India;* [2]*Landcare Research, Private Bag 3127, Hamilton, New Zealand*

Achatinidae are native to Africa. The family is represented by about 200 species in 13 genera. Several species have attained pest status within their native African range when the habitat is modified for human habitation and cropping. Furthermore, associated with the increased mobility of humans and globalization of travel and trade, several achatinids, the most notable of which is *Achatina fulica* Bowdich, have been accidentally or purposefully transported to areas outside their native range in Africa and further afield. In these new areas Achatinidae can cause significant economic and ecological impacts. This chapter provides a synopsis of Achatinidae as pests in tropical agriculture, focusing primarily on *A. fulica*, but also bringing together the relevant information on other pestiferous achatinid species.

Origins

The dominant features of the vegetation in Africa today are the tropical forest and the savannah. Most of the diversity in African terrestrial gastropods is concentrated in the forest and its isolated outliers, and indeed the forest is generally regarded as the centre of gastropod evolution on the continent. Van Bruggen (1986) recognized four sub-Saharan centres of endemism among African terrestrial gastropods, namely: (i) southern Africa; (ii) East Africa; (iii) north-east Africa; and (iv) Central/West Africa. Each centre was assumed to have functioned as an important refugium during periods of forest contraction in the Holocene.

The margins of the forest have never been permanent: throughout the climatic history of Africa the forest has waxed and waned in response to changing rainfall patterns. In the arid or interpluvial period *c.* 18,000 years BP the forest was reduced to a number of major blocks, which may have functioned as refugia for the tropical, forest-dependent biota (Diamond and Hamilton, 1980; Hamilton, 1981; van Zinderen Bakker, 1982; Mayr and O'Hara, 1986; van Bruggen, 1989). The East African forest in particular suffered the vicissitudes of climatic variation in the past due to the varied topography. Verdcourt (1984) and van Bruggen (1986) indicated that even fairly minor changes in rainfall and temperature have given repeated opportunities for vicariant speciation. However, this view may be an oversimplification of the evolutionary setting in Africa, particularly if the primary adaptive radiation(s) and the greater part of the speciation that led to the extant fauna predate the Holocene (which is highly probable in the case of the terrestrial gastropods) and if we accept that the savannah was or is not the biological desert that it is often purported to be.

The earliest fossil record for the Achatinidae is from the Pleistocene in Africa (Solem, 1979a,b) but the family clearly evolved much earlier. Mead (1950a, 1995, 1998) has postulated that the earliest achatinids originated north of the Zambezi, in the Lower Guinea of Cameroon and Gabon, with subsequent dispersive radiation into the southern parts of the subcontinent, in both the arid and the subarid areas, and in the moist parts east of the great watershed. Mead thus considered that the temperate species were in the main directly derived from tropical ancestors. Van Bruggen (1969, 1970, 1978) concurred with Mead in regarding the fauna of southern Africa as being derived from southward dispersal. None the less, the evolutionary history of the achatinids remains unknown. While much anatomical information is available (Pilsbry, 1906/7; Mead, 1950a, 1979b, 1988; van Bruggen, 1965, 1966, 1968, 1985; van Bruggen and Appleton, 1977; Sirgel, 1989) (much more is purported to be at hand but remains unpublished – A.R. Mead, personal communication, 2000), compelling data have yet to be presented to demonstrate that the nominal supraspecific taxa are in fact monophylogenetic units, and no quantitative character analysis has been presented to date to elucidate the phylogenetic relationships within the family. Thus the evolutionary history of the Achatinidae remains largely unkown.

Today the Achatinidae occupy practically all of sub-Saharan Africa, from Senegal (15°N) in the west, the region of Lake Chad (about 14°N) and the southern Egyptian Sudan (about 8°N) in the centre, and southern Ethiopia (about 7°30′ N) and Somaliland (about 5°N) in the east. They extend to South Africa, where species are to be found in the Orange River area on the west coast and in the District of George on the south coast of Cape Province. Central/West Africa is remarkably rich in achatinids, as is the East African centre. Achatinid diversity is considerably lower in southern Africa and north-east Africa (van Bruggen, 1969, 1986). São Tomé, a remote island off the Central/West African coast, has at least one endemic genus, the monotypic *Atopocochlis* Crosse & Fischer, while Príncipe and São Tomé share the monotypic subgenus *Archachatina* (*Archachatina*) Albers s.s. In contrast, the continental and little distant island of Fernando Poo (distance to continent 32 km) does not harbour any endemics among the Achatinidae.

Species richness among the achatinids is concentrated in two main genera, *Achatina* de Lamarck and *Archachatina*. J. Bequaert in Pilsbry (1919) considered *Archachatina* to be West African, being present on the islands of the Gulf of Guinea and in the coastal belt from Monrovia to the Kuilu River (Gabon). *Achatina* is widely distributed in sub-Saharan Africa. In West and Central Africa species of *Achatina* are confined to humid areas, while species of *Archachatina* are distributed in less humid areas (Hodasi, 1984). According to J. Bequaert in Pilsbry (1919), *Achatina* are essentially 'of the lowlands; in the mountains and on the plateaus of Central Africa the number of species and individuals decreases at about 1200 m and the genus is not found above 1500 m'. This points to a tropical origin. The majority of species in these genera are naturally confined to forested areas. However, as noted by H. Lang in Pilsbry (1919) in relation to species of the Belgian Congo, achatinids are often scarce in unmodified forest. Indeed, several achatinids, such as *Achatina achatina* (Linnaeus), have exhibited great adaptation to environmental change brought about by human encroachment and modification of the forests and in many of these modified areas Achatinidae occur in great numbers.

In East Africa a number of achatinid species are confined to humid, tropical forests. Further species are temperate forest dwellers, with *Achatina mulanjensis* Crowley & Pain, *Achatina tavaresiana* Morelet and *Archachatina bequaerti* Crowley & Pain occurring at high altitude in Malawi. However, as in West and Central Africa, a number of species are prevalent in forest-margin habitats. *A. tavaresiana* and *A. fulica*, for example, occur in large numbers along the margins of forest in East Africa (Crowley and Pain, 1964). *A. fulica* is present naturally from Natal and Mozambique in the south to Kenya and Italian Somaliland in the north. It extends 250–830 km from the coast, going farthest inland in the northern section of the range (Mead, 1949; J. Bequaert in Lange, 1950).

Numerous species of Achatinidae occur in humid, tropical–subtropical south-eastern Africa. The family is also well represented in temperate zones in South Africa: *Archachatina ustulata* (de Lamarck), *Archachatina marinae* Sirgel and *Achatina zebra* (Bruguière) are lowland species, while *Archachatina machachensis* (Smith), *Archachatina montistempli* van Bruggen and *Archachatina omissa* van Bruggen are confined to areas over 1300 m in the Drakensberg Range. *A. machachensis* occurs at altitudes of 1600–1800 m in Lesotho and the neighbouring plateau of southern Africa, areas that have cold winters with frosts and snow. The animals hibernate over winter. Elsewhere in Africa achatinids have a montane existence as well, but as rule only under temperate conditions. Furthermore, there are species of southern Africa that occur in less humid areas. The South African *Archachatina zuluensis* (Connolly) is restricted to dune and other coastal forests. *Achatina immaculata* de Lamarck is an example of a savannah-adapted species, occurring as large morphs in the savannah in southern Africa but as a somewhat smaller morph in the forests of the Rhodesian eastern escarpment (van Bruggen, 1978), possibly indicating a better adaptation to the savannah environment than to the forest. Bechuanaland of South Africa and the adjoining Botswanian deserts support five species of Achatinidae, namely *Achatina ampullacea* Böttger, *Achatina dammarensis* Pfeiffer, *Achatina passargei* von Martens, *Achatina schinziana* Mousson and *Achatina tracheia* Connolly (van Bruggen, 1969, 1978). These desert taxa have comparatively small shells compared with their more northern, forest-dwelling relatives, possibly indicating a gradient of selection pressure opposite to that operating on *A. immaculata*.

The genus *Limicolaria* Schumacher, represented by about 17 species (Crowley and Pain, 1970), extends from the southern limits of the Sahara south to the northern part of Malawi. According to Crowley and Pain (1970, p. 1), '*Limicolaria* are common everywhere on the west coast but are not found in maritime areas to the east.' Crowley and Pain (1970) regarded these animals as 'tropical' and to 'live equally in the forest and the veld country'. However, on both points these authors provide erroneous generalizations. First, a number of species are confined to montane habitats (to about 3000 m in the case of *Limicolaria turriformis* von Martens on Mt Mweru and *Limicolaria saturata* Smith on Kivu), and are more correctly to be regarded as temperate. Secondly, while a number of species occur in both forest and savannah, the majority are evidently confined principally to one or the other type of habitat. Many of the forest species occur in abundance in modified forest, at the forest edge and in plantations (e.g. Owen, 1965; Crowley and Pain, 1970; Tattersfield, 1996). In open country, *Limicolaria* spend long periods of time in soil, often at appreciable depths. These open-country species often also favour cultivated land and are found on the outskirts of settlements and farms.

Burtoa Bourguignat, as recognized by Crowley and Pain (1959) in their revision, comprises a single species (*Burtoa nilotica* (Pfeiffer)) widely distributed from the Sudan, south of 10°N, throughout the region of the Great Lakes to the Amanze Inyama River in the south, and into the upper Congo, upper Kasi and Lake Chad regions in the west. In Central Africa, *B. nilotica* occurs as a large, silvicolous form, often at high elevation, and is rarely seen in modified areas (H. Lang in Pilsbry, 1919). However, to the south, a smaller, savannah-adapted form is present (Crowley and Pain, 1959; van Bruggen, 1978). Crowley and Pain (1959) assigned subspecific status to seven regional variants of this taxon.

Among the minor achatinid genera, *Perideriopsis* Putzeys is restricted to the forests of the Congo basin. The genus *Limicolariopsis* d'Ailly occurs widely in Central and East Africa, represented by a small series of species of high-elevation forests. *Callistopepla* Ancey, a probable composite genus, has apparently been found only in the West African and equatorial rain-forest belt. *Callistopepla nyikaensis* (Pilsbry) occurs at high altitude in Malawi. J. Bequaert in Pilsbry (1919) considered *Cochlitoma* de Férussac to be restricted to South Africa, south of the Orange River on the west coast and of the Zambezi in the east. *Metachatina* Pilsbry is restricted to south-east Africa, predominantly in the coastal lowlands of Natal and southern Mozambique, but also occurring at 600–1300 m in the Lebombo escarpment and Drakensberg Range of Natal. Restricted to humid habitat in shrubland and forest vegetation, this monotypic genus, represented by *Metachatina kraussi* (Pfeiffer), is apparently a comparatively new development in a submarginal but none the less warm and humid area of the family.

Many achatinids are able to secrete a protective epiphragm in order to temporarily close the shell aperture, which for species living in the drier parts of Africa is considered of great survival value. H. Lang in Pilsbry (1919) notes, for example, that the species of the open-plain areas of Central Africa aestivate over the dry season, buried 'several inches below the surface, their aperture closed by a strong epiphragm some distance in from the edge of the shell'. None the less, the occurrence of aestivation varies among and within species (Hodasi, 1982) and even species living in moist rainforest, such as *A. achatina*, may aestivate during the drier months. Van Bruggen (1969) considered the absence of the capacity to produce an epiphragm, evident in a number of forest-dwelling achatinid species, to be a secondary phenomenon. This implies that at least some of the extant Achatinidae were derived from species that primarily inhabited the open veld country, which is contrary to the hypothesis of origin in the tropical forest.

Humans have long been part of the African biota and have had a profound influence on the African environment, particularly at the margins of the tropical rainforest (e.g. Boughey, 1963). As noted above, a number of achatinid species are evidently well adapted to this human-induced disturbance of the rainforest and can be locally abundant in plantations. There are occasional reports from various parts of Africa of achatinids causing damage to crop species (Table 3.1). However, many such situations are often short-lived, as the achatinids are collected for their meat, especially by peoples of West and Central Africa (Bequaert, 1950a). Hodasi (1989) reported that the increase in the human population in West Africa, coupled with the increasing cost of animal proteins, such as beef, pork and chicken, has meant that achatinid meat is an increasingly popular source of protein and iron for the rural poor.

Von Stanislaus *et al.* (1987) considered predation by humans as currently important in population regulation of forest-dwelling species, such as *A. achatina*, *Achatina monochromatica* Pilsbry, *Achatina balteata* Reeve, *Archachatina marginata* (Swainson), *Archachatina degneri* Bequaert & Clench and *Archachatina ventricosa* (Gould). The peoples in West Africa have different preferences for achatinids: in Nigeria the species of choice is *A. marginata*, while in Ghana *A. degneri* is preferred (Hodasi, 1989; Olufokunbi *et al.*, 1989). Coupled with habitat destruction through deforestation, the high rates of human predation are leading to a general decline in Achatinidae in West Africa (Hodasi, 1989). Consequently there is increasing interest in commercial production of achatinids to supply the lucrative urban gourmet trade (Elmslie, 1982;

Table 3.1. Crop plants in Africa recorded as being susceptible to feeding damage by Achatinidae.

Country in which damage was recorded	Crop species	References
Achatina achatina (Linnaeus)		
Ivory Coast	Cabbage (*Brassica oleracea* Linnaeus; Brassicaceae) Cassava (*Manihot esculenta* Crantz; Euphorbiaceae) Lettuce (*Lactuca* Linnaeus spp.; Asteraceae) Papaya (*Carica papaya* Linnaeus; Caricaceae) Sweet potato (*Ipomoea batatas* (Linnaeus)de Lamarck; Convolvulaceae) Yam (*Dioscorea alata* Linnaeus; Diascoreaceae)	Otchoumou *et al.* (1989/90), Tra (1994)
Ghana	Lettuce (*Lactuca* Linnaeus spp.; Asteraceae) Oil palm (*Elaeis guineensis* von Jacquin; Arecaceae) Orange (*Citrus sinensis* (Linnaeus) Osbeck; Rutaceae) Papaya (*Carica papaya* Linnaeus; Caricaceae) Pear (*Pyrus communis* Linnaeus; Rosaceae)	Hodasi (1975, 1979)
Achatina albopicta Smith		
Kenya	Papaya (*Carica papaya* Linnaeus; Caricaceae)	Williams (1951)
Achatina craveni Smith		
Tanzania	Coffee (*Coffea* Linnaeus spp.; Rubiaceae) Sesame (*Sesamum orientale* Linnaeus; Pedaliaceae)	Salaam (1938), van Dinther (1973)
Achatina fulica Bowdich		
Tanzania	Coffee (*Coffea* Linnaeus spp.; Rubiaceae)	Mead (1961)
Achatina zanzibarica Bourguignat		
Tanzania	Cotton (*Gossypium herbaceum* Linnaeus; Malvaceae) Sisal (*Agave sisalana* Perrine; Agavaceae)	Tomaszewski (1949), van Dinther (1973)
Archachatina marginata (Swainson)		
Nigeria	Banana (*Musa paradisiaca* Linnaeus; Musaceae) Lettuce (*Lactuca* Linnaeus spp.; Asteraceae) Papaya (*Carica papaya* Linnaeus; Caricaceae)	Imevbore & Ajayi (1993)
Limicolaria aurora (Jay)		
Cameroon	Oil palm (*Elaeis guineensis* von Jacquin; Arecaceae) Leguminous cover crops	Spence (1938)
Limicolaria flammea (Müller)		
Nigeria	Apple (*Malus* × *domestica* Borkhausen; Rosaceae)	Egonmwan (1991)
Limicolaria kambeul (Bruguière)		
Sudan	Maize (*Zea mays* Linnaeus; Gramineae) Groundunt (*Arachis hypogaea* Linnaeus; Fabaceae)	Salaam (1938), Godan (1983)
Limicolaria martensiana (Smith)		
Uganda	Cabbage (*Brassica oleracea* Linnaeus; Brassicaceae) Lettuce (*Lactuca* Linnaeus spp.; Asteraceae)	Owen (1965)
Nigeria	Carrot (*Daucus carota* Linnaeus; Apiaceae) Lettuce (*Lactuca* Linnaeus spp.; Asteraceae) Potato (*Solanum tuberosum* Linnaeus; Solanaceae)	Egonmwan (1991)
Limicolaria numidica (Reeve)		
Cameroon	Oil-palm (*Elaeis guineensis* von Jacquin; Arecaceae)	Spence (1938)
Limicolaria zebra Pilsbry		
Cameroon	Oil-palm (*Elaeis guineensis* von Jacquin; Arecaceae)	Spence (1938)

Awesu, 1988; Hodasi, 1989; Olufokunbi *et al.*, 1989; Awah, 1992; Monney, 1994). Most commercial interest in Africa is in *A. achatina*, *A. marginata*, *A. degneri* and *A. ventricosa*.

A number of Achatinidae are naturally restricted to virgin rainforest and decline markedly in abundance when the forest is replaced by second-growth vegetation. An example is the Liberian *Archachatina knorrii* (Jonas).

As Invasive Species in Africa

H. Lange in Pilsbry (1919, p. 55) remarks on the probable role of human agencies in the wide distribution of various Achatinidae in the Congo region of Africa. There is no reason to suspect that this does not also apply to other places on the continent. Bequaert (1950a, p. 41) raised the possibility that the disjunct distribution evident in *A. balteata* of Guinea was due to 'accidental or perhaps intentional introduction by man'. *A. zebra* occurs naturally in the south-eastern and southern coastal regions of South Africa. A colony of this species in the Hout Bay area of Cape Town, significantly further westwards, is believed to have been transported by humans (Sirgel, 1989).

A. marginata has evidently been dispersed by human agencies in West Africa, having recently invaded the south-west parts of Ghana (Monney, 1994). This species has also been introduced on to Annobón and São Tomé in the Gulf of Guinea (Gascoigne, 1994). On São Tomé it has become widespread and Gascoigne (1994) suggested that competitive interactions, along with habitat destruction, may have contributed to the decline in the indigenous *Archachatina bicarinata* (Bruguière).

The natural range of *A. fulica* is generally regarded to be the coastal area of East Africa, including its many islands (Pilsbry, 1904; Bequaert, 1950a), but at least part of this range in East Africa may be due to introductions by humans (Verdcourt, 1961). *A. fulica* now occurs in the southern part of Ethiopia and Somalia, throughout Kenya and Tanzania and into northern Mozambique. Very recently this species has been recorded in Morocco (van Bruggen, 1987), on the Ivory Coast (de Winter, 1988; Zong *et al.*, 1990) and in Ghana (Monney, 1994) of West Africa.

There is at present little information on the economic status of *A. fulica* in areas invaded in Africa. However, within a short period of its introduction, *A. fulica* achieved dominance in the achatinid community in Ivory Coast and Ghana and achieved significance as a crop pest (von Stanislaus *et al.*, 1987). *A. fulica* distributes in its faeces spores of *Phytophthora palmivora* (Butler) Butler, the cause of black pod disease in cacao (*Theobroma cacao* Linnaeus; *Sterculiaceae*) plants in Ghana (Evans, 1973). Since the local people do not accept *A. fulica* as an edible species, this alien species is allowed to go unchecked, while predation pressure is maintained on species such as *A. achatina*.

As Invasive Species Out of Africa

Pilsbry (1919) and van Bruggen (1981) treat the occurrence of Achatinidae in Madagascar as natural. None the less, Pilsbry (1919) admitted the possibility that the occurrence in Madagascar and several other islands off the African mainland were due to human importation. Van Bruggen (1981) considered *A. immaculata* (usually treated as *Achatina panthera* (de Férussac)) to be shared between south-eastern Africa and Madagascar, although the possibility was admitted that the occurrence in Madagascar is due to introduction through human agencies. Other authorities have considered that there exists no sound argument to consider Madagascar within the original geographical range of Achatinidae. Because of the absence of achatinid shells in Late Pleistocene deposits, both Dollfus (1899) and Germain (1921) considered the present occurrence of *Achatina* in Madagascar to be the consequence of introduction by human agency in the recent past. Bequaert (1950a) considered *A. fulica* to be an introduction to Madagascar. The presence of the East African *A. immaculata* in Rodrigues, Mauritius, Réunion, the Comores and the Seychelles (Bequaert, 1950a), clearly outside the realm of Africa, lends support to the idea that the Achatinidae have been dispersed to Madagascar and beyond by human agency.

The dispersal of *A. fulica* out of Africa has been discussed by a number of authors, including van Weel (1948/49), Lange (1950), Bequaert (1950a), Rees (1951), Mead (1961, 1979a), Wolfenbarger (1971), Lambert (1974), Srivastava (1992), Civeyrel and Simberloff (1996) and Cowie (2000). Bequaert (1950a, p. 73) concludes:

> that the spread of *Achatina fulica* from its original continental African home and Madagascar to the islands of the Indian Ocean, India, the Orient, the East Indies and the Pacific is entirely due to transport by man, usually deliberate, in a few cases accidental. Furthermore all later importations may be traced back ultimately to the first introduction from Madagascar into Mauritius, some 150 years ago.

A. fulica was evidently introduced to Madagascar prior to 1800 from Kenya, but was not accepted as an edible species. It assumed pest status through damage to crop plants. However, the species was attributed medicinal properties and, on these grounds, was introduced to Mauritius and thence to many island groups in the Indian Ocean. From there naturalists introduced them to India and Sri Lanka. By the 1930s *A. fulica* had been spread throughout tropical and subtropical East Asia. Subsequent further penetration of Asia and dispersal into the Pacific was aided by the Second World War and postwar commerce and by deliberate introductions for a variety of reasons. *A. fulica* had reached the outer islands of Papua New Guinea by 1946, New Ireland and New Britain by 1949 and mainland Papua New Guinea by 1976/77. *A. fulica* had invaded Tahiti by 1967 and New Caledonia and Vanuatu by 1972 and was reported from other areas in French Polynesia in 1978, the year in which it reached

American Samoa (Fig. 3.1). *A. fulica* continues to spread; for instance it was first reported on Upolu, Samoa, in 1990 and in Kosrae, Federated States of Micronesia, in 1998.

Small, incipient populations of *A. fulica* have been eradicated at various times from California, Florida, Queensland in Australia, Fiji, Western Samoa, Vanuatu and Wake Island (Abbott, 1949; Mead, 1961, 1979a; Colman, 1977, 1978; Muniappan, 1982; Waterhouse and Norris, 1987; Watson, 1985).

Bequaert (1950a) and Wolfenbarger (1971) had predicted the establishment of *A. fulica* in the New World tropics, based on the evident eastward dispersal of the species and the likely favourability of the Caribbean and American tropics as a habitat. This prediction was realized when, in 1984, *A. fulica* was found established in Guadeloupe, French West Indies (Frankiel, 1989). By 1987 it had spread to other parts of the island, and in 1988 was recorded in Martinique, about 200 km to the south of Guadeloupe (Schotman, 1990; Mead and Palcy, 1992).

With the advent of Achatinidae as a tradable commodity on the world market, captive breeding has been established for various species in different parts of the world (Mead, 1982; Upatham *et al.*, 1988; Runham, 1989; Monney, 1994), heightening the potential for further spread of *A. fulica* and related species. Considerable quantities of *Achatina* meat are exported to Europe and America from Taiwan, China and other Asian countries (Mead, 1982). Escapes and undoubtedly purposeful releases from these breeding facilities have certainly contributed to the naturalization of *A. fulica* in new areas in Asia. Furthermore, the continuing interest in achatinid meat has led to expansion of the industry into South America and was responsible for the very recent establishment of feral populations of *A. fulica* in many regions of Brazil, including São Paulo, Rio de Janeiro, Minas Gerais, Parana and Santa Catarina (Teles *et al.*, 1997; J. Coltro, personal communication, 2000).

Being of African origin, it has generally been assumed that *A. fulica* will be confined as an alien species to tropical environments. However, *A. fulica* exhibits wide environmental tolerances. The species is now well established in the temperate environs of Bonin and Ryūkyū Islands in the southern regions of Japan, and in the São Paulo region of Brazil. It also poses a serious threat to crops in the Coochbehar, Gauhati, Imphal, Nongpoh, Kumarghat, Chaibasa, Darbhanga, Dumka and Purnea districts of India, where temperatures down to 2°C occur during the winter months and the animals go into hibernation. Furthermore, published records indicate establishment in temperate environments imposed by altitude in low-latitude areas, such as at 350 m in Hawaii, 400 m in the Philippines, 600 m in Mauritius, 1166 m in India, 1200 m in Sri Lanka and 1500 m in Malaya (South, 1926; Mead, 1955, 1961, 1979a; Raut, 1983a). It is therefore apparent that *A. fulica* has the potential to occupy areas at 40° latitude, or the environmental equivalents at higher altitudes nearer the equator (Raut, 1983a).

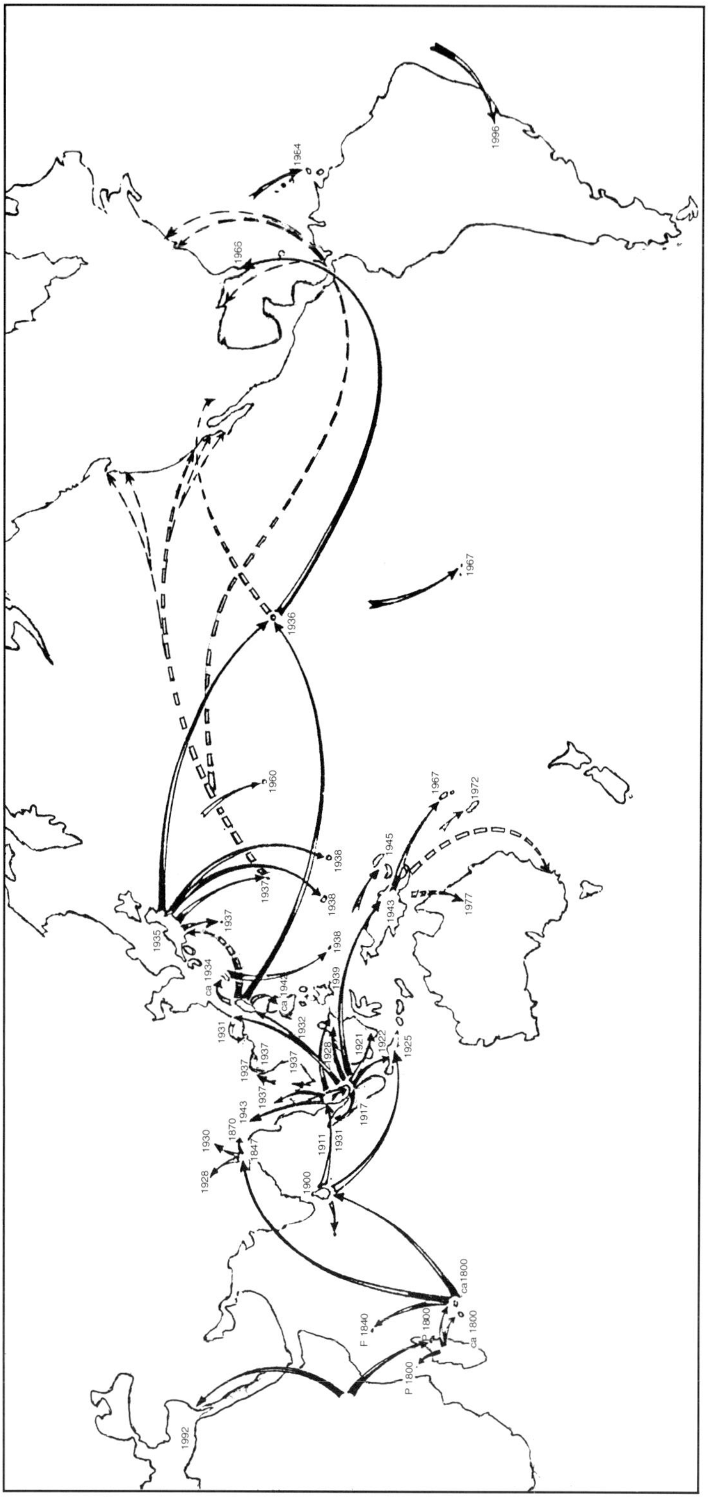

Fig. 3.1. Dispersal of *Achatina fulica* Bowdich (Achatinidae) out of Africa.

Over much of its introduced range, *A. fulica* has a predilection for modified environments, such as plantations and gardens. The emergence of *A. fulica* as an important crop pest within a decade or two of establishment has been repeated over much of its naturalized range. Van Benthem Jutting (1952) notes that *A. fulica* has not been found in truly undisturbed conditions in Java, or in tropical rainforest. Cowie (1998a) notes that *A. fulica* is primarily found in disturbed low- to mid-elevation sites in Hawaii. However, *A. fulica* has also been observed as an invader of primary or secondary forest in the Hawaiian Islands, Bonin Islands, India, Java, Sumatra and New Caledonia (Mead, 1979a; Tillier, 1982; Raut and Ghose, 1984; G.M. Barker, personal observation).

It is generally thought that animal species do not attain marked elevation in abundance and thus status as an environmental pest in their natural range, other than short periods of eruptive population behaviour. If this is so, in the case of introduced species, such as *A. fulica*, how many years are needed to develop an association with the local fauna such that the exogenous species can be regarded as endogenous with respect to the nature of its population dynamics? After an initial period of high abundance, do populations in their naturalized range decline due to the regulatory effects of natural enemies? Mead (1979a, p. 83) expressed the opinion that 'the phenomenon of decline in populations of *Achatina fulica* appears to be inevitable. The timing of its earliest manifestation, rate of progress and the ultimate degree of expression are functions of the environment.' Mead presented evidence for the principal role of disease in the decline. From information such as that presented in Fig. 3.1, it is possible to estimate the length of time that *A. fulica* has been resident in an area as an alien. It is evident that, in some areas of India, *A. fulica* has been thriving for a period of 100–150 years, with no clear evidence of abatement in its pest status (Raut and Ghose, 1984). None the less, there are situations where, after a period of remarkable abundance and environmental effects, *A. fulica* populations have declined. There is evidence, for example, that *A. fulica* became a lesser problem after only 20 years on Moorea in French Polynesia (Clarke *et al.*, 1984) and after some 60 years in Hawaii (Cowie, 1992) and Ogasawara (K. Takeuchi, personal communication, 2000).

Little information is currently available on the pest status of *A. immaculata* in its naturalized range in the islands of the Indian Ocean. It is of interest that this species has not been more widely dispersed by the human agencies responsible for the spread of *A. fulica*. Indeed, the great majority of achatinid species have not been dispersed to become feral outside Africa. A recent exception is the West African *Limicolaria aurora* (Jay), recorded for the first time outside Africa in 1989 when discovered in Martinique (Mead and Palcy, 1992; Palcy and Mead, 1993). According to Mead and Palcy (1992), the infestation probably arose from purposeful introduction as an edible species direct from Africa, some time after 1986. Mead and Palcy (1992) reported that *L. aurora* occurred in considerable numbers in the infested area of Martinique, causing damage to yam

(*Dioscorea alata* Linnaeus; Diascoreaceae), kidney bean (*Phaseolus vulgaris* Linnaeus; Fabaceae), black pepper (*Piper nigrum* Linnaeus; Piperaceae), Jerusalem artichoke (*Helianthus tuberosus* Linnaeus; Asteraceae), cucumber (*Cucumis sativus* Linnaeus; Cucurbitaceae), okra (*Abelmoschus esculentus* (Linnaeus) Moench; Malvaceae), rose-mallow (*Hibiscus* Linnaeus sp.; Malvaceae) and sweet potato (*Ipomoea batatas* (Linnaeus) de Lamarck; Convolvulaceae) within 8 months of being first recorded there.

In addition to farming for meat, several species of Achatinidae, including *A. fulica*, *A. achatina* and *A. marginata*, are maintained in temperate regions outside Africa as laboratory animals (e.g. Nisbet, 1974; Plummer, 1975; Pawson and Chase, 1984; Tranter, 1993).

Biology

The biology of some Achatinidae has been extensively studied. That of the great majority is hardly known at all. Two important books by Mead (1961, 1979a) bring together and appraise most of the literature on *A. fulica*. In this chapter we provide a synopsis of information relevant to the pest status and management of achatinids in tropical agriculture.

Achatinidae are nocturnal. Like other terrestrial gastropods they are dependent on the availability of moisture. Accordingly they are active under high-humidity conditions. In many tropical areas, activity is thus restricted to the monsoon season and the following moist summer period. Usually achatinids spend the daytime hours under protective cover. When populations are high, many *A. fulica* are to be found resting on exposed walls and tree trunks, indicating that under these conditions there may be a shortage of home sites. Activity generally commences with the approach of darkness at sunset. Takeda and Ozaki (1986) demonstrated an endogenous circadian rhythm in the activity of *A. fulica* that is independent of temperature and light conditions but regulated by hydration effects on haemolymph osmolality. Further, these authors showed that *A. fulica* only becomes active when the ambient relative humidity rises above 50%. In Calcutta, India, Panja (1995) found that foraging *A. fulica* spent on average 338 min (55%) of their nightly activity crawling, 95 min (15.5%) feeding and 180 min (29%) resting. Panja (1995) found that the distance travelled by *A. fulica* in a single night of activity decreased during the season irrespective of the age structure of the population, with an average of 1429 cm in June reducing to 912 cm by October. The distance travelled in a single night varies with the size of the animal. Tomiyama (1992) found that, in Chichi Jima, Japan, immature *A. fulica* dispersed up to a distance of 100 cm (standard deviation 34 cm), while mature animals moved an average distance of 161 cm (standard deviation 44 cm). It was observed that, in the course of searching for food, *A. fulica* typically moves some distance from the daytime resting site before commencing feeding. The animals may be active for over an hour before locating a favourable host plant. Foraging patterns are suspended when *A. fulica* is engaging in mating activity.

The length of feeding time at one site depends on the quality and quantity of the food, but invariably feeding is interspersed with periods of rest. Once the animals have satisfied their hunger or at the approach of dawn, *A. fulica* typically seek suitable daytime resting sites. In Chichi Jima, Tomiyama (1992) observed that mature *A. fulica* return to the same resting site after each night's activity, while immature *A. fulica* tend to use different resting sites each day. Similarly, Panja (1995), working in Calcutta, recorded the absence of homing activity in *A. fulica* of 20–29 mm shell size, but frequency of homing was 20% in animals of 40–49 mm shell size and 78% in animals of 70–79 mm shell size. Homing in terrestrial gastropods has been shown to be mediated by directional trail-following and chemoreception of airborne odours from the home site (Chelazzi, 1991; Cook, 2001). That animals are able to return to home sites despite being experimentally transplanted up to 30 m (Tomiyama, 1992) suggests that distant chemoreception is involved in the homing behaviour of *A. fulica*. Chase and Boulanger (1978) have shown that mucus trail-following behaviour can occur in *A. fulica* but, because snails do not crawl along old mucus trails on their way back to their home sites, Tomiyama (1992) concludes that mucus trail-following is not important in the homing of this species.

Any site that provides adequate protection from light and desiccation will be used by *A. fulica* for daytime sheltering and for aestivation. In the rain forest this need is evidently not so urgent and the animals will frequently rest on the bare ground or the litter (Dun, 1967). During the rainy season *A. fulica* will often ascend considerable distances up tree-trunks or the walls of buildings, embankments, etc. to rest during the day.

That many achatinid species aestivate during the dry season in Africa has been noted above. Throughout its naturalized range, *A. fulica* undergoes aestivation with the onset of dry weather. In the monsoonal tropics, such weather conditions occur in winter, when temperatures are typically 15–28°C but in some regions may fall below 10°C. As *A. fulica* is able to maintain activity at temperatures below 10°C (Mead, 1979a; Raut and Ghose, 1984), the cue for aestivation is evidently the humidity of the air. Raut and Ghose (1984) have reported aestivation when maximum temperatures reach 28–30°C at a humidity of 80–82%. Raut and Ghose (1983a) observed *A. fulica* feeding on fleshy and succulent food plants prior to aestivation, evidently as a body hydration strategy. *A. fulica* prefers to aestivate in moist soil, but will also aestivate at sites above the ground. Although aggregation (Chase *et al.*, 1980) and homing are well developed in *A. fulica*, there exists no affinity for particular aestivation sites in these animals. They aestivate singly or in aggregations of as many as 100 or more animals (Raut, 1978; Srivastava, 1992), with the shell aperture oriented downwards and sealed with an epiphragm (Raut and Ghose, 1984). In the aestivatory state there is considerable physiological change, including reduction of the heart rate from 52 to 8 beats min^{-1}

(Raut and Rahman, 1991). While the epiphragm functions to reduce loss of body water, the animals do gradually dehydrate during aestivation. The dehydrated animals periodically retract further into their shell and in doing so secrete a further epiphragm, which places further demands on body moisture. As many as six to 12 epiphragms may be produced (Raut and Ghose, 1984). The longer the aestivatory state is maintained, the greater the potential for body dehydration to reach a critical threshold for survival. In Calcutta, Raut and Ghose (1981) recorded 100% mortality over an aestivation period of 7 months (November to May) for *A. fulica* that were 10–15 days old at the onset of aestivation. These authors noted that the rate of mortality declined with increasing age of the animals at the commencement of aestivation, with 33–45% mortality in animals of the 100–105 day age-group. Aestivation in *A. fulica* lasts from 2 to 10 months, depending on the climatic zone (see Raut and Ghose, 1984).

Raut and Ghose (1984) state that more than 50 mm rainfall can terminate aestivation at any time. Often in the tropics the dry season can be interrupted by occasional, brief periods of rainfall. While these rains may not be sufficient to induce *A. fulica* to terminate their aestivation, the temporary restoration of humidity provides an opportunity for the animals to rehydrate. This rehydration can be critical to their survival over the long aestivatory period.

Nisbet (1974) found that achatinids exhibited a tendency to bury themselves in the soil, even in the absence of aestivation.

Achatinids are hermaphrodites. Mead (1949) recorded male sexual maturity in *A. fulica* before the animals are a year old; development of female organs and egg deposition takes a few months longer. Tomiyama (1991, 1993) demonstrated that *A. fulica* has determinate shell growth, with thickening of the shell peristome occurring after cessation of shell growth. During the shell growth phase the animals also develop sexually, but producing only male gametes. In the later part of the male phase, the animals begin to engage in copulation. At or shortly after cessation of shell growth, the animals complete reproductive development and enter a phase where both male and female gametes are produced. If there is no prolonged interruption by aestivation or hibernation, the animals mature within 1 year. *A. fulica* generally attains sexual maturity at the age of 5–8 months under field conditions (Leefmans, 1933; van Weel, 1948/49; Mead, 1949, 1961; van der Meer Mohr, 1949a; Bequaert, 1950a; Kondo, 1964; Pawson and Chase, 1984; Raut, 1991). Ghose (1959) reported that *A. fulica* attained sexual maturity within 6 months in the laboratory, consistent with the data of Pawson and Chase (1984), which indicated that this species laid the first eggs at the age of 5 months under controlled laboratory conditions of 20–24°C and 12 : 12 h light/dark photo regime. In subtropical areas, such as the Ryūkyū and Ogasawara Islands of Japan and certain regions in India, growth of *A. fulica* is interrupted by winter dormancy and the first eggs are not produced until the age of 12–15 months (Ghose, 1959; Sakae, 1968; Suzuki, 1981; Numasawa and Koyano, 1987; Tomiyama, 1993). *A. achatina* typically take 18 months to mature in

West Africa (Hodasi, 1979), while some animals that experience two intervening seasons of aestivation take 21 months to reach maturity (Hodasi, 1982).

Stievenart (1992) showed that *A. marginata* also has determinate shell growth, but, unlike *A. fulica*, the peristome is produced as a reflected lip after cessation of shell whorl growth. Furthermore, *A. marginata* was shown to reach sexual maturity and to produce eggs prior to peristome lip formation. *A. marginata* requires about 9–10 months under laboratory conditions (Plummer, 1975). Owen (1964) presented evidence for year-round reproductive activity of the Ugandan *Limicolaria martensiana* (Smith), but with peaks of activity in January–February and July. This bimodality was apparently associated with alternating wet and dry seasons.

Achatinidae are generally outcrossing and therefore require allosperm to produce fertile eggs. Olson (1973) summarizes the situation with respect to the possibility of self-fertilization in *A. fulica*: 'for all intents and purposes, cross fertilization is necessary for the laying of a sufficient quantity of eggs to ensure perpetuation of the species'. He states that self-fertilization does occur but that virgin animals provide clutches comprising fewer than ten eggs, that most of these eggs are sterile and that progeny arising from these eggs rarely survive through to sexual maturity.

In the case of *A. fulica*, individuals receptive to a mate can be distinguished by their dilated genital orifice and the occasional protrusion of the phallus (Raut and Ghose, 1984). Courtship is initiated by these animals immediately on encountering a prospective partner, and they often take an aggressive role in the courtship (Raut and Ghose, 1984; Tomiyama, 1994). The sequence of events in the courtship of *A. fulica* has been described by Raut and Ghose (1984) and Tomiyama (1994), and in that of *A. marginata* by Plummer (1975). Mating is generally reciprocal, and generally pairing occurs between animals of similar size. Mating generally occurs during the hours of darkness, although courtship may be initiated late in the afternoon (Lange, 1950). Tomiyama (1994) observed that, while 'young' adult *A. fulica* will initiate courtship at any time between 6.30 p.m. and 4.30 a.m., mating in the older animals was initiated only between 10 p.m. and 12.30 a.m. The duration of copulation in *A. fulica* is typically 6–8 h but can vary from 1 to 24 h (van Weel, 1948/49; van deer Meer Mohr, 1949a; Lange, 1950; Raut and Ghose, 1984; Tomiyama, 1994), and in *A. achatina* may continue for 12 h (Hodasi, 1979). Raut and Ghose (1984) reported that a small percentage of matings in *A. fulica* were not reciprocal.

In *A. fulica* one individual initiates courtship and the other may accept courtship. These initiators and acceptors exhibit different behaviours during the courtship process. Tomiyama (1994) describes the mating process. First, one animal (the initiator) approaches another from behind and mounts its shell. Generally, the phallus is extruded by the initiator during the shell-mounting phase. If the acceptor animal wishes to accept and proceed with courtship, it bends its head backward and rocks the whole body. Responding to this behaviour, the initiator bites the body of the acceptor in the cephalic region and then proceeds to rub its extruded phallus against the now extruded phallus of the acceptor animal. Finally, reciprocal intromission occurs with phallic penetration into the vagina of the partner. With intromission established, the conjoined animals fall to the ground side by side and remain in this position for the duration of copulation.

Tomiyama (1994) suggested that the courtship initiators are essentially 'male-behaving' and the courtship acceptors 'female-behaving'. Asami *et al.* (1998) have demonstrated that the shell-shape bimodality evident in stylommatophoran snails, where snails either carry a high-spired (height : diameter > 1) or a low-spired (height : diameter ≤ 1) shell (Cain, 1977), is associated with discrete mating behaviours. In general, flat-shelled species mate reciprocally, face to face, while tall-shelled species, such as Achatinidae, mate non-reciprocally: the 'male' copulates by mounting the 'female's' shell. Asami *et al.* (1998) categorized mating in achatinids as non-reciprocal, with one animal functioning as the 'male' and achieving copulation by mounting the 'female's' shell, consistent with Tomiyama's (1994) interpretation. The duration of courtship behaviour in *A. fulica* observed by Tomiyama (1994) was less than 5 min, i.e. less than *c.* 1.8% of the whole duration of successful mating. Copulation duration was much shorter in mating among young *A. fulica* than among relatively older animals.

Tomiyama (1994) found that, in *A. fulica*, courtship progressed successfully to copulation in only 10% of observed courtships. The rejection was usually made by the acceptor ('female-behaving') animal.

While eggs may be deposited within 8–20 days of mating in the case of *A. fulica* (Lange, 1950), the reproductive strategy of Achatinidae includes the capacity for long-term storage of allosperm. Owiny (1974) recorded production of viable eggs in *L. martensiana* 520 days after mating, while van deer Meer Mohr (1949a) and Raut and Ghose (1979b) record egg production 382 and 341 days, respectively, after mating in *A. fulica*. Allosperm viability is evidently maintained over lengthy periods of aestivation (Raut and Ghose, 1982). Allosperm storage provides achatinids with the capability to produce eggs at any time of the year given favourable environmental conditions. It is quite clear that introduction of a single allosperm-bearing specimen is sufficient for the establishment of a colony in a previously non-infested area.

A. fulica is oviparous, as evidently are most Achatinidae. Bequaert (1950a) presented information indicating that *Achatina zanzibarica* Bourguignat and *Achatina allisa* Reeve are ovoviviparous. Tompa (1979) indicated that all Achatinidae are egg retainers of one form or another. Delayed oviparity or ovoviviparity may, in some species, be associated with occupancy of a strongly seasonal habitat (van Bruggen, 1985). The reported duration of the egg stage in *A. fulica* varies from 1 to 17 days. Mead (1949) has reported retention of eggs in the spermoviduct so that hatching occurs within a few hours of oviposition. Ghose (1960, 1963) points out that eggs with embryos in different stages of development are laid; hence the period to hatching varies, with some eggs hatching within a few days of being laid. This has been confirmed by Pawson and Chase (1984) for *A. fulica* in laboratory culture. Such egg retention apparently has not been observed by others (e.g. Lange, 1950), indicating that egg retention in *A. fulica* may vary with environmental conditions. The incubation period for eggs of the oviparous *A. marginata* is approximately 30–40 days (Plummer, 1975; Plummer and Mann, 1983).

Achatinids produce shelled eggs. There is insufficient calcium in the albumen to allow for body-shell formation so the embryo utilizes calcium from the eggshell. Plummer and Mann (1983) found that *A. marginata* embryos use 33% of the calcium initially present in the eggshell. Eggs of *Archachatina* are larger than those produced by *Achatina* of comparable size. This is reflected in the bulbous, protoconch of *Archachatina* species. The eggs of Achatinidae are generally deposited in 'nests' excavated in the soil by the gravid animal, but occasionally may simply be deposited in moist crevices among plant litter, stones and other debris on the ground. The West African *Pseudachatina* Albers species, such as *P. downesii* (Sowerby) from Fernando Poo, deposit their eggs in the axils of the branches of the trees they inhabit. Tryon and Pilsbry (1904) mention a similar behaviour in *A. marginata*. The sites chosen by *A. fulica* for oviposition are similar to the resting sites on the ground, although if the cover is too sparse the gravid animals may turn some loose soil and deposit the eggs 25 mm or so below the surface.

The frequency of oviposition varies with the duration of the period favourable for activity. Mead (1961) stated that, in the field, *A. fulica* will lay a batch of eggs 'every few weeks' as long as favourable conditions prevail. In reality, however, the frequency of oviposition in the field does not approach this level. According to Dun (1967), egg laying by *A. fulica* in New Guinea occurs in two pronounced peaks each season, the first shortly after resumption of activity following the onset of the rainy season and the second 2–3 months later. Thus each reproductive animal typically produces two clutches of eggs each year. In Oahu, Hawaii, only five to six clutches of eggs are produced by *A. fulica* per season (Kekauoha, 1966). In Calcutta, India, where *A. fulica* is active for only 4 months in the year, 1.9, 4.2, 3.9 and 2.0 egg clutches were produced on average per animal in the first 4 years following attainment of reproductive maturity (Raut, 1991). Pawson and Chase (1984) showed that fecundity was maximal in *A. fulica* aged between 210 and 270 days under laboratory conditions. After that, the production of eggs declines markedly, with almost no clutches produced by animals older than 1 year. A similar pattern is evident in animals in the field, although the time to peak oviposition activity and the rate of subsequent decline is delayed commensurate with the slower growth rates. While data on *A. fulica* fecundity have not been collected by standardized methods for different regions, some estimates are available: 100 eggs in the first year and 500 eggs in the second year in Sri Lanka (Green, 1911); 100 eggs in the first year and 200–300 eggs in the second year in Hong Kong (Jarrett, 1931); 900–1200 eggs annually in Oahu (van Weel, 1948/49); 677–1817 eggs annually in Oahu (Kekauoha, 1966); 160–1024 eggs annually in Calcutta (Raut, 1991). Clutch size varies from ten to 400. Tomiyama and Miyashita (1992) demonstrated great variability in clutch size and egg size in *A. fulica*, with both parameters positively correlated with the size of the parent animals. The limited available data, summarized by Tomiyama and Miyashita (1992), indicate a considerably higher reproductive potential in *A. fulica* than in other *Achatina* species and in *Archachatina* species. Lange (1950) noted the discrepancy between high viability among eggs deposited in laboratory animals and the low rates of recruitment into feral populations. None the less, an enormous potential for recruitment into the population is indicated by the reproductive strategy in *A. fulica*.

Plummer (1975) reports an average longevity of 4.5 years for *A. marginata* kept in captivity in London, although specimens occasionally lived for 7.5–10 years. *A. fulica* can live as long as 9 years in captivity (van Leeuwen, 1932) but under field conditions maximum longevity is usually in the order of 3–5 years (Mead, 1979a; Suzuki and Yasuda, 1983; Tomiyama, 1993). Thus, these animals evidently persist long after their peak reproductive fitness. However, van Bruggen (1985) remarks that early maturity, possibly combined with a long life and a steady increase in clutch size, seems to be the key element in the reproductive strategy in *A. machachensis*.

After emerging from the egg, achatinids generally remain underground with other members of the clutch for several days. During this time the hatchlings consume their eggshells, sometimes the eggshells of unhatched siblings and soil organic matter. This eggshell-eating behaviour has been observed frequently, both in *A. fulica* (Rees, 1951; Pawson and Chase, 1984) and in other achatinids (e.g. Owiny, 1974; Plummer, 1975; Hodasi, 1979). Lange (1950) reported that the young of *A. fulica* feed on the eggshells for 3–4 days. In the field Rees (1951) determined that *A. fulica* hatchlings remain below the surface of the soil for 5–15 days, while for laboratory colonies of this species Pawson and Chase (1984) found hatchlings to remain in the soil for 4–7 days. Plummer (1975) stated that *A. marginata* hatchlings remain underground for 7–14 days before surfacing. On emergence from the soil the young snails display exploratory and voracious feeding behaviour.

Observations in India clearly indicate that emergent juvenile *A. fulica* typically do not disperse great distances. They initially remain near the site of hatching, feeding on decaying plant matter and preferred host plants. After about 2 weeks the juveniles begin to range further, but none the less still tend to be aggregated and forage on palatable plant species. While their small size limits the quantity of plant material consumed per animal, the aggregated nature of the infestations can lead to severe damage in infested plants. As the *A. fulica* grow, they progressively disperse, seeking out and inflicting substantial damage on susceptible plants. After about 2 months the snails establish home sites, from which they leave at dusk to forage and to which they return at or before dawn. A typical behavioural pattern is that, following emergence from their home sites, *A. fulica* move directly to the sites of preferred food plants in the garden or crop. Such behaviour in adults, possibly reinforced by entrained long-term memory (Croll and Chase, 1977) and selective feeding, can lead to severe damage in susceptible plant species. The behavioural sequence of sedentary early juveniles, dispersive juveniles and then home-site-bound adults can lead to progressive elimination of susceptible and vulnerable plant species from a localized area. This herbivory is most pronounced where the abundance of susceptible species is low and thus the selective pressure is greatest.

Achatinidae are generally regarded as herbivorous, feeding primarily on living and decaying vascular plant material. Van Weel (1948/49) reported that young of *A. fulica* feed on decaying matter and unicellular algae. Animals with shells between 5 and 30 mm height were observed to prefer living plants. It was during this period that *A. fulica* was found to be most injurious to plantations and gardens. Although not entirely neglecting living vegetation, the maturing snails were found to largely return to a scavenging, detritivorous habit. Olson (1973) refers to *A. fulica* as an opportunistic, omnivorous and carpophagous feeder. He considered this species to be basically a scavenger as 75% of its food is detritus. Das and Sharma (1984) comment on the necrophagous habit of *A. fulica*.

A considerable number of plant species susceptible to *A. fulica* are to be found listed in the popular and scientific literature. The information pertaining to economically important plant species is reviewed in a later section of this chapter. There are few reports on damage in indigenous plant species in areas where *A. fulica* has been introduced. This undoubtedly reflects a preoccupation with cultivated plants among investigators, rather than the absence of damage to the natural vegetation. Dun (1967) reports the virtual local extinction of the indigenous *Pipturus argenteus* (Forster) Weddell (Urticaceae) in parts of the Gazelle Peninsula, New Britain.

The literature on *A. fulica* is conspicuous for the scarcity of quantitative data on feeding preferences and impacts on plant communities. Generally, observations support the hypothesis of Waterhouse and Norris (1987) that the preference for particular food plants exhibited by *A. fulica* at a particular locality is dependent primarily on the composition of the plant communities, in respect to both the species present and the age of the plants belonging to the different species. Most severe damage is likely to be observed in susceptible species when they predominate in the plant community. In the absence of quantitative sampling methods, substantial damage to the less abundant plant species may often go undetected. Moreover, the extent of damage varies according to the age structure of the *A. fulica* population, which in turn will relate to the stages of the crop in relation to the phenology of *A. fulica* recruitment (Jaski, 1953; Raut, 1982).

The herbivory damage inflicted by *A. fulica* varies substantially between seasons, due to variation in plant occurrence in the habitat and variation in climatic favourability for gastropod activity. Of 22 plant species offered by Raut and Ghose (1983a) to *A. fulica* in outdoor cages in India, 13 plant species suffered damage during the monsoon and summer and only eight during the winter (Fig. 3.2).

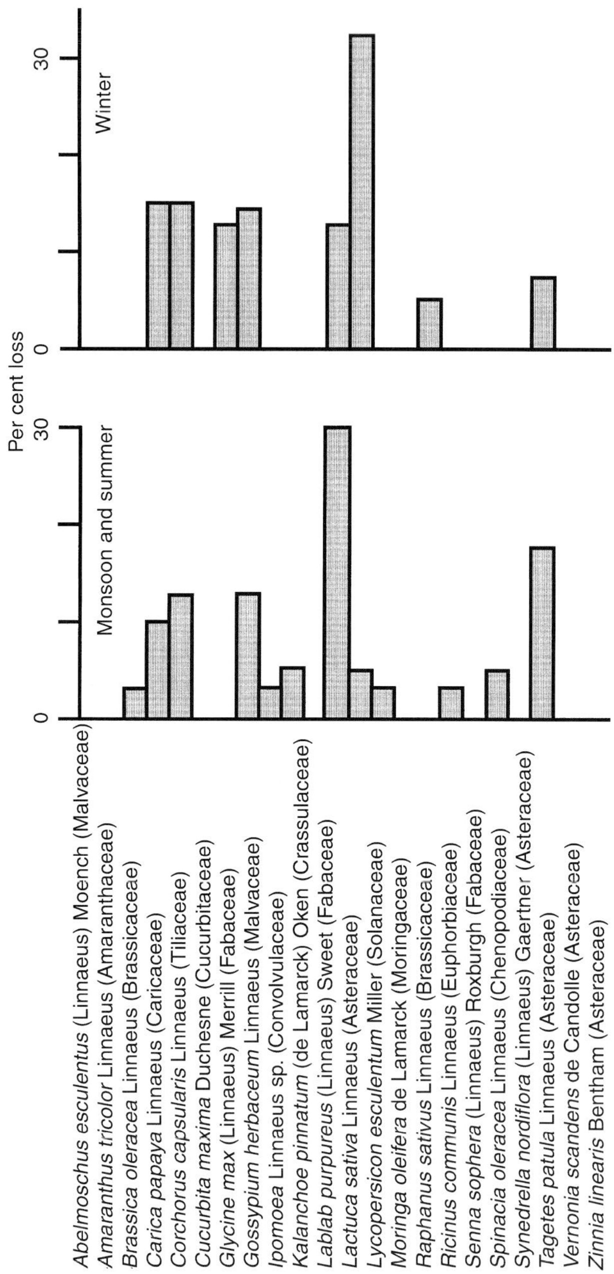

Fig. 3.2. Differential losses among 22 plant species due to feeding preferences by *Achatina fulica* Bowdich (Achatinidae) in field cages during two seasons in Calcutta, India (from Raut and Ghose, 1983a).

Ahmed and Raut (1991) demonstrated that *A. fulica* had higher growth rates on *Trichosanthes anguina* Linnaeus (Cucurbitaceae) and a mixed diet than when maintained on single-species diets comprising *Lactuca sativa* Linnaeus (Asteraceae), *Lablab purpureus* (Linnaeus) Sweet (Fabaceae), *Cucurbita maxima* Duchesne (Cucurbitaceae) or *Basella rubra* Linnaeus (Basellaceae). These differences occurred irrespective of the temperature regime at which the *A. fulica* were maintained (constant 20, 25 or 30°C; ambient 24.5–32.8°C), but were accentuated at 20°C. These results suggest that food-plant availability and feeding preferences may have important effects on the population dynamics of *A. fulica* by regulating growth rates and their subsequent effects on survival, fecundity and population recruitment. Egonmwan (1991) demonstrated that food preferences in *Limicolaria flammea* Müller varied between animals in somatic growth and those sexually active.

Ghose (1963) observed that young *A. fulica* denied access to soil 'did not thrive well'. He suggests that soil may be important in the provision of certain requirements of the juveniles in the early stages of postembryonic development. Nisbet (1974) subsequently found that ingestion of soil was important to the health of achatinids maintained in the laboratory.

A. fulica occurs across a range of soil pH and calcium conditions (summarized by Srivastava, 1992). By controlling the amount of available calcium in different soil types, Voelker (1959) was able to demonstrate environmentally induced variation in shell growth rate, size, weight, shape and colour in *A. fulica*. Schreurs (1963) conducted similar experiments in which he demonstrated the importance to normal development in *A. fulica* not only of calcium, but also of certain physical properties of the soil, the presence of adequate decaying organic material and the ample availability of green plant material. He found that, when many animals were kept together in a small space, the stress of 'crowding' was manifested in retarded growth, even though an abundance of food was available. This crowding effect is consistent with that observed in other terrestrial gastropod species (Cook, 2001).

According to Mead (1961), *A. fulica* persists but does not flourish at temperatures of 6–7°C. On the basis of observations in Hawaii, F.J. Olson (quoted in Mead, 1979a) established an optimal temperature for *A. fulica* of *c.* 26°C and predicted a maximum high temperature of *c.* 29°C and a minimum low temperature of 9°C for activity, and therefore feeding and growth, in this species. Singh and Birat (1969) recorded activity of *A. fulica* at a temperature of 8.8°C in Bihar. Raut and Ghose (1984) have stated that *A. fulica* will survive within the temperature range of 0 to 45°C, but for population increase a temperature range of 22–32°C is required. The latter authors found that hatching of *A. fulica* from eggs did not occur at temperatures below 15°C.

In the Bonin Islands winter temperatures are typically as low as 7°C and, according to Mead (1961), *A. fulica* persist there by winter hibernation 100–125 mm below the soil surface. Raut and Ghose (1984) reported that, despite favourable humidity during winter in India, at a temperature of 8°C about 58% of *A. fulica* go into hibernation. As a result of observation over several years, Raut and Ghose (1984) found that the timing and duration of hibernation and aestivation vary in different parts of India, reflecting seasonal variation in temperature and rainfall. Srivastava *et al.* (1987) observed that hibernation is initiated in New Delhi populations of *A. fulica* when temperatures declined to 11°C. By the time the temperature was down to 5.5°C and relative humidity was below 65%, all *A. fulica* had gone into hibernation. Larger animals were observed to hibernate earlier than small animals.

Modelling indicates that, under conditions of unrestricted growth, a group of 100 hatchling *A. fulica* is theoretically capable of producing a population in excess of 10^{12} individuals in the space of 2700 days (S.K. Raut, unpublished; Fig. 3.3). Under favourable field conditions, *A. fulica* can indeed reach high densities and biomasses. Tillier (1982), for example, recorded a biomass of up to 780 kg ha^{-1} in New Caledonia. Raut and Ghose (1984) record population densities of up to 46 m^{-2} in mainland India and up to 56 m^{-2} in Andaman and Nicobar. On the Philippine island of Bugsuk, Muniappan *et al.* (1986) estimated that 45 million *A. fulica* were collected and destroyed on 1600 ha (mean = 2.8 m^{-2}) over a 7-month period. In the Maldives, Muniappan (1987) reported 73 *A. fulica* m^{-2} for the island of Male. On Christmas Island, Lake and O'Dowd (1991) recorded a mean of 10 *A. fulica* m^{-2} in the heavily infested areas. As pointed out by Civeyrel and Simberloff (1996), there is almost invariably considerable variance in population density within infested areas.

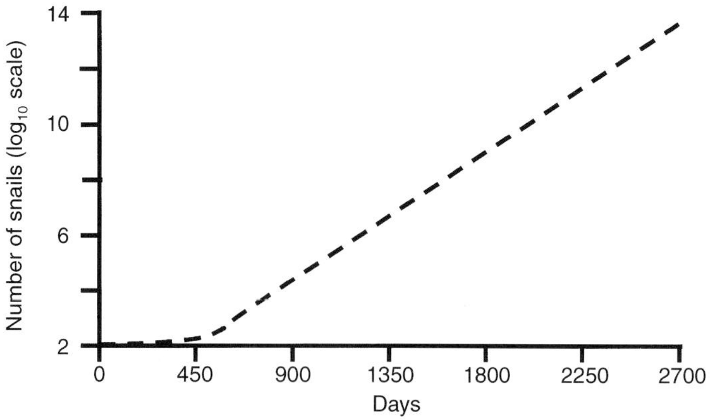

Fig. 3.3. Modelled growth rate in *Achatina fulica* Bowdich (Achatinidae) population size under abiotic environmental conditions pertaining to Calcutta, India. The model assumed an initial (Day 0) population of 100 0-day-old *A. fulica*, and incorporates growth, fecundity and mortality parameters derived from the literature and laboratory experimentation.

Pest Status

In tropical agriculture the cost of *A. fulica* is threefold. First, there is the loss of agricultural productivity caused by herbivory on crop plants, either through damage to the crop itself or to other plants that provide shade or soil enrichment in key elements such as nitrogen. Damage may also take the form of transmission of plant pathogens. Secondly, there is the cost of labour and materials associated with the management of the pest in such crop situations. Thirdly, there are the opportunity losses associated with enforced changes in agricultural practice, such as limiting the crop species to be grown in a region to those resistant to *A. fulica*. While outside the scope of this chapter, we may add the costs to the natural environment that arise from: (i) herbivory on native plant species; (ii) the altered nutrient cycling associated with the large volumes of plant material that pass through the achatinid gut under conditions of heavy infestations; (iii) the adverse effects on indigenous gastropods that may arise through competition for resource and fouling of the habitat with faeces and mucus; (iv) the adverse effects on indigenous gastropods that may arise through the non-target predation by malacophagous or generalist animals introduced as biological control agents of the achatinids; and (v) the adverse effects on indigenous gastropods that may arise through the non-target poisoning of chemical pesticides applied against the achatinids. Also beyond the scope of this contribution, but none the less a significant cost in many Asian, Pacific and American societies, is the role of achatinids in the transmission of the metastrongylid causative agents of eosinophilic meningoencephalitis, *Angiostrongylus cantonensis* (Chen) and *Angiostrongylus costaricensis* (Morera & Céspedes).

Estimates of costs to agricultural production associated with infestation by *A. fulica* are exceedingly scarce. Mead (1979a) argues that damage is characteristically localized and restricted to vegetable and flower gardens and that both the popular and scientific media have greatly exaggerated it. He expressed the opinion that the sheer numbers of snails, their slime trails, their excreta and even their decaying corpses have led observers to overestimate the threat to agriculture. Mead (1979a, p. 27) stated:

> by and large, the greatest damage caused by *Achatina fulica* is to be found either in new infestation sites or at the crest of expanding populations, with the amount of damage decreasing proportionately towards the epicentre. Even with the great numbers characteristic of young populations, however, the damage is fairly localized, and not catastrophic or devastating on a broad scale.

In a review of the economic importance of infestations, Mead (1979a) makes little mention of *A. fulica* as a crop pest. Civeyrel and Simberloff (1996) suggest that the apparent inevitable population decline that occurs in the wake of the invasion argues against a long-term threat to agricultural production. These views obviously do not accord with those of farmers whose land is infested by *A. fulica* and do not take into account the altered economy of farming that results from pest-enforced changes in agricultural practice. While it must be accepted that *A. fulica* populations have often declined after an initial period of severe infestation, we have already noted above that there are areas where *A. fulica* has persisted at pestiferous levels for many decades. While the constraints on agriculture imposed by infestations of *A. fulica* are well highlighted in the popular and scientific literature, there is little attention given in the literature to changes in farming practice following a decline in the pest. We thus have little information on the resilience of agricultural systems that have been subject to pest infestation for extended periods.

A. fulica has a reputation as a voracious herbivore. Schreurs (1963) determined that, in general, specimens up to 60 mm in shell length consume *c.* 10% of their own weight daily. From the literature it is well established that, in the agricultural landscapes of its naturalized range, this species feeds extensively, if not primarily, on cultivated and adventive, ruderal plant species. The species will persist on weeds and various indigenous vascular plants during periods in which cultivated plants are scarce. The list of cultivated plants reported to be susceptible to *A. fulica* is extensive, and is summarized in Tables 3.2 and 3.3 for economic and ornamental/medicinal plant species, respectively. Damage also extends to ground-cover and shade species grown in conjunction with cultivated shrub and tree species, such as cacao, tobacco (*Nicotiana tabacum* Linnaeus; Solanaceae), tea (*Camellia sinensis* (Linnaeus) Kuntze; Theaceae), rubber (*Hevea brasiliensis* (von Willdenow ex de Jussieu) Müller; Euphorbiaceae) and teak (*Tectona grandis* Linnaeus; Verbenaceae).

Irrespective of the crop, the seedling or nursery stage is most preferred and most vulnerable. In some situations, infestations of crops in the seedling or nursery stage are so severe as to demand changes in the crop species cultivated. In Guam, Indonesia and Malaysia, for example, *A. fulica* infestations made it uneconomic to grow vegetables, at least during the period of peak infestations (South, 1926; Kondo, 1950a; Mead, 1961). A similar situation was experienced by the growers of water melon (*Citrullus lanatus* (Thunberg) Matsumura & Nakai; Cucurbitaceae) in Mariana Islands and papaya (*Carica papaya* Linnaeus; Caricaceae) in Mariana Islands and India (Chamberlin, 1952; Raut and Ghose, 1984). Thus production of some crops has proved unsustainable in certain infested areas.

In more mature plants, the nature of the damage varies with the plant species, sometimes involving defoliation and in others involving damage to stems, flowers or fruit.

Waterhouse and Norris (1987) noted the differences in crop species reported to be susceptible in different regions. For example, in Sri Lanka (Green, 1910b), the Philippines (Pangga, 1949), Saipan (Lange, 1950), Rota (Kondo, 1952) and India (Raut and Ghose, 1984) it has proved difficult to produce yam and yet in Mariana Islands damage to this crop has proved

Table 3.2. Economically important plants recorded as being subject to losses through damage by *Achatina fulica* Bowdich (Achatinidae) in regions outside of Africa.*

Amaranth (Amaranthaceae)	*Amaranthus* Linnaeus spp., including *A. blitum* Linnaeus, *A. tricolor* Linnaeus and *A. viridis* Linnaeus
Banana (Musaceae)	*Musa* Linnaeus spp., particularly *M. acuminata* Colla and *M. paradisiaca* Linnaeus
Basella (Basellaceae)	*Basella alba* Linnaeus
Beans and peas (Fabaceae)	*Arachis hypogaea* Linnaeus; *Glycine max* (Linnaeus) Merill; *Lablab purpureus* (Linnaeus) Sweet; *Pisum* Linnaeus spp., particularly *P. sativum* Linnaeus; *Vigna radiatus* (Linnaeus) Wilczek and *V. unguiculata* (Linnaeus) Walpers
Blimbi (Oxalidaceae)	*Averrhoa bilimbi* Linnaeus and *A. carambola* Linnaeus
Breadfruits (Moraceae)	*Artocarpus* Forster & Forster spp., including *A. altilis* (Parkinson) Fosberg and *A. heterophyllus* de Lamarck
Brinjal/aubergine (Solanaceae)	*Solanum melongena* Linnaeus
Brassicas (Brassicaceae)	*Brassica oleracea* Linnaeus cultivars; *Raphanus sativus* Linnaeus
Cacao (Sterculiaceae)	*Theobroma cacao* Linnaeus
Carrot (Apiaceae)	*Daucus carota* Linnaeus
Cassava (Euphorbiaceae)	*Manihot esculenta* Crantz
Castor (Euphorbiaceae)	*Ricinus communis* Linnaeus
Chillies and peppers (Solanaceae)	*Capsicum* Linnaeus spp., particularly *C. annuum* Linnaeus and *C. baccatum* Linnaeus
Citrus (Rutaceae)	*Citrus* Linnaeus spp., particularly *C. sinensis* (Linnaeus) Osbeck and *C. reticulata* Blanco
Coffee (Rubiaceae)	*Coffea* Linnaeus spp., especially *C. arabica* Linnaeus and *C. canephora* Pierre ex Froehner
Corm (Araceae)	*Amorphophallus paeoniifolius* (Dennst.) Nicolson
Cotton (Malvaceae)	*Gossypium* Linnaeus spp., especially *G. herbaceum* Linnaeus
Drum stick (Moringaceae)	*Moringa oleifera* de Lamarck
Erythrina (Fabaceae)	*Erythrina* Linnaeus sp.
Eucalyptus (Myrtaceae)	*Eucalyptus* L'Héitier de Brutelle spp., especially *E. deglupta* Blume
Figs (Moraceae)	*Ficus hispida* Linnaeus
Gourd/pumpkins/ cucumber/melons (Cucurbitaceae)	*Citrullus lanatus* (Thunberg) Matsumura & Nakai; *Cucumis* Linnaeus spp., including *C. melo* Linnaeus and *C. sativus* Linnaeus; *Cucurbita* Linnaeus spp., including *C. maxima* Duchesne and *C. pepo* Linnaeus; *Edgaria darjeelingensis* Clarke; *Lagenaria* Seringe spp., including *L. siceraria* (Molina) Standley; *Luffa* Miller spp., including *L. acutangula* (Linnaeus) Roxburgh and *L. aegyptiaca* Miller; *Momordica* Linnaeus spp., principally *M. cochinchinensis* (de Loureiro) Sprengel
Jute (Tiliaceae)	*Corchorus capsularis* Linnaeus
Kokko (Fabaceae)	*Albizzia* Durazzini spp., including *A. lebbeck* (Linnaeus) Bentham; *Falcataria moluccana* (Miquel) Barneby & Grimes
Lettuce (Asteraceae)	*Lactuca* Linnaeus spp., including *L. sativa* Linnaeus and *L. indica* Linnaeus
Mahogany (Meliaceae)	*Swietenia mahagoni* (Linnaeus) von Jacquin
Mulberries (Moraceae)	*Broussonetia papyrifera* (Linnaeus) L'Héritier de Brutelle ex Ventenat; *Morus alba* Linnaeus

Table 3.2. Continued

Okra (Malvaceae)	*Abelmoschus esculentus* (Linnaeus) Moench
Onion (Liliaceae)	*Allium cepa* Linnaeus
Palm nuts (Arecaceae)	*Areca catechu* Linnaeus; *Elaeis quineensis* von Jacquin
Papaya (Caricaceae)	*Carica papaya* Linnaeus
Passion-fruit (Passifloraceae)	*Passiflora* Linnaeus sp.
Potato (Solanaceae)	*Solanum tuberosum* Linnaeus
Rubber (Euphorbiaceae)	*Hevea brasiliensis* (von Willdenow ex de Jussieu) Müller
Shishu (Fabaceae)	*Dalbergia sissoo* Roxburgh ex de Candolle
Soursop (Annonaceae)	*Annona muricate* Linnaeus
Spinach (Chenopodiaceae)	*Spinacia oleracea* Linnaeus
Sunflower (Asteraceae)	*Helianthus annuus* Linnaeus
Sweet potato (Convolvulaceae)	*Ipomoea batatas* (Linnaeus) de Lamarck
Taro (Araceae)	*Alocasia* (Schott) Don spp., including *A. macrorrhizos* (Linnaeus) Schott; *Colocasia esculenta* (Linnaeus) Schott; *Xanthosoma braziliense* (Desfontaines) Engler
Tea (Theaceae)	*Camellia sinensis* (Linnaeus) Kuntze
Teak (Verbenaceae)	*Tectona grandis* Linnaeus
Tobacco (Solanaceae)	*Nicotiana tabacum* Linnaeus
Tomato (Solanaceae)	*Lycopersicon esculentum* Miller
Vanilla (Orchidaceae)	*Vanilla* Miller sp.
Yam (Diascoreaceae)	*Dioscorea alata* Linnaeus

*Sources of information include: Green (1910b), Charmoy and Gébert (1922), South (1926), Bertrand (1928, 1941), Corbett (1933, 1937), Latif (1933), Leefmans and van der Vecht (1933a,b), Riel (1933), van Benthem Jutting (1934, 1952), Beeley (1935, 1938), Fairweather (1937), Heubel (1937, 1938), Cotton (1940), Feij (1940), Esaki and Takahashi (1942), Hatai and Kato (1943), Townes (1946), Anonymous (1947), Otanes (1948), van Weel (1948/49), Hes (1949, 1950), Pangga (1949), Rappard (1949), van der Meer Mohr (1949b), Altson (1950), Kondo (1950a,b, 1952), Lange (1950), Rees (1951), Chamberlin (1952), Holmes (1954), van Alphen der Veer (1954), Weber (1954a), Behura (1955), Mead (1961, 1979a), Chiu and Chou (1962), Dun (1967), Singh and Birat (1969), Ranaivosoa (1971), Olson (1973), Raut (1982), Raut and Ghose (1983a, 1984), Srivastavsa (1992), Jahan and Raut (1994).

negligible (Chamberlin, 1952). Similarly, Srivastava (1992) mentioned the bitter gourd (*Momordica charantia* Linnaeus: Cucurbitaceae) being grown free from *A. fulica* herbivory in the Andamans and yet there have been records of some damage to this crop species in various provinces in India (e.g. Raut and Ghose, 1984; Jahan and Raut, 1994). Other crop species for which there are conflicting reports of damage from different regions include tea, coffee (*Coffea* Linnaeus spp.; Rubiaceae) and various taro species (*Alocasia macrorrhizos* (Linnaeus) Schott, *Colocasia esculenta* (Linnaeus) Schott, *Xanthosoma brasiliense* (Desfontaines) Engler;

Table 3.3. Ornamental and medicinal crop species recorded as being subject to damage by *Achatina fulica* Bowdich (Achatinidae) in regions outside Africa.*

Aloe (Aloeaceae)	*Aloe indica* Royle
Alsophila (Cyatheaceae)	*Alsophila* Brown sp.
Amaranth (Amaranthaceae)	*Comphrena globosa* Linnaeus
Spleenwort (Aspleniaceae)	*Asplenium nidus* Linnaeus
Bauhinia (Fabaceae)	*Bauhinia acuminata* Linnaeus
Boatlily (Commelinaceae)	*Tradascantia spathacea* Swartz
Bouganvilles (Nyctaginaceae)	*Bougainvillea* Commerson ex de Jussieu spp., particularly *B. spectabilis* Willdenow
Buckhorn (Cactaceae)	*Opuntia* Miller sp.
Cactus (Cactaceae)	*Cereus* Miller sp.
Calophyllum (Clusiaceae)	*Calophyllum inophyllum* Linnaeus
Canna (Cannaceae)	*Canna* Linnaeus spp., particularly *C. indica* Linnaeus
Chrysanthemums (Asteraceae)	*Chrysanthemum* Linnaeus sp.
Clitoria (Fabaceae)	*Clitoria ternatea* Linnaeus
Cosmos (Asteraceae)	*Cosmos* Cavanilles spp.
Crinums (Liliaceae)	*Crinum* Linnaeus spp.
Dahlias (Asteraceae)	*Dahlia* Cavanilles sp.
Dumbcane (Araceae)	*Dieffenbachia seguine* (von Jacquin) Schott
Gardenias (Rubiaceae)	*Gardenia angusta* (Linnaeus) Merrill
Impatiens (Balsaminaceae)	*Impatiens balsamina* Linnaeus
Indian bark (Lauraceae)	*Cinnamonum tamala* (Buchanan-Hamlin) Nees & Eberm.
Jasmine (Oleaceae)	*Jasmin sambac* (Linnaeus) Aiton
Kalanchoe (Crassulaceae)	*Kalanchoe pinnatum* (de Lamarck) Oken
Marigold (Asteraceae)	*Tagetes* Linnaeus spp., including *T. erecta* Linnaeus and *T. patula* Linnaeus
Moth orchids (Orchidaceae)	*Phalaenopsis* Blume spp.
Oleander (Apocynaceae)	*Nerium* Linnaeus spp., including *N. indicum* Miller and *N. oleander* Linnaeus
Perwinkle (Apocynaceae)	*Catharanthus roseus* (Linnaeus) Don
Pothos (Araceae)	*Epipremnum pinnatum* (Linnaeus) Engler
Purslane (Portulacaceae)	*Portulaca grandiflora* Hooker
Rose-mallow (Malvaceae)	*Hibiscus* Linnaeus spp., including *H. rosasinensis* Linnaeus and *H. mutabilis* Linnaeus
Roses (Rubiaceae)	*Rosa* Linnaeus spp.
Sanseviera (Liliaceae)	*Sansevieria trifasciata* Prain
Snake gourd (Cucurbitaceae)	*Trichosanthes anguina* Linnaeus
Spiderwisp (Capparaceae)	*Cleome gynandra* Linnaeus
Sunflower (Asteraceae)	*Helianthus annuus* Linnaeus
Vanda (Orchidaceae)	*Vanda* Jones sp.
Zinnia (Asteraceae)	*Zinnia linearis* Bentham

*Sources of information include: Green (1910a), Jarrett (1923), South (1926), Dammerman (1929), Latif (1933), Leefmans and van der Vecht (1933a,b), Riel (1933), van Benthem Jutting (1934, 1952), Feij (1940), Otanes (1948), Pangga (1949), Lange (1950), Mead (1961), Olson (1973), Raut (1982), Raut and Ghose (1984), Manna and Raut (1986), Srivastava (1992), Jahan and Raut (1994).

Araceae). In the case of taro, part of the variance in damage reports undoubtedly relates to the different crop species grown in different regions. There are also several cases in the literature where reports from within one region are at variance. For example, Hutson (1920) reported no damage to cacao in Sri Lanka, but Mead (1961) reports damage to this crop there. Likewise, occasional damage to impatiens (*Impatiens balsamina*

Linnaeus; Balsaminaceae) has been recorded in India (e.g. Raut and Ghose, 1984; Jahan and Raut, 1994), but it is one of few ornamental species that have been reported to survive in infested gardens there (Srivastava, 1992).

That plant susceptibility can vary depending on the composition of the plant community suggests that the extent of damage to crop species will vary between agricultural systems based, at one extreme, on production in monocultures and those based, at the other extreme, on multiple land uses and crop species mixtures. This could well explain the feeding behaviour of *A. fulica* on the ornamental plant species *Canna indica* Linnaeus (Cannaceae). In the presence of many kinds of preferred food plants, Raut and Ghose (1984) noted that *A. fulica* rarely attack *C. indica*, but often use this species for daytime shelter. In contrast, *C. indica* was completely defoliated within a few days when the preferred host plants were no longer available (Manna and Raut, 1986).

Lange (1950) and Srivastava (1992) list observations on non-preferred plant species but, as noted earlier, there is no quantitative information available on the effect of *A. fulica* on the ecology of plant communities. At present there is little understanding of the chemical or physical traits that confer different levels of susceptibility among plant species or indeed as to whether any particular phylogenetic clades of vascular plants are more or less susceptible. As summarized by Schotman (1989), from the literature we may conclude that the economic crops generally suffering little damage from *A. fulica* include sugar cane (*Saccharum officinarum* Linnaeus; Gramineae), maize (*Zea mays* Linnaeus; Gramineae), rice (*Oryza sativa* Linnaeus; Gramineae), coconut (*Cocos nucifera* Linnaeus; Arecaceae), pineapple (*Ananas comosus* (Linnaeus) Merrill; Bromeliaceae) and screw pine (*Pandanus tectorius* Parkinson ex Zuccarini; Pandaceae). Onion (*Allium cepa* Linnaeus; Liliaceae), garlic (*Allium sativum* Linnaeus), yam-beans (*Pachyrhizus tuberosus* (de Lamarck) Sprengel; Fabaceae) and betel (*Piper betel* Linnaeus; Piperaceae) are particularly remarkable among crop species in that they are evidently immune to the attentions of *A. fulica* everywhere (Godan, 1983; Srivastava, 1992).

That *A. fulica* feed on a variety of plant species and the extent of damage varies temporally, spatially and with the compositional structure of the vegetation poses significant difficulties for the standardization of sampling and the development of economic thresholds in crops. This is accentuated by the generally small area of individual fields devoted to particular crops, the frequent intercropping within fields and the small-scale mosaic of dwellings, cultivated fields and primary and secondary forests that characterize much of the agriculture landscape in tropical regions. Undoubtedly the economics of infestations and appropriate action thresholds have been established for the more extensive crops, such as plantation banana (*Musa* Linnaeus spp.; Musaceae), but the relevant information is not available in the plant protection or malacological literature.

A. fulica has been implicated in transmission of plant diseases – *P. palmivora* in black pepper, betel pepper, coconut, papaya and vanilla (*Vanilla* Miller spp.; Orchidaceae), *Phytophthora colocasiae* Racib. in taro and *Phytophthora parasitica* Dastur in aubergine (brinjal; *Solanum melongena* Linnaeus: Solanaceae) and tangerine (*Citrus reticulata* Blanco; Rutaceae) (Mead, 1961, 1979a; Turner, 1964, 1967; Muniappan, 1983; Schotman, 1989). However, while the importance of these disease organisms is well established, the relative importance of *A. fulica* as a transmission agent in the epidemiology of these diseases under usual cropping conditions has not been well established.

While the pest status of achatinids has generally focused on *A. fulica* outside Africa, as outlined earlier in this chapter, various achatinid species can assume pest status in Africa. The achatinids feed on both dead and living plant tissues in their natural habitat, but, when that native habitat occurs adjacent to or is converted to sites of human habitation, they can assume pest status because of their predations on cultivated plants. Crop species damaged by Achatinidae under these circumstsances are listed in Table 3.1. Since some of these Achatinidae are edible, there is often a reluctance to regard them as pests (Hodasi, 1979, 1984; von Stanislaus *et al.*, 1987). Furthermore, the recent establishment of *L. aurora* as a crop pest in Martinique illustrates the potential for species in addition to *A. fulica* to adversely affect agricultural crops outside Africa (Mead and Palcy, 1992; Palcy and Mead, 1993).

Control

Physical, chemical and biological strategies have variously been used to manage infestations of *A. fulica*. However, the great variety of cropping and socio-economic environments in which infestations have occurred has prevented planned, coordinated and integrated approaches to the development of control methods. Most of the literature relating to the control or eradication of the pest predates the 1960s, primarily in relation to attempts to control infestations that developed as the pest was dispersed throughout the Indo-Pacific region. The published information pertaining to chemical control almost solely relates to that period. Mead (1979a, p. 8) noted in the Indo-Pacific:

> [a] growing attitude of resignation and even indifference – an acceptance of this pest as one of the many unfortunate facts of life. This attitude is explained in part by the fact that in most areas . . . where this snail is found, the people have learned to live with it.

Mead (1979a, pp. 8–9) goes on to suggest that the:

> overall picture that emerges . . . is one in which the snail continues to be a serious pest in the peripheral areas but is becoming less so in the older infested areas, to the point, in some cases, where it essentially ceases to be a

pest. In many areas, indeed if not most, there are virtually no organized efforts to control this snail.

Physical control strategies

Physical control relies primarily on the collection and destruction of the snails and their eggs from infested sites. The strategy has been effective in providing relief from *A. fulica* infestation in crops, albeit temporary, as reported from Guam (Peterson, 1957c), Hawaii (Olson, 1973), Japan and Sri Lanka (Mead, 1961). Schotman (1989) maintains that manual collection and destruction of the snails can be an effective control strategy when practised on a small scale or in organized campaigns involving the public or farmer groups. Collection and destruction of snails and their eggs have also played a significant part in eradication of incipient infestations in Japan (Mead, 1961), Australia (Colman, 1977), Arizona and Florida (Mead, 1961, 1979a).

The establishment of physical barriers that prevent or reduce movement of snails has long been practised as a control strategy for *A. fulica*. These barriers may simply be a strip of bare soil as a headland around the crop or may be a fence that comprises a screen of corrugated tin or security wire mesh. Schotman (1989) recommends that ditches be dug around the field and the snails collected and destroyed each day.

Protection of valuable horticultural plants can be provided during their vulnerable seedling stage by ringing them with a strip of cardboard that has been dipped in a suspension of metaldehyde, the dispersion of the latter being aided by the addition of a detergent (Bridgland and Byrne, 1956; Dun, 1967).

Chemical control strategies

Most early attempts at chemical control employed baits containing metaldehyde and/or calcium arsenate. A considerable number of toxicants and repellents have been evaluated at various times and locations for activity against *A. fulica* (summarized by Mead, 1979a; Raut and Ghose, 1984; Srivastava, 1992), but the great majority of these evaluations have not yielded significant advances over bran-based baits containing metaldehyde, which were initially developed in the 1930s for gastropod control in temperate regions. In many cases the evaluations were undertaken under experimental, laboratory conditions and the effectiveness of many materials under field conditions has not been demonstrated. Subsequently methiocarb baits also became available. In recent years the situation has not dramatically changed, although a number of new molluscicidal chemicals are now available, albeit rarely developed or registered specifically for use against *A. fulica*.

Bait formulations can be rendered ineffective by rain, which obviously poses constraints on the effectiveness of baits applied during the rainy season, when the gastropods are most active. Cement briquette formulations containing metaldehyde have provided for greater persistence and have enabled control in remote areas where repeated applications were not practicable (e.g. Dun, 1967; Watson, 1985). Many current commercial bait products have been formulated to persist, at least for a time, under moist field conditions. However, there is little published information on their effectiveness under tropical conditions.

Because a proportion of *A. fulica* occur arboreally and is thus not readily controlled by ground-applied baits, there has been interest in the efficacy of molluscicidal dusts or sprays. Nair *et al.* (1968), for example, demonstrated the effectiveness of kaolin dusts containing 1% metaldehyde and suspensions containing 1–4% metaldehyde.

Because of continuing concern about the environmental effects of synthetic chemicals, there is currently much interest in naturally occurring chemicals as molluscicides. Panigrahi and Raut (1994), for example, have demonstrated that an extract of the fruit of *Thevetia peruviana* (Persoon) Schumann (Apocynaceae) has activity against *A. fulica*. However, evaluations under field conditions are yet to be made.

Cropping strategies

Rees (1951, p. 585) noted that *A. fulica* 'does not appear to like aromatic plants, and it may be profitable to pursue this subject further to see whether judicious planting is likely to have some effect on its activity in gardens'. This strategy has not been seriously investigated. Relative to losses in monoculture crops, however, Raut and Ghose (1983b) demonstrated that planting selected non-crop species in headlands or guard rows can reduce economic losses within the crop (Fig. 3.4). As a strategy for the management of *A. fulica*, such mixtures of crop and non-crop species are not yet widely practised, although the approach is compatible with the current interest in the potential benefits of increased biological diversity in agriculture.

Biological control strategies

A. fulica, as with other Achatinidae, are subject to pathogens, parasites and invertebrate predators in their natural range in Africa. Those that are known are listed in Table 3.4. In addition, various vertebrates are recognized predators of Achatinidae in Africa (e.g. Rees, 1951; Williams, 1951, 1953; van Bruggen, 1978; Hodasi, 1989). None the less, the importance of these natural enemies in the regulation of *A. fulica* populations in Africa has not been studied, and much of the information on natural enemies stems from anecdotal observation made in the course of

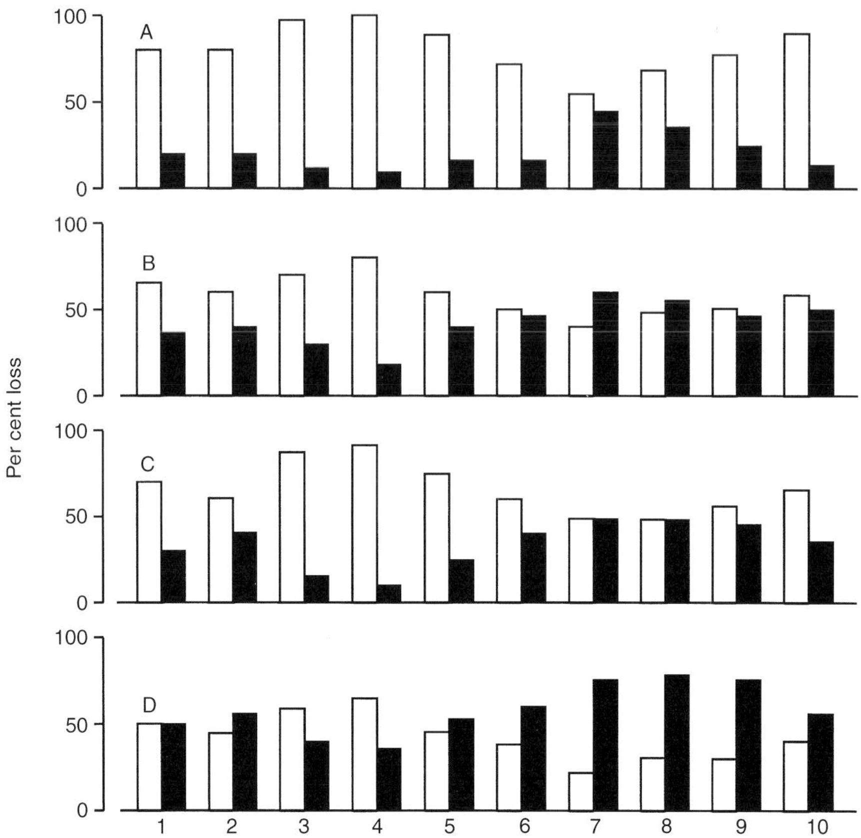

Fig. 3.4. Level of loss inflicted in ten crop plant species by *Achatina fulica* Bowdich (Achatinidae), in the presence of one of four non-crop species (black bar), namely: A. *Senna sophera* (Linnaeus) Roxburgh (Fabaceae). B. *Kalanchoe pinnatum* (de Lamarck) Oken (Crassulaceae). C. *Synedrella nordiflora* (Linnaeus) Gaertner (Asteraceae). D. *Tagetes patula* Linnaeus (Asteraceae). The crop species were: 1. *Lactuca sativa* Linnaeus (Asteraceae); 2. *Brassica oleracea* Linnaeus (Brassicaceae); 3. *Glycine max* (Linnaeus) Merrill (Fabaceae); 4. *Lablab purpureus* (Linnaeus) Sweet (Fabaceae); 5. *Cucurbita maxima* Duchesne (Cucurbitaceae); 6. *Carica papaya* Linnaeus (Caricaceae); 7. *Lycopersicon esculentum* Miller (Solonaceae); 8. *Gossypium herbaceum* Linnaeus (Malvaceae); 9. *Abelmoschus esculentus* (Linnaeus) Moench (Malvaceae); 10. *Ricinus communis* Linnaeus (*Euphorbiaceae*).

field surveys and searches for agents that may be employed in biological control outside Africa.

It is also evident that, when introduced into new areas, *A. fulica* is not without some level of population regulation from pathogens, parasites and predators naturally resident there, as evidenced by the suite of organisms reported to attack this gastropod species outside Africa (Table 3.5). That *A. fulica* almost invariably assumes pest status when introduced to areas of favourable climate clearly points to the lack of significant population regulation by pathogens, parasites and predators, at least in the early phases of invasion by the pest.

Table 3.4. Invertebrate natural enemies of Achatinidae in Africa, with observation on utilization in biological control programmes for *Achatina fulica* Bowdich outside Africa.*

Microspora	*Plistiphora husseyi* Michaud	Recorded from *Achatina zebra* (Bruguière)
Acari	Indeterminate sp.	Belgian Congo. Recorded from *Achatina scheinfurthi* von Martens and *Achatina stuhlmanni* von Martens
Decapoda	Undetermined sp.	East Africa. Recorded attacking *Achatina* de Lamarck sp.
Diptera: Phoridae	*Wandolleckia achatinae* Cook	Evidently widespread in Africa. Ectoparasitic. Recorded from *Achatina variegata* Roissy, *Achatina achatina* (Linnaeus), *Archachatina ventricosa* (Gould), *Achatina* de Lamarck sp. and *Lignus* Gray sp.
Diptera: Tachinidae	*Mydeae* sp. nr *bivittata* (Macquart)	Central Africa. *Burtoa nilotica* (Pfeiffer)
Diptera Muscidae	*Ochromusca trifaria* Big.	Malawi. Recorded from *Achatina craveni* Smith
Coleoptera: Carabidae	*Tefflus carinatus* Klug	Introduced into Hawaii. Apparently not established
	Tefflus zanzibaricus alluaudi Sternberg	Both adult and larval stages predacious on phytophagous terrestrial gastropods in Kenya. Introduced and established in Hawaii, but no demonstrated impact on *Achatina fulica* Bowdich
	Tefflus purpureipennis wituensis Kolbe	Kenya. Released in Hawaii. Apparently not established
	Tefflus raffrayi jamesoni Bates	Republic of the Congo. Released in Hawaii. Apparently not established
	Tefflus tenuicollis (Fairmaire)	Republic of the Congo. Released in Hawaii. Apparently not established
	Tefflus planifrons (Fabricius)	Nigeria. Released in New Britain but failed to establish
	Tefflus megerlei (Fabricius)	Nigeria. Introduced into Hawaii. Apparently not established
	Thermophilum hexastictum Gerstaecker	Kenya. Released in Hawaii. Apparently not established
Coleoptera: Drilidae	Undetermined species	West Africa
	Undetermined species	Morocco. Introduced into quarantine in Hawaii but evidently not released
	Undetermined species	Kenya. Introduced into quarantine in Hawaii but evidently not released
	Selasia unicolor (Guérin)	Nigeria. Introduced to New Britain. Apparently not established
Stylommatophora: Streptaxidae	*Gonaxis quadrilateralis* (Preston)	East Africa (Kenya). Introduced to India, parts of Asia, and many islands of the Pacific and Indian Oceans. Often failed to establish. Where established effect on *Achatina fulica* Bowdich when known, generally marginal. Generally preys on eggs and juveniles of *A. fulica*
	Gonaxis kibweziensis (Smith)	East Africa (Kenya). Established in Sri Lanka, Bermuda, and many islands of the Pacific, but impact on *Achatina fulica* Bowdich demonstrated only on Agiguan and Guam. Generally preys on eggs and juveniles of *A. fulica*
	Gonaxis vulcani Thiele	West Africa (Zaïre). Attempted introduction to Hawaii unsuccessful
	Gulella Pfeiffer sp.	South Africa. Released in Hawaii

Table 3.4. Continued

	Gulella Pfeiffer sp.	Republic of the Congo. Released in Hawaii
	Gulella Pfeiffer sp., nr *planti* (Pfeiffer)	South Africa
	Gulella bicolor (Hutton)[†]	Established as a tramp species in many tropical areas. Also introduced purposefully to Andaman Islands for control of *A. fulica* Bowdich but with no effect. Attempted introduction to Hawaii unsuccessful
	Gulella wahlbergi (Krauss)	South Africa. Established in Hawaii, but no demonstrated impact on *A. fulica*
	Edentulina affinis Boettger	East Africa (Kenya, Tanganyika). Released in Hawaii but failed to establish
	Edentulina obesa bulimiformis (Grandidier)	East Africa (Kenya, Tanzania). Attempted introduction to Hawaii unsuccessful
	Edentulina ovoidea (Bruguière)	Endemic to Mayotte. Preys on phytophagous gastropods. Introduced to Madagascar, Comores and Réunion. Attempted introduction to Hawaii unsuccessful
Stylommatophora: Rhytididae	*Ptychotrema* Mörch sp.	Belgian Congo. Introduced to Hawaii but establishment success and impact on *A. fulica* Bowdich unknown
	Ptychotrema walikalense Pilsbry	West Africa (Zaïre). Introduced to Hawaii but establishment success and impact on *A. fulica* Bowdich unknown
	Species complex	Twenty-two species, confined to eastern South Africa. Prey principally comprises achatinids and subulinids
	Natalina cafra (de Férussac)	South Africa. Predation on *Metachatina kraussi* (Pfeiffer). Attempted introduction to Hawaii unsuccessful

*Sources of information include: Stuhlmann (1894), Cook (1897), Wandolleck (1898), Brues (1903), Schmitz (1916, 1917, 1928, 1929, 1958), Bequaert in Pilsbry (1919), Bequaert (1925, 1926, 1950b), Pilsbry and Bequaert (1927), Williams (1951, 1953), Kondo (1952, 1956), Baer (1953), Weber (1953, 1954a,b, 1957), Davis (1954, 1958, 1959, 1960a,b, 1961, 1962, 1971, 1972), Pemberton (1954), Krauss (1955, 1964), Mead (1955, 1961, 1963a,b, 1979a), Peterson (1957b,c), Anon. (1961), Davis *et al.* (1961), Davis and Krauss (1962, 1963, 1964, 1965, 1967), Schreurs (1963), Simmonds and Hughes (1963), Davis and Butler (1964), Kim (1964), Dun (1967), Robinson and Foote (1968), Srivastava (1968b, 1976, 1992), Davis and Chong (1969), van Bruggen (1969, 1977, 1978), van der Schalie (1969), Ranaivosoa (1971), Etienne (1973), Lambert (1974, 1977), Sankaran (1974), Nakao *et al.* (1975), Nishida and Napompeth (1975), Srivastava *et al.* (1975), Lai *et al.* (1982), Muniappan (1982, 1983), Godan (1983), Backeljau (1984), Christensen (1984), Lionnet (1984), Nakamoto (1984), Raut and Ghose (1984), Howarth (1985, 1991), Nakahara (1985b), Waterhouse and Norris (1987), Eldredge (1988), Funasaki *et al.* (1988), Hodasi (1989), Nafus and Schreiner (1989), Naggs (1989), Napompeth (1990), Schreiner (1990), Herbert (1991), Cowie (1992, 1997, 1998a,b, 2000), Tillier (1992), Disney (1994), Civeyrel and Simberloff (1996), Sherley and Lowe (2000).
[†]Native range unknown. Possibly Africa or the Mascarene Islands (Solem, 1989) or Asia (Naggs, 1989).

Faced with infestation of *A. fulica*, many countries were eager to develop biological control strategies. Not only were natural enemies introduced from East Africa, in many cases introductions of polyphagous enemies were made from other parts of the world. Many introductions did not lead to the establishment of viable populations, as is typical for introduced species generally, but a great many of these introduced species were successful in naturalization. Unfortunately, the eagerness to effect biological control of *A. fulica* was not matched by consideration of environmental effects, particularly the impact on the indigenous

Table 3.5. Naturally occurring invertebrate enemies of terrestrial gastropods, utilizing the introduced species of Achatinidae in regions outside Africa, with observation on importance to regulation of achatinid populations.*

Bacteria	*Aeromonas hydrophila* (Chester) Stainer (= *Aeromonas liquifaciens* (Beijerinck))	Recorded in Sri Lanka, Singapore, Hong Kong, Thailand, Bangkok, Hawaii, India, Andaman Islands. Causing leucodermic lesions and epizootic disease in *A. fulica* Bowdich. Implicated as a causative agent in the decline of *A. fulica* observed in sectors many of the pests' naturalized range
Ciliophora: Peritrichida	*Trichodina* Ehrenberg spp.	India, recorded from *A. fulica* Bowdich. Germany, recorded from *A. zebra* (Brugière). Probably little effect on parasitized animals
	Pallitrichodina rogenae van As & Basson	Recorded from *A. fulica* Bowdich in Mauritius and Taiwan. No evidence of pathological effect. Regarded as a symbiont
	Pallitrichodina stephani van As & Basson	Recorded from *A. immaculata* de Lamarck in Mauritius. No evidence of pathological effect. Regarded as a symbiont
Nematoda: Rhabditidae	Unidentified sp.	Recorded from *A. fulica* Bowdich in India. Effect on *A. fulica* populations not known
Nematoda: Metastrongylidae	*Angiostrongylus cantonensis* (Chen)	Widespread in Asia and the Pacific. Definitive hosts are *Rattus* Fischer spp. (Muridae). Utilizes *A. fulica* Bowdich and other gastropods as intermediate hosts
	Angiostrongylus costaricensis (Morera and Céspedes)	Widespread in Americas. Definitive hosts are *Rattus* Fischer spp. (Muridae). Utilizes *A. fulica* Bowdich and other gastropods as intermediate hosts
	Anafilaroides rostratus Gerichter	Widespread. Definitive host *Felis* Linnaeus sp. (Felidae). Utilizes *A. fulica* Bowdich and other gastropods as intermediate hosts
Turbellaria: Geoplanidae	*Endeavouria septemlineata* (Hyman)	Hawaii. Important regulatory agent in *A. fulica* Bowdich. Also adversely affecting indigenous terrestrial gastropods, and the streptaxids and oleanicids introduced for biocontrol
	Undetermined sp.	Ogasawara. Observed attacking *A. fulica* Bowdich
Turbellaria: Rhynchodemidae	*Platydemus manokwari* de Beauchamp	New Guinea. Importance unknown but suspected as a contributory factor in decline in *A. fulica* Bowdich at some sites
Bipaliidae	*Bipalium indica* Whitehouse	India. Predation on juvenile *A. fulica* Bowdich. Effect on *A. fulica* populations not known
	Bipalium Stimpson sp.	Ogasawara. Observed attacking *A. fulica* Bowdich
Coleoptera: Lampyridae	*Lamprophorus tenebrosus* (Walker)	Sri Lanka and India. Important predator of *A. fulica* Bowdich
Hymenoptera: Formicidae	*Solenopsis geminata* (Fabricius)	Native to Central America. Invasive species, widely dispersed accidentally. Observations in New Britain, mainland New Guinea and Christmas Island suggest species can exert considerable mortality in young *A. fulica* Bowdich
	Oecophyllus Smith sp.	India. Predation on newly hatched *A. fulica* Bowdich. Importance in population regulation not known
	Pheidologeton affinis (Jerdon)	Sri Lanka, India. Mainly attacks the eggs of *A. fulica* Bowdich. Invasive species, widely dispersed accidentally. Importance of predation in *A. fulica* populations unknown
Diptera: Phoridae	*Megaselia javicola* (Beyer)	Asia. Recorded from *A. fulica* Bowdich

Table 3.5. Continued

	Spiniphora Malloch sp.	Recorded from *A. fulica* Bowdich
Diptera: Sarcophagidae	*Sarcophaga* Meigen sp.	India. Parasite of *A. fulica* Bowdich and other terrestrial gastropods
Diplopoda: Paradoxosomatidae	*Orthomorpha* sp.	Andaman Islands. Observed attacking *A. fulica* Bowdich
Chilopoda	Unidentified sp.	New Guinea. Occasional predation on *A. fulica* Bowdich
Decapoda: Coenobitidae	*Coenobita cavipes* Stimpson	East coast of Africa to Ryūkyū Island and Bismarck Archipelago. Confirmed predator of *A. fulica* Bowdich in Andaman Islands
	Coenobita perlatus Milne Edwards	Aldabra and Madagascar to Line and Gambier Islands. Confirmed predator of *A. fulica* Bowdich in various Pacific Islands
	Coenobita brevimanus Dana	East coast of Africa to Line and Tuamotu Archipelago. Confirmed predator of *A. fulica* Bowdich in Ogasawara, and member of a complex of *Coenobita* species implicated in control of *A. fulica* in Andaman Islands
	Coenobita purpreus Stimpson	Ogasawara. Predator of *A. fulica* Bowdich
	Coenobita rugosa Milne Edwards	East coast of Africa to Line Islands and Tuamotu Archipelago. Among a complex of *Coenobita* species implicated in control of *A. fulica* Bowdich in the Andaman Islands
	Birgus latro (Linnaeus)	East coast of Africa through to Malay Archipelago and Pacific Islands. Confirmed predator of *A. fulica* Bowdich, but level of control effected generally minimal
Decapoda: Grapsidae	*Geograpsus grayi* (Milne Edwards)	East coast of Africa to Japan and Society Islands. Confirmed predator of *A. fulica* Bowdich in Ogasawara
	Metopograpsus messor (Forskål)	Red Sea and east coast of Africa to Japan. Confirmed predator of *A. fulica* Bowdich in Ogasawara
	Sesarma dahaani (Milne Edwards)	Confirmed predator of *A. fulica* Bowdich in Ogasawara
Decapoda: Ocypodidae	*Ocypoda cordimana* Latreille	Red Sea and east coast of Africa to Japan and Society Islands. Confirmed predator of *A. fulica* Bowdich in Ogasawara
Decapoda: Gecarcinidae	*Gecarcoidea natalis* Pocock	Christmas Island. Confirmed predator of *A. fulica* Bowdich

*Sources of information include: Green (1910b, 1911), Annandale (1919), Paiva (1919), Hutson (1920), Austin (1924), Fantham (1924), Hutson and Austin (1924), South (1926), Jarrett (1931), Mead and Kondo (1949), Lange (1950), Mead (1950b, 1956, 1958a,b, 1961, 1963a, 1969, 1979a), Rees (1951), Kondo (1952), Davis (1954, 1971), van Zwaluwenburg (1955), Peterson (1957a), Seneviratna (1958), Beyer (1959), Ash (1962, 1976), Schreurs (1963), Alicata (1964, 1965a,b, 1966, 1969), Davis and Butler (1964), Davis and Krauss (1964), Cheng and Alicata (1965), Srivastava (1966, 1968b, 1970, 1976, 1992), Dun (1967), Srivastava and Srivastava (1967, 1968), Nair (1968), Robinson and Foote (1968), Davis and Chong (1969), van der Schalie (1969), Wallace and Rosen (1969a,b), Dean *et al.* (1970), Crook *et al.* (1971), Pradhan and Srivastava (1971), Raut and Ghose (1977, 1979a, 1984), Raut (1980, 1983b, 2001), Iga (1982), Godan (1983), Muniappan (1983), Nakahara (1985a), Higa *et al.* (1986), Waterhouse and Norris (1987), Raut and Panigrahi (1989), Schotman (1989), Kaneda *et al.* (1990), Lake and O'Dowd (1991), Raut (1993), van As and Basson (1993), Eldredge (1994), Ogren (1995), Teles *et al.* (1997), K. Takeuchi (personal communication, 1997), Kadirijan and Chauvet (1998), Cowie (2000), Sherley and Lowe (2000).

Fig. 3.5. A pair of *Achatina fulica* Bowdich (Achatinidae) in copulation.

molluscan biodiversity. Tests of host specificity preceding introductions of control agents have often been perfunctory or non-existent. More adverse effects on indigenous faunas, including species extinctions, can be attributed to species importation for biocontrol of *A. fulica* than can be attributed to the much more maligned chemical control. Despite some claims to the contrary (e.g. Tauili'ili and Vargo, 1993), the devastation wrought on indigenous terrestrial faunas by the polyphagous predator *Euglandina rosea* (de Férussac) (Oleacinidae) in the islands of the Pacific and India Oceans has been widely recognized and canvassed in the recent scientific literature (e.g. Tillier and Clarke, 1983; Civeyrel and Simberloff, 1996) and the popular media (e.g. Wells, 1988). Ironically there is no evidence that *E. rosea* or any other purposefully introduced pathogen, parasite or predator has effected population regulation in *A. fulica* (e.g. van der Schalie, 1969; Tillier, 1992; Tillier and Clarke, 1983; Clarke *et al.*, 1984; Pointier and Blanc, 1985; Cowie, 1992; Hopper and Smith, 1992; Griffiths *et al.*, 1993; Hadfield *et al.*, 1993; Civeyrel and Simberloff, 1996). The ecological effects of the great number of introduced agents remain to be investigated.

Populations of *A. fulica* have often been observed to pass through three phases following establishment in a new area (Mead, 1961, 1979a; Pointier and Blanc, 1985): (i) a phase of exponential increase, with the population typified by large, vigorous individuals; (ii) a stable phase of variable duration; and (iii) a phase of decline, with the population typified by small individuals. Thus naturalized populations of *A. fulica* often eventually decline greatly. There has been a widespread belief among local peoples that introduced biological control agents, particularly *E. rosea*, were responsible for the declines (Wells, 1988). The Hawaiian islands were often viewed as a pilot study that served as a model for other biological control projects and it is mainly from the Hawaiian islands that *E. rosea* and other predatory gastropods, such as

Table 3.6. Invertebrate enemies of terrestrial gastropods, naturally occurring outside Africa, introduced to different regions for biological control of *Achatina fulica* Bowdich (Achatinidae) and observations on importance to regulation of *A. fulica* populations.*

Turbellaria: Rhynchodemidae	*Platydemus manokwari* de Beauchamp	Native range not known. Accidentally introduced to Guam and northern Mariana Islands. Subsequent purposeful introduction to Bugsuk Island (Philippines) and Maldives and thence to many Pacific islands. Providing some regulation of *A. fulica*, and attributed with eradication in some areas. Probable adverse effect on indigenous gastropod faunas
Coleoptera: Lampyridae	*Lamprophorus tenebrosus* (Walker)	Native to Sri Lanka. Introduced to various Pacific and Indian Ocean islands, but did not establish
	Colophotia concolor (Olivier)	Native to the Philippines. Introduced to Hawaii, but not released
	Pyrophanes quadrimaculata bimaculata (Olivier)	Native to the Philippines. Introduced to Hawaii, but not released
	Diaphanes sp.	Native to Sri Lanka. Introduced to Hawaii, but perished in the laboratory prior to release
Coleoptera: Carabidae	*Damaster blaptoides* Kollar (includes subspecies *D. b. rugipennis* Motschulsky)	Native to Japan. Introduced to Hawaii but did not establish
	Scaphinotus striatopunctatus (Chaudoir)	Native to western North America. Introduced to Hawaii but did not establish
	Scaphinotus ventricosus (Dejean)	Native to western North America. Introduced to Hawaii but did not establish
Stylommatophora: Oleacinidae	*Euglandina rosea* (de Férussac)	Native to south-east USA. Introduced to India, parts of Asia and many islands of the Pacific and India oceans. Often failed to establish. Where established, no demonstrable regulatory effect on *A. fulica* but with adverse effect on indigenous fauna
	Euglandina singleyana (Binney)	Native to south-east USA. Introduced into quarantine in Hawaii but not released
	Salasiella Strebel sp.	Native to West Indies (Cuba). Introduced to Hawaii but did not establish
	Oleacina oleacea Deshayes	Native to West Indies (Cuba). Introduced to Hawaii but did not establish
	Oleacina Röding sp.	Native to West Indies (Cuba). Introduced to Hawaii but did not establish
Stylommatophora: Streptaxidae	*Streptaxis contundata* de Férussac	Native to South America (Brazil). Introduced to Hawaii but did not establish
Stylommatophora: Rhytididae	*Victaphanta compacta* (Cox & Hedley)	Native to Victoria, Australia. Imported into Hawaii but did not survive to be released
	Ptychorhytida ferreziana (Crosse)	Native to New Caledonia. Imported to Hawaii but evidently not released
	Ptychorhytida inaequalis (Pfeiffer)	Native to New Caledonia. Imported to Hawaii but evidently not released
	Austrorhytida capillacea (de Férussac)	Native to New South Wales and Victoria, Australia. Imported into Hawaii but did not survive to be released

Table 3.6. Continued

Stylommatophora: Zonitidae	*Oxychilus cellarius* (Müller)	Native to Europe. Purposefully introduced into New Britain. Early attempts at introduction into Hawaii unsuccessful, but accidentally introduced and established there. No evidence of effect on *A. fulica* populations
Stylommatophora: Haplotrematidae	*Haplotrema vancouverense* (Lea)	Native to north-west USA and western Canada. Imported into Hawaii but did not survive to be released
Decapoda: Coenobitidae	*Coenobita cavipes* Stimpson	East coast of Africa to Ryukyu Island and Bismarck Archipelago. Inundative releases in Andaman Islands provided control of *A. fulica*

*Sources of information include: Rees (1951), H. Macpherson in van Benthem Jutting (1952), Thistle (1953), Weber (1954b, 1956, 1957), Kondo (1956), Peterson (1957a,b,c), Davis (1958, 1959, 1960b, 1961, 1962, 1971, 1972, 1973), Chiu (1960), Mead (1961, 1979a), Chiu and Chou (1962), Schreurs (1963), Davis and Butler (1964), Krauss (1964), Davis and Krauss (1967), Dun (1967), Srivastava (1968a, 1976, 1992), Davis and Chong (1969), Mitchell (1969), van der Schalie (1969, 1970), Pradhan and Srivastava (1971), Ranaivosoa (1971), Etienne (1973), Lambert (1974, 1977), Sankaran (1974), Nishida and Napompeth (1975), Hart (1978), Hadfield and Mountain (1980), Hadfield and Kay (1981), Leehman (1981), Severns (1981, 1984), Whitten (1981), Muniappan (1982, 1983, 1987, 1990), Tillier (1982a), Godan (1983), Howarth (1983, 1985), Tillier and Clarke (1983), Wells *et al.* (1983), Backeljau (1984), Christensen (1984), Clarke *et al.* (1984), Nakamoto (1984), Raut and Ghose (1984), Pointier and Blanc (1985), Hadfield (1986), Muniappan *et al.* (1986), Waterhouse and Norris (1987), Eldredge (1988, 1992, 1994), Funasaki *et al.* (1988), Lai (1988), Gerlach (1989, 1993), Howarth and Medeiros (1989), Murray (1989), Nafus and Schreiner (1989), Schotman (1989), Cowie (1990, 1992, 1993, 1997, 1998a,b, 2000), Napompeth (1990), Schreiner (1990), Solem (1990), Hadfield and Miller (1992), Hopper and Smith (1992), Kawakatsu *et al.* (1992, 1993), Kinzie (1992), Smith (1992), Griffiths *et al.* (1993), Hadfield *et al.* (1993), Miller *et al.* (1993), US Congress (1993), Eldredge and Smith (1994), Griffiths (1994), Kobayashi (1994), Asquith (1995), Kay (1995), Obata (1995), Bauman (1996), Civeyrel and Simberloff (1996), Simberloff and Stiling (1996), K. Takeuchi (personal communication, 1997), Sherley and Lowe (2000).

Gonaxis quadrilateralis (Preston) (Streptaxidae), were introduced to other regions for control of *A. fulica*. There has thus been continued purposeful introduction of polyphagous enemies by people blissfully unaware of or blatantly dismissive of the ecological catastrophes unfolding in areas to which these same agents had earlier been introduced.

It is evident that lessons from the disastrous biological control effects of the past have not been well heeded. Generalist predators such as *E. rosea*, *G. quadrilateralis* and, more recently *Platydemus manokwari* de Beauchamp (Rhynchodemidae), continue to be dispersed to new areas in an attempt to control *A. fulica*.

The factor(s) causing the decline in *A. fulica* remains to be fully elucidated. Periods of high population densities of *A. fulica* are frequently followed by a high frequency of leucodermic lesions, evidently caused by the bacterium *Aeromonas hydrophila* (Chester) Stainer (Mead, 1979a). The disease has been considered a significant regulatory factor in declining *A. fulica* populations (Mead, 1961, 1979a; Raut and Ghose, 1984; Raut and Panigrahi, 1989). Exactly what triggers this epizootic disease is uncertain, but Mead (1979a) argues that various stresses associated with high populations lead to a breakdown in the natural resistance, while Civeyrel and Simberloff (1996) postulate that increasing density facilitates its transmission. Srivastava and Srivastava (1968) were successful in initiating disease outbreaks in *A. fulica* by spraying field populations with fluids derived from diseased animals. Undoubtedly, other natural enemies have also contributed to regulation of *A. fulica* in some areas, but the agents involved have not been studied.

Some island systems have evidently proved to be resistant to invasion by *A. fulica*. Schotman (1989) attributes the low abundance of *A. fulica* on some Pacific atolls to the sandy soils and predation by hermit crabs *Coenobita perlatus* Milne Edwards and *Birgus latro* (Linnaeus) (Coenobitidae). Lake and O'Dowd (1991) demonstrated that the omnivorous crab *Gecarcoidea natalis* Pocock (Gecarcinidae) provided biotic resistance to invasion by *A. fulica* on Christmas Island.

Future Prospects

A. fulica is a serious pest of agriculture in many tropical regions. Despite the decline in its abundance after long residence in many regions, *A. fulica* continues to impose severe economic constraints on agricultural productivity. Thus there is continuing demand for the development of effective, sustainable control strategies. There has been little advancement in the development of sustainable controls for *A. fulica* over the past 30 years. Further, this invasive species continues to spread. For those countries currently free from *A. fulica*, the most prudent control strategy is clearly the implementation of barriers to importation of unwanted organisms through apppropriate border security. Prevention of entry, rather than later control, is the most important means of mitigating the agricultural impacts of *A. fulica* and other invasive achatinids.

By reaching enormous numbers and invading native ecosystems *A. fulica* additionally poses a serious conservation problem. Not only do they eat native plants, modifying the environment, but they probably also outcompete native gastropods. However, the more insidious conservation problem they cause is that they tempt agricultural officials and individual farmers to initiate putative biological control measures. The best publicized of these measures is the introduction of generalist predators, most notably *E. rosea*. It cannot be stressed enough that these introductions of putative biological control agents against *A. fulica* are extremely adverse from the perspective of the conservation of native gastropod faunas. And, in any case, there is no good evidence that such generalist predators can indeed control *A. fulica* populations.

There is increasing awareness internationally of the adverse ecological and economic impacts of invasive species. Coupled with this is the recognition that mitigation of the effects of invasive species on biodiversity is best coordinated regionally, and agencies such as the International Union for the Conservation of Nature (IUCN) are coordinating development of biosecurity policies and operational procedures.

Examples are the *IUCN Guidelines for the Prevention of Biodiversity Loss caused by Alien Invasive Species* (IUCN, 2000) and the draft *Invasive Species Strategy for the Pacific Region* (Sherley *et al.*, 2000). There is a good case for integrating consideration of both agricultural and environmental pests in such strategy developments, given that impacts on agriculture result in a heavy demand for the introduction of biocontrol agents, which, by their very nature, involve further introductions of alien species. A coordinated effort among countries at the regional level is needed to prevent further spread of *A. fulica* and for the development of environmentally sustainable controls of current infestations.

References

Abbott, R.T. (1949) March of the giant African snail. *Natural History* 58, 68–71.

Ahmed, M. and Raut, S.K. (1991) Influence of temperature on the growth of the pestiferous land snail *Achatina fulica* (Gastropoda: Achatinidae). *Walkerana* 5, 33–62.

Alicata, J.E. (1964) *Parasitic Infections of Man and Animals in Hawaii*. Technical Bulletin 61, Hawaii Agricultural Experiment Station College of Tropical Agriculture, University of Hawaii, 138 pp.

Alicata, J.E. (1965a) Biology and distribution of the rat lungworm, *Angiostrongylus cantonensis*, and its relation to eosinophilic meningitis and other neurological disorders of man and animals. In: Dawes, B. (ed.) *Advances in Parasitology*, Vol. 3. Academic Press, New York, pp. 223–248.

Alicata, J.E. (1965b) Notes and observations on murine angiostrongylosis and eosinophilic meningoencephalitis in Micronesia. *Canadian Journal of Zoology* 43, 667–672.

Alicata, J.E. (1966) The presence of *Angiostrongylus cantonensis* in islands of the Indian Ocean and probable role of the giant African snail, *Achatina fulica*, in dispersal of the parasite to the Pacific Islands. *Canadian Journal of Zoology* 44, 1041–1049.

Alicata, J.E. (1969) Present status of *Angiostrongylus cantonensis* infection in man and animals in the tropics. *Journal of Tropical Medicine and Hygiene* 88, 65–73.

Altson, R.A. (1950) Giant snail. In: *Report for the Period January 1941 to August 1945*, p. 67. Rubber Research Institute, Malaya.

Annandale, N. (1919) Mortality among snails and the appearance of bluebottle flies. *Nature* 104, 412–413.

Anon. (1947) Giant snail numerous in Kokopo District, N.G. *Pacific Islands Monthly* 18, 33.

Anon. (1961) *Gulella wahlbergi* (Krauss). *Proceedings of the Hawaiian Entomological Society* 17, 325.

Asami, T., Cowie, R.H. and Ohbayashi, K. (1998) Evolution of mirror images by sexually asymmetric mating behavior in hermaphroditic snails. *American Naturalist* 152, 225–236.

Ash, L.R. (1962) The helminth parasites of rats in Hawaii and the description of *Capillaria traverae* sp. n. *Journal of Parasitology* 48, 66–68.

Ash, L.R. (1976) Observations on the role of mollusks and planarians in the transmission of *Angiostrongylus cantonensis* infection to man in New Caledonia. *Revista Biologia Tropical* 24, 163–174.

Asquith, A. (1995) Alien species and the extinction crisis of Hawaii's invertebrates. *Endangered Species Update* 12(6), 6–11.

Austin, G.D. (1924) The Indian glow-worm (*Lamprophorus tenebrosus* Wlk.). In: Year Book (1924) Ceylon Peradeniya Department of Agriculture, pp. 68–69.

Awah, A.A. (1992) Snail farming in mature rubber plantation: 1 Studies on aspects of specialized production techniques for farming *Archachatina marginata*. *Snail Farming Research* 4, 33–39.

Awesu, M.O. (1988) Observations on some aspects of reproduction of *Archachatina marginata* (Swainson) in captivity in a culture pen. *Snail Farming Research* 2, 39–44.

Backeljau, T. (1984) Introduction to the malacofauna of the Comoro Islands. *Africa-Tervuren* 30, 75–81.

Baer, J.G. (1953) Notes de faunistique éburnéene. III. Contribution à l'étude morphologique et biologique de *Wandolleckia achatinae* Cook, Phoridae (Diptera) commensal d'Achatines de la forêt tropicale. *Acta Tropica* 10, 73–79.

Bauman, S. (1996) Diversity and decline of land snails on Rota, Mariana Islands. *American Malacological Bulletin* 12, 13–27.

Beeley, F. (1935) Snails. In: *Annual Report (1934)*, Rubber Research Institute, Malaya. p. 109.

Beeley, F. (1938) The giant snail *Achatina fulica* (Fer): suggestions for control. *Tropical Agriculturalist* 90, 247–254.

Behura, B.K. (1955) Depredations of the giant African land snail, *Achatina fulica* (Ferussac) in Balasore (Orissa). *Journal of the Bombay Natural History Society* 54, 287.

Bequaert, J.[C.] (1925) The arthropod enemies of molluscs, with description of a new dipterous parasite from Brazil. *Journal of Parasitology* 11, 201–212.

Bequaert, J.C. (1926) Medical report of the Hamilton Rice Seventh Expedition to the Amazon, in conjunction with the Department of Tropical Medicine of Harvard University. XVIII. A dipterous parasite of a snail from Brazil, with an account of the arthropod enemies of mollusks. *Contributions from the Harvard Institute for Tropical Biology and Medicine* 4, 292–303.

Bequaert, J.C. (1950a) Studies on the Achatinidae, a group of African land snails. *Bulletin of the Museum of Comparative Zoology, Harvard* 105, 1–216.

Bequaert, J.C. (1950b) Enemies of *Achatina*. *Proceedings of the Hawaii Entomological Society* 14, 5–6.

Bertrand, H.W.R. (1928) Control of the Kalutara snail. *Tropical Agriculturalist* 71, 151–152.

Bertrand, H.W.R. (1941) Prevention of damage to young rubber by snails and slugs. *Tropical Agriculturalist* 97, 327.

Beyer, E.M. (1959) Gattung *Pericyclocera* Schmitz in Ostasien. *Entomologische Zeitschrift* 69, 167–169.

Boughey, A.S. (1963) Interaction between animals, vegetation and fire in southern Rhodesia. *Ohio Journal of Science* 63, 193–209.

Bridgland, L.A. and Byrne, P.N. (1956) Control of the giant snail (*Achatina fulica*) by baiting. *Papua and New Guinea Agricultural Journal* 11, 67–68.

Brues, C.T. (1903) A monograph of North American Phoridae. *Transactions of the American Entomological Society* 29, 331–404.

Cain, A.J. (1977) Variation in the spire index of some coiled gastropod shells, and its evolutionary significance. *Philosophical Transactions of the Royal Society of London, Series B* 277, 377–428.
Chamberlin, J.L. (1952) *Final Report on an Ecology and Population Study of the Giant African Snail on Tinian, Marianas Islands.* Pacific Science Board of the National Research Council, Invertebrate Consultants Committee for the Pacific, 27 pp.
Charmoy, D.d'E. and Gébert, S. (1922) Insect pests of various minor crops and fruit trees in Mauritius. *Mauritius Department of Agriculture Scientific Series Bulletin* 8, 14.
Chase, R. and Boulanger, C.M. (1978) Attraction of the snail *Achatina fulica* to extracts of conspecific pedal glands. *Behavioural Biology* 23, 107–111.
Chase, R., Croll, R.P. and Zeichner, L.L. (1980) Aggregation in snails, *Achatina fulica. Behavioural and Neural Biology* 30, 218–230.
Chelazzi, G. (1991) Eco-ethological aspects of homing behaviour in molluscs. *Ethological Ecology and Evolution* 2, 11–26.
Cheng, T.C. and Alicata, J.E. (1965) On the modes of infection of *Achatina fulica* by the larvae of *Angiostrongylus cantonensis. Malacologia* 2, 267–274.
Chiu, S.-C. (1960) The introduction and propagation of the natural enemy of *Achatina fulica* Bowd. *Bulletin of Plant Protection* 2, 39–43 [in Chinese].
Chiu, S.-C. and Chou, K.-C. (1962) Observations on the biology of the carnivorous snail *Euglandina rosea* Ferussac. *Bulletin of the Institute of Zoology, Academia Sinica* 1, 17–24.
Christensen, C.C. (1984) Are *Euglandina* and *Gonaxis* effective agents for biological control of the giant African snail in Hawaii? *American Malacological Bulletin* 2, 98–99.
Civeyrel, L. and Simberloff, D. (1996) A tale of two snails: is the cure worse than the disease? *Biodiversity and Conservation* 5, 1231–1252.
Clarke, B., Murray, J. and Johnson, M.S. (1984) The extinction of endemic species by a program of biological control. *Pacific Science* 38, 97–104.
Colman, P.H. (1977) An introduction of *Achatina fulica* to Australia. *Malacological Review* 10, 77–78.
Colman, P.H. (1978) An invading giant. *Wildlife in Australia* 15, 46–47.
Cook, A. (2001) Behavioural ecology: on doing the right thing, in the right place at the right time. In: Barker, G.M. (ed.) *The Biology of Terrestrial Molluscs.* CAB International, Wallingford, pp. 447–487.
Cook, O.F. (1897) A new wingless fly from Liberia. *Science* 6, 886.
Corbett, G.H. (1933) The giant snail (*Achatina fulica* Fér.) in Malaya. *Malayan Agricultural Journal* 21, 77–79.
Corbett, G.H. (1937) *Achatina fulica* Fér. In: *General Series* 26, Malaya Department of Agriculture, p. 40.
Cotton, B.C. (1940) The Kalutara snail, *Achatina fulica* (Fér.), attacks rubber trees in Ceylon. *South Australian Naturalist* 20, 3.
Cowie, R.H. (1990) Land snail extinction. *Hawaiian Shell News* 38(9), 5.
Cowie, R.H. (1992) Evolution and extinction of Partulidae, endemic Pacific island land snails. *Philosophical Transactions of the Royal Society of London B* 335, 167–191.
Cowie, R.H. (1993) Why tree snails are becoming scarce in Samoa. *Hawaiian Shell News* 41(3), 1, 9.

Cowie, R.H. (1997) *Catalog and Bibliography of the Nonindigenous Nonmarine Snails and Slugs of the Hawaiian Islands.* Occasional Papers 50, Bishop Museum, 66 pp.

Cowie, R.H. (1998a) Patterns of introduction of non-indigenous non-marine snails and slugs in the Hawaiian Islands. *Biodiversity and Conservation* 7, 349–368.

Cowie, R.H. (1998b) *Catalog of the Nonmarine Snails and Slugs of the Samoan Islands.* Bulletins in Zoology 3, Bishop Museum, 122 pp.

Cowie, R.H. (2000) Non-indigenous land and freshwater molluscs in the islands of the Pacific: conservation impacts and threats. In: Sherley, G. (ed.) *Invasive Species in the Pacific: a Technical Review and Draft Regional Strategy.* South Pacific Regional Environment Programme, Apia, pp. 143–172.

Croll, R.P. and Chase, R. (1977) A long-term memory for food order in the land snail, *Achatina fulica. Behavioural Biology* 19, 261–268.

Crook, J.R., Fulton, S.E. and Supanwong, K. (1971) The infectivity of third stage *Angistrongylus cantonensis* larvae shed from drowned *Achatina fulica* snails and the effects of chemical agents on infectivity. *Transactions of the Royal Society of Tropical Medicine and Hygiene* 65, 602–605.

Crowley, T.E. and Pain, T. (1959) A monographic revision of the African land snails of the genus *Burtoa* (Mollusca – Achatinidae). *Annales du Musée Royal de l'Afrique Centrale, Tervuren, Belgique, Sciences Zoologiques* 79, 1–35, 3 pls.

Crowley, T.E. and Pain, T. (1964) *Achatina* (*Lissachatina*) *tavaresiana* Morlet: its synonymy and distribution. *Revue de Zoologie et de Botanique Africaines* 69, 121–131.

Crowley, T.E. and Pain, T. (1970) A monographic revision of the African land snails of the genus *Limicolaria* Schumacher (Mollusca – Achatinidae). *Annales du Musée Royal de l'Afrique Centrale, Tervuren, Belgique, Sciences Zoologiques* 177, 1–61.

Dammerman, K.W. (1929) *The Agricultural Zoology of the Malay Archipelago.* J.H. de Bussy, Amsterdam, 437 pp.

Das, A.K. and Sharma, R.M. (1984) Necrophagous habit in the giant African snail, *Achaina fulica* Bowdich. *Journal of the Bombay Natural History Society* 81, 219–220.

Davis, C.J. (1954) *Report on the Davis Expedition to Agiguan, July–August, 1954. Ecological Studies, Island of Agiguan, Marianas Islands as Related to the African Snail,* Achatina fulica *Bowdich and its Introduced Predator,* Gonaxis kibweziensis *(E.A. Smith).* Invertebrate Consultants Commission for Micronesia, Pacific Science Board, Natural Resources Council, 24 pp.

Davis, C.J. (1958) Recent introductions for biological control in Hawaii – III. *Proceedings of the Hawaiian Entomological Society* 16, 356–358.

Davis, C.J. (1959) Recent introductions for biological control in Hawaii – IV. *Proceedings of the Hawaiian Entomological Society* 17, 62–66.

Davis, C.J. (1960a) *Gonaxis kibweziensis* (Smith) and *G. quadrilateralis* (Preston). *Proceedings of the Hawaiian Entomological Society* 17, 170.

Davis, C.J. (1960b) Recent introductions for biological control in Hawaii – V. *Proceedings of the Hawaiian Entomological Society* 17, 244–248.

Davis, C.J. (1961) Recent introductions for biological control in Hawaii – VI. *Proceedings of the Hawaiian Entomological Society* 17, 389–393.

Davis, C.J. (1962) African snail on Maui. *Proceedings of the Hawaiian Entomological Society* 18, 7.

Davis, C.J. (1971) Recent introductions for biological control in Hawaii – XV. *Proceedings of the Hawaiian Entomological Society* 20, 521–525.
Davis, C.J. (1972) Recent introductions for biological control in Hawaii XVI. *Proceedings of the Hawaiian Entomological Society* 21, 59–66.
Davis, C.J. (1973) Recent introductions for biological control in Hawaii XVII. *Proceedings of the Hawaiian Entomological Society* 21, 187–190.
Davis, C.J. and Butler, G.D. (1964) Introduced enemies of the giant African snail, *Achatina fulica* Bowdich, in Hawaii. *Proceedings of the Hawaiian Entomological Society* 18, 377–389.
Davis, C.J. and Chong, M. (1969) Recent introductions for biological control in Hawaii – VIII. *Proceedings of the Hawaiian Entomological Society* 20, 25–34.
Davis, C.J. and Krauss, N.L.H. (1962) Recent introductions for biological control in Hawaii – VII. *Proceedings of the Hawaiian Entomological Society* 18, 125–129.
Davis, C.J. and Krauss, N.L.H. (1963) Recent introductions for biological control in Hawaii – VIII. *Proceedings of the Hawaiian Entomological Society* 18, 245–249.
Davis, C.J. and Krauss, N.L.H. (1964) Recent introductions for biological control in Hawaii – IX. *Proceedings of the Hawaiian Entomological Society* 18, 319–397.
Davis, C.J. and Krauss, N.L.H. (1965) Recent introductions for biological control in Hawaii – X. *Proceedings of the Hawaiian Entomological Society* 19, 87–90.
Davis, C.J. and Krauss, N.L.H. (1967) Recent introductions for biological control in Hawaii – XI. *Proceedings of the Hawaiian Entomological Society* 19, 201–207.
Davis, C.J., Chock, Q.C. and Chong, M. (1961) Introduction of the liver fluke snail predator, *Sciomyza dorsata* (Sciomyzidae, Diptera), in Hawaii. *Proceedings of the Hawaiian Entomological Society* 17, 395–397.
Dean, W.W., Mead, A.R. and Northey, W.T. (1970) *Aeromonas liquifaciens* in the giant African snail, *Achatina fulica. Journal of Invertebrate Pathology* 16, 346–351.
de Winter, A.J. (1988) *Achatina fulica* in West Africa. *Basteria* 52, 2.
Diamond, A.W. and Hamilton, A.C. (1980) The distribution of forest passerine birds and Quaternary climatic change in tropical Africa. *Journal of Zoology* 191, 379–402.
Disney, R.H.L. (1994) *Scuttle Flies: the Phoridae.* Chapman & Hall, London, 467 pp.
Dollfus, G. (1899) Sur un coupe de Madagascar. *Bulletin de la Société Géologique de France* 27, 395.
Dun, G.S. (1967) The giant snail. *Papua and New Guinea Agricultural Journal* 18, 213–215.
Egonmwan, R.I. (1991) Food selection in the land snail *Limicolaria flammea* Müller (Pulmonata: Achatinidae). *Journal of Molluscan Studies* 58, 49–55.
Eldredge, L.G. (1988) Case studies of the impacts of introduced animal species on renewable resources in the U.S.-affiliated Pacific Islands. In: *Tropic Reviews in Insular Resource Development and Management in the U.S.-affiliated Islands.* Marine Laboratory Technical Report 88, University of Guan, pp. 118–146.
Eldredge, L.G. (1992) Unwanted strangers: an overview of animals introduced to Pacific islands. *Pacific Science* 46, 384–386.

Eldredge, L.G. (1994) Introductions and transfers of the triclad flatworm *Platydemus manokwari*. *Tentacle. Newsletter of the IUCN Species Survival Commission Mollusc Specialist Group* 4, 8.
Eldredge, L.G. and Smith, B.D. (1994) Introductions and transfers of the triclad flatworm *Platydemus manokwari*. *Tentacle* 3, 8.
Elmslie, L.J. (1982) Snails and snail farming. *World Animal Review* 41, 20–26.
Esaki, T. and Takahashi, K. (1942) Introduction of the African snail, *Achatina fulica* Ferussac into Japan, esp. Micronesia and subsequent developments. *Journal of the Palau Tropical Biological Station* 4, 16–25 [translation from Japanese by Toyohi Okada].
Etienne, J. (1973) Lutte biologique contre les escargots nuisibles aux cultures. In: *Rapport*. Institut de Recherche Agronomique Tropical et des Cultures Vivrières, Réunion, pp. 71–73.
Evans, H.C. (1973) Invertebrate vectors of *Phytophthora palmivora*, causing black pod disease of cocoa in Ghana. *Annals of Applied Biology* 75, 331–345.
Fairweather, J. (1937) *Pest and Diseases*. General Series 27, Malaya Department of Agriculture, 143 pp.
Fantham, H.B. (1924) Some parasitic Protozoa found in South Africa – VIII. *South African Journal of Science* 21, 435–444.
Feij, P.J. (1940) Enkele waarnemingen betreffende de agaatslak. *De Bergcultures* 14, 1112–1114.
Frankiel, L. (1989) *Les Ahatines aux Antilles*. Circular, Centre Départemental de Documentation Pédagogique, 10 pp.
Funasaki, G.Y., Lai, P.-Y., Nakahara, L.M., Beardsley, J.W. and Ota, A.K. (1988) A review of biological control introductions in Hawaii: 1890 to 1985. *Proceedings of the Hawaiian Entomological Society* 28, 105–160.
Gascoigne, A. (1994) The biogeography of land snails in the islands of the Gulf of Guinea. *Biodiversity and Conservation* 3, 794–807.
Gerlach, J. (1989) A report on the status of *Euglandina rosea* in Seychelles. *Papustyla* 3, 2–5.
Gerlach, J. (1993) Surveys of the status of *Euglandina rosea* in the Society Islands: its distribution, populations and effects on other mollusc species. The status of Partulidae and *Euglandina rosea* on Raiatea in 1992. *Papustyla* 7(3), 13–14.
Germain, L. (1921) Paléontologie de Madagascar ix – Mollusques Quaternaires terrestres et fluviatiles. *Annales de Paléontologie* 10, 21–36.
Ghose, K.C. (1959) Observations on the mating and oviposition of two land pulmonates, *Achatina fulica* Bowdich and *Macrochlamys indica*. *Journal of the Bombay Natural History Society* 56, 183–187.
Ghose, K.C. (1960) Observations on the gametes, fertilization and gonadal activities of two land Pulmonata *Achatina fulica* Bowdich and *Macrochlamys indica* Godwin-Austen. *Proceedings of the Zoological Society, Calcutta* 13, 91–96.
Ghose, K.C. (1963) The early stages of the development in *Achatina fulica* Bowdich (Mollusca: Gastropoda). *Journal of the Bombay Natural History Society* 60, 228–232.
Godan, D. (1983) *Pest Slugs and Snails*. Springer-Verlag, Berlin, 445 pp.
Green, E.E. (1910a) Introduction of an African snail. *Tropical Agriculturalist* 35, 311.
Green, E.E. (1910b) Report on the outbreak of *Achatina fulica*. *Circulars and Agricultural Journal of the Royal Botanic Gardens of Ceylon* 5, 55–64.
Green, E.E. (1911) The wanderings of a gigantic African snail. *Zoologist* 15, 41–45.

Griffiths, O. (1994) A review of the land snails of Rodrigues Island (Indian Ocean) with notes on their status. *Journal of Conchology* 35, 157–166.
Griffiths, O., Cook, A. and Wells, S.M. (1993) The diet of the introduced carnivorous snail *Euglandina rosea* in Mauritius and its implication for threatened island gastropod faunas. *Journal of Zoology* 229, 79–89.
Hadfield, M.G. (1986) Extinction in Hawaiian achatinelline snails. *Malacologia* 27, 67–81.
Hadfield, M.G. and Kay, E.A. (1981) The multiple villainies of *Euglandina rosea* (or its human proponents). *Hawaiian Shell News* 29(4), 5–6.
Hadfield, M.G. and Miller, S.E. (1992) Alien predators and decimation of endemic Hawaiian tree snails. Pacific Science 46, 395.
Hadfield, M.G. and Mountain, B.S. (1980) A field study of a vanishing species, *Achatinella mustella* (Gastropoda, Pulmonata), in the Waianae mountains of Oahu. *Pacific Science* 34, 345–358.
Hadfield, M.G., Miller, S.E. and Carwile, A.H. (1993) The decimation of endemic Hawaiian tree snails by alien predators. *American Zoologist* 33, 610–622.
Hamilton, A.C. (1981) The Quaternary history of African forests: its relevance to conservation. *African Journal of Ecology* 19, 1–6.
Hart, A.D. (1978) The onslaught against Hawaii's tree snails. *Natural History* 87, 46–57.
Hatai, S. and Kato, G. (1943) Observation upon growth of shells and ecology of *Achatina fulica* Ferussac in Palau. *Journal of the Palau Tropical Biological Station* 5, 1–19 [translation from Japanese by Toyohi Okada].
Herbert, D. (1991) South Africa's carnivorous snails. *African Wildlife* 45, 6–11.
Hes, J.W. (1949) Agaatslaken in suikerriet. *Chronica Naturae* 105, 226–227.
Hes, J.W. (1950) The African giant snail and sugar cane. *Sugar Journal* 12, 18.
Heubel, G.Ad. (1937) Enkele gegevens over de agaatslak (*Achatina fulica* Fer) in de Lampongsche Districten. *De Bercultures* 11, 1667–1670.
Heubel, G.Ad. (1938) Enkele gegevens over de agaatslak (*Achatina fulica* Fer) in de Lampongsche Districten. *De Bercultures* 11, 400–401.
Higa, H.H., Brock, J.A. and Palumbo, N.E. (1986) Occurrence of *Angiostrongylus cantonensis* in rodents, intermediate and paratenic hosts on the island of Oahu. *Journal of Environmental Health* 48, 319–323.
Hodasi, J.K.M. (1975) Preliminary studies on the feeding and burrowing habits of *Achatina achatina*. *Ghana Journal of Science* 15, 193–199.
Hodasi, J.K.M. (1979) Life history studies of *Achatina* (*Achatina*) *achatina* (Linné). *Journal of Molluscan Studies* 45, 328–339.
Hodasi, J.K.M. (1982) Some aspects of the biology of *Achatina* (*Achatina*) *achatina* (Linne). *Bulletin de l'Institut Fondamental d'Afrique Noire* 44, 100–114.
Hodasi, J.K.M. (1984) Some observations on the edible giant land snails of West Africa. *World Animal Review* 52, 24–28.
Hodasi, J.K.M. (1989) The potential for snail farming in West Africa. In: Henderson, I.F. (ed.) *Slugs and Snails in World Agriculture*. Monograph No. 41, British Crop Protection Council, Thornton Heath, pp. 27–31.
Holmes, C.H. (1954) Seed germination and seedling studies of timber trees of Ceylon. *Ceylon Forester* 1, 3–51.
Hopper, D.R. and Smith, B.D. (1992) Status of tree snails (Gastropoda: Partulidae) on Guam, with a resurvey of sites studied by H.E. Crampton in 1920. *Pacific Science* 46, 77–85.

Howarth, F.G. (1983) Classical biocontrol: panacea or Pandora's Box. *Proceedings of the Hawaiian Entomological Society* 24, 239–244.

Howarth, F.G. (1985) Impacts of alien land arthropods and mollusks on native plants and animals in Hawai'i. In: Stone, C.P. and Scott, J.M. (eds) *Hawai'i's Terrestrial Ecosystems: Preservation and Management.* University of Hawaii Cooperative National Park Resources Studies Unit, Honolulu, pp. 149–179.

Howarth, F.G. (1991) Environmental impacts of classical biological control. *Annual Review of Entomology* 36, 485–509.

Howarth, F.G. and Medeiros, A.C. (1989) Non-native invertebrates. In: Stone, C.P., Stone, D.B., Cuddihy, L.W. and Lane, M.E. (eds) *Conservation Biology in Hawai'i.* University of Hawaii Cooperative National Park Resources Studies Unit, Honolulu, pp. 82–87.

Hutson, J.C. (1920) The African snail (*Achatina fulica*). *Tropical Agriculturalist* 55, 217–225.

Hutson, J.C. and Austin, G.D. (1924) *Notes on the Habits and Life History of the Indian Glow Worm (an Enemy of the African or Kalutara snail).* Ceylon Department of Agriculture Bulletin 69, 16 pp.

Iga, M. (1982) Ecology and control of *Achatina fulica* Bowdich. *Japanese Journal of Applied Entomology and Zoology* 36, 24–28 [in Japanese].

Imevbore, E.A. and Ajayi, S.S. (1993) Food preference of the African snail (*Archachatina marginata*) in captivity. *African Journal of Ecology* 31, 265–267.

IUCN (2000) *IUCN Guidelines for the Prevention of Biodiversity Loss caused by Alien Invasive Species.* IUCN – the World Conservation Union, Gland.

Jahan, M.S. and Raut, S.K. (1994) Distribution and food preference of the giant African land snail, *Achatina fulica* Bowdich in Bangladesh. *Journal of Asiatic Society of Bangladesh, Science* 20, 111–115.

Jarrett, V.H.C. (1923) The occurrence of the snail *Achatina fulica* in Malaya. *Singapore Naturalist* 1, 73–76.

Jarrett, V.H.C. (1931) The spread of the snail *Achatina fulica* to South China. *Hong Kong Naturalist* 2, 262–264, 1 pl.

Jaski, C.J. (1953) *Achatina fulica. Tropische Natuur* 33, 91–98.

Kadirijan, T. and Chauvet, C. (1998) Distribution of the juvenile coconut crab, *Birgus latro* (L.), on the island of Lifou, New Caledonia. *Ecoscience* 5, 275–278.

Kaneda, M., Kitagawa, K.I. and Ichinohe, F. (1990) Laboratory rearing method and biology of *Platydemus manokwari* de Beauchamp (Tricladida: Terricola: Rhynchodemidae). *Applied Entomology and Zoology* 25, 524–528.

Kawakatsu, M., Ogren, R.E. and Munippan, R. (1992) Redescription of *Platydemus manokwari* de Beauchamp, 1962 (Turbellaria: Tricladida: Terricola), from Guam and the Philippines. *Proceedings of the Japanese Society for Systematic Zoology* 47, 11–25.

Kawakatsu, M., Oki, I., Tamura, S., Itô, H., Nagai, Y., Ogura, K., Shimabukuro, S., Ichinohe, F., Katsumata, H. and Kaneda, M. (1993) An extensive occurrence of a land planarian, *Platydemus manokwari* de Beauchamp, 1962, in the Ryûkyû Islands, Japan (Turbellaria, Tricladida, Terricola). *Biologiya Vnutrennikh Vod* 8, 5–14.

Kay, E.A. (ed.) (1995) *The Conservation Biology of Molluscs.* International Union for Conservation of Nature and Natural Resources, Gland, 81 pp.

Kekauoha, W. (1966) Life history and population studies of *Achatina fulica. The Nautilus* 80, 3–10, 39–46.

Kim, J. (1964) *Gonaxis quadrilateralis* (Preston). *Proceedings of the Hawaiian Entomological Society* 18, 328.

Kinzie, R.A. (1992) Predation by the introduced carnivorous snail *Euglandina rosea* (Ferussac) on endemic aquatic lymnaeid snails in Hawaii. *Biological Conservation* 60, 149–155.

Kobayashi, S.R. (1994) Saving Hawaii's landsnails. *Hawaiian Shell News* 42(6), 9, 12.

Kondo, Y. (1950a) *Report on the* Achatina fulica *Investigation on Palau, Pagan and Guam.* Pacific Science Board of the National Research Council, Invertebrate Consultants Committee for the Pacific, 94 pp.

Kondo, Y. (1950b) *The Giant African snail (*Achatina fulica*) in Palau, Pagan and Guam.* Pacific Science Board of the National Research Council, Invertebrate Consultants Committee for the Pacific, 13 pp.

Kondo, Y. (1952) *Report on Carnivorous Snail Experiment on Agiguan Island; Primary and Secondary* Achatina*-free Areas on Rota; and Gigantism among* Achatina *on Guam.* Pacific Science Board of the National Research Council, Invertebrate Consultants Committee for the Pacific, 50 pp., 2 pls.

Kondo, Y. (1956) Second *Helix aspersa* in Hawaii and data on carnivorous snails. *The Nautilus* 70, 71–72.

Kondo, Y. (1964) Growth rates in *Achatina fulica* Bowdich. *The Nautilus* 78, 6–15.

Krauss, N.L.H. (1955) *Tefflus zanzibaricus alluaudi* Sternberg. *Proceedings of the Hawaiian Entomological Society* 16, 1.

Krauss, N.L.H. (1964) Investigations on biological control of giant African (*Achatina fulica*) and other land snails. *The Nautilus* 78, 21–27.

Lai, P.-Y. (1988) Biological control: a positive point of view. *Proceedings of the Hawaiian Entomological Society* 28, 179–190.

Lai, P.-Y., Funasaki, G.Y. and Higa, S.Y. (1982) Introductions for biological control in Hawaii: 1979 and 1980. *Proceedings of the Hawaiian Entomological Society* 24, 109–113.

Lake, P.S. and O'Dowd, D.J. (1991) Red crabs in rainforest, Christmas Island: biotic resistance to invasion by an exotic snail. *Oikos* 62, 25–29.

Lambert, M. (1974) The African giant snail, *Achatina fulica*, in the Pacific islands. *South Pacific Bulletin* 24, 35–40.

Lambert, M. (1977) *The African Giant Snail.* South Pacific Commission Advisory Leaflet 6, 4 pp.

Lange, W.H. (1950) Life history and feeding habits of the giant African snail on Saipan. *Pacific Science* 4, 323–325.

Latif, S.M. (1933) Nog eens: het slakkandrama. *Die Orchidee* 2, 117, 123, 147, 151.

Leefmans, S. (1933) Voorloopige mededeeling inzake de bestrijding van *Achatina fulica* Fér. in Batavia. *Landbouw, Landbouwkundig Tijdscher voor Nederlandsch-Indie* 9, 289–298.

Leefmans, S. and van der Vecht, J. (1933a) De groote agaatslak (*Achatina fulica* Fér.) in Nederlandsche-Indie. *Landbouw, Landbouwkundig Tijdscher voor Nederlandsch Indië* 8, 668–677.

Leefmans, S. and van der Vecht, J. (1933b) De groote agaatslak (*Achatina fulica* Fér.) in Nederlandsche-Indië. *De Bergcultures* 7, 579–584.

Leehman, E.G. (1981) *Achatina fulica* vs. *Euglandina rosea*: which is the lesser villain? *Hawaiian Shell News* 29(2), 9.

Lionnet, G. (1984) Terrestrial testaceous molluscs of the Seychelles. In: Stoddart, D.R. (ed.) *Biogeography and Ecology of the Seychelles Islands*. Dr W. Junk, The Hague, pp. 239–244.

Manna, B. and Raut, S.K. (1986) Feeding adaptation of the giant African land snail *Achatina fulica*. *Environment and Ecology* 4, 158–159.

Mayr, E. and O'Hara, R.J. (1986) The biogeographic evidence supporting the Pleistocene forest refuge hypothesis. *Evolution* 40, 55–67.

Mead, A.R. (1949) The giant snails. *Atlantic Monthly* 184(2), 38–42.

Mead, A.R. (1950a) Comparative genital anatomy of some African Achatinidae (Pulmonata). *Bulletin of the Museum of Comparative Zoology, Harvard* 105, 217–291.

Mead, A.R. (1950b) *The Giant African Snail Problem (*Achatina fulica*) in Micronesia. Final report.* Invertebrate Consultants Commission for Micronesia, Pacific Science Board, Natural Resources Council, 55pp.

Mead, A.R. (1955) The proposed introduction of predatory snails into California. *The Nautilus* 69, 37–39.

Mead, A.R. (1956) Disease in the giant African snail *Achatina fulica* Bowdich. *Science* 123, 1130–1131.

Mead, A.R. (1958a) The continuing battle against the giant African snail. In: *Annual Report (1958)*. American Malacological Union, p. 37.

Mead, A.R. (1958b) Recent discoveries in the disease syndrome of the giant African snail. In: *34th Annual Meeting of the American Association for Advanced Science, Southwestern and Rocky Mountain Division*, p. 22.

Mead, A.R. (1961) *The Giant African Snail: a Problem in Economic Malacology*. University of Chicago Press, Chicago, 257 pp.

Mead, A.R. (1963a) A flatworm predator of the giant African snail *Achatina fulica* in Hawaii. *Malacologia* 1, 305–311.

Mead, A.R. (1963b) Disease, decline and predation in the giant snail populations of Hawaii. In: *Annual Report (1963)*. American Malacological Union, p. 22.

Mead, A.R. (1969) *Aeromonas liquefaciens* in the leukodermia syndrome of *Achatina fulica*. *Malacologia* 9, 43.

Mead, A.R. (1979a) Economic malacology with particular reference to *Achatina fulica*. In: Fretter, V. and Peake, J. (eds) *Pulmonates*, Vol. 2B. Academic Press, London, 150 pp.

Mead, A.R. (1979b) Anatomical studies in the African Achatinidae – a preliminary report. *Malacologia* 18, 133–138.

Mead, A.R. (1982) The giant African snails enter the commercial field. *Malacologia* 22, 489–493.

Mead, A.R. (1988) Anatomy of the South African *Archachatina ustulata* (Lamarck) (Pulmonata: Achatinidae). *Journal of Molluscan Studies* 54, 363–365.

Mead, A.R. (1995) Anatomy, phylogeny, and zoogeography in the African land snail family Achatinidae. In: *Proceedings of the 12th International Malacological Congress, Vigo*, pp. 422–423.

Mead, A.R. (1998) Comparative anatomy establishes correlativity in distributional direction and phylogenetic progression in the Achatinidae. In: Bieler, R. and Mikkelsen, P.M. (eds) *Abstracts, World Congress of Malacology, Washington D.C.* Field Museum of Natural History, Chicago, for Unitas Malacologica, p. 214.

Mead, A.R. and Kondo, Y. (1949) *Giant African Snail (*Achatina fulica*) Problem in Micronesia. Preliminary Report.* Invertebrate Consultants Commission for Micronesia, Pacific Science Board, Natural Resources Council, 6 pp.
Mead, A.R. and Palcy, L. (1992) Two giant African land snail species spread to Martinique, French West Indies. *The Veliger* 35, 74–77.
Miller, S.E., Cowie, R.H., Smith, B.D. and Rojek, N. (1993) The decline of partulid snail populations in American Samoa. *Species* 20, 65.
Mitchell, W.C. (1969) *Coptosoma xanthogramma* (White), *Euglandina rosea* (Férussac) and *Nezara viridula* (Linnaeus). *Proceedings of the Hawaiian Entomological Society* 20, 10.
Monney, K.A. (1994) Notable notes on giant African snails. *Snail Farming Research* 5, 1–13.
Muniappan, R. (1982) The giant African snail with special reference to its biological control. In: *Proceedings of Sub-regional Training Course on Methods of Controlling Diseases, Insects and Other Pests of Plants in the South Pacific, October 4–20, Vaini, Kingdom of Tonga*, pp. 229–237.
Muniappan, R. (1983) Biological control of the giant African snail. *Alafua Agricultural Bulletin* 8, 43–46.
Muniappan, R. (1987) Biological control of the giant African snail, *Achatina fulica* Bowdich, in the Maldives. *FAO Plant Protection Bulletin* 35, 127–133.
Muniappan, R. (1990) Use of the planarian *Platydemus manokwari*, and other natural enemies to control the giant African snail. In: *The Use of Natural Enemies to Control Agricultural Pests.* Book Series No. 40, FECT, pp. 179–183.
Muniappan, R., Duhamel, G., Santiago, R.M. and Acay, D.R. (1986) Giant African snail control in Bugsuk Island, Philippines, by *Platydemus manokwari. Oléagineux* 41, 183–186.
Murray, J.J. (1989) Extinction by intent. *The Virginia Explorer* 5(12), 8–9.
Nafus, D. and Schreiner, I. (1989) Biological control activities in the Mariana Islands from 1911 to 1988. *Micronesica* 22, 65–106.
Naggs, F. (1989) *Gulella bicolor* (Hutton) and its implications for the taxonomy of streptaxids. *Journal of Conchology* 33, 165–168, 1 pl.
Nair, K.R. (1968) Two sarcophagid parasites of phytophagous terrestrial snails in Mysore state, India. *Technical Bulletin of the Commonwealth Institute of Biological Control* 10, 113–121.
Nair, M.R.G.K., Das, N.M. and Jacob, A. (1968) Use of metaldehyde as dusts and sprays to control the giant African snail *Achatina fulica* Bowdich. *Indian Journal of Entomology* 30, 58–60.
Nakahara, L.M. (1985a) *Geoplana septemlineata* Hyman. *Proceedings of the Hawaiian Entomological Society* 25, 5.
Nakahara, L.M. (1985b) *Gonaxis quadrilateralis* (Preston). *Proceedings of the Hawaiian Entomological Society* 25, 5.
Nakamoto, K. (1984) A new menace in exotic snails. *Hawaiian Shell News* 32, 5.
Nakao, H.K., Funasaki, G.Y. and Davis, C.J. (1975) Introductions for biological control in Hawaii, 1973. *Proceedings of the Hawaiian Entomological Society* 22, 109–112.
Napompeth, B. (1990) Use of natural enemies to control agricultural pests in Thailand. In: *The Use of Natural Enemies to Control Agricultural Pests.* Book Series 40. Food and Fertilizer Technology Center, pp. 8–29.
Nisbet, R.H. (1974) The life of Achatinidae in London. *Proceedings of the Malacological Society of London* 41, 171–183.

Nishida, T. and Napompeth, B. (1975) Effect of age-specific predation on age distribution and survival of the giant African snail, *Achatina fulica*. *Proceedings of the Hawaiian Entomological Society* 22, 119–123.

Numasawa, K. and Koyano, S. (1987) Investigation of the ecology of giant African snail, *Achatina fulica*, in Ogasawara Islands. *Annual Report of the Ogasawara Subtropical Agriculture Centre*, pp. 106–128.

Obata, J. (1995) The decline of landshells – genus *Achatinella*. *Hawaiian Shell News* 43(4), 1, 3.

Ogren, R.E. (1995) Predation behaviour of land planarians. *Hydrobiologia* 305, 105–111.

Olson, F.J. (1973) The screening of candidate molluscicides against the giant African snail, *Archatina fulica* Bowdich (Stylommatophora: Achatinidae). Thesis, University of Hawaii.

Olufokunbi, B., Phillips, E.O., Omidiji, J.O., Ogbonna, U.O., Makinde, H.T. and Apansile, O.J. (1989) The economics of commercial domestication of the African giant land snail *Achachatina* (*Calachatina*) *marginata* (Swainson) in Nigeria. In: Henderson, I.F. (ed.) *Slugs and Snails in World Agriculture*. Monograph No. 41, British Crop Protection Council, Thornton Heath, pp. 41–48.

Otanes, F.Q. (1948) Notes on orchid pests and suggestions for their control. *Philippines Orchid Review* 1, 12–18.

Otchoumou, A., Zongo, D. and Dosso, H. (1989/90) Contribution á l'etude de l'escargot géant African *Achatina achatina* (Linné). *Annales d'Ecologie* 21, 31–58.

Owen, D.F. (1964) Bimodal occurrence of breeding in an equatorial land snail. *Ecology* 45, 862.

Owen, D.F. (1965) A population study of an equatorial land snail, *Limicolaria martensiana* (Achatinidae). *Proceedings of the Zoological Society of London* 144, 361–382.

Owiny, A.M. (1974) Some aspects of the breeding biology of the equatorial land snail *Limicolaria martensiana* (Achatinidae: Pulmonata). *Journal of Zoology, London* 172, 191–206.

Paiva, C.A. (1919) Notes on the Indian glow-worm (*Lamprophorus tenebrosus*). *Records of the Indian Museum, Calcutta* 16, 19–28.

Palcy, L. and Mead, A.R. (1993) Les deux redoutables escargots géants Africans à la Martinique. *Phytoma* 449, 48–50.

Pangga, G.A. (1949) A preliminary report on the biology, ecology, and control of the giant African snail (*Achatina fulica* Fér.). *Philippine Journal of Agriculture* 14, 337–347.

Panigrahi, A. and Raut, S.K. (1994) *Thevetia peruviana* (Family: Apocynaceae) in the control of slug and snail pests. *Memorias do Instituto Oswaldo Cruz* 89, 247–250.

Panja, U.K. (1995) Activity pattern in respect to homing of the giant African land snail, *Achatina fulica* Bowdich. PhD thesis, University of Calcutta, Calcutta.

Pawson, P.A. and Chase, R. (1984) The life-cycle and reproductive activity of *Achatina fulica* (Bowdich) in laboratory culture. *Journal of Molluscan Studies* 50, 85–91.

Pemberton, C.E. (1954) *Invertebrate Consultants Committee for Pacific. Report for 1949–54*. Pacific Science Board, National Academy of Science, Natural Resources Council, 56 pp.

Peterson, G.D. (1957a) *Lamprophorus tenebrosus* introduced into Guam to combat the giant African snail. *Journal of Economic Entomology* 50, 114.

Peterson, G.D. (1957b) An annotated check list of parasites and predators introduced into Guam during the years 1950–1955. *Proceedings of the Hawaiian Entomological Society* 16, 199–202.

Peterson, G.D. (1957c) Studies on control of the giant African snail on Guam. *Hilgardia* 26, 643–658.

Pilsbry, H.A. (1904) Family Achatinidae. In: Tryon, G.W. and Pilsbry, H.A. (eds) *Manual of Conchology*, Vol. 16. Academy of Natural Sciences, Philadelphia, pp. 205–329.

Pilsbry, H.A. (1906/7) *Manual of Conchology*. Second Series: *Pulmonata*. Vol. XVIII. *Achatinidae: Stenogyrinae and Coeliaxinae*. Academy of Natural Sciences, Philadelphia.

Pilsbry, H.A. (1919) A review of the land mollusks of the Belgian Congo, chiefly based on the collections of the American Museum Congo Expedition, 1909–1915. *American Museum of Natural History Bulletin* 40, 370 pp., 23 pls.

Pilsbry, H.A. and Bequaert, J.C. (1927) The aquatic mollusks of the Belgian Congo with geographical and ecological account on Congo malacology. *Bulletin of the American Museum of Natural History* 53, 69–602.

Plummer, J.M. (1975) Observations on the reproduction, growth and longevity of a laboratory colony of *Archachatina* (*Calachatina*) *marginata* (Swainson) subspecies *ovum*. *Proceedings of the Malacological Society of London* 41, 395–413.

Plummer, J.M. and Mann, O.V. (1983) Weight losses occurring during incubation of *Archachatina* eggs. *Journal of Molluscan Studies, Supplement* 12A, 223.

Pointier, J.-P. and Blanc, C. (1985) *Achatina fulica* en Polynésie Française. Répartition, caractérisation des populations et conséquences de l'introduction de l'escargot predateur *Euglandina rosea* en 1982–1983 (Gastropoda, Stylommatophora, Achatinacea). *Malakologische Abhandlungen* 11, 1–15.

Pradhan, S. and Srivastava, P.D. (1971) Role of distantly-related natural enemies in the integrate control of pests. *Entomologists' Newsletter* 1, 62–63.

Ranaivosoa, H. (1971) Lutte biologique contre les escargots phytophages à Madagascar et aux Comores. *L'Agronomie Tropicale* 26, 341–347.

Rappard, F.W. (1949) De agaatslak (*Achatina fulica* Fer.), een gevaar voor jonge djaticulturen. *Tectona* 40, 365–366.

Raut, S.K. (1978) Studies on the aestivating population of *Achatina fulica* Bowdich (Mollusca: Achatinidae) in West Bengal. *Bulletin of the Zoological Survey of India* 1, 243–246.

Raut, S.K. (1980) On a trichodin ciliate of the pestiferous land snail *Achatina fulica*. *Indian Journal of Animal Health* 19, 159–160.

Raut, S.K. (1982) The extent of damage inflicted by *Achatina fulica* Bowdich to agrihorticulture economic plants. *Journal of the Zoological Society of India* 34, 7–12.

Raut, S.K. (1983a) Altitudinal immobilization of the giant land snail *Achatina fulica* Bowdich. In: *Proceedings of a Workshop on High Altitude Entomology and Wildlife Ecology*. Zoological Survey of India, pp. 359–366.

Raut, S.K. (1983b) Epizootic disease of the giant African land snail, *Achatina fulica*. In: *Proceedings of the Symposium 'Host as an Environment'*. Zoological Survey of India, Calcutta, pp. 29–37.

Smith, C.W. (1992) Management of alien species in natural areas of Oceania. *Pacific Science* 46, 386–387.

Solem, A. (1979a) Biogeographic significance of land snails, Paleozoic to Recent. In: Gray, J. and Boucot, A.J. (eds) *Historical Biogeography, Plate Tectonics, and the Changing Environment.* Oregon State University Press, Corvallis, pp. 277–287.

Solem, A. (1979b) A theory of land snail biogeographic patterns through time. In: van der Spoel, S., van Bruggen, A.C. and Lever, J. (eds) *Pathways in Malacology.* Bohn, Scheltema and Holkema, Utrecht, and W. Junk, The Hague, pp. 225–249.

Solem, A. (1989) Non-camaenid land snails of the Kimberley and Northern Territory, Australia. I. Systematics, affinities and ranges. *Invertebrate Taxonomy* 2, 455–604.

Solem, A. (1990) How many Hawaiian land snail species are left? And what we can do for them. *Bishop Museum Occasional Papers* 30, 27–40.

South, F.W. (1926) The giant snail (*Achatina fulica*, Fer.) in Malaya. *Malayan Agricultural Journal* 14, 231–240.

Spence, G.C. (1938) *Limicolaria* as a pest. *Journal of Conchology* 21, 72.

Srivastava, P.D. (1966) Leucodermia like disease in the culture of giant African snail *Achatina fulica* Bowdich. *Indian Journal of Entomology* 28, 412–413.

Srivastava, P.D. (1968a) Role of hermit crabs in the biological control of *Achatina fulica* Bowdich on the Andamans. *Indian Journal of Entomology* 30, 217–219.

Srivastava, P.D. (1968b) *Gulella (Indoennea) bicolor* (Hutton), a predator of giant African snail *Achatina fulica* Bowdich. *Indian Journal of Entomology* 30, 240–241.

Srivastava, P.D. (1970) Integrated control of giant African snail. *Pesticides* 4, 92–96.

Srivastava, P.D. (1976) The giant African snail, *Achatina fulica* Bowdich and its control. *Proceedings of the National Academy of Sciences of India, Section B* 46, 60–64.

Srivastava, P.D. (1992) *Problem of Land Snail Pests in Agriculture: a Study of the Giant African Snail.* Concept Publishing Company, New Delhi, 234 pp.

Srivastava, P.D. and Srivastava, Y.N. (1967) *Orthomorpha* sp. – a new predatory millipede on *Achatina fulica* in Andamans. *Experientia* 23, 776.

Srivastava, P.D. and Srivastava, Y.N. (1968) Role of snails' disease in the biological control of *Achatina fulica* Bowdich, 1822 in the Andamans. *The Veliger* 10, 320–321.

Srivastava, P.D., Gupta, G.P. and Srivastava, Y.N. (1987) *Recent Advances in Entomology* 519–528.

Srivastava, Y.N., Srivastava, P.D. and Doharey, L.L. (1975) Predation preference of *Gulella (Indoenna) bicolor* (Hutton). *Entomologist's Newsletter* 5, 50.

Stievenart, C. (1992) Observations on shell lip formation and reproduction in the giant African snail *Archachatina marginata suturalis* (Philippi). *Snail Farming Research* 4, 43–45.

Stuhlmann, H. (1894) *Mit Emin Pascha ins Herz von Afrika.* Berlin.

Suzuki, H. (1981) Investigation of ecology and extirpation of giant African snail, *Achatina fulica*, in Okinawa. In: *Annual Report on Applied Zoology at the Okinawa Agricultural Centre*, pp. 47–51.

Suzuki, H. and Yasuda, K. (1983) Studies on ecology and control of the giant African snail, *Achatina fulica*, in Okinawa Island. (1) The optimal period for control with metaldehyde. *Annual Report of the Okinawa Agricultural Centre* 8, 43–50.

Takeda, N. and Ozaki, T. (1986) Induction of locomotor behaviour in the giant African snail, *Achatina fulica. Comparative Biochemistry and Physiology* 83A, 77–82.

Tattersfield, P. (1996) Local patterns of land snail diversity in a Kenyan rain forest. *Malacologia* 38, 161–180.

Tauili'ili, P. and Vargo, A.M. (1993) History of biological control in American Samoa. *Micronesica, Supplement* 4, 57–60.

Teles, H.M.S., Vaz, J.F., Fontes, L.R. and de Fátima Domingos, M. (1997) Registro de *Achatina fulica* Bowdich, 1822 (Mollusca, Gastropoda) no Brazil: caramujo hospedeiro intermediário da angiostrongilíase. *Revista de Saúde Pública* 31, 310–312.

Thistle, A.D. (1953) Chemical control of African snail. In: *Annual Report (1953).* Hawaii Board of Commissioners, Agriculture and Forestry, p. 28.

Tillier, S. (1982) Production et cycle réproducteur de l'escargot *Achatina fulica* Bowdich, 1822 en Nouvelle Calédonie (Pulmonata: Stylommatophora: Achatinidae). *Haliotis* 12, 111–122.

Tillier, S. (1992) Introduced land snails in New Caledonia: a limited impact in the past, a potential disaster in the future. *Pacific Science* 46, 396–397.

Tillier, S. and Clarke, B.C. (1983) Lutte biologique et destruction du patrimoine génétique: le cas des mollusques gastéropodes pulmones dans les territoires français du Pacifique. *Génétique, Sélection, Evolution* 15, 559–566.

Tomaszewski, W. (1949) Mollusca, Weichtiere. In: Sorauer, P. (ed.) *Handbuch der Pflanzewnkrankheiten, 4.* Paul Parey, Berlin and Hamburg, pp. 100–116.

Tomiyama, K. (1991) Reproductive behaviour of hermaphrodite land snail, *Achatina fulica.* In: *Proceedings of the 2nd International Ethological Conference, Otani University, Kyoto,* p. 43.

Tomiyama, K. (1992) Homing behaviour of the giant African snail, *Achatina fulica* (Ferussac) (Gastropoda; Pulmonata). *Journal of Ethology* 10, 139–147.

Tomiyama, K. (1993) Growth and maturation pattern in the African giant snail, *Achatina fulica* (Ferussac) (Stylommatophora: Achatinidae). *Venus* 52, 87–100.

Tomiyama, K. (1994) Courtship behaviour of the giant African snail, *Achatina fulica* (Férussac) (Stylommatophora: Achatinidae) in the field. *Journal of Molluscan Studies* 60, 47–54.

Tomiyama, K. and Miyashita, K. (1992) Variation of egg clutches in the giant African snail, *Achatina fulica* (Ferussac) (Stylommatophora: Achatinidae) in Ogasawara Islands. *Venus* 51, 293–301.

Tompa, A.S. (1979) Oviparity, egg retention and ovoviviparity in pulmonates. *Journal of Molluscan Studies* 45, 155–160.

Townes, H.K. (1946) *Results of an Entomological Inspection Tour of Micronesia.* United States Commercial Cooperative Economic Survey, U.S. Navy, Guam, 53 pp.

Tra, B.K.B. (1994) *Effets de la Densité et de Quelques Aliments sur les Performances de Croissance de l'Escargot Géant Africain* Achatina achatina *(Linné).* Réport de Stage, Ecole Nationale Supérieure Agronomique, Yamoussoukrere, 66 pp.

Tranter, J.A. (1993) The giant African land snail, *Achatina fulica*, and other species. *Journal of Biological Education* 27, 108–111.
Tryon, G.W. and Pilsbry, H.A. (1904) *Manual of Conchology*, Vol. 16. Academy of Natural Sciences, Philadelphia, 329 pp., 37 pls.
Turner, G.J. (1964) Transmission by snails of the species *Phytophthora* which causes foot rot of *Piper nigram* L. in Sarawak. *Nature* 202, 1133.
Turner, G.J. (1967) Snail transmission of the species of *Phytophthora* with special reference to foot rot of *Piper nigrum*. *Transactions of the British Mycological Society* 50, 251–258.
Upatham, E.S., Kruatrachu, M. and Baidikul, V. (1988) Cultivation of the giant African snail, *Achatina fulica*. *Journal of the Science Society of Thailand* 14, 25–40.
US Congress (1993) *Harmful Non-indigenous Species in the United States*. Office of Technology Assessment, US Government, Washington, DC, 391 pp.
van Alphen der Veer, E.J. (1954) De agaatslak (*Achatina fulica* Fer.), een gevaar voor jonge bosculturen. *Penggemar Alam* 34, 36.
van As, J.G. and Basson, L. (1993) On the biology of *Pallitrichodina rogenae* gen. n., sp. n. and *P. stephani* sp. n. (Ciliophora: Peritrichida), mantle cavity symbionts of the giant African snail *Achatina* in Mauritius and Taiwan. *Acta Protozoologica* 32, 47–62.
van Benthem Jutting (1934) *Achatina fulica* (Fér.) in the Netherlands East Indies. *Journal of Conchology* 20, 43–44.
van Benthem Jutting (1952) A snail farm in the Netherlands. *Basteria* 16, 25–30.
van Bruggen, A.C. (1965) Two new species of Achatinidae (Mollusca, Gastropoda, Pulmonata) from the Drakensberg Range, with general remarks on southern African Achatinidae. *Revue de Zoologie et de Botanique Africaines* 71, 79–91.
van Bruggen, A.C. (1966) Notes on non-marine molluscs from Mozambique and Bechuanaland, with a checklist of Bechuanaland species. *Annals Transvaal Museum* 25, 99–112.
van Bruggen, A.C. (1968) Additional data on the terrestrial molluscs of the Kruger National Park. *Annals Natal Museum* 20, 47–58.
van Bruggen, A.C. (1969) Studies on the land molluscs of Zululand with notes on the distribution of land molluscs in southern Africa. *Zoologische Verhandelingen Leiden* 103, 1–116.
van Bruggen, A.C. (1970) Notes on the distribution of terrestrial molluscs in southern Africa. *Malacologia* 9, 256–258.
van Bruggen, A.C. (1977) Studies on the ecology and systematics of the terrestrial molluscs of the Lake Sibaya area of Zululand, South Africa. *Zoologische Verhandelingen* 154, 1–44, 4 pls.
van Bruggen, A.C. (1978) Land molluscs. In: Werger, M.J.A. (ed.) *Biogeography and Ecology of Southern Africa*. Dr W. Junk, The Hague, pp. 877–923.
van Bruggen, A.C. (1981) The African element among the terrestrial molluscs of the island of Madagascar. *Proceedings of the Koninklijke Nederlandse Akademie van Wetenschappen, Series C, Zoology* 84, 115–129.
van Bruggen, A.C. (1985) The terrestrial molluscs of Lesotho (Southern Africa), a first contribution, with detailed notes on *Archachatina machachensis* (Mollusca, Gastropoda). *Proceedings of the Koninklijke Nederlandse Akademie van Wetenschappen, Series C, Zoology* 88, 267–296.
van Bruggen, A.C. (1986) Aspects of the diversity of the land molluscs of the Afrotropical Region. *Revue de Zoologie Africaines* 100, 29–45.

van Bruggen, A.C. (1987) *Achatina fulica* in Morocco, North Africa. *Basteria* 51, 66.
van Bruggen, A.C. (1989) The Dahomey Gap as evidenced by land molluscs, a preliminary report resulting from a reconnaissance of the literature. *Basteria* 53, 97–104.
van Bruggen, A.C. and Appleton, C.C. (1977) Studies on the ecology and systematics of the terrestrial molluscs of the Lake Sibaya area of Zululand, South Africa. *Zoologische Verhandelingen Leiden* 154, 1–44.
van deer Meer Mohr, J.C. (1949a) On the reproductive capacity of the giant African or giant land snail, *Achatina fulica* (Fer.). *Treubia* 20, 1–10.
van deer Meer Mohr, J.C. (1949b) *Achatina fulica* (Fer.) as a minor pest of tobacco. *Chronica Naturae* 104, 178–179.
van der Schalie, H. (1969) Man meddles with nature – Hawaiian style. *The Biologist* 51, 136–146.
van der Schalie, H. (1970) Snail control problems in Hawaii. In: *Annual Report of the American Malacological Union (1969)*. American Malacological Union, Hattiesburg, pp. 55–56.
van Dinther, J. (1973) Molluscs in agriculture and their control. *Mededeelingen Laboratorium Entomologie. Wageningen* 232, 281–286.
van Leeuwen, D. (1932) Notes and comments, conchology, *Achatina fulica. Hong Kong Naturalist* 3, 71.
van Weel, P.B. (1948/49) Some notes on the African giant snail, *Achatina fulica* Fer. I. On its spread in the Asiatic tropics. II. On its economic significance. III. On its biological balance and means of destruction. *Chronica Naturae* 104, 241–243, 278–280, 335–336.
van Zinderen Bakker, E.M. (1982) African palaeoenvironments 18 000 years BP. *Palaeoecology Africa* 15, 77–99.
van Zwaluwenburg, R.H. (1955) Minutes from the 7 Jan. 1955 meeting of the Hawaiian Entomological Society. *Proceedings of the Hawaiian Entomological Society* 16, 1.
Verdcourt, B. (1961) *Achatina fulica hamillei* (Petit) in the Kavirondo district of Kenya. *Journal of Conchology* 25, 34–35.
Verdcourt, B. (1984) Discontinuities in the distribution of some East African land snails. In: Solem, A. and van Bruggen, A.C. (eds) *World-wide Snails. Biogeographical Studies on Non-marine Mollusca.* E.J. Brill, Leiden, pp. 134–155.
Voelker, J. (1959) Der chemische Einfluß von Kalziumkarbonat auf Wachstum, Entwicklung und Gehäusebau von *Achatina fulica* Bowdich (Pulmonata). *Mitteilungen aus dem Hamburgischen Zoologische Museum und Institut, Hamburg*, 57, 37–78.
von Stanislaus, K., Morkramor, C., Potors, K.J. and Waitkuwait, E. (1987) Opportunities for utilizing the African giant snail. In: Siegmund, R. (ed.) *Untersuchungen für Wachstumsund Reprodurcktionsleitung Beider Achatschnecke.* Diplomabeit am Institute für Tierzucht und Hanstiergenetik der Universitat Gottingen, Gottingen, pp. 60–71.
Wallace, G.D. and Rosen, L. (1969a) Experimental infection of Pacific Island mollusks with *Angiostrongylus cantonensis. American Journal of Tropical Medicine and Hygiene* 18, 13–19.
Wallace, G.D. and Rosen, L. (1969b) Studies on eosinophilic meningitis V. Molluscan hosts of *Angiostrongylus cantonensis* on Pacific islands. *American Journal of Tropical Medicine and Hygiene* 18, 206–216.

Wandolleck, B. (1898) Die Stethopathidae, eine neue flügel- und schwingerlose Familie der Diptera. *Zoologische Jahrbücher Abteilung für Systematik* 11, 412–439, 2 pls.

Waterhouse, D.F. and Norris, K.R. (1987) *Achatina fulica* Bowdich, Mollusca: Achatinidae. Giant African snail. In: Waterhouse, D.F. and Norris, K.R. (eds) *Biological Control – Pacific Prospects*. Inkata Press, Melbourne, pp. 265–273.

Watson, B.J. (1985) The giant African snail in Australia: pest or nuisance. *Queensland Agricultural Journal* 111, 7–10.

Weber, P.W. (1953) Recent liberations of beneficial insects in Hawaii – II. *Proceedings of the Hawaiian Entomological Society* 15, 127–130.

Weber, P.W. (1954a) Studies of the giant African snail on Guam. *Hilgardia* 26, 643–658.

Weber, P.W. (1954b) Studies of the giant African snail. *Proceedings of the Hawaiian Entomological Society* 15, 363–367.

Weber, P.W. (1956) Recent introductions for biological control in Hawaii – I. *Proceedings of the Hawaiian Entomological Society* 16, 162–164.

Weber, P.W. (1957) Recent introductions for biological control in Hawaii – II. *Proceedings of the Hawaiian Entomological Society* 16, 313–314.

Wells, S.M. (1988) Snails going extinct at speed. *New Scientist* 117, 46–48.

Wells, S.M., Pyle, R.M. and Collins, N.M. (1983) *The IUCN Invertebrate Red Data Book*. International Union for the Conservation of Nature, Gland, Switzerland, and Cambridge, UK.

Whitten, H. (1981) Health threat to Samoa seen in *Achatina fulica. Hawaiian Shell News* 29(3), 3.

Williams, F.X. (1951) Life-history studies of East African *Achatina* snails. *Bulletin of the Museum of Comparative Zoology at Harvard College* 105(3), 295–317, 5 pls.

Williams, F.X. (1953) Some natural enemies of snails of the genus *Achatina* in East Africa. In: *Proceedings of the 7th Pacific Science Congress*, vol. 4. Pacific Science Association, Honolulu, pp. 277–278.

Wolfenbarger, D.O. (1971) Dispersion of the giant African snail, *Achatina fulica. Quarterly Journal of the Florida Academy of Sciences* 34, 48–52.

Zong, D., Coulibely, M., Diambra, O.H. and Adjiri, E. (1990) Note sur l'élévage de l'escargot géant African *Achatina achatina. Nature et Faune* 6, 32–44.

4 Vaginulidae in Central America, with Emphasis on the Bean Slug *Sarasinula plebeia* (Fischer)

ALFREDO RUEDA,[1] RAFAEL CABALLERO,[2] RINA KAMINSKY[3] AND KEITH L. ANDREWS[2]

[1]Escuela Agrícola Panamericana, Zamorano, PO Box 93, Tegucigalpa, Honduras, Central America; [2]Entomology Department, Forbes 410, Box 210036, University of Arizona, Tucson, AZ 85721-0036, USA; [3]Dirección de Investigación Científica, Universidad Nacional Autónoma de Honduras y Laboratorio de Parasitología, Hospital-Escuela, Apartado Postal 1587, Tegucigalpa, Honduras, Central America

Introduction

Central America has a population of about 30 million people. Of these, 55% live in rural areas and most of them live in poverty or extreme poverty. Dry beans (*Phaseolus vulgaris* Linnaeus) (Fabaceae) are the major source of protein for people in Central America, and 0.5 million ha are planted annually. Gastropods of the family Vaginulidae (= Veronicellidae) (Mollusca: Pulmonata: Soleolifera), especially *Sarasinula plebeia* (Fischer), are important pests on bean crops in the region. These gastropods feed on the leaves and stems of young dry-bean plants, defoliating and often killing them. Andrews (1987a) estimated that more than 400,000 farmers in Central America suffer economic losses due to these gastropods every year. The estimated losses range from US$27 million to US$45 million annually.

Vaginulids have a pantropical to subtropical distribution. Of the known *c.* 100 species in this family, more than half occur in the Americas. Vaginulids are important in Central America for two reasons. First, they are pests in several staple food and horticultural crops. Second, vaginulids are intermediate hosts of the metastrongylid nematode *Angiostrongylus costaricensis* (Morera & Cespedes), which parasitizes humans. Costs associated with the disease were estimated to be US$5 million annually (Andrews, 1987a).

The purpose of this chapter is to summarize the research carried out in Central America, principally over the past three decades, which focused on providing the understanding of the biology and ecology necessary for the development and implementation of effective management strategies for vaginulid pests.

History of Pest Infestations in Central America

Vaginulids are native to Central America, although several of the pest species have been introduced from other regions. Their presence in Central America was first recorded by Heynemann (1885) and Cockerell (1890, 1895). Before 1965, these gastropods were considered of minor importance on coffee (*Coffea arabica* Linnaeus) (Rubiaceae) in Guatemala (Alvarado, 1939), Nicaragua (Anon., 1953) and Costa Rica (Anon., 1960). Vaginulids were also reported occasionally on bananas (*Musa acuminata* Colla) (Musaceae) in Honduras (J. Osmark, personal communication 1984), on tobacco (*Nicotiana tabacum* Linnaeus) (Solanaceae) in Costa Rica (Anon., 1960) and on dry beans in Nicaragua (Anon., 1981) and Costa Rica (Anon., 1965). Most of these records of crop damage relate to *Diplosolenodes occidentalis* (Guilding). The Costa Rican and Nicaraguan references to damage on dry beans were for humid areas at high elevations. In these cases, the damage was attributed to *D. occidentalis* or the species responsible was not identified.

Andrews and Dundee (1987) reviewed the literature to reconstruct the apparent spread of *S. plebeia* in Central America (Fig. 4.1). Severe economic damage to dry-bean plants caused by vaginulids was reported first in two departments of El Salvador in 1967. These attacks were detected on bean crops around the experimental station of the Ministry of Agriculture in San Andrés, La Libertad. According to Mancía (1971), an unknown taxonomist identified the pest as *Vaginulus plebeius* Fischer. By 1970, damage to dry beans was common in all departments of the country (Mancía, 1971).

In Honduras, vaginulid damage in dry beans started in 1970 (Anon., 1976). By 1976, these infestations were present in the departments of El Paraiso, Francisco Morazan, Olancho, Ocotepeque and Comayagua. It is believed that at that time vaginulids were also present in the departments of Lempira and La Paz (Anon., 1976). In addition to the above departments, Córdova (1981) reported these gastropods in Cópan, Lempira, Yoro, Santa Bárbara, Cortés, Valle, Choluteca and La Paz. Caballero *et al.* (1991) presented a detailed survey of vaginulids in Honduras from field collections made between 1984 and 1988. In this study, four vaginulid species were encountered in Honduras: *Belocaulus angustipes* (Heynemann), *Leidyula moreleti* (Fischer), *D. occidentalis* and *S. plebeia* (Fig. 4.2).

Sarasinula-type vaginulids were first reported in the department of León, Nicaragua, in the early 1970s (R. Caballero, unpublished). Severe

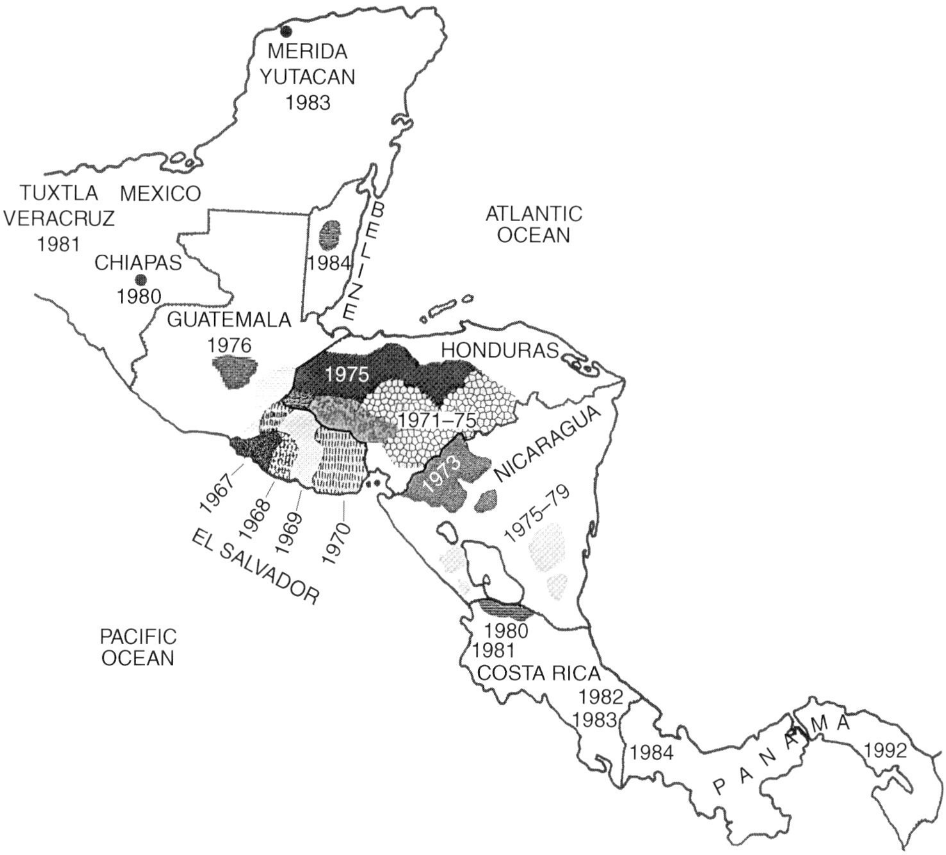

Fig. 4.1. Years in which Vaginulidae were first reported to cause chronic or severe damage to crops (Central America) or collected for the first time (Yucatán and Belize). Modified with permission from Andrews and Dundee (1987).

infestations on dry beans were reported during 1973 in the departments of Nueva Segovia, Madrid, Matagalpa, Nueva Guinea and Chinandega. By 1980, these gastropods were present in the department of Jinotega and, according to R. Daxl (personal communication, 1984), were causing considerable damage in the southern departments. Farmers reported *S. plebeia* as a severe pest in several crops, especially squash (*Cucurbita* Linnaeus spp.) (Cucurbitaceae) in Boaco, Nicaragua, in 1992.

As noted above, minor damage to dry beans has long been known from Costa Rica. This damage was attributed to *D. occidentalis*. However, collections made by R.A. Sequeira (personal communication, 1985) revealed *S. plebeia* in Costa Rica in 1981, close to the Nicaraguan border. By 1989 this species was widely distributed in Costa Rica (J. Saunders, personal communication, 1985). In Guatemala, vaginulids have been known since the early 1980s to cause damage on dry beans, particularly close to the El Salvador and Honduran borders (Salguero, 1981). To the

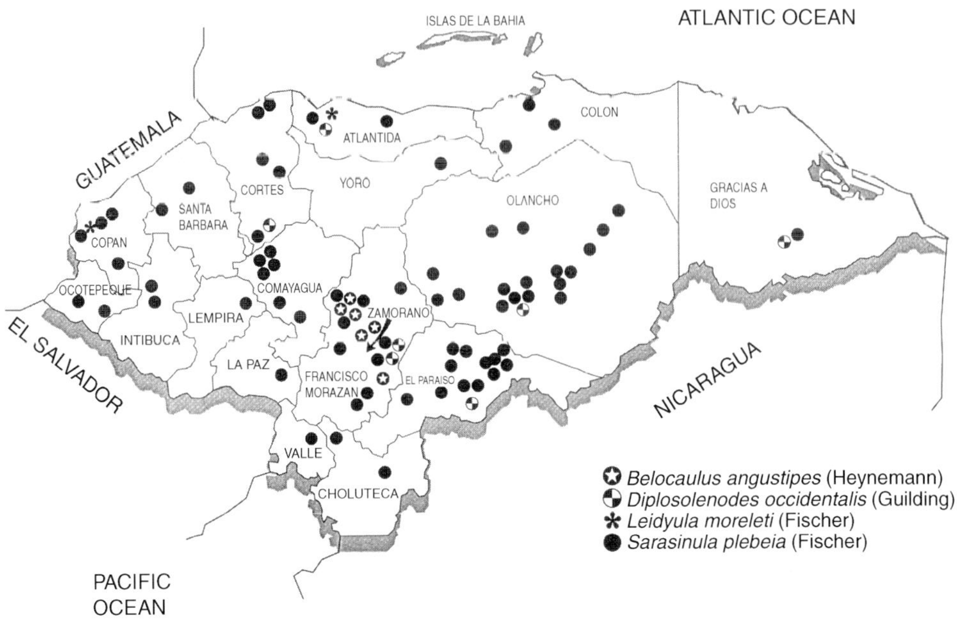

Fig. 4.2. Distribution of Vaginulidae in Honduras between 1984 and 1988. Modified with permission from Caballero *et al.* (1991).

east of Guatemala City farmers have reported severe infestations since 1976 (K.L. Andrews, unpublished data).

S. Dundee (Andrews and Dundee, 1987) collected *S. plebeia* in Orange Walk, Belize, during 1984. This is the first report of the species in Belize; however, some farmers reported damage to dry beans before that time. The pest status of *S. plebeia* in Belize is not well known (Andrews and Dundee, 1987). In Mexico, damage to dry beans caused by *S. plebeia* has been reported in Chiapas since 1980 (J. Gutiérrez, personal communication, 1985) and in Veracruz since 1981 (G. Arcos, personal communication, 1985). By 1983 the same species was present in Mérida, Yucatán (Andrews and Dundee, 1987). In Panama, damage produced by *S. plebeia* was first reported in 1984 (H. Iglesias, personal communication, 1985). R. Caballero (unpublished) observed considerable damage to sweet pepper (*Capsicum annuum* Linnaeus) and tomato (*Lycopersicon esculentum* Miller) (Solanaceae) in the province of Los Santos in 1992. *S. plebeia* is now known to occur throughout Panama.

The vaginulid species responsible for the widespread crop destruction in Central America is not native to the region, but was introduced through human commerce (Andrews, 1987a). The 1967 identification (Mancía, 1971) of vaginulids infesting crops in El Salvador as *V. plebeius* was subsequently confirmed by S. Dundee and J.W. Thomé (Andrews and Dundee, 1987). In the current systematic concepts of the Vaginulidae, *plebeia* is included in the genus *Sarasinula* (Grimpe & Hoffmann). However, there has been much confusion as to the identity of these gastropods, primarily due to the absence in the region of taxonomy experts. The taxonomy of vaginulids is based on the morphology of the hermaphroditic reproductive system, and a good knowledge of the comparative differences among species, the variation with stage of maturity and preservation artefacts is necessary to make the correct identifications.

Andrews and Dundee (1987) reviewed the literature and concluded that there were five species of vaginulids in Central America. Caballero *et al.* (1991) collected extensively in Honduras from 1984 to 1988 and determined four species: *B. angustipes*, *D. occidentalis*, *L. moreleti* and *S. plebeia*.

Belocaulus angustipes

This species has a black velvet notum and hyponotum and black ocular tentacles (Fig. 4.3A). The sole is blackish and narrower than any of the hyponotum. This vaginulid was introduced from South America. It is the smallest species in Central America, weighing not more than 1.2 g, and in Central America is known only from Francisco Morazan department in Honduras. It is not important as an agricultural pest, but has been found infected by *A. costaricensis* (Caballero *et al.*, 1991).

Diplosolenodes occidentalis

This species has a grey notum, varying somewhat in intensity and usually with very small, bright markings (Fig. 4.3B). The ocular tentacles are dark. The hyponotum is usually darkly pigmented, but this pigmentation varies with age and among individuals. Immature animals, weighing less than 1 g, have no pigmentation and occasionally there are adults without pigmentation. The sole is light grey and narrower than the hyponotum. This species reaches a weight in excess of 6 g. At present, this species is not considered a pest. It occurs primarily in low abundance in agricultural areas. However, *D. occidentalis* has been reported as causing damage in perennial crops in the past (Anon., 1953, 1960, 1965) and has been found infected with *A. costaricensis* (Caballero *et al.*, 1991). According to Thomé (1989), *D. occidentalis* is widely distributed in Central and South America.

Leidyula moreleti

This species has a light grey notum, with two parallel stripes, which are more defined in the anterior third (Fig. 4.3C). The integument at the perinotum is granulated. The sole is light yellow and wider than the hyponotum. Adults grow to a weight of 9 g. This species is not considered of economic or human health significance because of its low abundance.

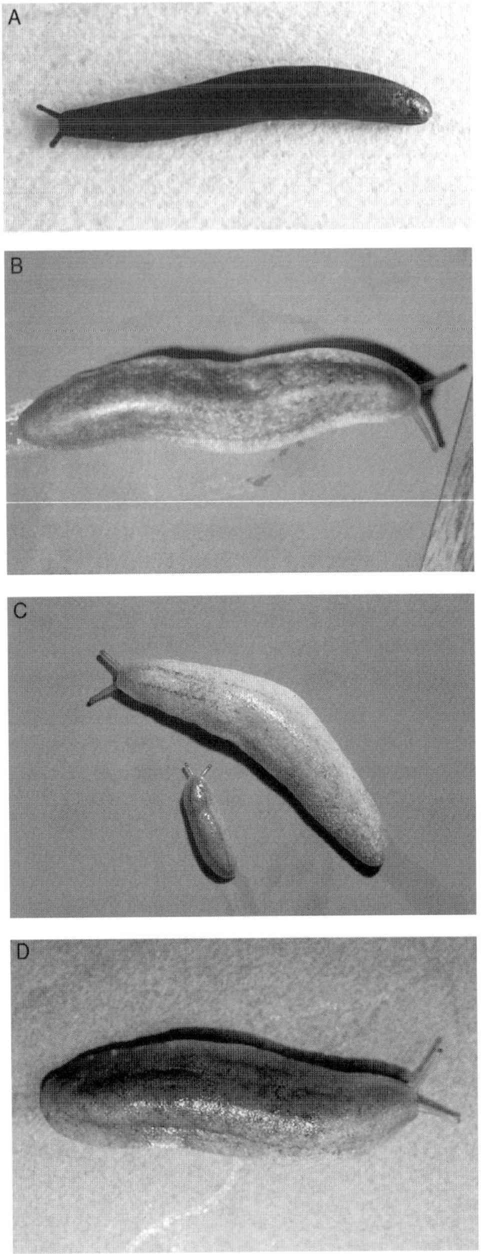

Fig. 4.3. A. *Belocaulus angustipes* (Heynemann), dorsal view. B. *Diplosolenodes occidentalis* (Guilding), dorsal view. C. *Leidyula moreleti* (Fischer), dorsal view. D. *Sarasinula plebeia* (Fischer), dorsal view.

It has not been found infected with *A. costaricensis* (Caballero *et al.*, 1991). *L. moreleti* is known from México, Guatemala and Honduras (Thomé, 1989).

Sarasinula plebeia

This species has a grey notum, with minute dark marks without a defined pattern (Fig. 4.3D). The sole is narrower than the hyponotum. Adults on average weigh about 3 g, but some individuals grow to 6 g. *S. plebeia* is economically the most important species in Central America. It is a key pest of bean crops in many regions, and also feeds on other horticultural and ornamental crops. In addition, because of its high abundance and wide distribution, it is the most important intermediate host of *A. costaricensis*. According to Thomé (1989), *Vaginula behni* Semper, *Sarasinula dubia* (Semper) and *Sarasinula lemei* Thomé are junior synonyms of *S. plebeia*.

Agricultural importance of *Sarasinula plebeia*

S. plebeia is the most important pest on bean crops in many regions of Central America. Andrews (1987a) considered the situation of *S. plebeia* to be one of the most significant mollusc infestations in the world. In recent years *S. plebeia* has assumed pest status on sweet pepper and tomato fruits in Panama and on cucurbits in Nicaragua (R. Caballero, unpublished). Despite the gravity of this situation, the geographical distribution and severity of damage by *S. plebeia* in dry-bean crops has gone largely undocumented over the past decade. The authors believe that the pest has declined in perceived importance, for farmers and scientists alike, due to the outbreak of the sweet-potato whitefly, *Bemisia tabaci* (Gennadius) (Aleyrodidae), and geminivirus in the region. As a result of experiences prior to 1990, farmers already know the appropriate control procedures for *S. plebeia*, but they have few effective alternatives to control *B. tabaci* and geminivirus.

It is important to consider the pest status of vaginulids in the light of the seasonal weather conditions and agronomic systems used for maize (*Zea mays* Linnaeus) (Gramineae) and dry-bean cultivation in Central America (Fig. 4.4). Most dry beans in the region are relay-planted with maize. The maize is planted at the beginning of the rainy season (May–June) and dry beans are manually planted between the maize rows when the maize reaches physiological maturity (August–September). *S. plebeia* abundance is generally very low at the beginning of the rainy season, due to the effects of the preceding dry season, which typically lasts 4–6 months. The surviving *S. plebeia* emerge at the onset of the rainy season, feed on broad-leaved weeds and reproduce during the maize-growing season. By the time of the bean-planting season, the *S. plebeia* populations are usually very large and the bean seedlings are subject to considerable damage (Rueda *et al.*, 1987).

Vaginulids are herbivorous, feeding mainly on leaf tissues, buds and soft stems. Sporadically they will also feed on bean pods and the fruits of sweet pepper and tomato. These gastropods have been considered a

Fig. 4.4. Seasonal changes in the activity of *Sarasinula plebeia* (Fischer) (Vaginulidae) in relation to weather and agronomic systems used for maize (*Zea mays* Linnaeus) (Gramineae) and dry-bean (*Phaseolus vulgaris* Linnaeus) (Fabaceae) cultivation in Central America. *S. plebeia* abundance is generally very low at the onset of the rainy season after a dry period of 4–6 months. The surviving *S. plebeia* feed on broad-leaved weeds and reproduce during the maize-growing season. By the time of the bean-planting season, *S. plebeia* populations may be very large and can destroy the bean crop at the seedling stage.

key pest in dry beans due to their attack on seedlings, where they defoliate the plants (Fig. 4.5) or sever the stems. Usually the young plants are irreversibly damaged and die (Caballero and Andrews, 1989).

Medical Importance of *Sarasinula plebeia* and Other Vaginulidae

Vaginulids are of medical importance as intermediate hosts of parasitic nematodes, namely *A. costaricensis*, a natural parasite occurring in the mesenteric arteries of rodents and other mammals (Morera, 1973). This metastrongylid nematode causes in humans a disease known as human abdominal angiostrongyliasis, which is primarily restricted to several countries in the Americas, particularly Costa Rica (Morera, 1973).

In rats, the first-stage larvae of *A. costaricensis*, measuring about 270 μm in length, are passed out in the faeces. Gastropods feeding on these faeces ingest the larvae, which migrate to the gastropod's mantle and foot tissues. There the nematodes moult twice before reaching the infective stage in about 16–19 days. Infection of rats (*Rattus* Fischer spp.; Muridae) and other mammals, including humans, occurs orally by ingestion of either the infected gastropods or viable, free-living larvae in mucus-contaminated food. Once ingested, the larvae migrate through the intestinal tissues to the lymphatic vessels in the abdominal cavity, moult

Fig. 4.5. Feeding damage on dry-bean (*Phaseolus vulgaris* Linnaeus) (Fabaceae) seedlings by *Sarasinula plebeia* (Fischer) (Vaginulidae). Note the severe defoliation.

twice and then migrate to the mesenteric arteries. This development is completed by about the tenth day post-infection. Oviposition by the mature female nematodes begins about 18 days after initial infection, with first-stage larvae appearing in faeces after 24 days (Morera, 1973).

A number of gastropod genera have been found infected in Latin America. In respect of vaginulids, *D. occidentalis* has been found infected by *A. costaricensis* in Colombia (Malek, 1981) and *S. plebeia*, *B. angustipes* and *D. occidentalis* are hosts in Honduras (Caballero *et al.*, 1991). In Nicaragua, the frequency of infection in *S. plebeia* varied from 4% in urban areas to 85% in rural areas (Duarte *et al.*, 1992). In Río Grande do Sul, Brazil, infections have been confirmed in the native vaginulids *Phyllocaulis variegatus* (Semper), *Phyllocaulis soleiformis* (Orbigny) and *B. angustipes*, as well as the introduced limacids *Limacus flavus* (Linnaeus) and *Limax maximus* Linnaeus (Graeff Teixeira *et al.*, 1993, 1994).

The first 31 cases of human abdominal angiostrongyliasis, spanning the period 1951–1967, were reported from Costa Rica in 27 children and four adults (Céspedes *et al.*, 1967). By 1980, 116 additional cases in children were reported in that country (Loría-Cortés and Lobo-Sanahuja, 1980). The first case of abdominal angiostrongyliasis from Honduras was reported in 1972 (Sierra and Morera, 1972). In 1983, six additional cases were reported (Zúniga *et al.*, 1983) and by 1995 there had been a further 23 cases (R. Kamisky, unpublished observation). Isolated cases of human infection by *A. costaricensis* have been reported from El Salvador (Sauerbrey, 1977), Nicaragua (Duarte *et al.*, 1991; Vásquez *et al.*, 1993) and Panama (Sánchez, 1992). Angiostrongyliasis is also known from the Caribbean and South America (Kamisky, 1997).

The usual clinical presentation is of appendicitis-like abdominal emergency, accompanied by eosinophilia of 40% or more. There are complaints of abdominal pain localized in the right iliac fossa. Pain is also present on palpation of the abdomen, and a tumour-like mass can be felt in the right lower quadrant. Rectal examination is painful. Anorexia, vomiting, diarrhoea and abdominal rigidity can also be found, alone or in combination (Céspedes *et al.*, 1967; Loría-Cortés and Lobo-Sanahuja, 1980). Other descriptions include yellow granulation of the subserosa of the intestinal wall, eosinophilic infiltration of the appendix and/or the lymph nodes (Loría-Cortés and Lobo-Sanahuja, 1980), extraintestinal lesions in the liver and testes (Céspedes *et al.*, 1967), a larva migrans-like syndrome (Morera *et al.*, 1982), an obstruction of the spermatic artery (Ruíz and Morera, 1983), a gangrenous ischaemic enterocolitis, acute appendicitis and Meckel's diverticulum-like presentation (Hulbert *et al.*, 1993).

It is unknown why there are about 300 new cases a year of angiostrongyliasis in Costa Rica, while in the rest of the Americas the infection is found sporadically or after diligent revision of pathology specimens (Morera, 1987). A 3-year study in Honduras indicated that the dynamics of human angiostrogyliasis in rural areas depends upon the presence of the parasites in rodents and the prevailing abiotic conditions, such as humidity and temperature. Without rain, farmers do not sow the crops, weeds do not germinate, gastropods such as vaginulids are inactive and the life cycle of the nematode is interrupted (Kaminsky *et al.*, 1995). This study does not indicate, however, what factors are critical in the regional variation in the prevalence of *A. costaricensis* in humans.

Biology, Ecology and Behaviour

Because of the economic importance of vaginulids as agricultural pests and as intermediate hosts of parasites, the management of these gastropods is critical for rural communities in Central America. Efficient management for any pest requires knowledge of its biology, ecology and behaviour. Therefore, when *S. plebeia* became a serious pest in Central America, research was initiated to provide basic information from which to develop control strategies.

B. angustipes has a subtropical native range in South America, and in Honduras occurs only above 800 m altitude in three different localities in the department of Francisco Morazan. The widely distributed *D. occidentalis* occurs from sea level up to 800 m. Normally found in low numbers, this species has a preference for undisturbed, shaded habitats and is rarely found in cultivated areas. In the north of Honduras, *L. moreleti* has been found at elevations up to 600 m above sea level. This species, too, occurs at low densities in humid, shaded habitats and is never found in cultivated areas. The most common and abundant species in Central America is *S. plebeia*. It has been found in all departments in Honduras, from near the sea up to 1000 m altitude. It prefers disturbed habitats, such as cultivated areas, backyards and gardens.

As indicated earlier, vaginulids are polyphagous herbivores. They are nocturnal, hiding during the day under stones, rotten logs and other plant residues on the ground. *S. plebeia* exhibits peak activity between 2 and 4 a.m. Individuals travel an average of 11 m in the course of a single night

(Andrews and López, 1987). They may also be active during cloudy, humid days. During the dry season these gastropods descend into soil crevices up to 1 m in depth to aestivate (Andrews *et al.*, 1987). They become active again at the beginning of the rainy season, normally in May.

Vaginulidae are most sexually active during the peak of the rainy season (Rueda *et al.*, 1987). They are hermaphroditic, with each individual having both female and male reproductive systems. Reproduction by self-fertilization is common in all species native to Central America, although outcrossing is known from the vaginulid species in Central America and species elsewhere in the Americas. *B. angustipes* and *D. occidentalis* are oviparous; they produce an average of seven and 33 eggs in each egg clutch, respectively (Caballero *et al.*, 1991). *D. occidentalis* becomes reproductive at about 7 months of age. *L. moreleti* is viviparous, giving birth to an average of 12 juveniles in each of one to three clutches per year. Reproduction in this species begins at an age of about 6 months, or when a weight of about 4 g is attained (Caballero *et al.*, 1991; R. Caballero, unpublished data).

S. plebeia is oviparous and produces clutches that average 37 eggs. This species lays one or two – occasionally four – egg clutches per year. Under laboratory conditions reproduction begins when the animals are about 6 months of age, or have attained a weight of 3 g on a bean-seedling diet (Caballero *et al.*, 1991) and up to 16 g on a dehydrated lucerne (*Medicago sativa* Linnaeus; Fabaceae) pellet diet (Rueda, 1989b). Self-fertilization is common in *S. plebeia* and outcrossing (mating) has not been observed in the field. Under laboratory conditions, however, mating occurs when the animals first reach sexual maturity but not subsequently (Rueda, 1989a). Rueda (1989a) observed that most individuals in a laboratory colony of *S. plebeia* copulated several times during the same night at the beginning of their reproductive life, suggesting that a sexual pheromone may be produced.

Studies of feeding behaviour in the laboratory (Andrews *et al.*, 1985b) demonstrated that the leaves of some plants species were highly acceptable to *S. plebeia*, while other plant were completely rejected by them (Table 4.1). They found that *Nicandra physalodes* (Linnaeus) Gaertner (Solanaceae), *Tithonia rotundifolia* (Miller) Blake (Asteraceae), *Melampodium divaricatum* (Cavanilles) Kunth (Asteraceae), *Ipomoea batatas* (Linnaeus) de Lamarck (Convolvulaceae), *Brassica oleracea* var. *capitata* Linnaeus (Brassicaceae) and *P. vulgaris* were consistently and readily accepted. Grasses, as well as a number of dicotyledonous species, were consistently rejected. Field observations made by Kaminsky *et al.* (1995) confirmed that more damage occurs on *M. divaricatum* than on *P. vulgaris*. Rueda *et al.* (1991) showed that *S. plebeia* has the capacity to regulate its food consumption substantially in response to water and the nutritional content of the diet. The experiment compared *S. plebeia* consumption and utilization of a poor-quality diet based on carrot, *Daucus carota* Linnaeus (Apiaceae), with those of a high-protein diet based on

Table 4.1. Average leaf area (mm^2 day^{-1}) consumed per *Sarasinula plebeia* (Fischer) (Vaginulidae) in a non-choice feeding experiment (obtained with permission from Andrews *et al.*, 1985b).

Plant species	Average consumption (mm^2 day^{-1})
Nicandra physalodes (Linnaeus) Gaertner (Solanaceae)	1064 a*
Melampodium divaricatum (Cavanilles) Kunth (Asteraceae)	1043 a
Phaseolus vulgaris Linnaeus (Fabaceae)	1008 a
Tithonia rotundifolia (Miller) Blake (Asteraceae)	695 b
Commelina diffusa Burman (Commelinaceae)	555 b
Brassica oleracea var. *capitata* Linnaeus (Brassicaceae)	382 c
Lactuca sativa Linnaeus (Asteraceae)	363 c
Glycine max (Linnaeus) Merrill (Fabaceae)	243 cd
Ageratum conyzoides Linnaeus (Asteraceae)	240 cde
Amaranthus hybridus Linnaeus (Amaranthaceae)	143 def
Ipomoea batatas (Linnaeus) de Lamarck (Convolvulaceae)	98 def
Medicago sativa Linnaeus (Fabaceae)	83 def
Ipomoea nil (Linnaeus) Roth (Convolvulaceae)	83 def
Citrus sinensis (Linnaeus) Osbeck (Rutaceae)	64 def
Manihot esculenta Crantz (Euphorbiaceae)	63 def
Beta vulgaris Linnaeus (Chenopodiaceae)	62 def
Rayjacksonia phyllocephalus (de Candolle) Hartman & Lane (Asteraceae)	30 def
Daucus carota Linnaeus (Apiaceae)	27 f
Lycopersicon esculentum Miller (Solanaceae)	26 f
Coffea arabica Linnaeus (Rubiaceae)	25 f
Portulaca oleracea Linnaeus (Portulacaceae)	18 f
Emilia sonchifolia (Linnaeus) de Candolle (Asteraceae)	12 f
Panicum maximum (Jacques) Robyns (Gramineae)	11 f
Pseudelephantopus spicatus (de Jussieu ex Aublet) Baker (Asteraceae)	10 f
Oxalis corniculata Linnaeus (Oxalidaceae)	5 f
Cyperus rotundus Linnaeus (Cyperaceae)	2 f
Sorghum bicolor (Linnaeus) Moench (Gramineae)	0 f
Nicotiana tabacum Linnaeus (Solanaceae)	0 f
Euphorbia heterophylla Linnaeus (Euphorbiceae)	0 f
Paspalum notatum Flueggé (Gramineae)	0 f

*Means followed by the same letter are not significantly different within each plant species ($P = 0.05$ Duncan mean separation test).

pellets of lucerne. For both diets, *S. plebeia* were able to increase their dry-matter intake as the diet's water content increased. The consumption of the lucerne-based diet was on average 5.5-fold higher than the carrot diet. This difference in consumption was attributed to a possible allochemical in the carrot diet. *S. plebeia* on the lucerne diet continued to grow and those on the carrot diet maintained or lost body weight. In another experiment in Honduras, *S. plebeia* were offered only jack bean, *Canavalia ensiformis* (Linnaeus) de Candolle (Fabaceae), leaves, to determine if this cover-crop species, known to contain allochemicals, had any detrimental effects on the gastropods. Results indicated that the survival and growth rates of *S. plebeia* on *C. ensiformis* were drastically reduced, compared with those fed on dry-bean leaves (Rizzo *et al.*, 1994).

Sampling and Economic Thresholds

Absolute sampling methods

From 1979 to 1987 intensive research efforts were made in Central America to develop sampling procedures to measure *S. plebeia* populations, for both scientific and management purposes. The sampling procedure adopted for experiments in Honduras was the visual inspection of 32 soil samples, each of 25 cm × 25 cm × 25 cm size (Rueda *et al.*, 1987). Data from a 2.5 year-long population-dynamics study showed that in 44% of the sample dates ($n = 123$), *S. plebeia* had an aggregated distribution in soil. The general problem with aggregated populations is that many samples are needed for a reliable population estimate. A further deficiency of this sampling procedure is that, when the soil is saturated (a common occurrence during the rainy season), it was extremely difficult to process the samples efficiently and some animals went undetected.

A. Rueda (unpublished data) found soil washing (Hunter, 1968) allowed extraction of *S. plebeia* from soil samples, with a recovery rate of *c.* 80%. This soil-washing technique has yet to be adequately evaluated for the processing of the 25 cm × 25 cm × 25 cm-sized sample units; however, Rueda *et al.* (1987) found that under dry conditions more than 85% of the vaginulids occurred in the upper 20 cm of the soil and thus were potentially recoverable.

In the slow-flotation method described by Hunter (1968), *S. plebeia* drown before they are able to escape to the surface of the soil sample (J. Garcia, unpublished data, 1985). A. Rueda (unpublished data) tested several methods to mark *S. plebeia* individuals with the idea of using mark–recapture sampling procedures (Southwood, 1978). Only hot-iron brandings of the notum and cutting pieces from the perinotum resulted in marks that remained visible for several weeks under laboratory conditions. Other possible markers were the albino *S. plebeia* found in Honduras (Andrews, 1987b). Unfortunately, such marked animals have not been tested in mark–recapture studies under field conditions.

Relative sampling methods

For the implementation of integrated pest management, relative sampling procedures are considered more pragmatic than absolute procedures, and have been used in conjunction with economic thresholds. The fact that *S. plebeia* is active at night and rests underground or beneath plant residues during the day makes monitoring difficult and costly, compared with other invertebrate pests of dry beans. Population estimates of

S. plebeia are also difficult to obtain because this gastropod is primarily active only on nights when the relative humidity is high: activity is negligible on nights when relative humidity drops below 75%. Daytime sampling of soil is also unreliable because under dry soil conditions *S. plebeia* seeks refuge in humid places under rocks or in soil crevices below practicable sampling depth. Thus, the spatial distribution of *S. plebeia* changes with soil humidity; under high soil humidity these gastropods are almost randomly distributed, but under dry soil conditions their dispersion is highly aggregated (Rueda *et al.*, 1987).

Crop-damage estimates from the extent of defoliation and yield reduction provide good estimates of the abundance of *S. plebeia* in the field. However, this approach cannot be used to predict pest abundance or to determine action thresholds because *S. plebeia* damage on dry-bean plants is not reversible. This sampling procedure is better suited to the evaluation of control practices (Wheeler and Peairs, 1980; Andrews, 1985; Meneses, 1985; Sobrado and Andrews, 1985).

Direct nocturnal observations with the help of a torch is the most common sampling procedure used to quantify *S. plebeia* infestations in Central America (Andrews, 1983, 1987b). For research purposes this method has been modified by using a 1 m square wooden frame; all animals present on the soil surface and on the plants inside the frame are counted. In a population-dynamics study this sampling procedure was used weekly for 2.5 years, taking 32 observations per sample (Rueda *et al.*, 1987). This procedure has several weaknesses: (i) *S. plebeia* activity depends on the environmental conditions, which may change from night to night and even within the same night (Andrews and López, 1987), and therefore counts are poor estimators of true abundance; (ii) small *S. plebeia* are difficult to observe; and (iii) a variable proportion of the active *S. plebeia* are not readily observed due to vegetative ground cover. However, the population-dynamics study of Rueda *et al.* (1987) provided data critical to the design of *S. plebeia* control programmes for farmers. In the nocturnal samples there were marked differences of *S. plebeia* abundance in the different seasons (Fig. 4.6). During the dry season (January–May), when the fields had little vegetation cover, there were 0.4 vaginulids m^{-2} on average. In this period these gastropods declined in number and activity due to the unfavourable soil conditions. During the early part of the rainy season, when the maize crop was planted (June–August), the *S. plebeia* population increased to 1.5 m^{-2} – this increment in population density resulted from the reproductive output of those animals that had survived the dry season and had resumed feeding (at or near the soil surface) on the broad-leaf weeds growing under the maize. During the dry-bean growing season (September–December) the *S. plebeia* population further increased to an average of 4.8 m^{-2}. At this population level, complete defoliation of the dry-bean seedlings typically occurs.

Baited traps have been used extensively to monitor *S. plebeia* populations in Central America. In preliminary studies (Andrews, 1983),

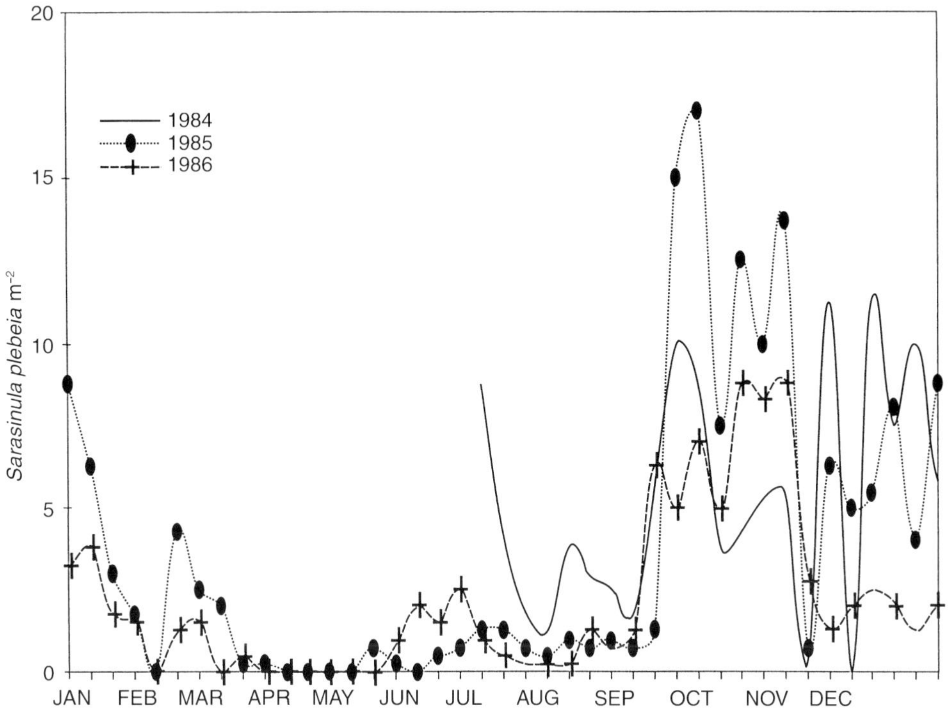

Fig. 4.6. Number of *Sarasinula plebeia* (Fischer) (Vaginulidae) observed during nocturnal sampling sessions between July 1984 and December 1986 in San Juan de Linaca, Danlí, El Paraíso, Honduras. Modified with permission from Rueda *et al.* (1987).

pitfall traps baited with a mixture of bran, molasses, beer and carbaryl were used. The trap consisted of a 1 quart oilcan with one of its ends removed. This can was buried in the soil and covered with a clay roof tile to avoid rainwater accumulation. With this trap design, seven *S. plebeia* were trapped for each *Sarasinula* per m^2 observed with a torch the night before. The ingredients of the bait were modified during the course of a series of experiments and in response to farmer comments. The bait now widely used in Central America comprises metaldehyde at 1% as the only active ingredient (Cáceres *et al.*, 1986). A quantity of 5 g of the bait was shown to be sufficient to capture 50% of the *S. plebeia* present in a radius of 9 m around the trap. For a series of experiments, Cáceres *et al.* (1986) concluded that the metaldehyde-based bait was equally effective when applied in pitfall traps, under pieces of plywood, fabric sacs or cut weeds, or simply as hills in the open field. As a consequence, the recommendation now made to farmers is to apply the bait as hills, without any cover.

For extension purposes it was recommended to use 20 hills each of 5 g of metaldehyde bait for plots up to 0.5 ha. Farmers have been encouraged to set out the baits after 4 p.m. to avoid drying of the bait in the afternoon sun. Counts of the captured *S. plebeia* should be made early the next morning; counts made later in the day are likely to underestimate the 'catch' as some of the gastropods may have recovered from a sublethal dose of metaldehyde and have escaped. Moreover, small *S. plebeia* may dry in the sun and thus be difficult to see, or may be removed by fire ants (*Solenopsis* Westwood sp.) (Formicidae). The major constraint in the use of the metaldehyde bait for monitoring has been the frequent unavailability of metaldehyde in the Central American market. Also, farmers were confused over the use of the same type of bait for monitoring vaginulid populations and later for control. They usually believed that the sampling was the control measure.

Economic thresholds

In the initial efforts to develop economic thresholds, Andrews and de Mira (1983) suggested an economic injury level (EIL) of 0.25 active *S. plebeia* m^{-2} or 0.4 *S. plebeia* per baited pitfall trap per night, at the time of dry-bean germination. They observed that each active *S. plebeia* m^{-2} per night resulted in a plant stand reduction of 20% and a yield reduction of 16%. For each animal captured in the baited pitfall trap per night, the subsequent plant stand was reduced by 14% and yield by 11% unless control procedures were instigated. Later, K.L. Andrews (unpublished data) determined that infestations of 0.12 *S. plebeia* m^{-2} $night^{-1}$ during the first 22–30 days of the crop was the EIL. However, under typical *S. plebeia* infestations of > 5 m^{-2}, control measures applied at the time of crop establishment did not provide adequate protection for the dry-bean seedlings.

Using baited pitfall traps, Andrews and Lema (1986) were able to correlate *S. plebeia* populations 10 weeks in advance of the dry-bean planting with populations present at plant emergence. Rueda *et al.* (1987) determined that numbers of active *S. plebeia* averaged 0.4 m^{-2} during the dry season, 1.5 m^{-2} during the maize season and 4.8 m^{-2} during the bean-growing season. These data suggested that control measures should be applied during the maize-growing season to reduce population recruitment and thus maintain *S. plebeia* at low numbers through to the time the dry beans are planted. In a validation study of the EIL on farmers' plots with varying degrees of infestation, Portillo *et al.* (1986) concluded that each *S. plebeia* captured at the baited hills represented a potential increase in bean yield of 10%. The EIL for *S. plebeia* during the maize season was set at 1 m^{-2} or 1 per bait (at least 10 weeks before dry-bean planting). This economic injury level for *S. plebeia* was reduced to 0.5 m^{-2} or 0.5 per trap during the establishment of the bean crop. The reduction of the EIL was made because observational studies suggested that chemical baits were less effective in attracting the gastropods in the presence of bean plants than when they were used in the maize crop.

Control of *Sarasinula plebeia*

Biological control

In Central America a limited number of investigations have been made of natural control agents for *S. plebeia*. Under laboratory conditions, Pinto (1988) observed a blistery disease on the notum of *S. plebeia*. *Citrobacter freundii* (Braak) Werkman & Gillen and an unidentified Gram-negative bacterium were suggested as the causative agents following Kosh postulates. The induced mortality was less than 50%.

In Nicaragua, a population of *S. plebeia* was found to be parasitized by a mermithid nematode (*Hexamermis* Steiner sp.) (van Huis, 1981). Later, a mermithid species was found to parasitize more than 50% of *S. plebeia* in the Olancho department in Honduras, but only 10% of the affected animals died under laboratory conditions. The parasitized *S. plebeia* had reduced growth and reproduction rates, relative to mermithid-free animals, but, after the mermithids (20 cm long) emerged, affected *S. plebeia* started to gain weight and reproduce (O. Ramírez, personal communication, 1985).

In El Paraiso, Honduras, A. Rueda (unpublished results, 1985) found the larvae of an unidentified lampyrid species (Coleoptera: Lampyridae) feeding on *S. plebeia* in the field. In the laboratory, this lampyrid was shown to be capable of killing *S. plebeia*, but the rate of predation was low at one *Sarasinula* every 5 days.

S. plebeia has been considered a recent introduction into Central America (Andrews, 1987a), where it has become a serious pest of dry beans and other crops. This situation contrasts sharply with its low abundance and status as a minor pest in South America, Florida and the Caribbean. Therefore it was considered possible that candidate biological-control agents may exist in these areas of low abundance. Bennett and Andrews (1985) and Bennett and Yaseen (1987) searched for natural enemies of Vaginulidae in Trinidad, West Indies, with a view to their introduction into Honduras. A nematode (*Panagrolaimus* Fuchs sp.; Panagrolaimidae) was isolated from vaginulids that had died in the laboratory, but this species could not be demonstrated to be parasitic or pathogenic in subsequent trials. The adult stage of *Scarites orientalis* (Fabricius), *Athrostictus* Bates sp. and *Pheropsophus aequinotialis* (Linnaeus) (Coleoptera: Carabidae) fed on *S. plebeia*, but breeding colonies of these beetles could not be established under laboratory conditions.

Exploration in Brazil, Bolivia, Colombia, Puerto Rico and Jamaica (F.D. Bennett, personal communication, 1989) and Florida, USA (Rueda, 1989a), yielded no evidence of attack on vaginulid species by natural enemies in the field. More than 1000 vaginulids collected from the Americas and observed under laboratory conditions for several generations similarly did not yield any natural enemies. Under laboratory conditions in Florida, Rueda (1989a) conducted a series of inoculation tests with bacteria, fungi, protozoa, free-living and entomogenous nematodes, carabid beetles, phorid flies, predatory gastropods and turbellarian flatworms. Again, no viable biological-control candidates were found.

Chemical control

With the invasion of dry-bean crops by *S. plebeia* in the 1970s, farmers adopted chemical control as the first and, in many cases, as the only control measure to protect their bean plantings. In a large extension campaign in Honduras, the government distributed a premixed chemical bait at a subsidized price. This bait was comprised of bran, molasses, metaldehyde and carbaryl (Barletta, 1987). Research on chemical control of *S. plebeia* was initiated in El Salvador by Mancía (1971), who recommended a bait with carbaryl as the active ingredient. Meneses (1985) tested and then recommended the use of granular applications of mephosfolan, but the product was discarded when a survey showed that more than 70% of farmers using this product exhibited symptoms of chemical poisoning.

From 1983 to 1986 a series of tests were made by the personnel of the malacology programme at Zamorano, Honduras, to optimize the recommendations for *S. plebeia* control. Based on a literature review and preliminary experiments in Central America, Andrews (1985) suggested that the basis for chemical control should be baits, but the type, active ingredient, dosages and application method of the baits to control *S. plebeia* needed to be developed. Sobrado *et al.* (1986) carried out two experiments to determine the effect of metaldehyde, carbaryl and their mixtures as active ingredients in a bait. In the first experiment, *S. plebeia* preferred the bait containing only metaldehyde as the active ingredient. In the second experiment under field conditions, baits with only metaldehyde killed three times the number of *S. plebeia* than did baits containing carbaryl alone and 1.5 times those with a mixture of active ingredients. Portillo *et al.* (1987) tested the effects on *S. plebeia* of several insecticides and herbicides that farmers were applying to bean foliage to control vaginulids and other pests. The only product that showed some activity as a molluscicide was carbaryl, but *S. plebeia* mortality was low and its use was not recommended.

By 1986 the most common bait formulation used in Honduras was a loose bait prepared with wheat (*Triticum aestivum* Linnaeus; Gramineae) bran (Fig. 4.7A). However, in the field the bait lost its efficacy in a day or two. Experiments with chemical preservatives demonstrated that calcium propanate at 1% improved the bait efficacy (Andrews *et al.*, 1986). The wheat bran used as the carrier was at times difficult to obtain on the market. For this reason, Andrews *et al.* (1986) compared locally available substitutes. They found that baits prepared from hay, rice shells and maize provided the same level of control as those prepared from wheat bran.

Fig. 4.7. A. *Sarasinula plebeia* (Fischer) (Vaginulidae) killed after ingesting a metaldehyde bait. B. Trash trap. Farmer killing vaginulids that had hidden under weed residues raked into hills.

Farmers were initially advised to apply bait in hills at the rate of 10 g m^{-2} (100 kg ha^{-1}) (Meneses, 1985). Field trials by Andrews (1985) indicated that effective control could be achieved with bait hills of 2 g every 4 m^2. For extension purposes the chemical bait recommendations were as follows: for application under maize plants before dry beans are planted, a bait pinch of 2 g every two steps in alternate rows (5 kg ha^{-1})

(Andrews and Rueda, 1987a); at the time of planting dry beans, a bait pinch of 2 g every step in every row of maize (20 kg ha^{-1}) (Andrews and Rueda, 1987b).

From 1975 to 1987 the government of Honduras imported metaldehyde to prepare the bait and sold it at a subsidized price to farmers. Chemical companies did not introduce any product to the country at that time because their products could not compete with the subsidized government baits. After the government stopped the direct importation of metaldehyde into the country, agrochemical companies introduced metaldehyde baits in a pellet formulation. By the 1997 dry-bean growing season, more than 13,000 kg of bait were sold in Honduras (F. Barahona, personal communication, 1998).

Coto and Saunders (1985) evaluated 60 species of exotic plants in the laboratory for control or feeding deterrence in *D. occidentalis*. Infusions and extracts were made with various solvents and applied in various doses to bean seedlings. Extracts of *Thevetia peruviana* (Persoon) Schumacher (Apocynaceae) leaves and jack-bean seeds provided 100% protection of dry-bean seedlings. Cantoral (1986) and Cordón (1986) reported that *Nerium oleander* Linnaeus (Apocynaceae) and *Jacquinnia macrocarpa* Cavanilles (Theophrastaceae) had some feeding deterrent effect on *S. plebeia*. Later, Sabillón *et al.* (1991) demonstrated that *N. oleander*, *T. peruviana*, *Solanum globiferum* Dunal (Solonaceae) and *Parthenium hysterophorus* Linnaeus (Asteraceae) deterred feeding by *S. plebeia* when extracts were applied to bean seedlings. None of the botanical products tested in these studies worked as molluscicides. To date, none of these plant extracts have been evaluated under field conditions.

Mechanical and cultural control

Soil preparation

S. plebeia populations have been found to be much higher in the maize–bean relay planting system than in monocropped dry beans planted in ploughed land (Fisher *et al.*, 1986). Soil preparation apparently destroys the vaginulids and their eggs mechanically and through exposure to the desiccating sun and to predators. However, most farmers in Central America do not use machinery to prepare their land because of the high cost and the topography of the fields.

Pitty and Andrews (1990) summarized the effects of tillage treatments, applied to the maize–bean relay, on *S. plebeia* in experimental plots at Zamorano, Honduras. In the first 3 years following conversion to zero tillage, *S. plebeia* abundance was greater than in plots with conventional tillage. However, in the fourth and subsequent years, populations were higher in conventional tillage plots. They attributed the initial differences to the absence of the mechanical disturbance and the abundance of broad-leaved weeds in the zero-tillage plots. The reduction of *S. plebeia* under zero tillage after the fourth year was attributed to a possible establishment of predators or pathogens, and the possible cumulative effect of chemical baits applied on those plots in the preceding 3 years.

Manual killing

One of the earliest recommendations to farmers was to search out and kill vaginulids at night. Farmers walked the whole plot in search of the pests with the help of a torch, and every *S. plebeia* found was killed with a stick or a machete. Some farmers preferred to collect the animals to show them to the neighbours and then kill them with salt. Sobrado and Andrews (1985) demonstrated that this control technique resulted in a threefold increase in bean yield compared with doing nothing.

Trash traps

A control technique that was based on the behaviour of *S. plebeia* is the trash trap. In this technique the plant residues obtained from hand weeding the maize fields are raked into hills of at least 30 cm × 30 cm × 30 cm and left as a daytime refuge for the *S. plebeia*. Farmers then destroyed the animals manually twice a week using a machete (Fig. 4.7B). Fisher *et al.* (1986) obtained an average bean-yield increment of 44% from this control practice in a multisite validation experiment in Honduras. Miranda *et al.* (1997) reported that Nicaraguan growers captured more than 6000 *S. plebeia* and 4 kg of eggs ha^{-1}, an estimated 95% reduction in the population, utilizing the trash-trap technique. If the removal of *S. plebeia* from these trash traps was infrequent, numbers of the pest tended to increase in the fields due to their reproduction within the trash, thus negating any benefit of this control technique.

Quick fire

Some farmers prepare the land for relay planting of the dry beans by setting fire to weeds and the dry leaves and tassels stripped from the mature maize plants. Following the fire, the land is raked clean to facilitate the manual sowing of the bean crop. During the first 2 weeks after the fire, Secaira *et al.* (1986) found that the abundance of active *S. plebeia* was reduced by 50% in burned plots compared with unburned plots. By the third week there were no differences in pest numbers between burned and control plots. *S. plebeia* cadavers were not found after the fire, and it is possible that the temporary suppression of vaginulid activity effected by the quick fire was due to the change in the microclimate forcing the animals to remain inactive underground. Another explanation is the possible recolonization of *S. plebeia* from adjacent plots. Secaira *et al.* (1986) thus recommended that the dry beans be planted immediately after the fire, to take advantage of the reduced vaginulid activity.

Errázuris (1985) reported that 63% of farmers in Honduras used this technique after they were given extension talks on this topic in 1985. These researchers found, however, that the nitrogen content of the maize plants declined by 10% as a consequence of the fires, and recommedation of this practice was discontinued because it conflicted with soil and nutrient conservation practices favoured by farmers.

Weed control

Andrews *et al.* (1985b) and Rueda (1989a) showed that *S. plebeia*, while polyphagous, has a strong preference for broad-leaved plants, many of which are common weeds in Central America. *S. plebeia* does not feed on maize or other grasses, even when other food is not available. Studies of *S. plebeia* population dynamics in Honduras (Andrews and Lema, 1986) corroborated the results of the laboratory feeding-preference trials. The highest populations were found in fields in which broad-leaved weeds were abundant. Since 1984, chemical control of broad-leaf weeds during the maize growth period has been advocated as an important means of minimizing *S. plebeia* damage in relay-planted dry beans (Fisher *et al.*, 1986; Andrews and Rueda, 1987a). Fisher *et al.* (1986) reported bean yield increments of 42% on farms that used herbicide treatment in the maize. Del Río *et al.* (1989) showed that maize fields treated with herbicides had 50% fewer vaginulids at the time of bean planting compared with fields that had no herbicide treatment.

Recent emphasis on less dependence on external inputs has resulted in the introduction of legume cover crops in the maize–bean relay systems. This is primarily to increase fertility and reduce weed infestations (Flores, 1994). Under laboratory conditions, Rizzo *et al.* (1994) found that *S. plebeia* was unable to grow or reproduce on a diet comprising the two most common legumes used as cover crops, velvet beans (*Mucuna pruriens* (Linnaeus) de Candolle) (Fabaceae) and jack beans. However, under field conditions *S. plebeia* populations increased more under a velvet-bean cover crop than in plots treated with herbicides. This suggests that the cover crop not only provides a good refuge for the vaginulids, but also does not eliminate the weeds that serve as food for these animals (Rizzo *et al.*, 1994).

Implementation of an Integrated Pest Management Programme

As a consequence of the introduction of *S. plebeia*, the area planted in dry beans had by 1975 declined by 50% in some areas of Central America. The first extension campaign to focus on *S. plebeia* control in Honduras was carried out by the Natural Resources Ministry in 1980 (Barletta, 1987). Farmers were by that time reluctant to plant dry beans because most of the crops had been destroyed by *S. plebeia* in the preceding years. The campaign was based on the application of a premixed, molluscicidal bait at the time dry beans were planted. An evaluation of the campaign concluded that most farmers achieved poor vaginulid control. The major technical problem identified was the inability of the bait to protect the highly susceptible dry beans in fields that were heavily infested with *S. plebeia*. The baits had to compete directly with the beans and broad-leaf weeds. We called these bait applications 'revenge applications' because, by the time control was achieved, the damage to the crop had already occurred and there was no compensatory response of the damaged dry beans.

The Integrated Pest Management Project in Honduras (MIPH) was initiated in 1983, based at Zamorano, Honduras. Before that date there had been virtually no systematic approach to the development of integrated pest management techniques for the maize–bean relay system in Honduras, or indeed anywhere in Central America. During the first 2 years of the project, several existing technologies were screened by a socio-economic filter to choose only those technologies that were appropriate to the resource-poor farmers of the region. These technologies were then tested in farmers' fields. Development of new technologies were only attempted when existing ones were found to provide inadequate control of the target pest (e.g. the 'revenge applications' of baits after bean-seedling emergence) or where they were unacceptable to farmers (e.g. highly labour-intensive methods) (Andrews *et al.*, 1985a).

In 1985, the 'best' technologies for control of *S. plebeia* before planting the dry beans were presented to the farmers in a menu format (Fig. 4.8). Farmers were able to choose from a range of technologies those that best fitted their particular economic resources and pest situation. The strategy behind the suggested technologies was that control of *S. plebeia* should be attempted when the populations are low due to the natural mortality during the dry season, and before the surviving vaginulids have started to reproduce in the maize. For the extension campaign the slogan 'one dead slug in the first season means 50 fewer slugs in the bean field' was used to motivate farmers to control the vaginulids during the maize-growing season. On 20 farmer cooperatives in Honduras, a demonstration plot was established to compare the projects technologies with traditional crop-management practices (Fisher *et al.*, 1986). For farmers with little capital, the manual-weeding, trash-trap and nocturnal-killing technologies were preferred, while farmers with some capital preferred the use of herbicides and chemical baits. In these demonstration plots, all the pest-management technologies evaluated reduced the number of *S. plebeia*, increased bean yields and produced an economic return to the farmer.

A large-scale study was carried out to compare different educational approaches to informing farmers about the biology and ecology of *S. plebeia* and appropriate control strategies and techniques (Barletta *et al.*, 1987). Farmers increased their knowledge of *S. plebeia* by 33–73%,

Fig. 4.8. Extension material made available to farmers in Central America, illustrating a menu of control strategies for *Sarasinula plebeia* (Fischer) (Vaginulidae) control. A. Control options to be applied in the maize (*Zea mays* Linnaeus) (Gramineae) crop, 3 months before planting dry beans (*Phaseolus vulgaris* Linnaeus) (Fabaceae). (i) Manual destruction of vaginulids found at night. (ii) Herbicide application in maize crop to reduce the abundance of broad-leaved weeds. (iii) Setting trash traps, made from raking weed residues into hills between the rows of maize. (iv) Application of molluscicidal baits for monitoring vaginulid abundance and their control. B. Control options to be applied at the time of planting dry beans. (i) Quick burning of weeds and maize leaves. (ii) Manual destruction of vaginulids found at night. (iii) Application of molluscicidal baits for monitoring vaginulid abundance and their control.

regardless of the teaching materials used in the presentations. The use of more printed or project material did not increase farmers' understanding. Moreover, the colourful cartoon extension publications (Andrews and Barletta, 1985) were modified to simple realistic drawings with short, direct, written messages (Andrews and Rueda, 1987a,b) in response to farmer feedback.

From 1987 to 1989, MIPH continued its research and extension efforts, working in pilot studies in different parts of the country (Secaira *et al.*, 1987; del Río *et al.*, 1988). In those years the main effort of MIPH was the dissemination of the information to public and private extension organizations. We estimate that more than 5000 farmers received some training in the control techniques developed by the project. This was possible by educating extension personnel through short courses, in-service training, seminars, congresses and alliances with extension organizations.

Pitty and Andrews (1990) concluded that the use of herbicides in maize was the most effective way to reduce *S. plebeia* populations, but it may not be the only technique needed to reduce vaginulid populations to a subeconomic level. They remarked that MIPH focused too much on the benefits of the herbicides for killing vaginulids, but failed to promote this practice as a means of increasing maize yields through the reduction of weed competition and the elimination of the lepidopteran maize pest, *Mocis latipes* (Guenée) (Noctuidae), which only lays its eggs on grassy weeds. They suggested that the herbicide technique should be promoted as 'the stone to kill three birds' (*S. plebeia*, weeds and *M. latipes*).

Acknowledgements

We thank Darlan Matute for assistance in preparing the line illustrations.

References

Alvarado, J.A. (1939) *Los Insectos Dañinos y los Insectos Auxiliares de la Agricultura en Guatemala*, 1st edn. Tipografía Nacional, Guatemala City, Guatemala, 301 pp.

Andrews, K.L. (1983) Trampa para determinar la densidad poblacional de la babosa, *Vaginulus plebeius*, plaga de frijol común. *Turrialba* 33, 209–211.

Andrews, K.L. (1985) Control químico de babosas especialmente la babosa del frijol. *Ceiba* 26, 90–102.

Andrews, K.L. (1987a) La importancia de las babosas veronicéllidos en Centro América. *Ceiba* 28, 149–153.

Andrews, K.L. (1987b) Técnicas de muestreo para la determinación de la densidad poblacional y actividad de las babosas veronicéllidos. *Ceiba* 28, 209–228.

Andrews, K.L. and Barletta, H. (1985) *Los Secretos de la Babosa. Parte II. Control en primera.* Extension Bulletin, Escuela Agrícola Panamericana, Tegucigalpa, Honduras, 12 pp.

Andrews, K.L. and de Mira, A.H. (1983) Relación entre densidad poblacional de la babosa *Vaginulus plebeius* y el daño en frijol común, *Phaseolus vulgaris*. *Turrialba* 33, 165–168.

Andrews, K.L. and Dundee, D.S. (1987) Las babosas veronicéllidos de Centro América con énfasis en *Sarasinula plebeia* (= *Vaginulus plebeius*). *Ceiba* 28, 163–172.

Andrews, K.L. and Lema, F. (1986) Dinámica poblacional de la babosa, *Vaginulus plebeius*, en lotes de maíz–frijol en relevo. *Turrialba* 36, 77–80.

Andrews, K.L. and López, J.G. (1987) Comportamiento nocturno de la babosa. *Ceiba* 28, 193–199.

Andrews, K.L. and Rueda, A. (1987a) *Control de la Babosa en Primera.* Publicación MIPH-EAP No. 126, Extension Bulletin, Escuela Agrícola Panamericana, Tegucigalpa, Honduras, 20 pp.

Andrews, K.L. and Rueda, A. (1987b) *Control de la Babosa en Postrera.* Publicación MIPH-EAP No. 137, Extension Bulletin, Escuela Agrícola Panamericana, Tegucigalpa, Honduras, 7 pp.

Andrews, K.L., Rueda, A., Fisher, R. and Barletta, H. (1985a) *Progreso del Proyecto MIPH en la Validación y Transferencia de Tecnologías para Productores de Maíz y Frijol en Honduras.* Publicación MIPH-EAP No. 63, Extension Bulletin, Escuela Agrícola Panamericana, Tegucigalpa, Honduras, 7 pp.

Andrews, K.L., Valverde, V.H. and Ramírez, O. (1985b) Preferencia alimenticia de la babosa, *Sarasinula plebeia* (Fischer). *Ceiba* 26, 59–65.

Andrews, K.L., López, J., Fisher, R. and Rueda, A. (1986) Avances en el control químico de babosas veronicéllidos en Honduras. In: *Memorias del IV Congreso de la Asociación Guatemalteca de Manejo Integrado de Plagas (AGMIP).* Guatemala City, Guatemala, pp. 143–154.

Andrews, K.L., López, J. and Rueda, A. (1987) Efecto de la humedad del suelo en la sobrevivencia de babosas, *Sarasinula plebia* (sensu-lato), durante la época seca. In: *Memoria IV Congreso de la Asociación Guatemalteca de Manejo Integrado de Plagas (AGMIP).* Guatemala City, Guatemala, pp. 129–134.

Anon. (1953) cited in Martínez, A. (1965) *Plagas Agrícolas de Cuba.* Empresa consolidada de Artes Gráficas, Havana, Cuba, 123 pp.

Anon. (1960) *Manual de Recomendaciones para el Cultivo de Productos Agrícolas en Costa Rica.* Boletín Técnico No. 32, Ministerio de Agricultura e Industrias, San José, Costa Rica, 23 pp.

Anon. (1965) *Cultivos Agrícolas de Costa Rica. Manual de Recomendaciones en Tabaco.* Ministerio de Agricultura y Ganadería, San José, Costa Rica, 95 pp.

Anon. (1976) Manual para el control de la babosa. In: *Proyecto Piloto de Maíz y Frijol.* Secretaria de Recursos Naturales, Tegucigalpa, Honduras, pp. 1–6.

Anon. (1981) Programa de actividades anuales de la División de Sanidad Vegetal para el presupuesto de 1981–1982. In: *Fotocopia del Documento Interno del Instituto Nicaragüense de Tecnología Agropecuaria (INTA)*, Managua, Nicaragua, pp. 80–93.

Barletta, H. (1987) Campaña nacional para el control de la babosa del frijol en Honduras en 1980. *Ceiba* 28, 285–304.

Barletta, H., Andrews, K.L. and Rueda, A. (1987) Dos modelos de comunicación para el control de la babosa del frijol, *Sarasinula plebeia* (Fischer) *sensu lato* en Honduras. In: *Memorias Resúmenes de la XXXIII Reunión Anual del Programa Cooperativo Centroamericano para el Mejoramiento de Cultivos Alimenticios (PCCMCA), Guatemala City, Guatemala*, p. 126.

Bennett, F.D. and Andrews, K.L. (1985) El control biológico clásico de veronicellidos en Centro América. Una Propuesta. *Ceiba* 26, 77–82.

Bennett, F.D. and Yassen, M. (1987) Investigations for possibilities of biological control of slugs in Honduras. *Ceiba* 28, 229–234.

Caballero, R. and Andrews, K.L. (1989) Daño causado por la babosa *Sarasinula plebeia* (Fischer), en diferentes etapas fenológicas del cultivo de frijol, *Phaseolus vulgaris* L. Publicación MIPH-EAP 211. In: *Memoria XXXV Reunión Anual del Programa Cooperativo Centroamericano para el Mejoramiento de Cultivos Alimenticios (PCCMCA), San Pedro Sula, Honduras*, p. 47.

Caballero, R., Thomé, J.W., Andrews, K.L. and Rueda, A. (1991) Babosas de Honduras (Soleolifera: Veronicellidae): biología, ecología, distribución, descripción, importancia económica, y claves para su identificación. *Ceiba* 32, 107–125.

Cáceres, O., Andrews, K.L. and Taylor, K. (1986) Estudio de trampas alternativas para el monitoreo de la babosa del frijol *Sarasinula plebeia* (Fischer) *sensu latu*. In: *Memorias Resúmenes de la XXXII Reunión Anual del Programa Cooperativo Centroamericano para el Mejoramiento de Cultivos Alimenticios (PCCMCA), San Salvador, El Salvador*, p. L-1.

Cantoral, O. (1986) Evaluación de diferentes técnicas de control de la babosa (Veronicellidae) en el cultivo del frijol, en San José La Aranda, Chiquimula. Thesis, Universidad de San Carlos de Guatemala, Guatemala City, Guatemala.

Céspedes, R., Salas, J., Mekbel, S., Troper, L., Mullner, F. and Morera, P. (1967) Granulomas entéricos y linfáticos con intensa eosinofilia tisular producidos por un estrongilídeo (*Strongylata*). *Acta Médica Costarricence* 10, 235–255.

Cockerell, T.D.A. (1890) Notes on the slugs, chiefly in the collection at the British Museum. *Annals and Magazine of Natural Hist*ory 6, 380–390.

Cockerell, T.D.A. (1895) Notes on the species of *Veronicella* found in Central América. *The Nautilus* 8, 140–143.

Cordón, F.A. (1986) Evaluación de 14 tratamientos para el control de la babosa (Mollusca; Veronicellidae) en el cultivo de frijol, San Jacinto, Chiquimula. Ing. Agr. thesis. Universidad de San Carlos de Guatemala, Guatemala City, Guatemala.

Córdova, W. (1981) *Informe del Seminario-taller Contra la Babosa in 1981.* Fotocopia del reporte interno, Ministerio de Recursos Naturales, Tegucigalpa, Honduras, 18 pp.

Coto, D.A. and Saunders, J.L. (1985) Prevención alimenticia de la babosa, *Diplosolenodes occidentale*, Soleolifera: Veronicellidae, con repelentes botánicos. *Ceiba* 26, 66–76.

del Río, L., Andrews, K.L. and Martínez, B. (1988) Evaluación de diversas modalidades de transferencia de tecnología para el manejo de *Sarasinula plebeia* (*sensu latu*) y su importancia en la extensión integral. In: *Memoria de la Sexta Semana Científica UNAH, Tegucigalpa, Honduras*, p. 12.

Del Río, L., Rubio, J. and Andrews, K.L. (1989) *Dinámica Poblacional de la Babosa del Frijol* Sarasinula plebeia *Fischer, en tres Sistemas de Manejo de la Plaga en Honduras.* Publicación MIPH-EAP No. 206, Extension Bulletin, Escuela Agrícola Panamericana, Tegucigalpa, Honduras, 9 pp.

Duarte, Z., Morera, P. and Vuong, P.N. (1991) Abdominal angiostrongyliasis in Nicaragua: a clinico-pathological study on a series of 12 cases reports. *Annales de Parasitologie Humaine et Comparé* 66, 259–262.

Duarte, Z., Morera, P., Dávila, P. and Gantier, J.C. (1992) *Angiostrongylus costaricensis* natural infection in *Vaginulus plebeius* in Nicaragua. *Annales de Parasitologie Humaine et Comparé* 67, 94–96.

Errázuris, C.F.L. (1985) *Estudio del Proceso de Transferecia y Adopción de Tecnología en dos Cooperativas Campesinas en Honduras.* Escuela Agrícola Panamericana, El Zamorano, Honduras, 29 pp.

Fisher, R.W., Andrews, K.L., Rueda, A. and Sobrado, C.E. (1986) Impacto económico de prácticas culturales y químicos en el control de la babosa del frijol *Sarasinula plebeia* (*sensu lato*), en Honduras. In: *Memoria del IV Congreso de la Asociación Guatemalteca de Manejo Integrado de Plagas (AGMIP).* Guatemala City, Guatemala, pp. 136–142.

Flores, M. (1994) The use of legumes cover crops in traditional farming systems in Central America. In: Thurston, H.D., Smith, M., Abawi, G. and Kearl, S. (eds) *Tapado. Slash/Mulch: How Farmers Use it and what Researchers Know about it.* Cornell International Institute for Food, Agriculture and Development (CIIFAD), Ithaca, New York, pp. 149–157.

Graeff Teixeira, C., Thiengo, S., Thomé, J.W., Bueno Medeiros, A., Camillo-Courra, L. and Agostini, A. (1993) On the diversity of mollusc intermediate hosts of *Angiostrongylus costarricensis* Morera and Céspedes 1971 in southern Brazil. *Memorias do Instituto Oswaldo Cruz* 88, 487–489.

Graeff Teixeira, C., Pinto, V., Busato Junior, E. and Agostini, A. (1994) Natural infection of *Phyllocaulis soleiforms* with larvae morphologically similar to L2 of *Angiostrongylus costarricensis. Memorias do Instituto Oswaldo Cruz, Rio de Janeiro* 89, 121.

Heynemann, D.F. (1885) Die Nackten Landpulmonaten des Erdbodens. *Jahrbücher der Deutschen Malakozooligischen Gesellschaft* 12, 236–330.

Hulbert, T.V., Larsen, R.A. and Chandrasoma, T.P. (1993) Abdominal angiostrongyliasis mimicking acute appendicitis and Meckel's diverticulum: report of a case in the United States and review. *Clinical Infectious Diseases* 14, 836–840.

Hunter, P.J. (1968) Studies on slugs on arable ground. I Sampling methods. *Malacología* 17, 161–164.

Kaminsky, R.G. (1997) Situación actual de *Angiostrongylus costaricensis* y la infeccción en humanos y animales en las Américas. *Revista Médica Hondureña* 64, 139–147.

Kaminsky, R.G., Caballero, R. and Andrews, K.L. (1995) Presencia de *Angiostrongylus costaricensis* en Honduras y sus relaciones agro-ecológicas y humanas. *Parasitología al Día* 19, 81–90.

Loría-Cortés, R. and Lobo-Sanahuja, J.F. (1980) Clinical abdominal angiostrongylosis: a study of 116 children with intestinal eosinophilic granuloma caused by *Angiostrongylus costaricensis. American Journal of Tropical Medicine and Hygiene* 29, 538–544.

Malek, E. (1981) Presence of *Angiostrongylus costaricensis* Morera and Cespedes 1971 in Colombia. *American Journal of Tropical Medicine and Hygiene* 30, 81–83.

Mancía, J.E. (1971) Combate de la babosa del frijol *Vaginulus plebeius* (Fischer) en El Salvador. Documento de discusión. In: *Memoria XVIII Reunión Anual del Programa Cooperativo Centroamericano para el Mejoramiento de Cultivos Alimenticios (PCCMCA), 2–6 marzo, 1971, Panamá, Panamá*, pp 46–61.

Meneses, R.R. (1985) Control químico de la babosa (*Sarasinula plebeia*) en El Rosario, Comayagua, Honduras, 1982. *Ceiba* 26, 103–109.

Miranda, P.A., Inestroza, J.R., Torres, G. and Pérez, M. (1997) Utilización del rastrojo como trampa para babosa (*Sarasinula plebeia*). In: Cáceres, O., Pilarte, F. and Tondeur, F. (eds) *Memorias del Primer Congreso Regional de las Segovias de Manejo Integrado de Plagas.* Proyecto MIP/Zamorano/ COSUDE, Estéli, Nicaragua, 18 pp.

Morera, P. (1973) Life history and redescription of *Angiostrongylus costaricensis* Morera and Céspedes, 1971. *American Journal of Tropical Medicine and Hygiene* 22, 613–621.

Morera, P. (1987) Angiostrongilosis abdominal. ¿Un problema de salud pública? *Revista de la Asociación Guatemalteca de Parasitología Medicina Tropical* 2, 98–111.

Morera, P., Pérez, F., Mora, F. and Castro, L. (1982) Viceral larva migrans-like syndrome caused by *Angiostrongylus costaricensis. American Journal of Tropical Medicine and Hygiene* 31, 67–70.

Pinto, C.P.C. (1988) Microorganismos asociados con la babosa común del frijol *Sarasinula antillarum* (Becker) que causan mortalidad en el laboratorio. Thesis, Escuela Agrícola Panamericana, Tegucigalpa, Honduras.

Pitty, A. and Andrews, K.L. (1990) Efecto del manejo de malezas y labranza sobre la babosa del frijol. *Turrialba* 40, 272–277.

Portillo, H., Rueda, A. and Andrews, K.L. (1986) Comprobación de un nivel crítico para la babosa del frijol, *Sarasinula plebeia* (*sensu lato*), en Honduras. In: *Memoria del IV Congreso de la Asociación Guatemalteca del Manejo Integrado de Plagas (AGMIP).* Guatemala City, Guatemala, pp. 136–140.

Portillo, H., Andrews, K.L., Valverde, V.H., Rueda, A. and Wheeler, G. (1987) Prueba de campo de la toxicidad de algunos plaguicidas sobre poblaciones de la babosa del frijol. *Ceiba* 28, 235–238.

Rizzo, R.B., del Río, L., Rueda, A. and Pitty, A. (1994) Effect of using two diets based on leguminous cover crops on weight gain and reproductive capacity of the slug *Sarasinula plebeia* Fischer. In: Thurston, H.D., Smith, M., Abawi, G. and Kearl, S. (eds) *Tapado. Slash/Mulch: How Farmers Use it and what Researchers Know about it.* Cornell International Institute for Food, Agriculture and Development (CIIFAD), Ithaca, New York, pp. 109–114.

Rueda, A. (1989a) Biology, nutritional ecology and natural enemies of the slug *Sarasinula plebeia* (Fischer, 1868) (Soleolifera: Veronicellidae). MSc thesis, University of Florida, Gainesville, Florida.

Rueda, A. (1989b) Artificial diet for laboratory maintenance of the veronicellid slug, *Sarasinula plebeia* (Fischer). In: Henderson, I.F. (ed.) *Slugs and Snails in World Agriculture.* Monograph No. 41, British Crop Protection Council, Thornton Heath, pp. 361–366.

Rueda, A., Valdivia, A. and Andrews, K.L. (1987) Dinámica poblacional de la babosa del frijol *Sarasinula plebeia* (Fischer) *sensu-latu* en Danlí, el Paraíso, Honduras. In: *Memorias Resúmenes de la XXXIII Reunión Anual del Programa Cooperativo Centroamericano para el Mejoramiento de Cultivos Alimenticios (PCCMCA), Guatemala City, Guatemala*, p. 118.

Rueda, A., Slansky, F. and Wheeler, G.S. (1991) Compensatory feeding response of the slug *Sarasinula plebeia* to dietary dilution. *Oecologia* 88, 181–188.

Ruíz, P.J. and Morera, P. (1983) Spermatic artery obstruction caused *by Angiostrongylus costaricensis. American Journal of Tropical Medicine and Hygiene* 32, 1958–1959.

Habitat

Ampullariids are freshwater inhabitants. Some may be able to tolerate low levels of salinity (Prashad, 1925b; Hunt, 1961; Santos *et al.*, 1987; Fujio *et al.*, 1991a), but they generally do not live in brackish-water habitats. Most species are amphibious, able to spend significant lengths of time out of water breathing air. Many species, especially species of *Pomacea*, *Marisa*, *Pila* and *Lanistes*, inhabit slow-moving or stagnant water in low-land swamps, marshes, ditches, lakes and rivers (e.g. Pain, 1950a, 1960; Andrews, 1965b; Robins, 1971; Louda and McKaye, 1982; Keawjam, 1986). Some species could be considered preadapted for living in rice paddies, taro (*Colocasia esculenta* (Linnaeus) Schott; Araceae) patches and other similar artificial habitats in which aquatic crops are grown. There may be differences in habitat among closely related species. For instance, in Argentina, Scott (1957) reported that *Pomacea canaliculata* (de Lamarck) inhabited relatively still water, while the almost indistinguishable *Pomacea insularum* (d'Orbigny) was found in rivers. In Lake Malawi, five species of *Lanistes* occupy marshy areas and the lake edges, with the lake-edge species being generally found at slightly different depths (Louda and McKaye, 1982). In Thailand, the five native species of *Pila* have overlapping distributions but slight differences in habitat preferences (Keawjam, 1986). The Asian genus *Turbinicola* Annandale & Prashad (treated as a synonym of *Pila* by Berthold (1991)) is found in fast-flowing hill streams, and exhibits a somewhat more terrestrial existence than other ampullariids (Prashad, 1925a; Andrews, 1965b; Keawjam, 1986; Berthold, 1991). *Limnopomus* Dall, a subgroup of *Pomacea* treated as a distinct genus or subgenus by some authors but synonymized by Berthold (1991), also inhabits swiftly flowing mountain streams (e.g. Pain, 1950a,b, 1960).

Reproduction, growth and demographics

Breeding system

Ampullariids are dioecious, internally fertilizing and oviparous (not reciprocally fertilizing hermaphrodites as stated by Chang (1985)). There is evidence that females are larger than males, at least in some species (Prashad, 1925b; Robins, 1971; Keawjam, 1987; Marwoto, 1988; Lum-Kong and Kenny, 1989; Cazzaniga, 1990a; Perera and Walls, 1996; Wada, 1997; Estebenet and Cazzaniga, 1998). The extent of dimorphism may vary among populations within species: it may be slight (Cazzaniga, 1990a) or dramatic (Fig. 5.1); and it is possible that this has caused considerable nomenclatural and taxonomic confusion. Preliminary study of *P. canaliculata* in Hawaii found no size dimorphism and an approximately 1 : 1 sex ratio in the wild, although in the laboratory females grew faster than males under certain feeding regimes (H. Ako and T. Nishimura, Honolulu,

Fig. 5.1. A mating pair of *Pomacea canaliculata* (de Lamarck) (Ampulariidae), collected by the author from a domestic rearing facility in Phnom Penh, Cambodia, in November 1995; female on the left, shell height 84 mm.

personal communication, 1997). In addition to size dimorphism there appears to be some slight dimorphism in shape of the aperture and operculum. In *P. canaliculata*, females have a broader shell aperture (Cazzaniga, 1990a) and a concave operculum (convex in males) (Adalla and Morallo-Rejesus, 1989; Anon., 1989; Guerrero, 1991; Schnorbach, 1995) (H. Ako (personal communication, 1997), says it is the other way round: females have convex opercula). In *M. cornuarietis*, the shell aperture of males is said to be more round compared to the more oval shape in females, and males are said to produce thicker, more heavily ridged shells (Robins, 1971). Species of *Pila*, and perhaps *Pomacea*, have been reported to change sex (Keawjam, 1987; Keawjam and Upatham, 1990). The sex change is from male to female (protandry) and takes place during aestivation (*Pila*) or without aestivation (*Pomacea*). The larger size of females in *Pila* has therefore been attributed to age, with continued growth following protandric sex change (Keawjam, 1987). The ubiquity and significance of this phenomenon need further investigation; it was not reported by Andrews (1964) in her detailed account of the functional anatomy and development of the reproductive system in *P. canaliculata* or by Estebenet and Cazzaniga (1998) in their study of dimorphism in the same species.

Mating, oviposition, eggs and fecundity

Breeding in many species is seasonal and apparently related to latitude, temperature and rainfall (Andrews, 1964). In equatorial regions, many species aestivate during the dry season as their habitats dry up (see below), breeding in the rainy season; in subtropical regions, ampullariids may only breed during summer, once temperatures reach a certain level (Scott, 1957; Andrews, 1964). Local variation in reproductive regime may be related to local climatic variation, especially availability of water (Bourne and Berlin, 1982).

Species of *Pomacea* generally lay their eggs above water on exposed substrates, such as vegetation and rocks, perhaps to avoid aquatic predators or in response to low oxygen tension in their often near-stagnant aquatic habitats (Snyder and Snyder, 1971). The eggs are individually enclosed in a calcium carbonate shell, which may or may not be used as a source of calcium for the developing embryo (Andrews, 1964; Tompa, 1980; Turner and McCabe, 1990). The eggs of *Pomacea scalaris* (d'Orbigny) have been reported to lack the calcareous shell and to be laid under water (Scott, 1957). In fact, as for other species of *Pomacea*, they are laid out of water and do have a calcareous coating; they are salmon-pink (A. Castro-Vazquez, personal communication, 1997).

The eggs of *Pila* species are also laid out of water, but in depressions made by the gravid females on banks or mud-flats (Michelson, 1961; Andrews, 1964; Keawjam, 1986). They have a calcareous coating (Prashad, 1925b; Keawjam, 1986). In India, *Pila* begin laying their eggs at the start of the rainy season (Prashad, 1925b; Andrews, 1964). The eggs of *Turbinicola* have a calcareous coating and are laid out of water attached to stones (Prashad, 1925a). Those of *Marisa*, *Lanistes* and *Asolene* lack the calcareous coating (Keawjam, 1986) and are deposited under water on submerged vegetation or other surfaces (Michelson, 1961; Robins, 1971; Albrecht *et al.*, 1996).

In *P. canaliculata*, and indeed other *Pomacea* species (e.g. *Pomacea dolioides* (Reeve) (van Dinther, 1956); *Pomacea haustrum* (Reeve) (Guimarães, 1981a,b); *Pomacea paludosa* (Perry, 1974)), oviposition (above water) takes place predominantly at night or in the early morning or evening (Andrews, 1964; Chang, 1985; Halwart, 1994a; Schnorbach, 1995; Albrecht *et al.*, 1996), about 24 h after copulation (up to 2 weeks after mating, according to Chang (1985)). Andrews (1964) and Albrecht *et al.* (1996) described copulation in *P. canaliculata* in detail; it occurs at any time of the day or night (Naylor, 1996; R.H. Cowie, personal observations), although there may be some diurnal rhythm, and it takes 10–18 h. Individuals mate repeatedly, about three times per week (Albrecht *et al.*, 1996). On each oviposition occasion, a single clutch is laid, of highly variable egg number (Table 5.1). The interval between successive ovipositions has been reported as 12–14 days (Chang, 1985) and about 5 days (Albrecht *et al.*, 1996) for *P. canaliculata* and 8–16 days for an unidentified '*Ampularius* sp.', presumably *P. canaliculata* (Lacanilao, 1990). Hatching generally takes place about 2 weeks after oviposition, but this period varies greatly (Table 5.1). Newly hatched *Pomacea* immediately fall or crawl into the water. On average, one animal can produce 4375 (maximum observed 8680) eggs per year (Mochida, 1988a, 1991), which, if clutch size is about 200 (Table 5.1), translates into about 22 clutches (see also Anon. (1989), which gave an even higher figure of up to 1200 eggs per month). Development is highly dependent on temperature (e.g. Robins, 1971; Demian and Yousif, 1973; Aldridge, 1983; Mochida, 1988a,b; Estebenet and Cazzaniga, 1992; Schnorbach, 1995), and therefore locality, which probably largely accounts for the variability in the data for *P. canaliculata* in Table 5.1. The eggs of *M. cornuarietis* take 8 days to hatch at 25–30°C and 20 days at 15–20°C; those of *Pila globosa* (Swainson) take 10–14 days at 32–38°C and 3 weeks at 21–27°C (Demian and Yousif, 1973).

Table 5.1. Reproduction and growth in Ampullariidae. Asterisks indicate laboratory studies. Some data have been reported without indicating whether they were obtained from laboratory cultures or from the wild. Some of the references listed undoubtedly simply reiterate data from others that are also listed. Some, at least, of the variability exhibited (especially by *Pomacea canaliculata*) may be due to taxonomic confusion (see text).

Species	Locality	Clutch size	Hatchability (%)	Time to hatching	Time from hatching to maturity	Longevity	References
Pomacea canaliculata (de Lamarck)							
	Argentina	120–478					Cazzaniga and Estebenet (1988)
	Argentina			28 days			Scott cited by Demian and Yousif (1973)
	Argentina*			12–15 days			Thiengo *et al.* (1993)
	Argentina*				7 months–2 years	13 months–4 years	Estebenet and Cazzaniga (1992)
	Argentina	Mean 101	43				Albrecht *et al.* (1996)
	Argentina*				100–150 days		Estebenet and Cazzaniga (1998)
	Philippines	25–320		10–15 days	60 days	2–3 years	Adalla and Morallo-Rejesus (1989)
	Philippines	25–500		7–14 days	59–84 days	119 days– > 3 years	Anon. (1989); Olivares *et al.* (1992)
	Philippines	25–500		10–15 days	60–85 days	2–3 years	Halwart (1994a)
	Philippines*	50–400	20.9–35.0	9–12 days			Lacanilao (1990) (as '*Ampularius* sp.')
	Philippines	50–500		10–15 days	< 1 year		Guerrero (1991)
	Philippines	25–500	80	8–15 days			Rondon and Callo (1991)
	Philippines		80				Bombeo-Tuburan *et al.* (1995)
	Philippines	25–320	90–100	8–21 days		3 years	Schnorbach (1995)
	Taiwan*	200–300	60	Average 12.4 days	55 days		Chang (1985)
	Japan*	Average 200–700	41.9	12 days	6 months		Fujio and von Brand (1990); Fujio *et al.* (1991a,b)
	Japan	30–700	7–90		2 months	> 2 years	Wada (1997)
	Thailand	800–1000					Keawjam and Upatham (1990)

Table 5.1. (*Cont'd*)

Species	Locality	Clutch size	Hatchability (%)	Time to hatching	Time from hatching to maturity	Longevity	References
	Asia	321 (average)	7–90	9–37 days	2–3 months	2–5 years	Halwart (1994a); Mochida (1988a,b, 1991)
	Asia	320	7–90		60–90 days		Naylor (1996)
	Asia	25–500	7–90	8–15 days	60–90 days	4 years	Heidenreich and Halwart (1995); Heidenreich *et al.* (1997)
	Hawaii	200–500		7–14 days	3–4 months	2–5 years	Glover and Campbell (1994)
	Hawaii	*c.* 200		3 weeks	3–4 months	3–4 years	Kobayashi *et al.* (1993)
	Hawaii	350		2–3 weeks			Tamaru and Hun (1996)
	Hawaii				10 months		Laeh *et al.* (2001)
	England*	100 +			< 12 months		Andrews (1964)
Pomacea dolioides (Reeve)							
	Surinam	200–300 (max. 437)		13–16 days	8–12 months	c. 18 months?	van Dinther (1956, 1962) (as '*P. lineata*')
Pomacea gigas (Spix)							
	Not known			2–6 weeks			Köhler cited by Demian and Yousif (1973)
Pomacea glauca (Linnaeus)							
	Surinam	30–90		14–17 days	8–12 months	*c.* 18 months?	van Dinther (1956, 1962)
	Guadeloupe				13.5 months	*c.* 3 years	Pointier *et al.* (1988)
Pomacea haustrum (Reeve)							
	Brazil*			15–23 days	374–529 days		Guimarães (1981a)
	Brazil	236		9–30 days			Guimarães (1981b)
	Brazil				< 1 year		Estebenet and Cazzaniga (1992)

Table 5.1. *(Cont'd)*

Species	Locality	Clutch size	Hatchability (%)	Time to hatching	Time from hatching to maturity	Longevity	References
Pomacea lineata (Spix)							
	Brazil*	Average 100		15 days			Thiengo (1987)
Pomacea paludosa (Say)							
	Florida, USA	Mean 26.7 (max. 141)		18–28 days			Hanning (1978)
	Florida, USA	3–50, 80		15–20 days			Perry (1974)
Pomacea urceus (Müller)							
	Trinidad	21–93	0–84*	22–30 days*			Lum-Kong and Kenny (1989); Lum-Kong (1989)
	Venezuela	50–200			6–7 months	2.5–3.5 years	Burky (1973, 1974)
Marisa cornuarietis (Linnaeus)							
	Egypt*			8–20 days			Demian and Yousif (1973)
	Florida	Max. 210		11–24 days			Robins (1971)
Pila globosa (Swainson)							
	India	200–300		*c.* 1 month			Prashad (1925b); Andrews (1964)
	India			10 days – 3 weeks			Ranjah cited by Demian and Yousif (1973)
Pila polita (Deshayes)							
	Not known			< 14 days			Semper cited by Demian and Yousif (1973)
Pila spp.							
	Thailand					At least 3 years	Keawjam (1987)
Lanistes nyassanus Dohrn							
	Lake Malawi					5–10 years	Louda and McKaye (1982)

A very different strategy is adopted by *Pomacea urceus* (Burky, 1973, 1974; Lum-Kong and Kenny, 1989). This species mates towards the end of the rainy season. Females then bury themselves into the muddy substrate and aestivate as their marshy habitat dries out. *P. urceus* living in permanent rivers also aestivate, burying themselves into the river banks above the water level. Eggs are laid at the start of the aestivation period and are maintained within the shell, between the operculum and the aperture. Development takes place during the dry season, within the female's shell (even if she dies). The young hatch but remain and aestivate within the female's shell until the start of the rainy season. Burky *et al.* (1972) argued that this strategy protects the newly hatched juveniles from high (possibly lethal) temperatures and water loss during the dry season. *P. urceus* is the only ampullariid known to adopt this strategy.

The eggs of most species of *Pomacea* are brightly coloured, perhaps as a warning of unpalatability since the eggs appear to be distasteful, at least to vertebrates although perhaps not to invertebrates (Snyder and Snyder, 1971; Kushlan, 1978). They are various shades of pink, orange and red in *Pomacea australis* (d'Orbigny), *Pomacea bridgesii* (Reeve), *P. canaliculata*, *P. dolioides* (possibly a synonym of *Pomacea lineata* (Spix) but considered distinct by Geijskes and Pain, 1957), *P. hanleyi* (Reeve), *P. insularum*, *P. lineata*, *Pomacea megastoma* (Sowerby), *P. paludosa* and *Pomacea sordida* (Swainson) (van Dinther, 1956, 1973; Snyder and Snyder, 1971; Thiengo, 1987, 1989; Winner, 1989, 1996; Keawjam and Upatham, 1990; Thiengo *et al.*, 1993; Perera and Walls, 1996). The eggs are green in *Pomacea glauca* (Linnaeus), *Pomacea pyrum* (Philippi), *Pomacea decussata* (Moricand) and *Pomacea nais* Pain (Pain, 1950b; van Dinther, 1956, 1973; Snyder and Snyder, 1971; Perera and Walls, 1996). Snyder and Snyder (1971) hesitated to consider green eggs to have evolved for camouflage because they generally remained distinctly visible, at least to humans. The eggs of *Pomacea falconensis* Pain & Arias, *Pomacea flagellata* (Say), *Pomacea gossei* (Reeve), *Pomacea fasciata* (Roissy) and *Pomacea cuprina* (Reeve) are white (Andrews, 1933; Snyder and Snyder, 1971; Perera and Walls, 1996). Eggs of *P. urceus* are reported as either white (Burky, 1973, 1974) or orange (Lum-Kong and Kenny, 1989) or to vary from orange to pale green (Perera and Walls, 1996); those of *P. haustrum* have been reported as both green (Winner, 1989, 1996) and red/pink (Snyder and Snyder, 1971; Guimarães, 1981a); in both cases this may represent taxonomic confusion. Comfort's (1947) report that the eggs of *P. glauca* are red is based on a misidentification of a species (perhaps *P. canaliculata*) from Argentina. Egg colour may change somewhat as the egg surface dries following oviposition and subsequently as the dark-coloured embryo develops inside (Snyder and Snyder, 1971).

The eggs of *Pila* and *Lanistes* are not brightly coloured. The colour of the eggs of *Asolene* have not been reported. Those of *Marisa* (at least *M. cornuarietis*) are orange when laid but soon lose this colour (Michelson, 1961); they were reported as 'grayish-white' by Robins (1971).

Growth and longevity

Little is known of these aspects of the biology of ampullariids (Table 5.1; see also Estebenet and Cazzaniga, 1992). A number of studies have investigated growth in the laboratory (Table 5.1), but it is difficult to relate these studies to growth in the wild. One laboratory study, on *P. canaliculata* in its native Argentina (Estebenet and Cazzaniga, 1992), did, however, demonstrate the crucial role of temperature in growth and reproduction. At a constant 25°C, *P. canaliculata* matured in 7 months, then bred continuously for a single 'season' of about 4 months and then died. In contrast, under seasonally fluctuating temperatures (7–28°C), *P. canaliculata* took 2 years to reach maturity; they then bred for two distinct annual breeding seasons, for a lifespan of about 4 years. In the wild in Argentina, *P. canaliculata* breeds only during the summer (Scott, 1957), and the life cycle under the fluctuating laboratory temperature regime may indeed approximate the life cycle under natural field conditions. Under semi-artificial conditions in Japan (an outdoor pond but with food provided), *P. canaliculata* grew to maturity in less than 2 months (Chang, 1985). In tropical regions of South-East Asia, release from the seasonality of its natural range may be at least one reason why *P. canaliculata* is so prolific; rapid growth and breeding, and hence rapid succession of generations, are permitted year-round (Naylor, 1996), leading to rapid population expansion and high abundance. A short life might then be predicted, but longevity in the field under such tropical conditions has not been reported.

A multitude of other biotic and abiotic factors may influence growth. For instance, the growth rate of *P. dolioides* in its native South America is determined by food availability, as well as by the quantity of water and the duration and intensity of the dry season (van Dinther, 1956, 1962; Donnay and Beissinger, 1993). And, as for other gastropod species, population density (Cameron and Carter, 1979; Cazzaniga and Estebenet, 1988) and both inter- and intraspecific competition may be important in growth regulation.

Maximum size varies greatly among populations (e.g. Keawjam, 1986; Estebenet and Cazzaniga, 1992; Donnay and Beissinger, 1993) and may be related to a number of environmental factors, including habitat size (Johnson, 1958), microclimatic variation and differing water regimes (Donnay and Beissinger, 1993) and population density. The maximum size of *P. canaliculata* in Hawaii is about 30 mm, but in Asia it can reach at least 65 mm (Schnorbach, 1995) or even 90 mm (Heidenreich *et al.*, 1997).

Population dynamics and abundance

Few studies have addressed ampullariid population dynamics directly. It is clear, however, that seasonal changes in water availability are important. In Florida, recruitment of *P. paludosa* was dramatically enhanced during years when the water-table remained high, allowing the animals to remain active instead of having to aestivate; larger animals were also thought to be more able to withstand dry periods (Kushlan, 1975). *P. urceus* will only enter aestivation once a shell length of 85 mm has been reached (Burky *et al.*, 1972). In Venezuela, habitats with more permanent standing water (rice-fields, as opposed to natural wetlands) allowed *P. dolioides* to grow larger and achieve higher abundance, essentially because the period spent in aestivation (no growth) was shorter; densities ranged from 3 100 m^{-2} in natural wetlands to 33 100 m^{-2} in rice-fields (Donnay and Beissinger, 1993). In Hawaii, recorded abundances of *P. canaliculata* in taro patches have exceeded 130 m^{-2} (12.9 ft^{-2}) (Tamaru, 1996; Tamaru and Hun, 1996). In rice paddies in the Philippines *P. canaliculata* abundance is generally 1–5 m^{-2} but densities up to 150 m^{-2} have been reported (Halwart, 1994a; Schnorbach, 1995). Anderson (1993), perhaps mistakenly, reported '1,000 mature snails per square metre' in the Philippines. In rice in Japan, population densities of 3–7 m^{-2} (Okuma *et al.*, 1994) and 12–19 m^{-2} (Litsinger and Estano, 1993) have been reported. Clearly, in irrigated systems with water present longer than would naturally be the case, *Pomacea* can reach maturity in a shorter period. Fluctuations in abundance in *P. haustrum* in Brazil (up to a maximum of 215 m^{-2}) were reported by Freitas *et al.* (1987), although the underlying reasons for the fluctuations were unclear. Again, numerous biotic and abiotic factors may be involved (Bryan, 1990; Perera and Yong, 1990). Over the reproductive season in Florida, *P. paludosa* populations were estimated to produce 1.2–1.5 million individuals ha^{-1} (= 120–150 m^{-2}) (Hanning, 1978).

M. cornuarietis, introduced to Florida, has been reported at densities over 200 m^{-2} (20.9 ft^{-2}) (Robins, 1971). *Lanistes nyassanus* Dohrn is found at densities of about 1 m^{-2} in Lake Malawi (Louda and McKaye, 1982). It is listed as an endangered species by the International Union for the Conservation of Nature (IUCN, 1996).

Clearly, following hatching, abundance in the immediate vicinity of the clutch will be high. However, few of the above reports are sufficiently detailed to assess the impact of survivorship on population densities, although clearly some species, especially when introduced, are able to achieve remarkably high adult abundances.

Natural enemies

Predators

Perhaps the best-known predator of New World ampullariids is the kite *Rostrhamus sociabilis* d'Orbigny (Accipitridae) (in Florida, known as the Everglade snail kite and considered endangered in the USA), which has a long, slender, hooked bill adapted for extracting gastropods from their shells (Pain, 1950b; van Dinther, 1956; Snyder and Snyder, 1971; Kushlan, 1975; Beissinger *et al.*, 1994). In Florida, the kite's natural prey is *P. paludosa*; in South America it preys on a number of other species:

P. lineata, *P. dolioides*, *Pomacea papyracea* (Spix) and *M. cornuarietis*. Another important New World avian predator is the limpkin (*Aramus guarauna* d'Orbigny) (Aramidae), a large wading bird similar to an ibis. Limpkins are found virtually throughout the New World distribution of ampullariids, which constitute a major part of their diet (Peterson, 1980). Caiman lizards (*Dracaena guianensis* Daudin) feed almost exclusively on ampullariids; their teeth have evolved into rounded knobs for crushing gastropod shells (Perera and Walls, 1996). Other predators include various birds, crocodilians, fishes, turtles, crayfish and aquatic insects (Robins, 1971; Snyder and Snyder, 1971; Kushlan, 1975; Donnay and Beissinger, 1993). However, insufficient work has been done to evaluate whether any of these predators have a significant impact on ampullariid population dynamics (Donnay and Beissinger, 1993). In Africa, species of *Lanistes* are a major food resource of various species of fish, which may exert significant selection pressure on shell morphology (Louda and McKaye, 1982), but their impact, if any, on population dynamics is unknown.

Some species (*P. paludosa* and, to a lesser degree, *P. glauca* and *P. dolioides*) exhibit an alarm response on detecting chemical stimuli in the water from predators (turtles) or the juices of damaged conspecifics (Snyder and Snyder, 1971) and in response to mechanical disturbance of vegetation they are sitting on (Perry, 1974). The animals drop to the substrate (if they are not already on it) and bury themselves in it.

Parasites and pathogens

Little has been published in this area. Various ampullariid species are vectors of metastrongylid nematodes, including *Angiostrongylus cantonensis* (Chen), and various trematodes, including schistosomes (see below). The importance of these parasites in ampullariid population dynamics is at present unknown. Three species of temnocephalid flatworms are reported to live symbiotically in the mantle cavity of species of *Pomacea* and *Asolene* in South America (de León, 1989). Whether any of these parasites cause harm to the ampullariid host is unknown. There appears to be no knowledge of natural microbial pathogens in apple snails, although other gastropods are known to be associated with microorganisms such as protozoa, both as parasites and as symbionts or commensals (Godan, 1983). Again, whether any of these parasites or pathogens play (or could play) a role in population regulation is unknown.

Food and feeding

The feeding habits of ampullariids are microphagous, zoophagous and macrophytophagous, none being mutually exclusive (Estebenet, 1995). Ciliary feeding on particulate matter on the water surface has been described for some species (McClary, 1964). Some species will feed on insects, crustaceans, small fish, etc., mostly as carrion but not always so (McLane, 1939; Estebenet, 1995). Some species (e.g. *M. cornuarietis*, *P. canaliculata*) will attack other gastropods and their eggs (Demian and Lutfy, 1966; Robins, 1971; Aldridge, 1983; Cedeño-León and Thomas, 1983; Cazzaniga, 1990b). The predominant habit, however, is macrophytophagous, which from an agricultural pest standpoint is also the most significant. Andrews (1965a) described the feeding behaviour of *P. canaliculata* in detail. This species shows preferences among different food plants; its rate of growth correlates with its feeding on a preferred plant; and it is able to detect its food plants from some distance, using chemical cues in the water (Estebenet, 1995; Lach *et al.*, 2001), as can *P. paludosa* (McClary, 1964). However, despite exhibiting such preferences, *P. canaliculata* appears to be somewhat of a generalist and indiscriminate (e.g. Schnorbach, 1995), and, as suggested for *M. cornuarietis* by Robins (1971), it may be 'more pertinent to determine what the animal does not eat than what it will eat'. In fact, anecdotal comments suggest that *P. canaliculata* is particularly voracious compared with other ampullariids (Neck, 1986).

Respiration

Many ampullariids are amphibious, in both physiology and behaviour. The mantle cavity contains both a ctenidium ('gill') and a portion modified as a pulmonary sack or 'lung' (Andrews, 1965b). In *P. urceus* in Venezuela (Burky and Burky, 1977), ventilation of the lung by extending the siphon to the water surface occurs periodically, and more frequently under conditions of low oxygen tension. Ventilation of the lung, as well as being used for respiration, is also used to adjust buoyancy levels, such that *P. urceus* can float at the water surface under periods of low oxygen tension. Burky and Burky (1977) reported similar patterns of ventilation for *P. falconensis*, *Pomacea luteostoma* (Swainson) and *M. cornuarietis*, with smaller animals (including juvenile *P. urceus*) ventilating more frequently than larger ones, probably reflecting differences in lung capacity relative to body weight and metabolic rate. Lung ventilation is obligatory, but the animals can nevertheless survive extended periods without active intake of air: up to 6 h in *P. lineata* (van Dinther, 1956). Similar respiratory behaviour has been described for other ampullariids (McClary, 1964; Freiburg and Hazelwood, 1977), with differences among species in the relative significance of aerial and aquatic respiration (Andrews, 1965b). Work on respiration rate and its relation to temperature and oxygen tension has been reviewed by Aldridge (1983) and Santos *et al.* (1987). The ability to use the ctenidia and the 'lung' for respiration allows many ampullariids to survive significant periods out of water and to disperse significant distances over land. This is clearly of adaptive value for species that live in marshy or other habitats that dry out periodically. It also means that when introduced to new habitats, such as rice paddies or taro patches, the animals may be difficult to contain within circumscribed areas as they may easily cross the raised burms between paddies (although this has been reported as not occurring (Eversole, 1992)).

Aestivation

Many ampullariids aestivate during dry periods (Lum-Kong and Kenny, 1989). When the animals' habitat dries out, they bury themselves in the mud. Some species (e.g. *P. urceus*) bury themselves only superficially, with part of the shell remaining above the surface of the hardened mud (Burky *et al.*, 1972); others bury up to 1 m deep (e.g. *Pila ampullacea* (Linnaeus), *Pila pesmei* (Morelet)) (see Keawjam, 1986)). They can survive in this state in some cases for extended periods (in laboratory experiments), far longer than are likely to be necessary in the wild, e.g. 8 months for *P. glauca* (van Dinther, 1956), 13 months for *P. lineata* (Little, 1968), 17 months for *P. urceus* (Burky *et al.*, 1972), at least a year for *P. ampullacea* and *P. pesmei* (Keawjam, 1986) and 25 months for *P. globosa* (Chandrasekharam *et al.*, 1982). *P. canaliculata* is reported to survive buried in the earth only up to 3 months (Schnorbach, 1995). They can withstand significant loss of soft-tissue weight during aestivation; in *P. lineata* up to 50% (Little, 1968) and in *P. urceus* up to 62% (Burky *et al.*, 1972). *Pila virens* (de Lamarck) and *P. globosa* lose considerably less weight (5%) but can nevertheless aestivate for at least 6 months; the shell and operculum appear to be effective barriers to water loss, especially as the operculum is sealed in the shell aperture with dried mucus (Meenakshi, 1964). Metabolism during aestivation is anaerobic in *P. virens* and *P. globosa* (Meenakshi, 1964; Aldridge, 1983), but aerobic in *P. ovata* (Olivier) and *P. urceus* (Burky *et al.*, 1972). *P. virens* and *P. globosa* (and other species of *Pila*) aestivate buried very deep in the ground; their anaerobic aestivation metabolism may be an adaptation to this, in contrast to the aerobic respiration of the shallow-burying *P. urceus*. Metabolism during aestivation in *M. cornuarietis* has been described by Horne (1979); this species tolerates anaerobic conditions only for about 48 h. However, adult *M. cornuarietis* can withstand at least 30 days out of water in 20% relative humidity or 120 days at 80% relative humidity, although juveniles have little resistance to desiccation (Robins, 1971).

Physiology

Lethal temperatures

During aestivation, *P. urceus* regulates its body temperature below 41°C, in part through evaporative cooling; this species has an upper lethal temperature between 40 and 45°C (Burky *et al.*, 1972). Adult *P. urceus* are more tolerant of high temperature than are juveniles (Burky *et al.*, 1972), perhaps because adults can afford to lose more water through evaporative cooling (cf. Cowie, 1985). In both *P. paludosa* and *M. cornuarietis*, 40°C is lethal when animals are exposed for 1–4 h (Freiburg and Hazelwood, 1977), although Thomas (1975) reported that they could withstand temperatures up to 45°C. Robins (1971) gave 39°C as the 'upper limit of short-term heat tolerance' for *M. cornuarietis*, with juveniles more tolerant than adults at 37°C and both adults and juveniles feeding normally between 33.5 and 35.5°C; eggs did not develop normally at 35–37°C. Mochida (1991) reported for *P. canaliculata* that mortality is high at water temperatures above 32°C (35°C in Mochida (1988b) and Eversole (1992)). *P. virens* and *P. globosa* cannot survive for 2 days at 40°C (Meenakshi, 1964). *P. lineata* survived 1 h exposure at 40°C (Santos *et al.*, 1987).

Regarding low-temperature tolerance, Robins (1971) (see also Neck, 1984) reported that *M. cornuarietis* could survive for over 24 h at 11°C (although egg development ceased at this temperature), but succumbed in 5 h when exposed to 8°C – although Thomas (1975) reported that it could withstand 6°C. *P. paludosa* can survive exposure at 5°C (Freiburg and Hazelwood, 1977). Mochida (1991) reported *P. canaliculata* survival for 15–20 days at 0°C, 2 days at – 3°C, but only 6 h at – 6°C (see also Neck and Schultz, 1992; Wada, 1997). *P. lineata* survived 1 h exposure at 5°C (Santos *et al.*, 1987). *P. virens* and *P. globosa* cannot survive for 4 days at 20°C and die within 1 day at 10°C (Meenakshi, 1964).

Differences among species in both their upper and lower lethal limits may reflect adaptation to their natural climatic environment. There appears to be less variability in the upper limit, which in general appears to be around 40°C for many aquatic organisms. Comparability among the studies mentioned above is poor, however, largely because experimental procedures differed. Nevertheless, *P. canaliculata*, of more temperate habitats (Argentina), seems to have lower tolerance to high temperatures than more tropical species, such as *P. urceus*, *M. cornuarietis* and the two *Pila* species. Lower temprerature limits seem more variable, with *P. canaliculata* able to tolerate freezing temperatures, in marked contrast to the *Pila* species, which are unable to survive at 20°C beyond a few days. These differences probably have significant consequences for the potential establishment, reproduction, growth and population dynamics of apple snails when they are introduced to new regions with climates differing from those in their natural ranges.

Salinity tolerance

M. cornuarietis can withstand up to about 30% salt water (Hunt, 1961; Robins, 1971; Santos *et al.*, 1987). *P. globosa* can 'live in salt water of low salinity' (Prashad, 1925b). Fujio *et al.* (1991a) indicated differences in salinity tolerance among three 'strains' of *P. canaliculata*. Preliminary observations in Hawaii suggest that *P. canaliculata* is sufficiently tolerant of sea water to survive long enough to be carried by currents from one stream mouth to another, thereby providing for expansion of its distribution (F. Reppun and N. Reppun, personal communication, 1997). However, although exhibiting some tolerance of salinity, ampullariids generally live only in fresh water, and brackish water may limit the spread of some species (e.g. *M. cornuarietis* (Robins, 1971)).

Dispersal

Adult *P. lineata* travel several metres an hour (van Dinther, 1956). *L. nyassanus* moves an average 2.8 m day^{-1} (Louda and McKaye, 1982). *P. globosa* makes long excursions on land both for going from one source of water to another and for the purpose of laying eggs (Prashad, 1925b). Short-term dispersal activity, however, does not necessarily translate into long-term, long-distance dispersal. There is little documentation of the spread of ampullariids from a focus of introduction. In a canal in Florida, an introduced population of *M. cornuarietis* expanded by at least 1.5 km downstream in 6–8 months (Hunt, 1958) and within 12 years had become distributed in virtually the entire freshwater canal system in the Miami area, dispersal being predominantly by floating downstream on vegetation (Robins, 1971). Floating downstream (unattached to vegetation) has been seen in Hawaii and no doubt facilitates rapid dispersal, but crawling upstream is also possible (H. Ako, personal communication, 1997). However, the rapid dispersal of *P. canaliculata* within countries in South-East Asia, following its initial introduction, has been predominantly human-mediated.

Introductions

Distribution as a food item

A number of ampullariids are used as human food in their native ranges. For instance, *P. urceus* is eaten in Trinidad (Lum-Kong, 1989; Lum-Kong and Kenny, 1989) and in times of food scarcity in Guyana (Pain, 1950b). In India, *P. globosa* is eaten (Thomas, 1975), as are *Pila conica* (Wood) in Malaysia (Johnson, 1958), *Pila luzonica* (Reeve) in the Philippines (Palomino and Jueco, 1983) and native species of *Pila* in Thailand (Keawjam, 1986, 1990). Elsewhere in South-East Asia, ampullariids may be only a minor part of the local diet (e.g. in Cambodia (Cowie, 1995a)). In Africa, *Pila congoensis* Pilsbry & Bequaert was eaten, although only by older people because eating snails was thought to cause 'infecundity' (Pilsbry and Bequaert, 1927).

In both Guam (first reported in 1984) and Hawaii (first recorded in 1966) *P. conica* was introduced without authorization, either accidentally or deliberately as a food item (Smith, 1992; Cowie, 1995b). It was introduced to Palau in 1984 or 1985 but was thought to have been eradicated by 1987 (Eldredge, 1994).

Between 1979 and 1981 a species of *Pomacea*, usually referred to as *P. canaliculata*, was introduced to South-East Asia, initially from Argentina to Taiwan (e.g. Mochida, 1988a,b, 1991; Cheng, 1989 (as '*Pomacea lineata*')). The initial introduction to Taiwan was illegal (Cheng, 1989). The purpose of the introduction was both for local consumption and for development for export to the gourmet restaurant trade, with the expectation of high profits (Acosta and Pullin, 1991; Naylor, 1996). Prior marketing research had not been undertaken. The subsequent spread of this ampullariid, distributed for the same purposes and still with no market research, has been summarized by Mochida (1991), Litsinger and Estano (1993), Halwart (1994a), Naylor (1996) and Vitousek *et al.* (1996). In 1981 they were taken from Taiwan to Japan (they were present in Okinawa by at least 1984 (Fujio *et al.*, 1991a; Wada, 1997)), and by 1983 about 500 apple snail businesses had been established in various parts of Japan (Wada, 1997). In either 1980 or 1982 these apple snails were introduced from Taiwan to the Philippines (Mochida, 1991; Anderson, 1993; Halwart, 1994a; Joshi *et al.*, 2001), and introductions to the Philippines continued from various sources (and possibly including more than one species (Mochida, 1988b)) as snail-farming was promoted by both governmental and non-governmental organizations (Anderson, 1993). Later, *Pomacea* were taken to China (1985), Korea (probably 1986), parts of Malaysia (Sarawak and Peninsular Malaysia, 1987), Indonesia (Java and Sumatra, 1989), Thailand (1989), Vietnam (1988 or 1989) and Laos (1992). They have also been introduced to Hong Kong (Laup, 1991), Cambodia (Cowie, 1995a), Singapore (Ng *et al.*, 1993), Guam (from Taiwan (Smith, 1992; Eldredge, 1994), Papua New Guinea (from the Philippines (Laup, 1991 (as '*P. lineata*'); Anon., 1993; Eldredge, 1994)) and Hawaii (Cowie, 1995b, 1997b).

However, the ampullariids' economic potential seems to have been overestimated and no major trade based on aquaculture operations has developed (Acosta and Pullin, 1991). In Taiwan, failure of the local market was said to be because consumers did not like the 'tough meat and repulsive taste' (Chang, 1985; see also Cheng, 1989; Naylor, 1996; Vitousek *et al.*, 1996). In addition, developed nations maintain stringent health regulations that have largely precluded importation of *Pomacea* products (Anderson, 1993; Naylor, 1996). Animals escaped or were deliberately released from captivity and *Pomacea* has since become widespread and abundant in many countries. Expansion of their distribution has been assisted by, among other things, floods and typhoons, movement of infested soil during new paddy preparation, deliberate release of animals for weed control and their use for fishing bait (Anderson, 1993; Wada, 1997).

Distribution by the aquarium trade

Ampullariids are popular domestic aquarium animals (Perera and Walls, 1996) and a number of species have been introduced to many parts of the world via the aquarium trade. *M. cornuarietis* of northern South America has been introduced to several countries (e.g. the USA) by aquarists (Hunt, 1958; Robins, 1971; Neck, 1984; Perera and Walls, 1996). *P. bridgesii* of South America became established in Florida, either accidentally or by intent, probably introduced in the early 1960s (Clench, 1966). The shell of *P. bridgesii* is usually dark greenish-brown with or without spiral banding patterns. However, when a bright yellow shell-colour variant was discovered, it was selectively bred by the aquarium trade and subsequently transported to many parts of the world. Bright yellow, orange and other shell-colour variants have also been found in other species of *Pomacea*. These brightly coloured forms are favoured by the aquarium trade and the market for ampullariids for aquariums has expanded since their discovery (Perera and Walls, 1996); they have become established or have been intercepted by customs officials as far from their natural range as Singapore (Chan Sow-Yan, personal communication 1996; Bishop Museum collections) and Australia (P. Colman, personal communication, 1994; Bishop Museum collections). *P. canaliculata* (including brightly coloured forms) has recently been reported in California (Cerutti, 1998). The promulgation of these brightly coloured varieties has unfortunately led to the use of the terms 'golden snail', 'golden apple snail', 'golden mystery snail', as if referring to a single, particular species. Indeed, the terms 'golden apple snail' and 'golden snail' have become the most frequently used common names for the major pest species in South-East Asia, which has been distributed almost entirely for use as food (above) rather than via the aquarium trade. Use of these common names has led to immense confusion and misunderstanding, both within the aquarium trade and, more seriously, among managers, politicians and other non-biologists trying to cope with the agricultural problems that have arisen as these ampullariids have escaped or been released into the wild far beyond their natural range.

Keawjam and Upatham (1990) considered the *Pomacea* species known in Thailand to have been imported by the aquarium trade, but it is also probable that they were introduced for food, as elsewhere in South-East Asia. In Hawaii, *P. canaliculata*, as well as being introduced for food, has been available in aquarium stores, as has *P. bridgesii*. Purchase in aquarium stores, followed by release for culture as food items, appears to have been one reason for the spread of *P. canaliculata* in Hawaii (H. Ako, personal communication, 1997). *P. lineata* has been introduced to South Africa (Berthold, 1991), perhaps via the aquarium trade.

Introduction for biological control

As a snail competitor/predator

Ampullariids have been introduced in attempts to control the gastropod vectors of schistosomes, either by outcompeting them and/or by predation (especially on eggs and juveniles). In Guadeloupe, introduced *P. glauca* and *M. cornuarietis* caused the decline of the planorbid *Biomphalaria glabrata* (Say) through competition (Pointier *et al.*, 1988, 1991). In Puerto Rico, *M. cornuarietis* caused a decline in *B. glabrata* and the lymnaeid *Lymnaea columella* Say through predation (Robins, 1971; Peebles *et al.*, 1972). *M. cornuarietis* is said to have had a similar effect in the Dominican Republic (Perera and Walls, 1996) and in Egypt (Demian and Kamel, cited by Cedeño-León and Thomas, 1983; Berthold, 1991).

For aquatic weed control

Ampullariids such as *M. cornuarietis* are voracious feeders on aquatic plants; this is partly the reason for their reported success in controlling other gastropod species: they reduce the available food supply (Perera and Walls, 1996). They have therefore been used or suggested for control of aquatic weeds. For instance, in Florida and Puerto Rico, *M. cornuarietis* has been deliberately introduced in attempts to control aquatic plant nuisances (Robins, 1971; Simberloff and Stiling, 1996a); *P. globosa* in India has been tested as a control agent for the aquatic weed *Salvinia molesta* Mitchell (Salviniaceae) (Thomas, 1975); and, in Japan, introduced *P. canaliculata* has been suggested as a possible agent for weed control (Okuma *et al.*, 1994; Wada, 1997).

Ampullariids as Agricultural Pests

Pests in their native range

In general, ampullariids are not serious agricultural pests in their native ranges. A major exception is in rice-growing areas of Surinam (van Dinther, 1956, 1962, 1973; van Dinther and Stubbs, 1963). Here, since the development in the 1930s–1950s of large-scale rice farming involving mechanized sowing, *P. dolioides* (also referred to as *P. lineata*) in particular, but also *P. glauca*, have caused serious problems. The seeds, planted directly into the irrigated fields, are attacked by the ampullarrids as they germinate. Prior to the development of direct sowing as the preferred crop-establishment method, seedlings were germinated in nurseries and transplanted in the field. Although some damage occurred in the nurseries and after planting out, larger seedlings were more able to withstand attack.

In El Salvador, 'snails' (presumably ampullariids) necessitated treatment of rice with pesticides (Efferson, 1968). Damage has also been reported in Barbados, Bolivia, Brazil, Colombia, Guadeloupe and Trinidad (Litsinger and Estano, 1993; Bombeo-Tuburan *et al.*, 1995), but no details are available. Some of these reports may be of non-native species. In Africa, *Lanistes ovum* Peters and *Lanistes carinatus* (Olivier) have been reported damaging seedlings (van Dinther, 1973). *Pila* has occasionally been implicated as an agricultural pest. In Burma, *P. pilosa* has been reported to damage paddy rice (van Dinther and Stubbs, 1963). *P. globosa* has been reported as a paddy pest in India (Singh and Agarwal, 1981), although Thomas (1975) stated that this species did not feed on rice. Van Dinther (1973) mentioned *P. globosa* and *Pila polita* (Deshayes) in India and Burma.

Pests when introduced

In Puerto Rico, *M. cornuarietis* has been reported destroying rice seedlings (van Dinther, 1973) following its introduction. Species of *Pila* have been introduced outside their natural range and have become pests, including *P. conica* in Hawaii (Cowie, 1995b). However, it is species of *Pomacea* that have become by far the most serious ampullariid pests, attacking a wide range of crops, with their most serious impacts being on rice in South-East Asia.

Rice

The history of the introduction of non-native apple snails into South-East Asia (above) and the damage they cause to rice farming have been reviewed by Halwart (1994a) and Naylor (1996), among others. One or more species of *Pomacea* (usually identified as *P. canaliculata*) have become pests of paddy rice in many countries, including Thailand, Vietnam, parts of Malaysia and Indonesia, China, Taiwan and Japan, but probably most seriously in the Philippines.

In Taiwan, where *Pomacea* was first introduced between 1979 and 1981, 17,000 ha of rice and other crops had been infested by 1982, increasing rapidly to 171,425 ha by 1986 (Mochida, 1991). In the Philippines the spread of *Pomacea* has been even more rapid, from 9500 ha of rice in 1986 to over 400,000 ha in late 1988, 500,000 ha by 1989 and occurring in most provinces (Adalla and Morallo-Rejesus, 1989; Mochida, 1991; Olivares *et al.*, 1992; Anderson, 1993; Litsinger and Estano, 1993) and 800,000 ha by 1995 (Palis *et al.*, 1996). Apple snails have now become the most important pest of rice in the Philippines (Mochida, 1988a, 1991; Cheng, 1989; Acosta and Pullin, 1991; Halwart, 1994a; Naylor, 1996; Vitousek *et al.*, 1996). In Japan, where it was introduced in 1981, *Pomacea* had spread to 35 out of 47 prefectures by 1989 (Mochida, 1991) and by 1995 occurred in over 50,000 ha of paddy fields (Wada, 1997). Details of the *Pomacea*'s spread in other countries are not readily available.

Problems caused by *P. canaliculata* are not totally confined to South-East Asia. In 1990/91 infestation of rice-growing areas in the Dominican Republic by a species tentatively identified as *P. canaliculata* was reported, and by 1997 40% of this country's rice-growing areas were infested, with losses up to 75% in some areas (D. Robinson, personal communication, 1997).

Other crops

Other crops reported as being attacked by ampullariids are listed in Table 5.2. In about 1989 *P. canaliculata* was introduced to the Hawaiian Islands. Within 3 years it had been taken deliberately to most of the main islands in the archipelago and had escaped or been deliberately released into taro patches (which are similar ecologically to rice paddies), where it is now a major pest (Cowie, 1995b, 1996, 1997b). Taro is the main traditional staple for native peoples not only in Hawaii but on many Pacific islands. A number of other introduced ampullariids have been present in the Hawaiian Islands since before the arrival of *P. canaliculata*, but do not seem to cause major agricultural problems (Cowie, 1995b, 1997b), although there have been no formal economic yield-loss assessments.

Table 5.2. Crops other than rice that are attacked by Ampullariidae. This is not a comprehensive list, but simply a brief compilation from some of the more accessible literature, illustrating the susceptibility of a wide range of plants.

Plant	Region	Reference
Water hyacinth (*Eichornia crassipes* (Martius) Solms) (Pontederiaceae)	Hong Kong	Laup (1991)
Water spinach/swamp cabbage (*Ipomoea aquatica* Forsskaol) (Convolvulaceae)	Hong Kong	Laup (1991)
	Japan	Mochida (1991)
	Hawaii	Nishimura *et al.* in Tamaru (1996)
Lotus (*Nelumbo nucifera* Gaertner) (Nelumbonaceae)	Hong Kong	Laup (1991)
	Japan	Mochida (1991); Wada (1997)
Water cress (*Rorippa* Scopoli spp.) (Brassicaceae)	Hong Kong	Laup (1991)
	Hawaii	Nishimura *et al.* in Tamaru (1996)
Taro (*Colocasia esculenta* (Linnaeus) Schott) (Araceae)	Japan	Mochida (1988b, 1991); Wada (1997)
	Hawaii	Cowie (1995b); Nishimura *et al.* in Tamaru (1996)
Mat rush (*Juncus decipiens* (Buchenau) Nakai) (Juncaceae)	Japan	Mochida (1988b, 1991); Wada (1997)
Chinese mat rush (*Cyperus monophyllus* Vahl) (Cyperaceae)	Japan	Mochida (1991)
Wild rice (*Zizania latifolia* (Grisebach) Stapf) (Gramineae)	Japan	Mochida (1991)
Japanese parsley/dropwort (*Oenanthe javanica* (Blume) de Candolle) (Apiaceae)	Japan	Mochida (1991); Wada (1997)
Water chestnut (*Trapa bicornis* Osbeck) (Trapaceae)	Japan	Mochida (1991); Wada (1997)
Azolla (*Azolla* de Lamarck spp.) (Azollaceae)	Philippines	Adalla and Morallo-Rejesus (1989)
	Japan	Mochida (1991)

Yield loss and economic cost of ampullariid damage

Yield loss can be massive but variable. In rice in the Philippines (Naylor, 1996), where probably the most serious damage occurs, losses vary from 5% to 100%, depending on locality and the level of infestation. Yield loss is related to the abundance and size of the apple snails. In experimental studies, one apple snail per m^2 was shown to reduce the rice crop by 20%, but eight per m^2 reduced it by over 90% (see also Olivares *et al.*, 1992; Schnorbach, 1995); a single apple snail eats seven to 24 rice seedlings day^{-1} (Litsinger and Estano, 1993). Densities in infested rice paddies in the Philippines are generally 1–5 m^{-2} but densities up to 150 m^{-2} have been reported (Halwart, 1994a). In Japan, 3–19 m^{-2} have been reported in rice (Litsinger and Estano, 1993; Okuma *et al.*, 1994).

However, little quantitative yield-loss or economic information is available. In Taiwan, the cost of the loss of rice was estimated as US$2.7 million in 1982, increasing rapidly to US$30.9 million in 1986 (Mochida, 1991). Huge areas were treated with pesticides (103,350 ha in 1986) at an additional enormous cost. In Japan, control in just 176 ha cost US$64,385 (Mochida, 1991). In the Philippines, between 1987 and 1990, farmers spent US$10 million on pesticides (Anderson, 1993).

The most detailed published economic analysis so far is that of Naylor (1996), reported also by Vitousek *et al.* (1996), for the Philippines. This analysis included not only the cost of loss of rice, but also the costs of replanting, application of pesticides and manually removing the apple snails. Total costs in 1990 due to *Pomacea* infestation were estimated as US$28–45 million. This was 25–40% of what the Philippines spent on rice imports in 1990. Naylor (1996) also compared the costs of control measures (pasturing ducks (*Anas* Linnaeus spp., Anatidae) in the paddies, hand-picking the apple snails and applying pesticides) in the Philippines with costs in Vietnam. She showed that the relative implementation of each of these techniques differed because of different cost structures in the two countries, but also because infestation in Vietnam had not reached the extreme levels that it had in the Philippines, and hand-picking combined with duck pasturing was relatively more feasible and effective in keeping ampullariid numbers low. As ampullariid populations increase and spread in Vietnam this may change. No doubt the suite of management practices will need to be varied to meet local needs; research will be necessary.

Environmental Impacts

In addition to the serious agricultural problems caused by introduced ampullariids, there are also potential concerns for the natural environment, including impacts on native mollusc species. As noted above, purposefully introduced *Pomacea* and *Marisa* have effected decline in populations of the gastropod hosts of schistosomes. While the implication

origin' (Lacanilao, 1990). Keawjam and Upatham (1990) recognized three species of *Pomacea* introduced in Thailand: *P. canaliculata*, *P. insularum* and an unidentified species of *Pomacea*. Mochida (1991) indicated that, as well as *P. canaliculata* (which he considered frequently to have been misidentified as *P. insularum*), two other species of *Pomacea* have also been introduced to the Philippines: *P. gigas* and *P. cuprina* (the latter possibly a misidentification of *P. bridgesii*, a species that has been carried all over the world by the domestic aquarium trade (Cowie, 1995b)). In the Philippines the apple snails have even been identified as species of *Pila* (see Guerrero, 1991). In Japan, three 'strains of *Pomacea canaliculata*' have been identified, differing in shell colour and pattern, salinity tolerance and aspects of reproduction and growth (von Brand *et al.*, 1990; Fujio *et al.*, 1991a). In Hawaii, where four ampullariid species are recorded (Cowie, 1995b), ampullariids in an aquaculture project have been reported as hybrids of *P. canaliculata* and *P. paludosa* (Nishimura *et al.* in Tamaru, 1996).

Ampullariid species-level taxonomy has been heavily reliant on shell morphology, yet gastropod shells, and especially those of ampullariids, exhibit much intraspecific variation. The taxonomy and systematics of most species have not been adequately worked on since their original descriptions. The pest species (even if it turns out to be more than one species) in South-East Asia nevertheless appears to belong to a relatively well-circumscribed group of more or less closely related species from South America. However, within this group, the species and their relationships are very poorly understood. The group comprises about 15 nominal species, including *P. canaliculata*. I refer to this group as the '*canaliculata* group'. From time to time, some of the species within the '*canaliculata* group' have been formally synonymized, informally linked together, distinguished as separate species and so on. This confusion was discussed but not resolved by Alderson (1925), the most recent author to comprehensively revise (as '*Ampullaria*') taxa now assigned to the genera *Pomacea* and *Pila*. He implicitly recognized most of the species in the '*canaliculata* group' as a more or less closely knit group. Within this group Alderson further recognized a number of rather vaguely defined associations of species – for instance, explicitly linking *Pomacea immersa* (Reeve), *Pomacea amazonica* (Reeve) and *P. haustrum*, although without formally synonymizing them, and informally referring to another subset of the group as 'the *lineata* group'. However, he did retain most species as valid. It is quite possible that, just as for the large number of Central American species synonymized under *P. flagellata* by Pain (1964), many other 'species' of Ampullariidae, including those in the '*canaliculata* group', do not deserve distinct specific status (Pain, 1960; Cazzaniga, 1987). A modern revision, involving not only conchology but also internal anatomy and molecular characters, might reduce the '*canaliculata* group' to as few as three species, possibly *P. canaliculata*, *P. lineata* and *P. gigas*. Until such work is undertaken, however, the status of these various nominal species will remain obscure.

The predominant pest may indeed be *P. canaliculata*. However, *P. canaliculata* is a widespread and morphologically highly variable species in South America (Cazzaniga, 1987). How distinct it really is from the various nominal South American species also reported from South-East Asia is not at all clear. In particular, the distinction between *P. canaliculata* and *P. insularum* is especially unclear, despite attempts to clarify it (Scott, 1957; Keawjam and Upatham, 1990). In many populations, the maximum shell size of *P. canaliculata* is around 30 mm (e.g. in Hawaii), but, in other populations, females can reach 70–80 mm (Wada, 1997) and some individuals may even reach sizes over 90 mm (Estebenet and Cazzaniga, 1992). Even in populations in adjacent ponds and with one population derived from the other, there can be enormous differences of this kind (e.g. in Cambodia (R.H. Cowie, personal observations)). Sexual dimorphism in size (Fig. 5.1), combined with this overall morphological variability and the existence of considerable variation in shell colour and banding pattern, no doubt confounds attempts at accurate identification.

These problems of identification cast doubt on the correctness of the names applied both in the pest-related literature and in much of the ecological, physiological, behavioural and other literature on species in the '*canaliculata* group'. Comparability among studies is unreliable. The problems are compounded when no precise natural-provenance information is available, as is the case with studies dealing with introduced populations in South-East Asia and elsewhere. Because the taxonomy of the '*canaliculata* group' of species is so confused, it is difficult to assess natural species distributions. For instance, *P. canaliculata* itself has been considered to be extremely widely distributed, from the Plata River system in Argentina to Guyana and from the easternmost tip of Brazil to Bolivia (Scott, 1957), although its presence in northern South America depended on synonymizing it with *P. dolioides*. Study of museum material (R.H. Cowie, personal observations) suggests that it does not extend north of Brazil and that perhaps *P. dolioides* (possibly a synonym of *P. lineata*) replaces it in Surinam, Guyana and Venezuela.

It is therefore clear that any sound attempt to develop integrated pest management of apple snails requires a modern assessment of their basic taxonomy.

Ecology, behaviour, physiology

A taxonomic re-evaluation of the '*canaliculata* group' would lay the foundation for the study of the ecology and behaviour of the species then identified as the major, if not only, pest species in South-East Asia. Necessary ecological information as a background to implementing integrated pest-management strategies would include data on life history (lifespan, age at maturity, growth, fecundity, etc.), population dynamics, predators, parasites and habitat characteristics (water temperature, flow rate, etc.). Behavioural and physiological data on aestivation, thermal and desiccation relations, seasonality of growth and reproduction, dispersal, etc. would also be important. This information needs to be obtained in the native range of the pest species (South America) as well as in the localities in South-East Asia where the species has become a serious pest. In this way, changes in the species' ecology and behaviour – for instance, niche expansion following relaxation of competition – that have arisen as a result of the different ecosystems in which they now function can be understood. Ultimately such studies would focus on aspects of ecology and behaviour that can be manipulated as possible routes to eventual control. If biological control is being considered, selection of biocontrol agents must be done with an understanding of their potential to operate and remain host-specific in a novel ecological context.

Yield-loss assessments

There is still far too little information on economic losses and how these can be offset by the various control measures applied singly or in different combinations. Frequently, achieving a reduction, even a large reduction, in apple snail numbers is assumed to lead to increased crop yield, but the relationship is unlikely to be simple. Furthermore, levels of yield loss are often assumed to reflect directly levels of damage early in crop development, but compensatory effects may be important (Okuma *et al.*, 1994) and accurate yield-loss assessment may be highly complex (see Wood and Cowie, 1988). Such analyses are almost entirely lacking for all but a few regions where apple snails have assumed pest status. Yet detailed yield-loss and economic analyses are a necessary basis for developing recommendations on management options, now and in the future. Economic yield-loss assessments also provide ammunition to convince political authorities of the seriousness of the problem and the need for funding to address it.

Control

Even with the diversity of control and management approaches that have been developed or tried, no fully adequate strategy has surfaced that can be widely recommended. Despite huge effort on the part of farmers, losses continue to be severe, especially where infestations are heavy. Control is still dominated by the use of pesticides (Halwart, 1994a), with their often extremely harmful environmental and human health effects. Biological control using predators, parasites and pathogens that occur in the native range of the ampullariids has barely been addressed. The search for possible predators/pathogens in South America has been minimal: preliminary surveys have not identified species-specific agents that could be easily manipulated and that have the potential to have a significant impact on pest populations. The idea of promoting practices that might enhance populations of native competitors and/or predators (Settle *et al.*, 1996) has not been investigated and no such competitors/predators have been identified. Research, based on a sound understanding of the pest species' biology, must then be directed towards reduction in the use of pesticides, identifying new approaches to cultural control and combining both new and existing strategies into fully integrated pest-management protocols that can be tailored to the needs of particular situations.

Prevention

It is crucial that further spread of ampullariids be minimized. Quarantine regulations need to be put in place (if they are not in place already) and applied strictly. Such measures are far more cost-effective than attempting to address the problems once the apple snails have invaded and become pests. But probably most important is raising the awareness of the threat of alien species. Publicity must be directed at the general public, the aquarists and those seeking quick profit from marketing apple snails as food. In particular, in South-East Asia, the full import of the spread of apple snails to rice-growing areas so far uninfested must be clearly communicated to farmers in a serious and comprehensive education campaign that has the backing of politicians and agricultural managers. It is only with public cooperation and a real public understanding of the potential problems that apple snails can cause that disasters on the scale that has befallen the Philippines can be avoided.

Conclusions

Frequently there is a conflict between potential environmental harm and economic benefit when species are considered for introduction (Baskin, 1996; McNeely, 1996; Williamson, 1996). In the case of apple snails, the potential for short-term economic gain (which has not materialized) has been foremost in people's minds, while they have been blind (and in some cases continue to be so) to the potential (and realized) long-term environmental and agricultural destruction.

There is a rapidly increasing literature on invasive species: factors causing their success or failure, the dynamics of establishment, the ecological and agricultural problems they cause and what can be done about them (e.g. Niemelä and Mattson, 1996; Williamson, 1996; Williamson and Fitter, 1996; Carlton, 1999). In many cases, new invaders exhibit a lag phase before their populations expand rapidly (Crooks and Soulé, 1996), when they become pests. In the case of *Pomacea* introduced to South-East

Asia and other areas, the lag phase seems virtually non-existent, giving authorities minimal time to make a decision to act.

For regions not as yet infested, prevention of the introduction of apple snails must then be the primary strategy. Awareness must therefore be raised so that officials know the potential problems that will overcome them should the apple snails be introduced, rather than only becoming aware of the problems when it is too late. Officials must also be prepared to act quickly if an introduction is detected. Eradication at this early stage might still be possible, but there will be only a very narrow window of opportunity.

For areas already infested and with little hope of eradicating the apple snails, control must rely on integration of various management strategies. These strategies will differ from region to region, depending on the levels of infestation, potential environmental consequences, the specific needs of the local farmers and the options open to them, and local economics. Development of widely implemented and successful strategies has to be based on a thorough understanding of relevant aspects of the pest's biology.

Acknowledgements

I thank Harry Ako, Geoff Baker, Gary Barker, Lu Eldredge, Neal Evenhuis, Matthias Halwart, Don Heacock, Frank Howarth and Rosamond Naylor for reviewing all or part of the manuscript. For additional information, I also thank Alfredo Castro-Vazquez, Phil Colman, Gary Jahn, Frederick and Nicholas Reppun, David Robinson, Barry Smith, Chan Sow-Yan and Fred Thompson. Clyde Imada and George Staples helped with names of crop plants. Melanda Hoque and Yoichi Yusa helped with obtaining literature. My own work on ampullariids has been funded in part by the Food and Agriculture Organization (FAO), International Rice Research Institute (IRRI) and US Agency for International Development (USAID), and I thank Peter Kenmore, Gary Jahn and K.L. Heong for their support.

References

Acosta, B.O. and Pullin, R.S.V. (1991) *Environmental Impact of the Golden Snail* (Pomacea *sp.) on Rice Farming Systems in the Philippines.* Freshwater Aquaculture Center, Central Luzon State University, Munoz, Nueva Ecija, and ICLARM, Manila, vi + 34 pp.

Adalla, C.B. and Morallo-Rejesus, B. (1989) The golden apple snail, *Pomacea* sp. a serious pest of lowland rice in the Philippines. In: Henderson, I.F. (ed.) *Slugs and Snails in World Agriculture.* Monograph 41, British Crop Protection Council, Thornton Heath, pp. 417–422.

Albrecht, E.A., Carreño, N.B. and Castro-Vazquez, A. (1996) A quantitative study of copulation and spawning in the South American apple-snail, *Pomacea canaliculata* (Prosobranchia: Ampullariidae). *The Veliger* 39, 142–147.

Alderson, E.G. (1925) *Studies in* Ampullaria. Heffer, Cambridge, xx + 102 pp., 19 pls.

Aldridge, D.W. (1983) Physiological ecology of freshwater prosobranchs. In: Russell-Hunter, E.D. (ed.) *The Mollusca*, Vol. 6, *Ecology*. Academic Press, London, pp. 329–358.

Anderson, B. (1993) The Philippine snail disaster. *The Ecologist* 23, 70–72.

Andrews, E.A. (1933) Eggs of *Ampullaria* in Jamaica. *The Nautilus* 46, 93–97.

Andrews, E.B. (1964) The functional anatomy and histology of the reproductive system of some pilid gastropod molluscs. *Proceedings of the Malacological Society of London* 36, 121–139.

Andrews, E.B. (1965a) The functional anatomy of the gut of the prosobranch gastropod *Pomacea canaliculata* and of some other pilids. *Proceedings of the Zoological Society of London* 145, 19–36.

Andrews, E.B. (1965b) The functional anatomy of the mantle cavity, kidney and blood system of some pilid gastropods (Prosobranchia). *Journal of Zoology* 146, 70–94.

Anon. (1989) *Integrated 'Golden' Kuhol Management.* Philippines Department of Agriculture and Food and Agriculture Organization of the United Nations, Manila, 44 pp.

Anon. (1993) Golden apple snail present in Papua New Guinea. *FAO Plant Protection Bulletin* 41(3), 35–36.

Arthur, J.S., Taylor, E.J. and Bowen, I.D. (1996) Optimising the chemical control of the rice pest *Pomacea canaliculata* (Lamarck). In: Henderson, I.F. (ed.) *Slug and Snail Pests in Agriculture.* Symposium Proceedings No. 66, British Crop Protection Council, Farnham, pp. 389–396.

Baker, G.H. (1998) The golden apple snail, *Pomacea canaliculata* (Lamarck) (Mollusca: Ampullariidae), a potential invader of freshwater habitats in Australia. In: Zalucki, M.P., Drew, R.A.I. and White, G.G. (eds) *Pest Management – Future Challenges. Proceedings of the Sixth Australasian Applied Entomological Research Conference, Brisbane, Australia, 29 September–2 October, 1998*, Vol. 2. University of Queensland Printery, Brisbane, pp. 21–26.

Baskin, Y. (1996) Curbing undesirable aliens. *BioScience* 46, 732–736.

Beissinger, S.R., Donnay, T.J. and Walton, R. (1994) Experimental analysis of diet specialization in the snail kite: the role of behavioral conservatism. *Oecologia* 100, 54–65.

Berthold, T. (1988) Anatomy of *Afropomus balanoides* (Mollusca, Gastropoda, Ampullariidae) and its implications for phylogeny and ecology. *Zoomorphology* 108, 149–159.

Berthold, T. (1989) Comparative conchology and functional morphology of the copulatory organ of the Ampullariidae (Gastropoda, Monotocardia) and their bearing upon phylogeny and palaeontology. *Abhandlungen des Naturwissenschaftlichen Vereins in Hamburg (nf)* 28, 141–164.

Berthold, T. (1991) Vergleichende Anatomie, Phylogenie und historische Biogeographie der Ampullariidae (Mollusca, Gastropoda). *Abhandlungen des Naturwissenschaftlichen Vereins in Hamburg (nf)* 29, 1–256.

Bieler, R. (1992) Gastropod phylogeny and systematics. *Annual Review of Ecology and Systematics* 23, 311–338.

Bieler, R. (1993) Ampullariid phylogeny – book review and cladistic re-analysis. *The Veliger* 36, 291–299.

Bombeo-Tuburan, I., Fukumoto, S. and Rodriguez, E.M. (1995) Use of the golden apple snail, cassava, and maize as feeds for the tiger shrimp, *Penaeus monodon*, in ponds. *Aquaculture* 131, 91–100.

Bourne, G.R. and Berlin, J.A. (1982) Predicting *Pomacea dolioides* (Reeve) (Prosobranchia: Ampullariidae) weights from linear measurements of their shells. *The Veliger* 24, 367–370.

Brönmark, C. (1989) Interactions between epiphytes, macrophytes and freshwater snails: a review. *Journal of Molluscan Studies* 55, 299–311.

Bryan, D.C. (1990) Apple snail densities at Alexander Springs, Lake County, and observations on snail ecology. *Florida Scientist* 53 (suppl. 1), 13.

Burky, A.J. (1973) Organic content of eggs and juveniles of an amphibious snail, *Pomacea urceus* (Müller), from the Venezuelan savannah and their ecological significance. *Malacological Review* 6, 59.

Burky, A.J. (1974) Growth and biomass production of an amphibious snail, *Pomacea urceus* (Müller), from the Venezuelan savannah. *Proceedings of the Malacological Society of London* 41, 127–143.

Burky, K.A. and Burky, A.J. (1977) Buoyancy changes as related to respiratory behavior in an amphibious snail, *Pomacea urceus* (Müller), from Venezuela. *The Nautilus* 91, 97–104.

Burky, A.J., Pacheco, J. and Pereyra, E. (1972) Temperature, water, and respiratory regimes of an amphibious snail, *Pomacea urceus* (Müller), from the Venezuelan savannah. *Biological Bulletin* 143, 304–316.

Cameron, R.A.D. and Carter, M.A. (1979) Intra- and interspecific effect of population density on growth and activity in some helicid snails (Gastropoda: Pulmonata). *Journal of Animal Ecology* 48, 237–246.

Carlton, J.T. (1999) A journal of biological invasions. *Biological Invasions* 1, 1.

Cazzaniga, N.J. (1987) *Pomacea canaliculata* (Lamarck, 1801) en Catamarca (Argentina) y un comentario sobre *Ampullaria catamarcensis* Sowerby, 1874 (Gastropoda, Ampullariidae). *Iheringia, Série Zoologia* 66, 43–68.

Cazzaniga, N.J. (1990a) Sexual dimorphism in *Pomacea canaliculata* (Gastropoda: Ampullariidae). *The Veliger* 33, 384–388.

Cazzaniga, N.J. (1990b) Predation of *Pomacea canaliculata* (Ampullariidae) on adult *Biomphalaria peregrina* (Planorbidae). *Annals of Tropical Medicine and Parasitology* 84, 97–100.

Cazzaniga, N.J. and Estebenet, A.L. (1988) Effects of crowding on breeding *Pomacea canaliculata* (Gastropoda: Ampullariidae). *Comparative Physiology and Ecology* 13, 89–96.

Cedeño-León, A. and Thomas, J.D. (1983) The predatory behaviour of *Marisa cornuarietis* on eggs and neonates of *Biomphalaria glabrata*, the snail host of *Schistosoma mansoni*. *Malacologia* 24, 289–297.

Cerutti, R. (1998) An infestation of *Pomacea canaliculata* (Lamarck, 1804) in Lake Miramar, San Diego, California. *The Festivus* 30, 25–27, 29.

Chandrasekharam, V., Satyam, P., Srikanth, N.S., Naidu, K.A., Raghavaiah, K. and Ramamurthi, R. (1982) Retention of label of leucine-U-^{14}C in the haemolymph of *Pila globosa* (Swainson) as a function of sex and long-term aestivation. *The Veliger* 24, 373–374.

Chang, W.-C. (1985) The ecological studies on the *Ampullaria* snails (Cyclophoracea: Ampullaridae [*sic*]). *Bulletin of Malacology, Republic of China* 11, 43–51.

Chao, D., Lin, C.-C. and Chen, Y.-A. (1987) Studies on growth and distribution of *Angiostrongylus cantonensis* larvae in *Ampullarium canaliculatus*. *Southeast Asian Journal of Tropical Medecine and Public Health* 18, 248–252.

Cheng, E.Y. (1989) Control strategy for the introduced snail *Pomacea lineata*, in rice paddy. In: Henderson, I.F. (ed.) *Slugs and Snails in World Agriculture.* Monograph 41, British Crop Protection Council, Thornton Heath, pp. 69–75.
Civeyrel, L. and Simberloff, D. (1996) A tale of two snails: is the cure worse than the disease? *Biodiversity and Conservation* 5, 1231–1252.
Clench, W.J. (1966) *Pomacea bridgesi* (Reeve) in Florida. *The Nautilus* 79, 105.
Comfort, A. (1947) Lipochromes in the ova of *Pila. Nature* 160, 333.
Cowie, R.H. (1985) Microhabitat choice and high temperature tolerance in the land snail *Theba pisana. Journal of Zoology, London (A)* 207, 201–211.
Cowie, R.H. (1994) *Freshwater Snail Surveys in Relation to Eradication of Apple Snails with Copper Sulphate on Kauai. Bishop Museum Technical Reports 5*, 26 pp.
Cowie, R.H. (1995a) Report on a visit to Cambodia to advise on apple snails as potential rice pests. Unpublished report to Cambodia–IRRI–Australia Project, Phnom Penh, Bishop Museum, Honolulu, 8 + [2] pp.
Cowie, R.H. (1995b) Identity, distribution and impacts of introduced Ampullariidae and Viviparidae in the Hawaiian Islands. *Journal of Medical and Applied Malacology* 5[1993], 61–67.
Cowie, R.H. (1996) New records of introduced land and freshwater snails in the Hawaiian Islands. *Bishop Museum Occasional Papers* 46, 25–27.
Cowie, R.H. (1997a) *Pila* Röding, 1798 and *Pomacea* Perry, 1810 (Mollusca, Gastropoda): proposed placement on the Official List, and AMPULLARIIDAE Gray, 1824: proposed confirmation as the nomenclaturally valid synonym of PILIDAE Preston, 1915. *Bulletin of Zoological Nomenclature* 54, 83–88.
Cowie, R.H. (1997b) Catalog and bibliography of the non-indigenous non-marine snails and slugs of the Hawaiian Islands. *Bishop Museum Occasional Papers* 50, 1–66.
Cowie, R.H. (2001) Can snails ever be effective and safe biocontrol agents? *International Journal of Pest Management* 47, 23–40.
Crooks, J. and Soulé, M.E. (1996) Lag times in population explosions of invasive species: causes and implications. In: Sandlund, O.T., Schei, P.J. and Viken, A. (eds) *Proceedings of the Norway/UN Conference on Alien Species. Trondheim, 1–5 July 1996.* Directorate for Nature Management and Norwegian Institute for Nature Research, Trondheim, pp. 39–46.
de León, R.P. (1989) Description of *Temnocephala haswelli* n. sp. (Platyhelminthes) from the mantle cavity of *Pomacea canaliculata* (Lamarck). *Journal of Parasitology* 75, 524–526.
Demian, E.S. and Lutfy, R.G. (1966) Factors affecting the predation of *Marisa cornuarietis* on *Bulinus* (*B.*) *truncatus*, *Biomphalaria alexandrina* and *Lymnaea caillaudi. Oikos* 17, 212–230.
Demian, E.S. and Yousif, F. (1973) Embryonic development and organogenesis in the snail *Marisa cornuarietis* (Mesogastropoda: Ampullariidae) I. General outlines of development. *Malacologia* 12, 123–150.
Demian, E.S. and Yousif, F. (1975) Embryonic development and organogenesis in the snail *Marisa cornuarietis* (Mesogastropoda: Ampullariidae). V. Development of the nervous system. *Malacologia* 15, 29–42.
Donnay, T.J. and Beissinger, S.R. (1993) Apple snail (*Pomacea dolioides*) and freshwater crab (*Dilocarcinus dentatus*) population fluctuations in the llanos of Venezuela. *Biotropica* 25, 206–214.
Efferson, J.N. (1968) A rice seed farm in Central America. *The Rice Journal* 71, 6, 8–10.

Eldredge, L.G. (1994) *Perspectives in Aquatic Exotic Species Management in the Pacific Islands*. Vol. 1. *Introductions of Commercially Significant Aquatic Organisms to the Pacific Islands*. South Pacific Commission, Noumea, v + 127 pp.

Estebenet, A.L. (1995) Food and feeding in *Pomacea canaliculata* (Gastropoda: Ampullariidae). *The Veliger* 38, 277–283.

Estebenet, A.L. and Cazzaniga, N.J. (1992) Growth and demography of *Pomacea canaliculata* (Gastropoda: Ampullariidae) under laboratory conditions. *Malacological Review* 25, 1–12.

Estebenet, A.L. and Cazzaniga, N.J. (1998) Sex-related differential growth in *Pomacea canaliculata* (Gastropoda: Ampullaridae). *Journal of Molluscan Studies* 64, 119–123.

Eversole, A.G. (1992) Golden snail problem on rice: report for short- and long-term steps for management. Unpublished report. IRRI, Los Baños, 12 pp.

Freiburg, M.W. and Hazelwood, D.H. (1977) Oxygen consumption of two amphibious snails: *Pomacea paludosa* and *Marisa cornuarietis* (Prosobranchia: Ampullariidae). *Malacologia* 16, 541–548.

Freitas, J.R., Bedé, L.C., de Marco, P., Rocha, L.A. and Santos, M.B.L. (1987) Population dynamics of aquatic snails in Pampulha reservoir. *Memorias do Instituto Oswaldo Cruz, Rio de Janeiro* 82 (suppl. 4), 299–305.

Fujio, Y. and von Brand, E. (1990) Two linked loci with a null allele for leucine aminopeptidase isozymes in apple snail *Pomacea canaliculata. Nippon Suisan Gakkaishi* 56, 1039–1043.

Fujio, Y., Kurihara, H. and von Brand, E. (1991a) Differences metric traits among three strains of apple snail, *Pomacea canaliculata. Tohuku Journal of Agricultural Research* 41(3–4), 61–68.

Fujio, Y., von Brand, E. and Kobayashi, M. (1991b) Apparent differential hatchabilities associated with degrees of heterozygosity at leucine aminopeptidase isozyme loci in the apple snail *Pomacea canaliculata. Nippon Suisan Gakkaishi* 57, 459–461.

Gammon, E.T. (1943) Helicid snails in California. *Bulletin of the State of California Department of Agriculture* 32, 173–187.

Geijskes, D.C. and Pain, T. (1957) Suriname freshwater snails of the genus *Pomacea. Studies on the Fauna of Suriname and Other Guyanas* 1(3), 41–48, 2 pls.

Glover, N. and Campbell, C. (1994) *Apple Snails in Wetland Taro*. Pacific Islands Farm Manual, Taro Pest and Disease Leaflet 5, 4 pp.

Godan, D. (1983) *Pest Slugs and Snails. Biology and Control*. Springer Verlag, Berlin, 445 pp.

Guerrero, L. (1991) The biology of golden snail in relation to Philippine conditions. In: Acosta, B.O. and Pullin, R.S.V. (eds) *Environmental Impact of the Golden Snail (*Pomacea *sp.) on Rice Farming Systems in the Philippines*. Freshwater Aquaculture Center, Central Luzon State University, Munoz, Nueva Ecija, and ICLARM, Manila, pp. 10–11.

Guimarães, C.T. (1981a) Algumas observações de laboratória sobre biologia e ecologia de *Pomacea haustrum* (Reeve, 1856). *Memorias do Instituto Oswaldo Cruz, Rio de Janeiro* 76, 33–46.

Guimarães, C.T. (1981b) Algumas observações de campo sobre biologia e ecologia de *Pomacea haustrum* (Reeve, 1856) (Mollusca, Pilidae). *Memorias do Instituto Oswaldo Cruz, Rio de Janeiro* 76, 343–351.

Halwart, M. (1994a) The golden apple snail *Pomacea canaliculata* in Asian rice farming systems: present impact and future threat. *International Journal of Pest Management* 40, 199–206.

Halwart, M. (1994b) The ecology of fish in rice fields. *Annals of Tropical Research* 10 (*Special Issue on Tropical Ecology* 1), 1–15.

Hanning, G.W. (1978) Oviposition of *Pomacea paludosa* (Say) in Lake Okeechobee, Florida. *Bulletin of the American Malacological Union* 1977, 90–91.

Hanning, G.W. and Leedom, W.S. (1978) Schistosome dermatitis from *Pomacea paludosa* (Say) (Prosobranchia: Pilidae). *The Nautilus* 92, 105–106.

Heidenreich, A. and Halwart, M. (1995) Fish farming and other integrated pest management strategies against pomacea snails in rice fields: a computer simulation. In: *Proceedings of the 4th Asian Fisheries Forum, 16–20 October 1995, Beijing, P.R. China*. Beijing, pp. 1–4.

Heidenreich, A., Poethke, H.-J., Halwart, M. and Seitz, A. (1997) Simulation der Populationsdynamik von *Pomacea canaliculata* (Prosobranchia) zur Bewertung von Managementmaßnahmen. *Verhandlungen der Gesellschaft für Ökologie* 27, 441–446.

Hendarsih, S., Suriapermana, S., Fagi, A. and Manwan, I. (1994) Potential of fish in rice–fish culture as a biological control agent of rice pests. In: dela Cruz, C.R. (ed.) *Role of Fish in Enhancing Ricefield Ecology and in Integrated Pest Management*. Agency for Agricultural Research and Development, Bogor, and ICLARM, Manila, pp. 32–33.

Horne, F.R. (1979) Comparative aspects of estivating metabolism in the gastropod, *Marisa*. *Comparative Biochemistry and Physiology* 64A, 309–311.

Howarth, F.G. (1991) Environmental impacts of classical biological control. *Annual Review of Entomology* 36, 485–509.

Hunt, B.P. (1958) Introduction of *Marisa* into Florida. *The Nautilus* 72, 53–55.

Hunt, B.P. (1961) Tolerance of a fresh-water snail, *Marisa cornuarietis* L., to sea water. *Quarterly Journal of the Florida Academy of Sciences* 23, 278–284.

IUCN (1996) *1996 IUCN Red List of Threatened Animals*. IUCN, Gland, 70 + 368 + 10 pp.

ICZN (1999) Opinion 1913. *Pila* Röding, 1798 and *Pomacea* Perry, 1810 (Mollusca, Gastropoda): placed on the Official List, and Ampullariidae Gray, 1824: confirmed as the nomenclaturally valid synonym of *Pilidae* Preston, 1915. *Bulletin of Zoological Nomenclature* 56, 74–76.

Jepson, P.C. (ed.) (1989) *Pesticides and Non-target Invertebrates*. Intercept, Wimborne, xiv + 240 pp.

Johnson, D.S. (1958) Size variation in the edible pond-snail. *Malayan Nature Journal* 12, 185–186.

Joshi, R.C., Delacruz, M.S., Martin, E.C., Cabigat, J.C., Bahatan, R.G., Bahatan, A.D., Abayao, E.H., Choy-Awan, J., Chilagan, N.P. and Cayong, A.B. (2001) Current status of the golden apple snail in the Ifugao rice terraces, Philippines. *Journal of Sustainable Agriculture* 18, 71–90.

Keawjam, R.S. (1986) The apple snails of Thailand: distribution, habitats and shell morphology. *Malacological Review* 19, 61–81.

Keawjam, R.S. (1987) The apple snails of Thailand: aspects of comparative anatomy. *Malacological Review* 20, 69–89.

Keawjam, R.S. (1990) The apple snails of Thailand: molecular genetics. *Journal of Medical and Applied Malacology* 2, 1–48.

Keawjam, R.S. and Upatham, E.S. (1990) Shell morphology, reproductive anatomy and genetic patterns of three species of apple snails of the genus *Pomacea* in Thailand. *Journal of Medical and Applied Malacology* 2, 49–62.

Kobayashi, W., Cowie, R.H. and Glover, N. (1993) Escargot has got to go. *The Taro Tattler* 5, 2.

Kushlan, J.A. (1975) Population changes in the apple snail, *Pomacea paludosa*, in the southern Everglades. *The Nautilus* 89, 21–23.

Kushlan, J.A. (1978) Predation on apple snail eggs (*Pomacea*). *The Nautilus* 92, 57–58.

Lacanilao, F. (1990) Reproduction of the golden apple snail (Ampullaridae [*sic*]): egg mass, hatching, and incubation. *Philippine Journal of Science* 119, 95–105.

Lach, L. and Cowie, R.H. (1999) The spread of the introduced freshwater apple snail *Pomacea canaliculata* (Lamarck) (Gastropoda: Ampullariidae) on O'ahu, Hawai'i. *Bishop Museum Occasional Papers* 58, 66–71.

Laup, S. (1991) Golden apple snail and its eradication in Papua New Guinea. In: Kumar, R. (ed.) *Proceedings of a Seminar on Pests and Diseases of Food Crops – Urgent Problems and Practical Solutions.* Department of Agriculture and Livestock, Konedobu, pp. 55–62.

Leedom, W.S. and Short, R.B. (1981) *Cercaria pomaceae* sp. n., a dermatitis-producing schistosome cercaria from *Pomacea paludosa*, the Florida apple snail. *Journal of Parasitology* 67, 257–261.

Litsinger, J.A. and Estano, D.B. (1993) Management of the golden apple snail *Pomacea canaliculata* (Lamarck) in rice. *Crop Protection* 12, 363–370.

Little, C. (1968) Aestivation and ionic regulation in two species of *Pomacea* (Gastropoda, Prosobranchia). *Journal of Experimental Biology* 48, 569–585.

Little, C. (1981) Osmoregulation and excretion in prosobranch gastropods. Part I: physiology and biochemistry. *Journal of Molluscan Studies* 47, 221–247.

Louda, S.M. and McKaye, K.R. (1982) Diurnal movements in populations of the prosobranch *Lanistes nyassanus* at Cape Maclear, Lake Malawi, Africa. *Malacologia* 23, 13–21.

Lum-Kong, A. (1989) The potential of *Pomacea urceus* as a culture species in Trinidad. In: Henderson, I.F. (ed.) *Slugs and Snails in World Agriculture.* Monograph 41, British Crop Protection Council, Thornton Heath, pp. 33–39.

Lum-Kong, A. and Kenny, J.S. (1989) The reproductive biology of the ampullariid snail *Pomacea urceus* (Müller). *Journal of Molluscan Studies* 55, 53–65.

McClary, A. (1964) Surface inspiration and ciliary feeding in *Pomacea paludosa* (Prosobranchia: Mesogastropoda: Ampullariidae). *Malacologia* 2, 87–104.

McLane, W.A. (1939) *Pomacea paludosa* a predator on the brown darter. *The Nautilus* 52, 141–142.

McNeely, J.A. (1996) Costs and benefits of alien species. In: Sandlund, O.T., Schei, P.J. and Viken, A. (eds) *Proceedings of the Norway/UN conference on Alien Species, Trondheim, 1–5 July 1996.* Directorate for Nature Management and Norwegian Institute for Nature Research, Trondheim, pp. 176–181.

Madhu, C., Rao, K.J., Rao, K.S. and Rao, K.V.R. (1982) Comparative evaluation of the toxicity of lindane and sumithion on freshwater snail, *Pila globosa* (Swainson). *Geobios* 9, 181–183.

Maini, P.M. and Morallo-Rejesus, B. (1992) Toxicity of some volatile oils against golden snail (*Pomacea* spp.). *Philippine Journal of Science* 121, 391–397.

Maini, P.M. and Morallo-Rejesus, B. (1993) Molluscicidal activity of *Derris elliptica* (Fam. *Leguminosae*). *Philippine Journal of Science* 122, 61–75.

Marwoto, R.M. (1988) The occurrence of a freshwater snail *Pomacea* sp. in Indonesia. *Treubia* 29, 275–276.

Meenakshi, V.R. (1964) Aestivation in the Indian apple snail *Pila* – 1. Adaptation in natural and experimental conditions. *Comparative Biochemistry and Physiology* 11, 379–386.

Michelson, E.H. (1961) On the generic limits in the family Pilidae (Prosobranchia: Mollusca). *Breviora* 133, 1–10.

Mochida, O. (1988a) Nonseedborne rice pests of quarantine importance. In: *Rice Seed Health*. IRRI, Los Baños, pp. 117–129.

Mochida, O. (1988b) The rice water weevil (*Lissorhoptrus oryzophilus* Kuschel) and the freshwater snail (*Pomacea canaliculata* (Lamarck)) as important pests of crops for plant quarantine in Asia. *Asean Planti* 1988, 71–80.

Mochida, O. (1991) Spread of freshwater *Pomacea* snails (Pilidae, Mollusca) from Argentina to Asia. *Micronesica, Supplement* 3, 51–62, 2 pls.

Naylor, R. (1996) Invasions in agriculture: assessing the cost of the golden apple snail in Asia. *Ambio* 25, 443–448.

Neck, R.W. (1984) Occurrence of the striped ram's horn snail, *Marisa cornuarietis*, in central Texas (Ampullariidae). *The Nautilus* 98, 119–120.

Neck, R.W. (1986) A second record of an introduced apple snail, *Pomacea canaliculata*, from the lower Rio Grande valley of Texas. *Texas Conchologist* 22(3), 54–57.

Neck, R.W. and Schultz, J.G. (1992) First record of a living channeled applesnail, *Pomacea canaliculata* (Pilidae), from Texas. *Texas Journal of Science* 44, 115–116.

Ng, P.K.L., Chou, L.M. and Lam, T.J. (1993) The status and impact of introduced freshwater animals in Singapore. *Biological Conservation* 64, 19–24.

Niemelä, P. and Mattson, W.J. (1996) Invasion of North American forests by European phytophagous insects. *BioScience* 46, 741–753.

Okuma, M., Tanaka, K. and Sudo, S. (1994) Weed control method using apple snail (*Pomacea canaliculata*) in paddy fields. *Weed Research, Japan* 39, 114–119.

Olivares, F.M. Jr, Medina, J.R. Jr and Obien, S.R. (1992) Integrated pest management activities on the golden snail in the Philippines: a progress report. Unpublished report presented to the Crop and Resource management Networks Joint Meeting of INSURF, ARFSN, and IPMN, 12–17 October 1993, Vietnam. 36 pp.

Pain, T. (1950a) A new species of *Pomacea* (*Limnopomus*) from Venezuela. *Journal of Conchology* 23, 109–111.

Pain, T. (1950b) *Pomacea* (Ampullariidae) of British Guiana. *Proceedings of the Malacological Society of London* 28, 63–74, 2 pls.

Pain, T. (1960) *Pomacea* (Ampullariidae) of the Amazon River system. *Journal of Conchology* 24, 421–432.

Pain, T. (1964) The *Pomacea flagellata* complex in Central America. *Journal of Conchology* 25, 224–231, 1 pl.

Pain, T. (1972) The Ampullariidae, an historical survey. *Journal of Conchology* 27, 453–462.

Palis, F.V., Macatula, R.F. and Browning, L. (1996) Niclosamide, an effective molluscicide for the golden apple snail (*Pomacea canaliculata* Lamarck) control in Philippine rice production systems. In: Henderson, I.F. (ed.) *Pests in Agriculture*. Symposium Proceedings No. 66, British Crop Protection Council, Farnham, 213–230.

Palomino, M.L.P. and Jueco, N.L. (1983) Radular pattern in three edible freshwater snails. *Philippine Journal of Biology* 12, 174–176.

Pedini, M. and Shehadeh, Z.H. (1996) Projects and other activities. *FAO Aquaculture Newsletter* 14, 24–30.

Peebles, C.R., Oliver-González, J. and Ferguson, F.F. (1972) Apparent adverse effect of *Marisa cornuarietis* upon *Lymnaea columella* and *Biomphalaria glabrata* in an ornamental pond in Puerto Rico. *Proceedings, Hawaiian Entomological Society* 21, 247–256.

Perera, G. and Walls, J.G. (1996) *Apple Snails in the Aquarium.* T.F.H. Publications, Neptune City, New Jersey, 121 pp.

Perera, G. and Yong, M. (1990) Seasonal studies on *Pomacea paludosa* in Cuba. *Florida Scientist* 53 (suppl. 1), 15.

Perry, M.C. (1974) Ecological studies of the apple snail at lake Woodruff National Wildlife Refuge. *Florida Scientist* 36, 22–30.

Peterson, R.T. (1980) *A Field Guide to the Birds*, 4th edn. Houghton Mifflin, Boston, 384 pp.

Pilsbry, H.A. and Bequaert, J. (1927) The aquatic mollusks of the Belgian Congo. With a geographical and ecological account of Congo malacology. *Bulletin of the American Museum of Natural History* 53, 69–602, 78 pls.

Pleadwell, B.A. (1997) Raising apple snails for escargot vs. getting rid of them. *Ka Wai Ola o OHA* 14(4), 3, 6.

Pointier, J.P. and Guyard, A. (1992) Biological control of the snail intermediate hosts of *Schistosoma mansoni* in Martinique, French West Indies. *Tropical Medicine and Parasitology* 43, 98–101.

Pointier, J.P., Théron, A. and Imbert-Establet, D. (1988) Decline of a sylvatic focus of *Schistosoma mansoni* in Guadeloupe (French West Indies) following competitive displacement of the snail host *Biomphalaria glabrata* by *Ampullaria glauca*. *Oecologia* 75, 38–43.

Pointier, J.P., Théron, A., Imbert-Establet, D. and Borel, G. (1991) Eradication of a sylvatic focus of *Schistosoma mansoni* using biological control by competitor snails. *Biological Control* 1, 244–247.

Ponder, W.F. and Warén, A. (1988) Classification of the Caenogastropoda and Heterostropha – a list of the family-group names and higher taxa. *Malacological Review, Supplement* 4, 288–326.

Prashad, B. (1925a) Revision of the Indian Ampullariidae. *Memoirs of the Indian Museum* 8, 69–89, 3 pls.

Prashad, B. (1925b) Anatomy of the common Indian apple-snail, *Pila globosa. Memoirs of the Indian Museum* 8, 91–151, 3 pls.

Rao, M.R., Rao, K.S.P., Reddy, M.S., Rao, K.R.S.S. and Reddy, N.M. (1983) Toxicity of phenthoate and changes in the organic constituents of the snail, *Pila globosa* (Swainson) under sublethal and lethal impacts. *Geobios* 10, 250–253.

Robins, C.H. (1971) Ecology of the introduced snail, *Marisa cornuarietis* (Ampullariidae) in Dade County, Florida. *The Biologist* 53, 136–152.

Rondon, M.B. and Callo, D.P. Jr. (1991) Distribution and mode of infestation of golden snail in rice farming. In: Acosta, B.O. and Pullin, R.S.V. (eds) *Environmental Impact of the Golden Snail (*Pomacea *sp.) on Rice Farming Systems in the Philippines.* Freshwater Aquaculture Center, Central Luzon State University, Munoz, Nueva Ecija, and ICLARM, Manila, p. 12.

Santos, C.A.Z., Penteado, C.H.S. and Mendes, E.G. (1987) The respiratory responses of an amphibious snail *Pomacea lineata* (Spix, 1827), to temperature and oxygen tension variations. *Comparative Biochemistry and Physiology* 86A, 409–415.

Schnorbach, H.-J. (1995) The golden apple snail (*Pomacea canaliculata* Lamarck), an increasingly important pest in rice, and methods of control with Bayluscid. *Pflanzenschutz-Nachrichten Bayer* 48, 313–346.

Scott, M.I.H. (1957) Estudio morfologico y taxonomico de los ampullaridos de la Republica Argentina. *Revista del Museo Argentino de Ciencias Naturales 'Bernardino Rivadavia' e Instituto Nacional de Investigacion de las Ciencias Naturales. Ciencias Zoologicas* 3, 231–333, 23 pls.

Settle, W.H., Ariawan, H., Astuti, E.T., Cahyana, W., Hakim, A.L., Hindayana, D., Lestari, A.S. and Sartanto, P. (1996) Managing tropical rice pests through conservation of generalist natural enemies and alternative prey. *Ecology* 77, 1975–1988.

Simberloff, D. and Stiling, P. (1996a) Risks of species introduced for biological control. *Biological Conservation* 78, 185–192.

Simberloff, D. and Stiling, P. (1996b) How risky is biological control? *Ecology* 77, 1965–1974.

Singh, O. and Agarwal, R.A. (1981) Toxicity of certain pesticides to two economic species of snails in northern India. *Journal of Economic Entomology* 74, 568–571.

Smith, B.D. (1992) Introduction and dispersal of apple snails (Ampullariidae) on Guam. *Pacific Science Association Information Bulletin* 44, 12–14.

Snyder, N.F.R. and Snyder, H.A. (1971) Defenses of the Florida apple snail *Pomacea paludosa. Behaviour* 40, 175–215, 4 pls.

Tamaru, C.S. (1996) Control of the apple snail (*Pomacea canaliculata*), planning project. Six month report: contract 40785. Unpublished Sea Grant Miscellaneous Report, University of Hawaii Sea Grant College Program, Honolulu, 18 + [5] + 24 pp.

Tamaru, C.[S.] and Hun, F. (1996) Survey of apple snails for Hui Na'auao. Unpublished report, University of Hawaii Sea Grant College Program, Honolulu, 13 pp.

Taylor, E.J., Arthur, J.S. and Bowen, I.D. (1996) Kill or cure? Control of aquatic mollusc pests. In: Henderson, I.F. (ed.) *Pests in Agriculture*, Symposium Proceedings No. 66, British Crop Protection Council, Farnham, 199–204.

Thiengo, S.C. (1987) Observations on the morphology of *Pomacea lineata* (Spix, 1827) (Mollusca, Ampullariidae). *Memorias do Instituto Oswaldo Cruz, Rio de Janeiro* 82, 563–570.

Thiengo, S.C. (1989) On *Pomacea sordida* (Swainson, 1823) (Prosobranchia, Ampullariidae). *Memorias do Instituto Oswaldo Cruz, Rio de Janeiro* 84, 351–355.

Thiengo, S.C., Borda, C.E. and Araújo, J.L.B. (1993) On *Pomacea canaliculata* (Lamarck, 1822) (Mollusca; Pilidae: Ampullariidae). *Memorias do Instituto Oswaldo Cruz, Rio de Janeiro* 88, 67–71.

Thomas, K.J. (1975) Biological control of *Salvinia* by the snail *Pila globosa* Swainson. *Biological Journal of the Linnean Society* 7, 243–247.

Tompa, A.S. (1980) Studies on the reproductive biology of gastropods: part III. Calcium provision and the evolution of terrestrial eggs among gastropods. *Journal of Conchology* 30, 145–154.

Turner, R.L. and McCabe, C.M. (1990) Calcium source for protoconch formation in the Florida apple snail, *Pomacea paludosa* (Prosobranchia: Pilidae): more evidence for physiologic plasticity in the evolution of terrestrial eggs. *The Veliger* 33, 185–189.

van Dinther, J.B.M. (1956) Control of *Pomacea* (*Ampullaria*) snails in rice fields. *Agricultural Experiment Station, Paramaribo, Surinam, Bulletin* 68, 1–20.

van Dinther, J.[B.M.] (1962) Levenswijze en bestrijding van slakken in de rijstvelden van Suriname. *Mededelingen van de Landbouwhogeschool en de Opzoekingsstations van de Staat te Gent* 27, 670–676.

van Dinther, J.B.M. (1973) Molluscs in agriculture and their control. *World Crops* 25, 282–286.

van Dinther, J.B.M. and Stubbs, R.W. (1963) Summary of research on the control of rice snails in Surinam. *Bulletin of the Agricultural Experiment Station, Paramaribo, Surinam* 82, 415–420.

Vitousek, P.M., D'Antonio, C.M., Loope, L.L. and Westbrooks, R. (1996) Biological invasions as global environmental change. *American Scientist* 84, 468–478.

von Brand, E., Yokasawa, T. and Fujio, Y. (1990) Chromosome analysis of apple snail *Pomacea canaliculata. Tohuku Journal of Agricultural Research* 40, 81–89.

Wada, T. (1997) Introduction of the apple snail *Pomacea canaliculata* and its impact on rice agriculture. In: *Proceedings of an International Workshop on Biological Invasions of Ecosystems by Pests and Beneficial Organisms.* National Institute of Agro-Environmental Sciences, Ministry of Agriculture, Forestry and Fisheries, Tsukuba, pp. 170–180.

Wallace, G.D. and Rosen, L. (1969) Studies on eosinophilic meningitis V. Molluscan hosts of *Angiostrongylus cantonensis* on Pacific Islands. *American Journal of Tropical Medicine and Hygiene* 18, 206–216.

Williamson, M.H. (1996) *Biological Invasions.* Chapman & Hall, London, xii + 244 pp.

Williamson, M.H. and Fitter, A. (1996) The varying success of invaders. *Ecology* 77, 1661–1666.

Winner, B.E. (1989) A comparison of three species of *Pomacea* and their spawn. *American Conchologist* 17(3), 15–16.

Winner, B.E. (1996) A comparison of four species of *Pomacea* and their spawn. *Hawaiian Shell News* 44(7–8), 5–7.

Wood, T.G. and Cowie, R.H. (1988) Assessment of on-farm losses in cereals in Africa due to soil insects. *Insect Science and its Application* 9, 709–716.

Yusa, Y. (2001) Predation on eggs of the apple snail *Pomacea canaliculata* (Gastropoda: Ampullariidae) by the fire ant *Solenopsis geminata. Journal of Molluscan Studies* 67, 275–279.

6 Helicidae and Hygromiidae as Pests in Cereal Crops and Pastures in Southern Australia

Geoff H. Baker

CSIRO Entomology, GPO Box 1700, Canberra, ACT 2601, Australia

Introduction

Four species of gastropod snails, *Theba pisana* (Müller) (Helicidae), *Cernuella virgata* (da Costa), *Cochlicella acuta* (Müller) and *Cochlicella barbara* (Linnaeus) (Hygromiidae), are introduced pests of grain crops and pastures in southern Australia (Baker, 1986; Fig. 6.1). *T. pisana* and *C. virgata* are known locally as white snails. *Cochlicella* species are called conical snails. All four species are native to the western Mediterranean and were first reported in southern Australia in the 1920s (Baker, 1986).

The distribution and abundance of white and conical snails has increased in southern Australia in recent years. Prior to the 1980s, infestations causing economic losses were confined to South Australia (SA) and Western Australia (WA). However, large numbers of these pests are now also being reported in western Victoria and southern New South Wales. The increase in abundance of white and conical snails seems to be due in part to changing agricultural practices, especially those aimed at improving soil conservation. Greater adoption of reduced tillage, with associated retention of stubble and pasture residues (instead of burning them prior to sowing new crops), has resulted in more gastropods surviving throughout crop rotations than did under previous conventional tillage practices. Research over the past 15 years has focused on improving our understanding of the biology of white and conical snails as a basis for developing sustainable strategies for their control. This chapter provides an overview of this research.

(ed. G.M. Barker)

Fig. 6.1. Distribution of white snails (*Theba pisana* (Müller), Helicidae; *Cernuella virgata* (da Costa), Hygromiidae) in Australia.

Pest Status in Australia

In early summer, white snails and *C. acuta* climb on to the heads, pods and stalks of cereals and legumes to aestivate. These gastropods often occur at sufficient abundance to clog machinery during the harvesting of infested crops, resulting in significant labour costs to farmers, who have to clear blockages and/or repair breakages in the machinery. The gastropods contaminate harvested grain, resulting in a product that is unacceptable to grain-handling authorities upon delivery to the silo or is downgraded in quality (e.g. malting to animal feed in barley, *Hordeum vulgare* Linnaeus (Gramineae)). Such downgrading represents a significant loss to the farmer (e.g. payment reduced from A$160 to $120 t^{-1}). Contaminated grain may be cleaned mechanically, but this is costly, and difficult when gastropod shells are comparable in size to cereal grains. This is particularly so for small conical snails. During seasons with abundant gastropods, significant areas of crops are not harvested, due to the likelihood of product rejection because of contamination. Shipments of barley originating in southern Australia have been rejected overseas, with substantial compensation payments paid by the Australian grain industry to overseas customers. Gastropod-contaminated produce has damaged Australia's international reputation as a supplier of good-quality grain and poses a serious threat to the marketing of future exports.

White snails and *C. barbara* feed on legume-based pastures, such as those based on annual medics and lucerne (*Medicago* Linnaeus spp.,

Fabaceae) and clovers (*Trifolium* Linnaeus spp., Fabaceae). In a field trial in a permanent pasture in the south-east of SA, *T. pisana* reduced herbage yield by 23% when confined to cages at normal field densities for 1 month in spring. Notably, the clover component was reduced by 75% (Baker, 1989a). In a similar experiment in the pasture phase of a cereal–pasture rotation on Yorke Peninsula, 83% of the herbage was lost after exposure to *T. pisana* for 2 months in winter (Baker, 1989a). Re-establishment of pastures in gastropod-infested areas is particularly difficult. Furthermore, domestic livestock reject pasture and hay that are heavily contaminated with gastropods (because of the mucus). Losses due to gastropods in permanent pastures in SA have been estimated, by farmers, to be 0.5–2.5 dry sheep equivalents (d.s.e.) ha^{-1} (normal carrying capacity approximately 10 d.s.e. ha^{-1}). These pests also cause significant losses in wheat (*Triticum aestivum* Linnaeus, Gramineae), barley and oilseed rape (*Brassica napus* Linnaeus var. *oleifera* Linnaeus, Brassicaceae) crops through the destruction of seedlings during crop establishment. A trial on Yorke Peninsula, SA, demonstrated that *T. pisana* and *C. virgata* can substantially reduce the numbers of wheat seedlings (control, *T. pisana*, *C. virgata*: 11.6, 0.2, 5.7 plants, respectively) (Baker, 1989a).

In addition to agricultural ecosystems, white and conical snails have invaded natural systems (e.g. sclerophyllous woodland and sand-dunes) in southern Australia, but their effects on the indigenous flora and fauna, and in particular the abundance of native gastropods, is essentially unknown. Serventy and Storr (1959) reported that the native *Bothriembryon melo* (Quoy & Gaimard) (Orthalicidae) was rare or extinct in areas invaded by *T. pisana* on Rottnest Island, WA, but quite common where *T. pisana* had yet to establish. White and conical snails are also intermediate hosts for a flukeworm, *Brachylaima* Dujardin sp. (Digenea: Brachylaimidae) (Cribb, 1990). Young children who have ingested infected gastropods in SA have suffered severe gut disorders.

Life History and Abundance

The life histories and population dynamics of white and conical snails have been intensively investigated in permanent pastures and pasture–cereal rotations in SA (Baker, 1988a, 1989a,b, 1996; Baker and Vogelzang, 1988; Baker and Hawke, 1990; Baker *et al.*, 1991). Measurements of the size of the albumen gland, as an indicator of reproductive condition, suggest that breeding in all species occurs from autumn through to spring (e.g. Fig. 6.2). Gastropods can commonly be found depositing clutches of eggs 1–2 cm below the surface of the soil in the field in winter. Laboratory studies suggest that white snails are capable of laying many clutches (Baker, 1991). For example, individuals of *T. pisana* and *C. virgata* taken from a pasture–cereal rotation in summer subsequently produced averages of 19 and 42 clutches, respectively in glasshouse cultures. Average clutch sizes were 70 eggs per clutch for *T. pisana* and 60 eggs per clutch

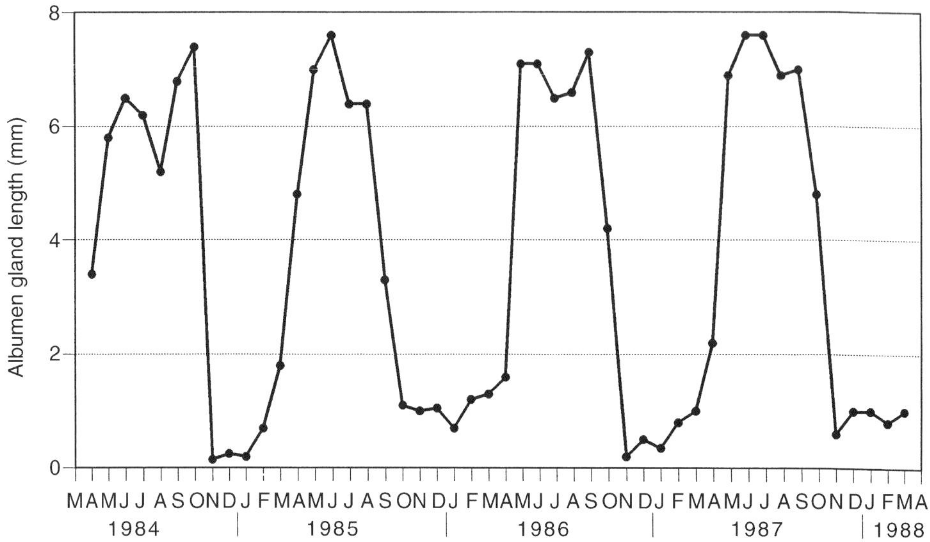

Fig. 6.2. Lengths of albumen glands in *Cernuella virgata* (da Costa) (Hygromiidae), collected from a pasture–cereal rotation at Weetulta, Yorke Peninsula, South Australia. The *C. virgata* were 12 mm in shell diameter. (Adapted from Baker, 1989b.)

for *C. virgata*. Largest clutches and most eggs were laid early in the season. In a similar glasshouse study, Baker and Hawke (1991) found that *C. acuta* individuals produced on average only seven clutches, with an average clutch size of 36 eggs. This suggests that the fecundity of conical snails is much less than that of the white snails. How many clutches any of these three species lay in the field is, however, unknown.

The life cycles of white snails in southern Australia may be annual or biennial. Baker and Vogelzang (1988) found two distinct cohorts of *T. pisana* were produced during the breeding season in a permanent pasture. One cohort reproduced when 1 year old while the other cohort, which was less numerous than the first, did not become reproductive until 2 years old. A small proportion of *T. pisana* lived for more than 2 years. The mechanism by which the two cohorts with different reproductive phenologies were generated in the pasture has not been explained. Possibly, oviposition and/or survival of young was reduced during the middle of the breeding season when conditions were coldest and wettest. The young produced earlier in the breeding season may simply have had more time to grow and/or more suitable food than the young produced at the end of the breeding season. Alternatively, genetic differences in growth rates may have existed between cohorts within the population.

In a roadside habitat and in a pasture–cereal rotation, Baker and Hawke (1990) found that *T. pisana* was predominantly biennial. Large numbers of *T. pisana* were produced during the pasture phase of the rotation, but production was limited during the crop phase (Fig. 6.3). Reasons for the relatively poor production in crops is not understood.

One-year-old *T. pisana* in crops certainly have large albumen glands (Baker and Hawke, 1990) indicating reproductive activity. Possibly the tilled soil beneath the crop is unsuitable for oviposition, but *T. pisana* oviposits readily in disturbed soil in the laboratory (Baker, 1991). Perhaps food is in short supply for the newly hatched young or the herbicides used to control weeds (in this case trifluralin and 2,4-dichlorophenoxyacetic acid (2,4-D) (ester)) are lethal to young *T. pisana*. Other herbicides have been shown to kill several gastropod species of (Godan, 1983). In contrast, *T. pisana* living near fence lines dividing fields in opposite phases of a pasture–cereal rotation can have annual life cycles, presumably because they have the opportunity for access to pastures (and perhaps high-quality food) in consecutive years.

Comparable variations in the life cycle of *C. virgata* have also been reported by Baker (1988a, 1989a,b). In permanent pastures the life cycle of *C. virgata* is annual, while in pasture–cereal rotations it is predominantly biennial. Similarly, *C. acuta* has a biennial life cycle in pasture–cereal rotations (Baker *et al.*, 1991). Like *T. pisana*, both these species recruit juveniles in the pasture phase of the rotation much more successfully than in the crop phase.

The life history and population dynamics of *T. pisana* have been extensively studied in Europe, Israel and South Africa, mostly in non-agricultural habitats (Durr, 1946; Nevo and Bar, 1976; Swart *et al.*, 1976; Heller, 1982; Cowie, 1984a,b,c; Moran, 1989; Avivi and Arad 1993; see Baker, 1986, for several other references). Both annual and biennial life cycles have been reported, occasionally within the same populations. There has been much debate as to the factors controlling the expression of these different life cycles, especially the role of climate. Recently, both Moran (1989) and Avivi and Arad (1993) have identified the production of two cohorts of young being produced during the winter months in Israel. One cohort grows fast and reproduces when 1 year old. The growth of the other cohort is inhibited and its reproduction is delayed a year. These findings on the life cycle of *T. pisana* in Israel are similar to those for *T. pisana* in permanent pastures in southern Australia. The variability in life cycle displayed by *T. pisana* probably serves as an insurance against a poor reproductive season, i.e. a spreading of risk (Den Boer, 1968).

Less is known of the life histories and population dynamics of *C. virgata* and *Cochlicella* de Férussac species in their native European and Mediterranean habitats (Mienis, 1969; Lewis, 1977; De Smet, 1983; see Baker, 1986, for other references). In general, these gastropods have been attributed with annual life cycles in Europe.

White and conical snails are patchily distributed within fields in SA. During summer, *T. pisana* and *C. virgata* aggregate on robust weeds, such as horehound (*Marrubium vulgare* Linnaeus; Lamiaceae), cut-leaf mignonette (*Reseda lutea* Linnaeus) and turnip weed (*Rapistrum rugosum* (Linnaeus); Allioni). As well as providing opportunities for the gastropods to climb and thus escape the hot soil surface in summer, some of these plants, by their leafy nature, provide shade (e.g. horehound). Living weeds probably provide more humid microclimates through their transpiration. When deprived of access to horehound in summer, *T. pisana* suffered high mortality (Baker and Vogelzang, 1988). Conical snails also aggregate on robust weeds (e.g. on cut-leaf mignonette), as well as at the bases of grass tussocks and beneath loose rocks (Baker *et al.*, 1991).

White and conical snails also aggregate in winter. For example, *T. pisana* aggregates on false caper (*Euphorbia terracina* Linnaeus; Euphorbiaceae) in pastures in winter (Baker and Vogelzang, 1988). The young are more aggregated in this way than older animals. The reasons for these winter aggregations are not understood, but they may in part reflect feeding preferences, mating attractions, distributions of oviposition sites or limited dispersals of young from nests.

Numbers of *T. pisana* are much higher in spring–early summer than at other times of the year, reflecting the recruitment of young (Fig. 6.3). Large numbers of these young gastropods die during summer, especially if the autumn rains commence a month or two late. This mortality is believed to be more related to starvation than to extreme weather conditions (Pomeroy, 1966, 1968; Hodson, 1969). Brief rain showers during summer may stimulate gastropod activity, but may reduce rather than increase survival if the gastropod expend more energy moving to and from aestivation sites than they gain through feeding. Butler (1972, 1976) recognized that the availability of food in winter and the accumulation of sufficient reserves for the coming summer could be crucial in determining the abundance of *C. virgata*.

It is widely recognized by farmers that contamination by gastropods is more of a problem with grain legumes than with cereal crops. However, G.H. Baker (unpublished data) has monitored *C. virgata* numbers in wheat–barley–bean rotations on Yorke Peninsula, SA, for 12 years and failed to find conclusive evidence that gastropod snails are more abundant in the legume crops. When grain legumes are harvested, sieve openings in the machinery are set much wider than when cereals are harvested and the crop is cut much closer to the ground. Presumably, greater problems are experienced with gastropods in legume crops simply because there is more chance for the snails to be included with the harvest.

The abundance of gastropods varies markedly between years (Baker, 1996). Figure 6.4 illustrates the numbers of *C. virgata* recorded in a field on Yorke Peninsula, SA, from 1985 to 1995. *C. virgata* numbers were much higher in 1992 than in the other years in which populations were sampled. Figure 6.5 gives tonnages of barley downgraded in SA because of gastropod contamination from 1990 to 1995 (data are not available for earlier years). These data also suggest that gastropods were extremely abundant in SA during 1992. Annual rainfall was much higher in 1992 than in other years (Fig. 6.4), reflecting higher than average precipitation in autumn and spring. Such weather should assist the survival of

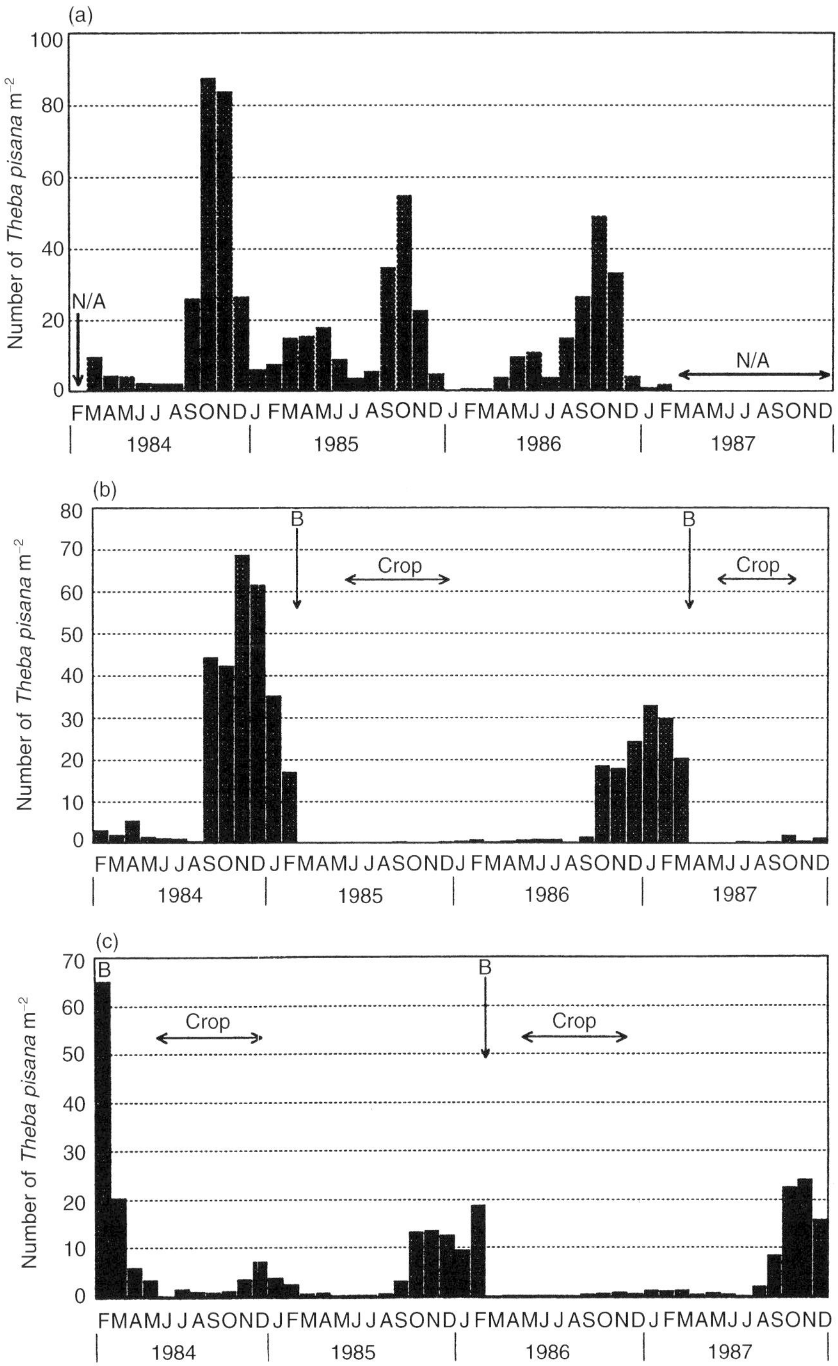

Fig. 6.3. Abundance of *Theba pisana* (Müller) (Helicidae) (m^{-2}) in (a) a permanent pasture at Mt Benson in the south-east of South Australia and (b and c) two adjacent fields in opposite phases of a barley–pasture rotation at Hardwicke Bay, Yorke Peninsula, South Australia. *T. pisana* sampled were > 6 mm in shell diameter. B indicates when fields were burnt prior to sowing crops. N/A, data not available.

gastropods through the end of the dry summer, stimulate breeding earlier in the year, thus giving rise to a longer breeding season, and exacerbate fouling of crops through extended activity in spring and greater invasion into crops from adjacent areas with high numbers (see following section on dispersal).

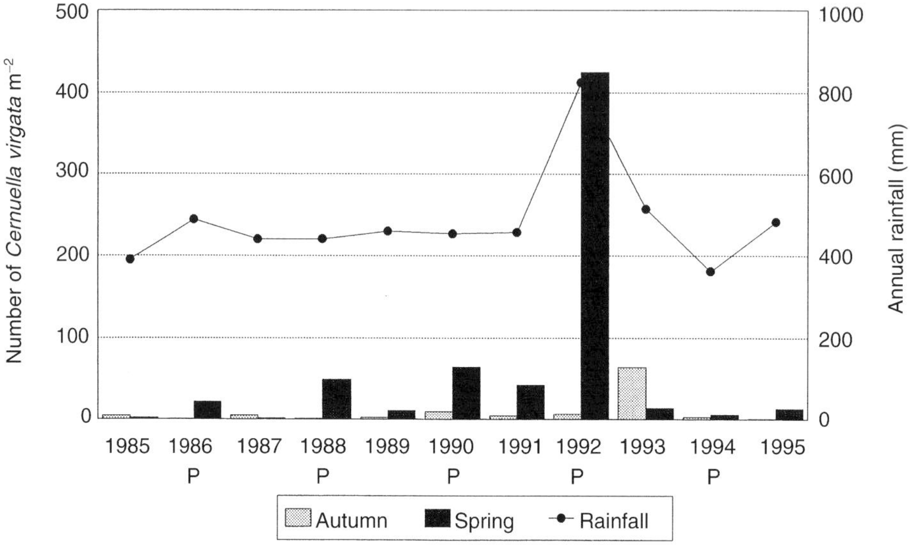

Fig. 6.4. Abundance of *Cernuella virgata* (da Costa) (Hygromiidae) (m^{-2}) during autumn and spring (bars) in a barley–pasture rotation at Weetulta, Yorke Peninsula, South Australia, from 1985 to 1995. The *C. virgata* sampled were > 6 mm in shell diameter. Annual rainfalls are also included (line), for the years 1985–1995. P, pasture phase of rotation.

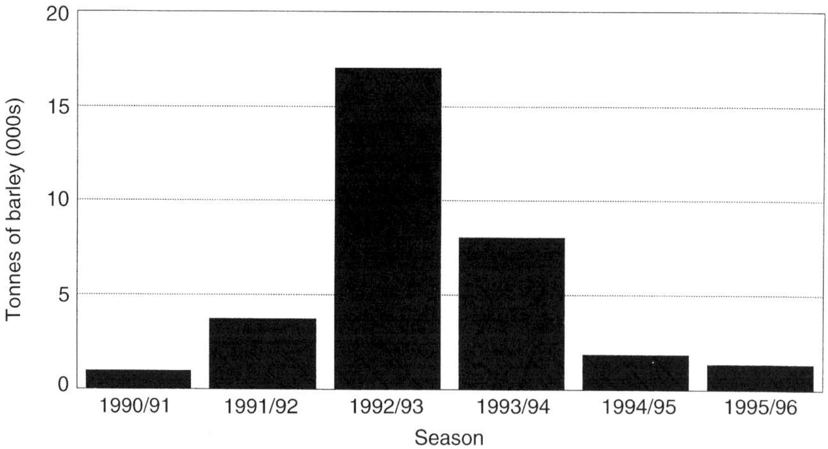

Fig. 6.5. Tonnages of barley (*Hordeum vulgare* Linnaeus, Gramineae) grain downgraded at South Australian silos because of contamination by gastropods during the period 1990–1995. Data kindly provided by Ken Saint, Australian Barley Board. (Adapted from Baker, 1996.)

Dispersal

The abundance of white snails in cereal fields is generally higher at the field margins, adjacent to pastures, than in the centres of the fields (Baker, 1989a,b). The same pattern occurs in well-grazed pastures that are bordered by weedy or shrubby wasteland (Baker, 1988a; Baker and Vogelzang, 1988). These patterns in abundance can be explained by seasonal movements of white snails. Baker (1988b,c, 1989a, 1992) and Baker and Hawke (1990) marked large numbers of white snails with different-coloured paints. They then released them and later recaptured them along transects that spanned pastures and adjacent cereal crops or roadside habitats. These authors were able to demonstrate that white snails were capable of moving relatively large distances and that the directions of movement were not random. For example, *T. pisana* and *C. virgata* moved out of a well-grazed pasture in late spring–early summer to aestivate upon tall weeds and small trees along a roadside (Baker, 1988b,c). Some white snails moved > 55 m in 1 month. In autumn–early winter, the white snails moved similar distances in the opposite direction (roadside to pasture). In another example, *C. virgata* moved from newly sown crops to adjacent pasture in winter and then back from pasture to maturing crop in spring (Baker, 1989a).

The cues for white snail movement are poorly understood. Perhaps the animals in the well-grazed pasture cited above could detect the taller vegetation at the roadside and oriented towards it, searching for aestivation sites. Peake (1978) suggested that gastropods can move towards shapes such as trees and shrubs that are silhouetted against the sky at night, and Zanforlin (1976) showed that the movement of *T. pisana* is biased towards large objects in the laboratory. Chase and Croll (1981) have argued that olfaction is the primary sense that terrestrial gastropods use for the detection and location of objects at a distance. Anemotaxis (moving upwind) is one suggested means by which gastropods orient to olfactory stimuli (Farkas and Shorey, 1976; Goodfriend, 1983). Perhaps the painted white snails released in the roadside in autumn detected preferable food odours in the pasture and moved towards them.

White snails move more in autumn and winter as adults than they do in spring as juveniles (Baker, 1992). They also move more when vegetation is sparse (Baker, 1988c). Baker (1992) mowed a weedy pasture to reduce the height of vegetation and showed that this increased the numbers of *T. pisana* invading an adjacent barley crop (Fig. 6.6). Crowding has been suggested by some authors to either increase or decrease the rate of dispersal by gastropods, while other authors could find no effect (Cain and Currey, 1968; Greenwood, 1974; Cameron and Williamson, 1977; Oosterhoff, 1977; Baur, 1993; Baur and Baur, 1994). There is as yet no evidence that crowding influences the distances moved by *T. pisana* and *C. virgata* in Australia (Baker, 1988b,c). However, dispersal has only been measured in experiments where white snails have temporarily been grouped in large numbers and then allowed to move

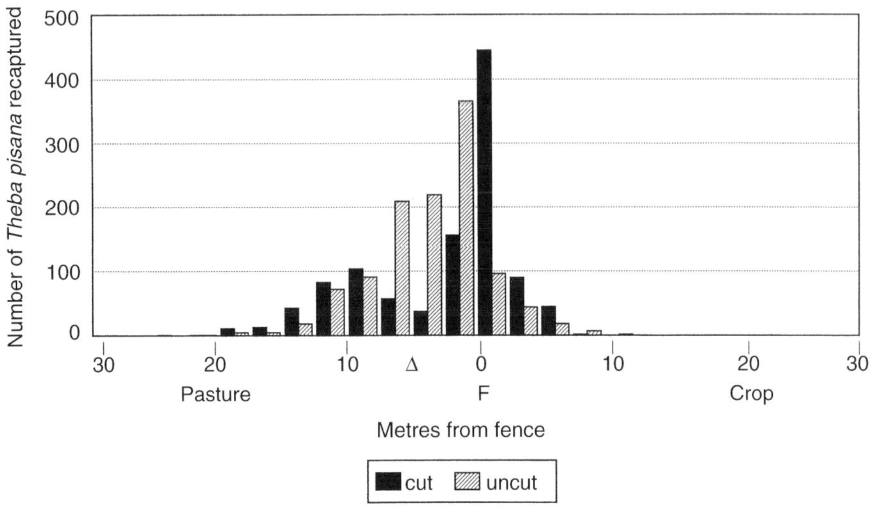

Fig. 6.6. Recoveries of marked *Theba pisana* (Müller) (Helicidae) at varying distances from a fence (F) dividing two fields at Hardwicke Bay, Yorke Peninsula, South Australia, during spring 1988. The fence was made of wire mesh, through which gastropods could easily pass. Four thousand *T. pisana* were released 5 m into the pasture (Δ) and were searched for after 44 days. Two thousand of these *T. pisana* were released in a plot within the pasture that had been mown to a height of 3 cm; the remaining snails were released in a nearby plot that had not been cut and was very weedy (pasture height = 30–40 cm). (Adapted from Baker, 1992.)

away to less crowded habitats. It would be of interest to measure the distances moved by white snails in naturally occurring dense and sparse populations.

Polymorphism

Both *T. pisana* and *C. virgata* are polymorphic with regard to shell-banding patterns. Both species can be completely white through to heavily marked with dark bands. The adaptive advantages of shell-banding patterns of helicid snails, including *T. pisana*, have been extensively debated (Nevo and Bar, 1976; Jones *et al.*, 1977; Heller, 1981; Heller and Gadot, 1984). Arguments have particularly centred on climatic tolerances (e.g. banded shells absorb more heat than unbanded ones) and cryptic coloration and avoidance of predators. Johnson (1980, 1981) and Hazel and Johnson (1990) reported that the frequency of unbanded *T. pisana* in sand-dunes in WA is higher in exposed habitats (e.g. low grasses) than in nearby more sheltered ones (e.g. thickets of *Acacia* Miller; Fabaceae). They found that banded *T. pisana* were more likely to move to sheltered habitats in summer than unbanded ones. Baker (1988a) and Baker and Vogelzang (1988) surveyed throughout the Australian distributions of *T. pisana* and *C. virgata*. They could show no relationship between the level of banding and climatic variables at the regional scale for either species. There was some evidence to suggest that banded white snails were more closely associated with sheltered habitats within fields and roadside vegetation. The majority of white snails in Australia are unbanded – hence their common name.

Intra- and Interspecific Interactions

The shell sizes of adult gastropod snails, including *T. pisana*, *C. virgata* and *C. acuta*, vary inversely with population size (Pomeroy, 1969; Butler, 1976; Williamson *et al.*, 1976; Cameron and Carter, 1979; Tattersfield, 1981; Heller, 1982; Baker, 1988a; Baur and Baur, 1990; Heller and Ittiel, 1990; Baker *et al.*, 1991). This negative relationship has been attributed to a shortage of food and to stimuli transmitted through the mucus laid down during locomotion. The fecundity of white snails is positively correlated with shell size (Baker, 1991), which suggests a density-dependent mechanism for the regulation of population numbers, which has been proposed for other gastropod species (Cameron and Carter, 1979; Carter and Ashdown, 1984). However, the positive relationship between shell size and fecundity is only weak in white snails (see also Cowie, 1984a) and may therefore not be particularly important in determining field abundance (Baker, 1991). Laboratory studies with *C. acuta* have failed to show a relationship between shell size and fecundity (Baker and Hawke, 1991).

The numbers of juvenile *T. pisana* and *C. virgata* (spring populations) produced per adult are inversely related to the abundance of those adults at the start of the autumn–winter breeding season (Baker, 1991, 1996; Fig. 6.7). If differences in adult sizes cannot fully explain these variations in the production of young (see above), perhaps greater behavioural interference between adults or between the young they produce can (e.g. inhibitory signals in the mucus/faeces, competition for food) (Baker, 1991). Baur (1988) argued that egg cannibalism can act as a population regulator in the helicid, *Arianta arbustorum* (Linnaeus). Whether such cannibalism occurs in *T. pisana* or *C. virgata* is not known.

Lim and Jenkins (1972) rarely found *T. pisana* and *C. virgata* together in high numbers when they surveyed roadside sites in the south-east of SA. Baker (1988a) also showed that, although both *T. pisana* and *C. virgata* were widespread and very abundant in patches, they rarely coexisted in high numbers at microsites (e.g. within quadrats of 0.25 m^2) within a pasture in the south-east of SA. The possibility that such patterns in distribution might be explained by interspecific competition was investigated by Smallridge and Kirby (1988), who demonstrated, in the laboratory, that *T. pisana* reduced the survival, growth and activity of *C. virgata*. They proposed that stimuli in the mucus trails of *T. pisana* were responsible for these effects, as has been argued for interactions between other species (Cameron and Carter, 1979; Tilling, 1985a,b). However, Bull *et al.* (1992) noted that Smallridge and Kirby (1988) had not isolated mucus from faeces in their experiments. Bull *et al.* (1992)

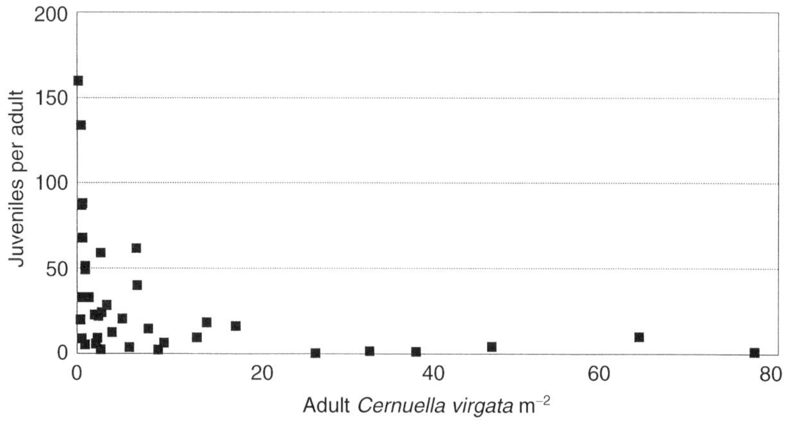

Fig. 6.7. Ratio of juvenile *Cernuella virgata* (da Costa) (Hygromiidae) (spring populations) produced per adult (autumn populations) as a function of adult density (numbers m^{-2}). Data are taken from surveys in several pastures (permanent and in rotations with cereals) throughout South Australia between the years 1984 and 1995.

could not demonstrate consistently that the mucus of *T. pisana* inhibited the activity of *C. virgata*, nor was it avoided by *C. virgata*. However, *C. virgata* ate the faeces of *T. pisana*. Bull *et al.* (1992) suggested that this feeding behaviour might provide an alternate mechanism for interspecific interactions. Field studies in SA have also demonstrated that *T. pisana* can reduce the survival of *C. virgata* (Fig. 6.8).

Prospects for Control

Cultural

Early attempts at control employed outriggers attached to harvesting machinery to act as rakes and knock white and conical snails off cereal plants before they foul the machinery (Rimes, 1968). However, they are now rarely used, principally because they tend to dislodge the grain (especially barley), as well as the gastropods! Some farmers resort to heavy rollers to crush white and conical snails, but this is only effective if the soil is hard and flat. In SA, white and conical snails are particularly abundant in sandy soils where undulations on the soil surface make rolling ineffective. Windrowing crops (cutting the stalks before the grain is fully ripe, raking the fallen crop into rows across the field, leaving the crop to mature on the ground for a week or so and then harvesting it) can reduce contamination. Many conical and white snails aestivate on the stubble between rows and hence do not contaminate the grain when harvested. Windrowing is more effective in reducing contamination by white snails than conical snails, but the reasons for this difference are not known. There is a cost of approximately A$10 ha^{-1} associated with windrowing.

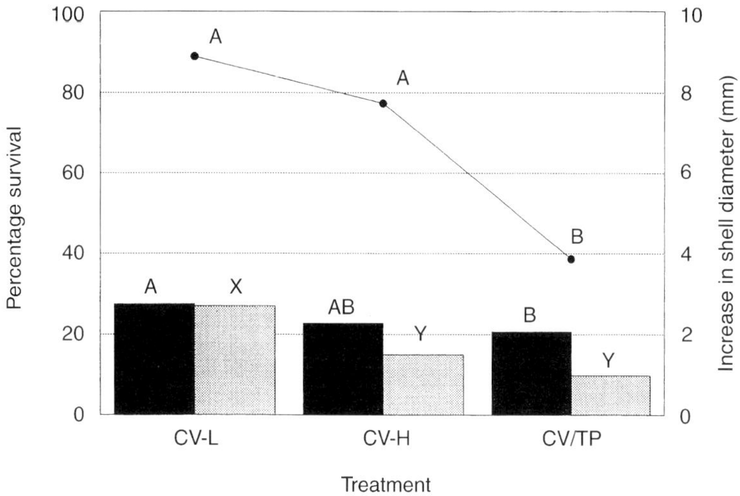

Fig. 6.8. Percentage survival (line) and growth increments (bars) of *Cernuella virgata* (da Costa) (Hygromiidae) when caged at densities of 35 per cage (CV-L), 70 per cage (CV-H) and 35 per cage + 35 *Theba pisana* (Müller) per cage (CV/TP) in a pasture on Yorke Peninsula, South Australia. The experiment ran for 6 weeks in spring 1987. Cages were made of metal rings, 60 cm in diameter and 8 cm high, pushed into the soil surface and covered on top with fly mesh. Shell growth increments are given for *C. virgata* that were either living ■ or dead ▦ at the end of the experiment. Where letters above bars or points on the line graph differ, values are significantly different ($P < 0.05$).

For many years, farmers have relied on tillage and the burning of pasture residues and stubbles before sowing new crops to kill white and conical snails. However, with recent general acceptance by farmers of the need to conserve organic matter and improve soil structure and fertility, there is a conflict of interest between striving for soil conservation and achieving gastropod control through burning and cultivating fields. Unfortunately, white and conical snails tend to be most abundant in fields where soils are light and the risks of wind erosion are high following burning. While burning can be very effective in killing white snails, it is less effective against *C. acuta*, many of which escape the fire by sheltering beneath loose rocks. If the vegetation is sparse, a hot, even burn across fields is not achieved and large numbers of gastropods escape in unburnt patches.

Some farmers drag ‘prickle’ chains or iron bars across fields in summer to dislodge aestivating gastropods from the vegetation. The dislodged animals often die, apparently from exposure to extreme climatic conditions on the soil surface or starvation (food reserves are exhausted during the search for new aestivation sites). Intensive grazing and improved control of weeds (especially the tall robust ones that provide aestivation sites) should reduce gastropod numbers. Cutting the weeds that occur in pastures adjacent to crops may, however, encourage invasion into the crops (see above), unless large numbers of the animals are killed *in situ* (Baker, 1992). Removal of weeds within crops may also reduce available food and increase feeding damage to crops, at least in the short term (Godan, 1983).

Chemical

Many molluscicides are available in a range of formulations (Godan, 1983; Bowen and Antoine, 1995). Methiocarb and metaldehyde baits have been traditionally used to kill white and conical snails in southern Australia. In the context of the Australian agricultural industry, these molluscicides are expensive at currently recommended rates (5 kg ha^{-1}, A$10–40 ha^{-1}) and not very effective at reducing numbers of *C. acuta*. The use of some molluscicides is of concern because of their effects on non-target fauna (e.g. methiocarb effects on earthworms) (Bieri *et al.*, 1989; Baker, 1998). The recognition that gastropods can migrate into and out of fields in certain seasons offers the possibility that molluscicides might be used as strategic barriers to protect 'clean' crops.

Chemicals released by aquatic gastropods into water can influence the growth and reproduction of other gastropod species (e.g. Thomas and Benjamin, 1974; Thomas *et al.*, 1974, 1975; Chaudhry and Morgan, 1986). Thomas (1995) has suggested that chemicals of aquatic gastropod origin might be used as molluscicides to reduce the incidence of schistosomiasis, which is vectored by gastropods. Isolation of inhibitory substances (or organisms) in the faeces of *T. pisana* (see above) or the mucus trails of other terrestrial gastropods might also provide alternative molluscicides to those currently used (Baker, 1989a) and are being investigated, along with molluscicides of plant origin.

Fumigants (e.g. CO_2, phosphine) can kill white and conical snails contaminating grain stored in silos, but the dosages required for efficient kills are higher than those recommended for insect control and would be very difficult to achieve in commercial silos (P. Annis, personal communication). While fumigation may reduce quarantine implications for grain exports, it would not, of course, solve the problem of contamination with shells from dead gastropod snails.

Biological

Few animals are known to prey upon or parasitize white and conical snails in Australia. Some birds, lizards and mice feed upon them, but their impact on pest numbers is considered trivial. On the other hand, a wide range of invertebrate and vertebrate predators (molluscs, beetles, lizards, birds, small mammals) and insect parasitoids (sarcophagid, sciomyzid, phorid and calliphorid flies) feed upon white and conical snails in Europe and the Mediterranean (Baker, 1986; Hopkins and Baker, 1993; Coupland, 1994). The diseases of white and conical snails are poorly known but bacterial epizootics have reduced the numbers of other gastropod species (*Helix* Linnaeus species, Helicidae) in Mediterranean France (Meynadier *et al.*, 1964). Coupland (1995) isolated a rhabditid nematode (*Phasmarhabditis hermaphrodita* (Schneider)) from *T. pisana* and the hygromiid (*Trochoidea elegans* (Gmelin) in southern France which proved lethal to *T. pisana*, *C. virgata* and *C. acuta* in laboratory cultures. This same nematode is being used as a biological control agent against pestiferous gastropods (principally Agriolimacidae) in Europe (Wilson *et al.*, 1993).

Sciomyzid flies have most commonly been considered as potential biological control agents of aquatic and semi-aquatic gastropods (e.g. Chock *et al.*, 1961; Gormally, 1987; Maharaj *et al.*, 1992). However, Knutson *et al.* (1970), Berg and Knutson (1978), Godan (1983) and Baker (1986) have also suggested that they could prove useful as biological control agents of terrestrial gastropods. In particular, Knutson *et al.* (1970) reported that *Salticella fasciata* Meigen occurred as a parasite of *T. pisana* and was widely distributed throughout the Mediterranean in dry habitats with low, sparse vegetation such as sand-dunes and wheat stubble. Such habitats are very similar to those in which white snails are abundant in southern Australia. The possibility of using sciomyzids, and also sarcophagid flies, as biological control agents against white and conical snails has recently been further investigated (Hopkins and Baker, 1993; Coupland and Baker, 1994, 1995; Coupland *et al.*, 1994).

While *S. fasciata* lays large numbers of eggs upon the shells of adult *T. pisana* and *C. virgata*, it is not a true parasitoid (Hopkins and Baker, 1993; Coupland *et al.*, 1994). Rather, the larvae of *S. fasciata* feed upon the decaying remains of dead gastropod snails. Larvae that hatch on the shells of live gastropods abandon their hosts. *S. fasciata* is therefore no longer considered as a potential biological control agent.

Other sciomyzids, in particular *Pherbellia cinerella* Fallén, do attack white and conical snails (Coupland, 1994; Coupland and Baker, 1995). *P. cinerella*, a multivoltine species, is widespread in western Europe, incuding the Mediterranean, and its larvae feed on a range of gastropod species (Bratt *et al.*, 1969; Gormally, 1987; Vala, 1989; Coupland and Baker, 1995). Adults are active for most of the year (except midsummer in the Mediterranean). *P. cinerella* occurs in a wide range of habitats, but in the Mediterranean it is particularly abundant in open pasture and dune systems and rare in wooded habitats (Coupland and Baker, 1995). A culture of *P. cinerella* was established in quarantine in Australia in 1993 to test its host specificity against native Australian gastropods. Unfortunately, *P. cinerella* fed readily upon several Australian species (D. Hopkins, personal communication) and for this reason it is no longer regarded as a potential biological control agent. However, it is worth commenting that little is known about the ecologies of native gastropods in southern Australia; with improved knowledge on seasonal activity and habitat preferences, it may be possible to make a more informed assessment of which native species would actually be at risk if *P. cinerella* were to be released, rather than having to simply rely on attack rates against gastropods in artificial enclosures in quarantine.

Current research interest in the biological control of white and conical snails is focused on the use of sarcophagid flies, most notably *Sarcophaga penicillata* Villeneuve, which attacks conical snails, and *Sarcophaga uncicurva* Pandelle and *Sarcophaga balanina* Pand., which attack white snails (Coupland and Baker, 1994; Carter and Baker, 1997). These flies are multivoltine, summer-active and more host-specific in Europe than *P. cinerella*. Species of *Sarcophaga* Meigen can kill large numbers of their hosts (e.g. parasitism levels of up to 88% recorded for *S. penicillata* in *C. acuta*) (J.B. Coupland and G.H. Baker, unpublished data). They, in turn, are subject to hyperparasitism, especially by the pteromalid wasp, *Novitzkyanus cryptogaster* Boucek (up to 90%). In the absence of hyperparasites in Australia, these sarcophagids are predicted to have a large impact on the abundance of white and conical snails. Cultures of all three *Sarcophaga* species are currently being evaluated for host specificity under quarantine, using a range of gastropod species indigenous to Australia.

Conclusions

Introduced white and conical snails cause significant feeding and fouling damage to pastures and cereal crops in southern Australia. Traditional methods for controlling them, such as burning fields prior to sowing crops or using molluscicides, are either ecologically unacceptable or expensive. Recent research has focused on improving our understanding of the biologies of the pests in order to develop appropriate methods for their control. While the life histories and ecologies of white snails and *C. acuta* are now reasonably well understood, little is known about *C. barbara*.

The life histories of white snails are clearly very plastic and enable these animals to cope well with a range of agricultural managements, such as permanent pastures and pasture–cereal rotations. Nevertheless, the growing of cereals inhibits in some way the production of young in white snails. Perhaps food is less available for the young in a cereal crop, but the reason(s) for the poor recruitment there, compared with pastures, are not well understood and deserve more attention.

Continuous cereal cropping reduces white and conical snail abundance (G.H. Baker, unpublished data). However, many farmers include legume-based pastures in rotations, primarily to diversify their incomes, but also to limit the development of herbicide resistance in weed species, improve soil structure, increase soil organic matter and replenish soil nutrients, such as nitrogen. The inclusion of grain legumes in rotations with cereals seems not to increase the abundance of gastropods to the same extent as the inclusion of legume-based pastures. Especially with the adoption of minimum tillage and stubble retention, the use of grain legumes can achieve many of the benefits in terms of soil structure and fertility that are gained by using pastures. Unfortunately, reduced tillage and increased organic matter on the soil surface increase gastropod pest numbers! In addition, white and conical snails cause more problems when harvesting legumes compared with cereals, probably because of the different methods used.

Access to aestivation sites above the surface of the ground is critical to the survival of these pest species, especially *T. pisana*. Improved management of robust weeds in pastures should therefore greatly reduce gastropod numbers. Simple methods, such as dragging chains across pastures in the heat of summer, can dislodge gastropods from their aestivation sites on above-ground vegetation and kill them through exposure to the hot soil surface.

White snails move substantial distances, both as adults in autumn–winter and juveniles in spring–early summer. Movements are biased in direction. For example, white snails move into cereal crops from adjacent pastures in spring. This behaviour offers the prospect of strategic placement of chemicals as traps or barriers to protect fields against invasion. Such use of chemicals should be encouraged where possible to reduce costs and reduce impacts on non-target organisms. Whether or not chemicals are best used in autumn to kill adult white snails prior to breeding or in spring to kill juveniles before they foul harvests is as yet unknown.

Although white and conical snails are found in temperate regions of southern Australia, they are pests principally in areas with Mediterranean-type climates. By analysing allozyme patterns in separate populations, Johnson (1988) concluded that the *T. pisana* now found in southern Australia probably originated from the Mediterranean rather than further north in Europe. The centre of radiation of the genus *Theba* Risso seems to have been Morocco and southern Spain, where several species now occur (Gittenberger and Ripken, 1987). Surveys for natural enemies of white and conical snails that might prove useful as biological control agents have therefore been focused within the Mediterranean, particularly in southern France, Spain, Portugal and Morocco (Coupland, 1994). This maximizes the chance that the selected agents are optimally matched for the climate and hosts they will face in Australia. The most promising agents for use against white and conical snails appear to be sarcophagid flies. Whether or not these will prove to be sufficiently host-specific in quarantine studies against native Australian gastropods, such that the impact on non-target fauna is deemed unlikely to be significant and permission for release in Australia is granted, remains to be seen.

White and conical snails have invaded native ecosystems as well as agricultural fields in southern Australia. Although information is scarce, it seems highly probable that these introduced gastropods are having a significant impact on the native flora and fauna in such ecosystems. Biological control of white and conical snails, while principally targeted at reducing pest numbers in agricultural habitats, could perhaps also help reduce damage in native ecosystems.

Acknowledgements

I am indebted to several colleagues who have helped gather many of the data that are presented here. I especially wish to thank James Coupland, Bonnie Vogelzang, Bruce Hawke, Penny Carter and Vicki Barrett. Much of the research has been funded by the Australian Grains Research and Development Corporation and the International Wool Secretariat.

References

Avivi, T.R. and Arad, Z. (1993) Population dynamics of the snail *Theba pisana* (Helicidae) in the sand dunes of northern Israel. *Israel Journal of Zoology* 39, 245–254.

Baker, G.H. (1986) *The Biology and Control of White Snails (Mollusca: Helicidae), Introduced Pests in Australia.* Division of Entomology Technical Paper No. 25, Commonwealth Scientific and Industrial Research Organization, Canberra, 31pp.

Baker, G.H. (1988a) The life history, population dynamics and polymorphism of *Cernuella virgata* (Mollusca: Helicidae). *Australian Journal of Zoology* 36, 497–512.

Baker, G.H. (1988b) The dispersal of *Cernuella virgata* (Mollusca: Helicidae). *Australian Journal of Zoology* 36, 513–520.

Baker, G.H. (1988c) Dispersal of *Theba pisana* (Mollusca: Helicidae). *Journal of Applied Ecology* 25, 889–900.

Baker, G.H. (1989a) Damage, population dynamics, movement and control of pest helicid snails in southern Australia. In: Henderson, I.F. (ed.) *Slugs and Snails in World Agriculture.* Monograph No. 41, British Crop Protection Council, Thornton Heath, pp. 175–185.

Baker, G.H. (1989b) Population dynamics of the white snail, *Cernuella virgata* (Mollusca: Helicidae), in a pasture–cereal rotation in South Australia. In: *Proceedings of the 5th Australasian Conference on Grassland Invertebrate Ecology.* D and D Printing, Melbourne, pp. 177–183.

Baker, G.H. (1991) Production of eggs and young snails by adult *Theba pisana* (Müller) and *Cernuella virgata* (da Costa) (Mollusca: Helicidae) in laboratory cultures and field populations. *Australian Journal of Zoology* 39, 673–679.

Baker, G.H. (1992) Movement of introduced white snails between pasture and cereal crops in South Australia. In: Del Fosse, E. (ed.) *Pests of Pastures: Weed, Invertebrate and Disease Pests of Australian Sheep Pastures.* Commonwealth Scientific and Industrial Research Organisation Information Services, Melbourne, pp. 115–120.

Baker, G.H. (1996) Population dynamics of the mediterranean snail, *Cernuella virgata*, in a pasture–cereal rotation in South Australia. In: Henderson, I.F. (ed.) *Slug and Snail Pests in Agriculture.* Proceedings No. 66, British Crop Protection Council, Farnham, pp. 117–124.

Baker, G.H. (1998) Recognising and responding to the influences of agriculture and other land-use practices on soil fauna. *Applied Soil Ecology* 9, 303–310.

Baker, G.H. and Hawke, B.G. (1990) Life history and population dynamics of *Theba pisana* (Mollusca: Helicidae) in a cereal–pasture rotation. *Journal of Applied Ecology* 26, 16–29.

Baker, G.H. and Hawke, B.G. (1991) Fecundity of *Cochlicella acuta* (Müller) (Mollusca: Helicidae) in laboratory cultures. *Invertebrate Reproduction and Development* 20, 243–247.

Baker, G.H. and Vogelzang, B.K. (1988) Life history, population dynamics and polymorphism of *Theba pisana* (Mollusca: Helicidae) in Australia. *Journal of Applied Ecology* 25, 867–887.

Baker, G.H., Hawke, B.G. and Vogelzang, B.K. (1991) Life history and population dynamics of *Cochlicella acuta* (Müller) (Mollusca: Helicidae) in a pasture–cereal rotation. *Journal of Molluscan Studies* 57, 259–266.

Baur, A. (1993) Effects of food availability and intra- and interspecific interactions on the dispersal tendency in the land snail *Chondrina clienta*. *Journal of Zoology, London* 230, 87–100.

Baur, B. (1988) Population regulation in the land snail *Arianta arbustorum*: density effects on adult size, clutch size and incidence of egg cannibalism. *Oecologia* 77, 390–394.

Baur, B. and Baur, A. (1990) Experimental evidence for intra- and interspecific competition in two species of rock-dwelling land snails. *Journal of Animal Ecology* 59, 301–315.

Baur, B. and Baur, A. (1994) Dispersal of the land snail *Chondrina avenacea* on vertical rock walls. *Malacological Review* 27, 53–59.

Berg, C.O. and Knutson, L.V. (1978) Biology and systematics of the Sciomyzidae. *Annual Review of Entomology* 23, 239–258.

Bieri, M., Schweizer, H., Christensen, K. and Daniel, O. (1989) The effect of metaldehyde and methiocarb slug pellets on *Lumbricus terrestris* Linne. In: Henderson, I.F. (ed.) *Slugs and Snails in World Agriculture*. Monograph No. 41, British Crop Protection Council, Thornton Heath, pp. 237–244.

Bowen, I.D. and Antoine, S. (1995) Molluscicide formulation studies. *International Journal of Pest Management* 41, 74–78.

Bratt, A.D., Knutson, L.V., Foote, B.A. and Berg, C.O. (1969) *Biology of* Pherbellia *(Diptera: Sciomyzidae)*. Memoirs of the New York State College of Agriculture, No. 404, Ithaca, New York, 246 pp.

Bull, C.M., Baker, G.H., Lawson, L.M. and Steed, M.A. (1992) Investigations of the role of mucus and faeces on interspecific interactions of two land snails. *Journal of Molluscan Studies* 58, 433–441.

Butler, A.J. (1972) The food of the terrestrial snail *Helicella virgata* (da Costa) in South Australia. PhD thesis, University of Adelaide.

Butler, A.J. (1976) A shortage of food for the terrestrial snail *Helicella virgata* in South Australia. *Oecologia* 25, 349–371.

Cain, A.J. and Currey, J.D. (1968) Studies on *Cepaea*. III. Ecogentics of a population of *Cepaea nemoralis* (L.) subject to strong area effects. *Philosophical Transactions of the Royal Society of London Series B* 253, 447–482.

Cameron, R.A.D. and Carter, M.A. (1979) Intra- and interspecific effects of population density on growth and activity in some helicid land snails (Gastropoda: Pulmonata). *Journal of Animal Ecology* 48, 237–246.

Cameron, R.A.D. and Williamson, P. (1977) Estimating migration and the effects of disturbance in mark–recapture studies on the snail *Cepaea nemoralis* L. *Journal of Animal Ecology* 46, 173–179.

Carter, M.A. and Ashdown, M. (1984) Experimental studies on the effects of density, size and shell colour and banding phenotypes on the fecundity of *Cepaea nemoralis*. *Malacologia* 25, 291–302.

Carter, P.J. and Baker, G.H. (1997) Biological control of white and conical snails. *Farming Systems Developments* 1997, 113–114.
Chase, R. and Croll, R.P. (1981) Tentacular function in snail olfactory orientation. *Journal of Comparative Physiology* 143, 357–362.
Chaudhry, M.A. and Morgan, E. (1986) Factors regulating oviposition in *Bulinus tropicus* in snail-conditioned water. *Malacologia* 27, 249–263.
Chock, Q.C., Davis, C.J. and Chong, M. (1961) *Sepedon macropus* (Diptera: Sciomyzidae) introduced into Hawaii as a control of the liver fluke snail, *Lymnaea ollula*. *Journal of Economic Entomology* 54, 1–4.
Coupland, J.B. (1994) Diptera associated with snails collected in south-western and west-Mediterranean Europe. *Vertigo* 3, 19–26.
Coupland, J.B. (1995) Susceptibility of helicid snails to isolates of the nematode *Phasmarhabditis hermaphrodita* from southern France. *Journal of Invertebrate Pathology* 66, 207–208.
Coupland, J.B. and Baker, G.H. (1994) Host distribution, larviposition behaviour and generation time of *Sarcophaga penicillata* (Diptera: Sarcophagidae): a parasitoid of conical snails. *Bulletin of Entomological Research* 84, 185–189.
Coupland, J.B. and Baker, G.H. (1995) The potential of several species of terrestrial Sciomyzidae as biological control agents of pest helicid snails in Australia. *Crop Protection* 14, 573–576.
Coupland, J.B., Espiau, A. and Baker, G.H. (1994) Seasonality, longevity, host choice, and infection efficiency of *Salticella fasciata* (Diptera: Sciomyzidae), a candidate for the biological control of pest helicid snails. *Biological Control* 4, 32–37.
Cowie, R.H. (1984a) The life-cycle and productivity of the land snail *Theba pisana* (Mollusca: Helicidae). *Journal of Animal Ecology* 53, 311–325.
Cowie, R.H. (1984b) Ecogenetics of *Theba pisana* (Pulmonata: Helicidae) at the northern edge of its range. *Malacologia* 25, 361–380.
Cowie, R.H. (1984c) Density, dispersal and neighbourhood size in the land snail *Theba pisana*. *Heredity* 52, 391–401.
Cribb, T.H. (1990) Introduction of a *Brachylaima* species (Digenea: Brachylaimidae) to Australia. *International Journal for Parasitology* 20, 759–796.
Den Boer, P.J. (1968) Spreading of risk and stabilisation of animal numbers. *Acta Biotheoretica* 18, 165–194.
De Smet, W.H.O. (1983) Notes on the distribution and the shell polymorphism of *Cochlicella acuta* (Gastropoda: Helicidae) in Belgium. *Biologisch Jaarboek* 51, 23–35.
Durr, H.J.R. (1946) *A Contribution to the Morphology and Bionomics of* Theba pisana *(Muller) (Gastropoda: Helicidae)*. Elsenburg Scientific Series No. 47, South African Department of Agriculture, Stellenbosch, 34 pp.
Farkas, S.R. and Shorey, H.H. (1976) Anemotaxis and odour-trail following by the terrestrial snail *Helix aspersa*. *Animal Behaviour* 24, 686–689.
Gittenberger, E. and Ripken, T.E.J. (1987) The genus *Theba* (Mollusca: Gastropoda: Helicidae), systematics and distribution. *Zoologische Verhandelingen* 241, 3–59.
Godan, D. (1983) *Pest Slugs and Snails*. Springer-Verlag, Berlin, 445 pp.
Goodfriend, G.A. (1983) Anemotaxis and its relation to migration in the land snail *Cepaea nemoralis*. *American Midland Naturalist* 109, 414–415.

Gormally, M.J. (1987) Effect of temperature on the duration of larval and pupal stages of two species of sciomyzid flies, predators of the snail *Lymnaea truncatula*. *Entomologia Experimentalis et Applicata* 43, 95–100.

Greenwood, J.J.D. (1974) Effective population numbers in the snail *Cepaea nemoralis*. *Evolution* 28, 513–526.

Hazel, W.N. and Johnson, M.S. (1990) Microhabitat choice and polymorphism in the land snail *Theba pisana* (Müller). *Heredity* 65, 449–454.

Heller, J. (1981) Visual versus climatic selection of shell banding in the land snail *Theba pisana* in Israel. *Journal of Zoology, London* 194, 85–101.

Heller, J. (1982) Natural history of *Theba pisana* in Israel (Pulmonata: Helicidae). *Journal of Zoology, London* 196, 475–487.

Heller, J. and Gadot, M. (1984) Shell polymorphism of *Theba pisana* – the effects of rodent distribution. *Malacologia* 25, 349–354.

Heller, J. and Ittiel, H. (1990) Natural history and population dynamics of the land snail *Helix texta* in Israel (Pulmonata: Helicidae). *Journal of Molluscan Studies* 56, 189–204.

Hodson, A.C. (1969) Adaptations that permit the terrestrial snail *Helicella virgata* (da Costa) to survive in dry places. PhD thesis, University of Adelaide.

Hopkins, D.C. and Baker, G.H. (1993) Biological control of white and conical snails. In: Corey, S.A., Dall, D.J. and Milne, W.M. (eds) *Pest Control and Sustainable Agriculture*. Commonwealth Scientific and Industrial Research Organization Information Services, Melbourne, pp. 246–249.

Johnson, M.S. (1980) Association of shell banding and habitat in a colony of the land snail *Theba pisana*. *Heredity* 45, 7–14.

Johnson, M.S. (1981) Effects of migration and habitat choice on shell banding frequencies in *Theba pisana* at a habitat boundary. *Heredity* 47, 121–134.

Johnson, M.S. (1988) Founder effects and geographic variation in the land snail *Theba pisana*. *Heredity* 61, 133–142.

Jones, J.S., Leith, B.H. and Rawlings, P. (1977) Polymorphism in *Cepaea*: a problem with too many solutions? *Annual Review of Ecology and Systematics* 8, 109–143.

Knutson, L.V., Stephenson, J.W. and Berg, C.O. (1970) Biosystematic studies *of Salticella fasciata* (Meigen), a snail-killing fly (Diptera: Sciomyzidae). *Transactions of the Royal Entomological Society, London* 122, 81–100.

Lewis, G. (1977) Polymorphism and selection in *Cochlicella acuta*. *Philosophical Transactions of the Royal Society of London Series B* 276, 399–451.

Lim, Y.P. and Jenkins, R.B. (1972) The Italian white snail *Theba pisana* (Müller) in South Australia with particular reference to its distribution in the south-east of South Australia. *Experimental Records of the Department of Agriculture, South Australia* 6, 23–25.

Maharaj, R., Appleton, C.C. and Miller, R.M. (1992) Snail predation by larvae *of Sepedon scapularis* Adams (Diptera: Sciomyzidae), a potential biocontrol agent of snail intermediate hosts of schistosomiasis in South Africa. *Medical and Veterinary Entomology* 6, 183–187.

Meynadier, G., Bergoin, M. and Vago, C. (1964) Bacteriose epizootique chez les Helicides (Mollusques). *Antonie van Leewenhoek* 30, 76–80.

Mienis, H.K. (1969) *Cernuella virgata* (da Costa, 1778) in Nederland. *Basteria* 33, 31–49.

Moran, S. (1989) Weather- and population density-induced infantilism in the landsnail *Theba pisana* in a semi-arid climate. *International Journal of Biometeorology* 33, 101–108.

Nevo, E. and Bar, Z. (1976) Natural selection of genetic polymorphisms along climatic gradients. In: Karlin, S. and Nevo, E. (eds) *Population Genetics and Ecology*. Academic Press, New York, pp. 159–184.

Oosterhoff, L.M. (1977) Variation in growth rate as an ecological factor in the land snail *Cepaea nemoralis* (L.). *Netherlands Journal of Zoology* 27, 1–132.

Peake, J. (1978) Distribution and ecology of Stylommatophora. In: Fretter, V. and Peake, J. (eds) *Pulmonates*, Vol. 2A. *Systematics and Ecology*. Academic Press, London, pp. 429–526.

Pomeroy, D.E. (1966) The ecology of *Helicella virgata* and related species of snails in South Australia. PhD thesis, University of Adelaide.

Pomeroy, D.E. (1968) Dormancy in the land-snail *Helicella virgata*, in South Australia. *Australian Journal of Zoology* 16, 857–869.

Pomeroy, D.E. (1969) Some aspects of the ecology of the land snail, *Helicella virgata*, in South Australia. *Australian Journal of Zoology* 17, 495–514.

Rimes, G.D. (1968) Snail investigations – a progress report. *Journal of Agriculture of Western Australia* 9, 584–587.

Serventy, D.L. and Storr, G.M. (1959) The spread of the Mediterranean snail on Rottnest Island – Part II. *West Australian Naturalist* 6, 193–196.

Smallridge, M.A. and Kirby, G.C. (1988) Competitive interactions between the land snails *Theba pisana* (Müller) and *Cernuella virgata* (da Costa) from South Australia. *Journal of Molluscan Studies* 54, 251–258.

Swart, P.L., Barnes, B.N. and Myburgh, A.C. (1976) Pest of table grapes in the Western Cape. *Deciduous Fruit Grower* 26, 169–195.

Tattersfield, P. (1981) Density and environmental effects on shell size in some sand dune snail populations. *Biological Journal of the Linnean Society* 16, 71–82.

Thomas, J.D. (1995) The snail hosts of schistosomiasis – some evolutionary and ecological perspectives in relation to control. *Memorias do Instituto Oswaldo Cruz* 90, 195–204.

Thomas, J.D. and Benjamin, M. (1974) The effects of population density on growth and reproduction of *Biomphalaria glabrata* (Say) (Gasteropoda: Pulmonata). *Journal of Animal Ecology* 43, 31–50.

Thomas, J.D., Goldsworthy, G.J. and Benjamin, M. (1974) Chemical conditioning of the environment by the fresh water pulmonate snail (*Biomphalaria glabrata*) and its effect on growth and natality rates. *Journal of Zoology, London* 172, 443–467.

Thomas, J.D., Goldsworthy, G.J. and Aram, R.H. (1975) Studies on the chemical ecology of snails: the effect of chemical conditioning by adult snails on the growth of juvenile snails. *Journal of Animal Ecology* 44, 1–28.

Tilling, S.M. (1985a) The effects of density and interspecific interaction on mortality in experimental populations of adult *Cepaea* (Held). *Biological Journal of the Linnean Society* 24, 61–70.

Tilling, S.M. (1985b) The effect interspecific interaction on spatial distribution patterns in experimental populations of *Cepaea nemoralis* (L.) and *C. hortensis* (Müll.). *Biological Journal of the Linnean Society* 24, 71–81.

Vala, J.-C. (1989) *Diptères Sciomyzidae Euro-Mediterranéens*. Faune de France 72, Fédération Française des Sociétés de Sciences Naturelles, Paris, 300 pp.

Williamson, P., Cameron, R.A.D. and Carter, M.A. (1976) Population density affecting adult shell size of the snail *Cepaea nemoralis. Nature (London)* 263, 496–497.

Wilson, M.J., Glen, D.M. and George, S.K. (1993) The rhabditid nematode *Phasmarhabditis hermaphrodita* as a potential biological control agent for slugs. *Biocontrol Science and Technology* 3, 503–511.

Zanforlin, M. (1976) Observations on the visual perception of the snail *Euparipha pisana* (Muller). *Bolletino di Zoologia* 43, 303–315.

7 Planorbidae and Lymnaeidae as Pests of Rice, with Particular Reference to *Isidorella newcombi* (Adams & Angus)

Mark M. Stevens

NSW Agriculture and Cooperative Research Centre for Sustainable Rice Production, Yanco Agricultural Institute, PMB Yanco, NSW 2703, Australia

Introduction

Rice (*Oryza sativa* Linnaeus) (Gramineae) is the staple food of approximately half the world's population. The world rice harvest exceeded 596 million t in 1999, with 90% of production occurring in Asia. China and India alone account for 56% of global rice production (FAO, 2000). Primarily a subsistence crop, less than 5% of world production is traded on international markets. Rice is grown across a wide range of agricultural environments, from irrigated lowland fields to upland areas entirely dependent on seasonal rainfall.

Invertebrate pests impose serious constraints on rice production throughout the world. Pest problems are generally more severe in tropical and subtropical areas, where leafhopppers, planthoppers (Homoptera), stemborers and leaf-rollers (Lepidoptera) are among the most important insect pests of the crop. Gastropods have generally been regarded as minor pests; however, since its introduction into Asia, the apple snail, *Pomacea canaliculata* (de Lamarck) (Ampullariidae), has become one of the most serious invertebrate rice pests in the region. Planorbids and lymnaeids are less frequently recognized as rice pests. Although they may be responsible for substantial crop damage in particular areas, this damage is often sporadic in occurrence and appears to be strongly influenced by changes to farming practices and/or unusual seasonal conditions.

Planorbidae and Lymnaeidae are ubiquitous throughout most of the world's rice-growing areas, with high populations occurring not only in the rice-fields themselves, but also in the extensive networks of irrigation and drainage canals often associated with rice cultivation. As pests, planorbids and lymnaeids can have an impact on rice growing in three principal ways: as primary crop pests, as secondary crop pests and as vectors of disease in both livestock and humans. Primary pest activity involves gastropods feeding directly on the rice crop itself. Secondary pest activity can be characterized as gastropods detrimentally affecting the agronomic environment of the rice crop, typically by feeding on blue-green algae (BGA) (Cyanophyceae) or the aquatic fern *Azolla* de Lamarck (Azollaceae), which enhance crop nutrition through biological nitrogen fixation. The role of planorbids and lymnaeids as vectors of schistosomiasis and other parasitic diseases is well known, and has been reviewed elsewhere (see Madsen, 1992, and references therein).

Planorbids and Lymnaeids as Primary Pests of Rice

The majority of gastropods recognized as pests of rice belong to the family Ampullariidae, including species in the genera *Pomacea* Perry, *Pila* Röding and *Lanistes* de Montfort. Of these, *P. canaliculata* is undoubtedly the most serious gastropod pest of rice, and has been dealt with by Cowie (Chapter 5, this volume). Although reports of planorbids and lymnaeids as primary pests are not widespread, they may be significant pests of rice in particular geographical areas. Reports of planorbids and lymnaeids as

Table 7.1. Published reports of Planorbidae and Lymnaeidae as primary pests of rice.

Species	Country	Pest status	References
Planorbidae			
Biomphalaria pfeifferi (Krauss)	Mali	Potential	Madsen, 1992
Helisoma duryi (Wetherby)	Florida strain tested in Mali	Potential	Madsen, 1992
Indoplanorbis exustus (Deshayes)	India	Actual	Cherian and Krishnamurthy, 1940; Tirumala Rao *et al.*, 1953; Butani and Jotwani, 1976
Isidorella newcombi (Adams & Angus) *sensu lato*	Australia	Actual	Stevens *et al.*, 1996; Stevens and Coombes, 2001
Isidorella Tate sp.	Australia	Actual	Grist and Lever, 1969
Physastra Tapparone-Canefri sp. (now *Glyptophysa* Crosse)	Australia	Actual	Jones, 1975, 1976, 1979, 1984
Planorbis Müller sp.	Italy	Actual	Grandori, 1952; Foschi 1975
Unidentified species	Italy	Actual	Chiappelli, 1952
Lymnaeidae			
Lymnaea acuminata de Lamarck	India	Actual*	Butani and Jotwani, 1976
Lymnaea acuminata de Lamarck form *rufescens* Gray	India	Actual	Cherian and Krishnamurthy, 1940
Lymnaea de Lamarck sp.	Italy	Actual	Foschi, 1973, 1975

*Ittyaverah *et al.* (1979a) state that this species does not feed on rice.

primary rice pests, based on either laboratory or field assessments, are summarized in Table 7.1.

Rice production in New South Wales, Australia

Although trials with upland rice were conducted in northern New South Wales (NSW) as early as 1891, extensive rice production did not begin until the development of the Murrumbidgee Irrigation Area in south-western NSW during the early part of the 20th century. The first commercial crop of about 225 t was produced by eight farmers in 1924 (Lewis, 1994). Since that time, and with the development of further irrigation districts, the area sown each year has increased to more than 120,000 ha, and production has increased to over 1 million t of paddy rice per annum. Climatic constraints limit production to one crop per year, which is sown in October/November and harvested in March/April.

In NSW rice is sown either by drilling dry seed into prepared seed-beds or uncultivated annual pastures, or by pregerminating the seed and sowing into flooded bays using fixed-wing aircraft. Aerial sowing became popular in the mid-1960s (Lewis, 1994; McDonald, 1994), and in 1997 well over 90% of the crop was sown using this method. Although fast and cost-effective, aerial sowing has increased the severity of invertebrate pest problems in the NSW rice crop. Crops drilled into seed-beds or pastures prior to flooding need to be 'flushed' several times to stimulate germination, a process that involves the repeated flooding and draining of fields (Woodlands *et al.*, 1984). A permanent aquatic environment is not created in the fields until the plants are 15 cm or more in height and able to tolerate reasonable levels of damage from gastropods and chironomid midge larvae. In contrast, aerial sowing involves dropping pregerminated seed into a permanently flooded field, where aquatic pests are often already active. Aerially sown rice seed rests on the soil surface during the crop-establishment period, where it is much more vulnerable to attack from both vertebrate and invertebrate pests.

Isidorella ***and*** **Glyptophysa** ***as pests of rice***

The first published report of a planorbid attacking rice in southern NSW was provided by Grist and Lever (1969). Despite this report referring to a species of *Isidorella* Tate, subsequent reports and studies between 1975 and 1984 dealt with an unidentified species of *Physastra* Tapparone-Canefri, a genus now considered a junior synonym of *Glyptophysa* Crosse (Walker, 1988). Jones (1975) provided an illustration which confirms that early work was conducted on *Glyptophysa*, rather than *Isidorella*. No confirmed reports of *Glyptophysa* causing significant crop damage have been received since 1989, although small populations of *Glyptophysa* are often found in rice-fields, sometimes in association with much larger *Isidorella* populations. All recent instances of planorbid damage have been attributed to a species of *Isidorella*, which, following Walker (1988), has been referred to as *Isidorella newcombi* (Adams & Angus) *sensu lato*.

Planorbids were considered minor and sporadic rice pests until 1973, when crop damage increased substantially, particularly in the Murray Valley (Jones, 1975, 1979). High levels of crop damage were recognized as being related to both higher than usual rainfall in the months preceding crop establishment and the practice of repeat cropping, the planting of rice in individual fields in consecutive years without a summer fallow period. These observations were consistent with the gastropods apparently 'overwintering' in the soil (Jones, 1975). Repeat cropping would be expected to favour the survival of gastropods between crops, since the dry period (between draining one crop and flooding the next) is only 6 months and occurs during the cooler, wetter part of the year. Aestivation in *Glyptophysa cosmeta* (Iredale) was subsequently observed by Smith and Burn (1976) and confirmed in *I. newcombi* by Stevens and Coombes (2001).

I. newcombi (Plate 1) and *Glyptophysa* sp. (Plate 2) both feed on the roots of rice until plants reach the late tillering stage (Plate 3). Typically, gastropod damage retards plant growth, reduces tillering and can lead to asynchronous crop maturity. Severe damage to plant root systems at the seedling stage often results in the loss of whole plants, which detach from the soil and are blown to one corner of the field. Aerially sown crops are particularly vulnerable, since the young plants developing under this management regime have shallower root systems than those established by drill sowing. In addition, aerially sown crops are usually sown into well-prepared cultivated fields with limited alternative food sources for the gastropods. The progressive trend towards aerial sowing over the last three decades appears to be a major factor in the increased importance of *I. newcombi* as a primary crop pest.

Chemical control of Isidorella *and* Glyptophysa

Attempts have been made to control both *I. newcombi* and *Glyptophysa* sp. with copper sulphate since they assumed significant pest status in 1973. Application rates of up to 12 kg ha^{-1} for $CuSO_4.5H_2O$ have been used. However, gastropod mortality arising from these treatments has been inconsistent. The efficacy of copper sulphate is reduced by precipitation of insoluble compounds in response to aspects of water chemistry, by adsorption on to sediments and through uptake by aquatic plants (Malek and Cheng, 1974). All of these factors are significant within rice-fields, and their variability between crops contributes to differences in the efficacy of copper applications. Although copper sulphate usage may enhance rice yields in areas deficient in copper (Foschi, 1973), pasture plants grown on land previously subjected to repeated copper sulphate application may become toxic to livestock, particularly sheep. This problem is exacerbated by the use of *Trifolium* Linnaeus (Fabaceae) as the main constituent of pastures on rice farms. Grazing on these pastures increases copper retention in sheep (*Ovis aries* Linnaeus; Bovidae) and may induce secondary copper poisoning even if the pastures contain only small amounts of copper. The risk is further increased if sheep have been exposed to certain pasture weeds containing hepatotoxic alkaloids that lead to reduced copper tolerance (Radostits *et al.*, 1994). Although molybdenum fertilizers can be used to reduce copper retention in livestock, the continued use of copper sulphate as a molluscicide must be considered environmentally unsustainable in NSW rice-cropping systems.

Jones (1976) reported that in bioassays with niclosamide (2′,5-dichloro-4′-nitrosalicylanilide), trifenmorph (4-tritylmorpholine) and ziram (zinc bis(dimethyldithiocarbamate)) against *Glyptophysa* sp., niclosamide was the most effective, causing 100% mortality at a concentration of 0.1 mg l^{-1}. He found ziram to be effective at 7 kg active ingredient (a.i.) ha^{-1} in field trials, although residual activity against gastropods was regarded as limited.

Stevens *et al.* (1996) evaluated 27 agricultural chemicals at 3 mg l^{-1} in experimental microcosms containing soil, water and aquatic plants, and found that niclosamide, nicotinanilide (pyridine-3-carboxanilide) and trifenmorph were the most toxic to *I. newcombi*. Ziram and sodium pentachlorophenol exhibited moderate levels of toxicity, while carbamates, pyrethroids and organophosphorus compounds were shown to have low toxicity to these gastropods. Niclosamide represents the best potential alternative to copper sulphate, since it is more toxic than nicotinanilide (Stevens *et al.*, 1996) and has stronger residual activity than trifenmorph (Fig. 7.1). Unlike trifenmorph, niclosamide is known to have ovicidal activity against aquatic gastropods (Malek and Cheng, 1974), although its activity against the eggs of *I. newcombi* has yet to be evaluated. While *n*-tritylmorpholine is classed as a superseded compound (Tomlin, 1994), niclosamide is currently being formulated by its manufacturers for use against *P. canaliculata* in South-East Asia. Concerns about the small market size and high registration costs have so far precluded niclosamide becoming available for *Isidorella* control in Australian rice-fields.

Cultural control of Isidorella newcombi

Most rice areas in NSW are infested by *I. newcombi*, with substantial populations commonly present by the time the crop is drained prior to harvest. Populations that develop late in the growing season are generally regarded as non-damaging to the crop. Anecdotal evidence indicates that *I. newcombi* feed on algae and broad-leaf aquatic weeds in preference to rice. The herbicides applied to the crop for general weed control during the establishment period markedly reduce these food sources. When the residual effects of the herbicides decline and these plants re-establish, *I. newcombi* will more commonly be found in weedy areas along field

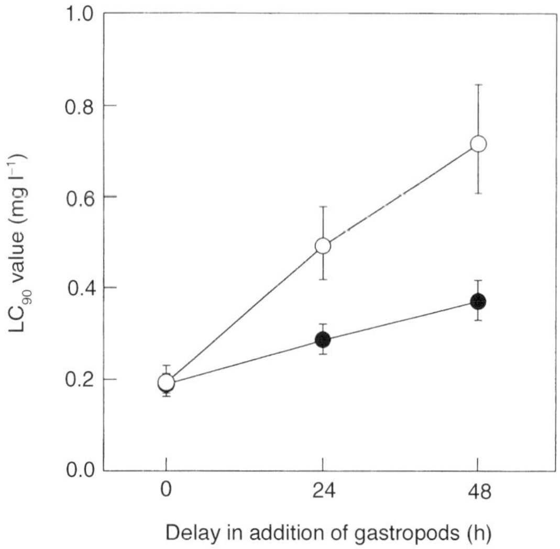

Fig. 7.1. Toxicity of niclosamide (●) and trifenmorph (○) to mature *Isidorella newcombi* (Adams & Angus) (Planorbidae) in laboratory microcosms. Error bars represent 95% confidence intervals. Gastropods were added to microcosms either when treatments were applied or at 24 or 48 h after treatment application. LC_{90} values were calculated on the basis of initial application rates. A slower increase in LC_{90} values with increasing delays prior to the addition of gastropods reflects stronger residual control from niclosamide. (After Stevens *et al.*, 1996, with permission.)

margins and adjacent to irrigation stops, with relatively few under the crop canopy itself. Fields that develop large gastropod populations in the first 2–3 months after sowing are, however, prone to severe damage and, in some cases, total crop loss. These early infestations are often characterized by moderate numbers of large, post-aestivation individuals and very high numbers of their smaller progeny. It is these young, actively growing animals that are often responsible for the majority of crop damage.

While many farmers rotate rice with alternative summer crops or pastures, others choose to grow consecutive rice crops on individual fields. Jones (1975, 1979, 1984) found that *Glyptophysa* sp. damaged rice more severely in fields that had been sown to rice in the previous year and that this damage tended to be most severe after wet winters. These observations led Stevens and Coombes (2001) to evaluate the capacity of *I. newcombi* to aestivate in fallow fields. They found that mature *I. newcombi* could enter dormancy when fields were drained prior to harvest, and that survival over 180 days (from autumn to spring) can exceed 40% if there is adequate ground cover and the gastropods are able to burrow beneath the soil surface before entering dormancy. Survival was found to decline to negligible levels after 300 days, suggesting that farmers could eliminate aestivating *I. newcombi* populations by limiting rice production to every second season in individual fields. Implementation of this strategy by rice farmers has led to a reduction in the incidence of damaging *I. newcombi* infestations and a decline in molluscicide use. Some farmers still grow consecutive rice crops, primarily because of limited land availability, and these growers continue to rely on chemical control.

Planorbids and lymnaeids as primary pests of rice in other regions

Records of planorbids and lymnaeids as primary pests of rice in other regions of the world are relatively uncommon. Generally, those records that do exist mention only chemical control recommendations and/or observations on the nature of damage to the plants. Butani and Jotwani (1976) recorded *Indoplanorbis exustus* (Deshayes) (Planorbidae) and *Lymnaea acuminata* de Lamarck (Lymnaeidae) as primary rice pests in India and advocated the use of copper sulphate or metaldehyde (*r*-2,*c*-4, *c*-6,*c*-8-tetramethyl-1,3,5,7-tetroxocane; acetaldehyde homopolymer) bait pellets for their control. Madsen (1992) studied the feeding habits of the schistosomiasis vector *Biomphalaria pfeifferi* (Krauss) (Planorbidae) and two potential biocontrol agents, *Helisoma duryi* (Wetherby) (Planorbidae) and the *Marisa cornuarietis* (Linnaeus) (Ampullariidae). He found that both *B. pfeifferi* and *H. duryi* will attack exposed rice roots under laboratory conditions, but that when the root systems were covered by a substrate of coarse sand the rice plants were no longer attractive to these species. In contrast, *M. cornuarietis* caused significant damage to rice by feeding on the roots and lower parts of the stems and leaf sheaths and was not deterred by the presence of a substrate. Although there have been no reports of *B. pfeifferi* or *H. duryi* damaging rice in the field, they must be regarded as having the potential to assume pest status, particularly if cropping practices such as aerial sowing are adopted in the areas where they occur. Species of *Planorbis* Müller and *Lymnaea* de Lamarck have been reported attacking rice in Italy (Grandori, 1952; Foschi, 1973, 1975), although Foschi (1975) regards them as not constituting a ‘very serious threat’ to the crop under Italian conditions.

Planorbids and Lymnaeids as Secondary Pests of Rice

Rice yields are strongly influenced by the amount of nitrogen available to the crop. In rice-growing regions it is common practice to supplement soil-available nitrogen by the application of nitrogenous fertilizers (e.g. urea, anhydrous ammonia, etc.) and by rotating rice with leguminous crops. Alternative methods of enhancing nitrogen availability include green manuring with legumes or *Azolla*, dual-cropping *Azolla* and rice simultaneously and encouraging the growth of BGA.

The genus *Azolla* contains seven species (Roger, 1996) of aquatic fern which have symbiotic associations with the nitrogen-fixing BGA *Anabaena azollae* Strasburger. *Azolla* is generally inoculated into rice-fields and incorporated into the soil prior to transplanting. However, its cultivation as a dual crop with rice can also provide substantial crop benefits, either with or without periodic soil incorporation (Kannaiyan, 1987; Mabbayad, 1987; Roger, 1996). *Azolla* crops have the potential to assimilate the full nitrogen requirements of a high-yielding rice crop (50–100 kg N ha^{-1}) in 10–20 days (Lumpkin, 1987), although this potential is seldom achieved in practice. *Azolla* is exposed to both terrestrial and aquatic herbivores that can substantially reduce yields and therefore reduce the amount of nitrogen available to the rice crop after *Azolla* incorporation. The principal herbivores that limit *Azolla* productivity include larval stages of Chironomidae (Diptera), Curculionidae (Coleoptera) and Pyrallidae (Lepidoptera) and a range of gastropod species, including some lymnaeids and planorbids (De Lara, 1986; Mochida, 1987; Mochida *et al.*, 1987). In addition to the direct effects of herbivory, gastropod damage is known to exacerbate black-rot disease of *Azolla*, which is caused by the pathogen *Rhizoctonia solani* Kuhn (Agonomycetales) (Kannaiyan and Nandabalan, 1987b).

Roger *et al.* (1987) identified three categories of photodependent nitrogen-fixing organisms in wetland rice-fields: *Azolla*, photosynthetic bacteria and free-living BGA. The contribution of photosynthetic bacteria to rice-field nitrogen levels is considered to be very minor (Roger and Watanabe, 1985). In contrast, the high abundance of free-living BGA in rice-fields provides for low but consistent rice yields in fields that do not receive any supplementary applications of synthetic or organically-derived nitrogenous fertilizer (Grant *et al.*, 1986; Roger *et al.*, 1987). Although there are technical difficulties associated with measuring nitrogen fixation by BGA, Roger *et al.* (1987) have estimated that, where BGA growth is visible, a figure of 30 kg N ha^{-1} per crop is supported by the available data. A full cover of BGA within a rice crop can contain between 5 and 20 kg N ha^{-1}, depending on which algal species are present. Control of invertebrate grazers, particularly microcrustacea and Gastropoda, has generally been shown to enhance the growth of BGA (Grant *et al.*, 1986; Roger *et al.*, 1987; Roger, 1991). Records of Planorbidae and Lymnaeidae as secondary pests of rice (primary pests of *Azolla* or BGA) are listed in Table 7.2.

Chemical control of secondary rice pests

Utilization of *Azolla* or a reliance on BGA for biological nitrogen fixation is common only where rice is grown as a subsistence crop. In these situations economics precludes or limits the use of synthetic fertilizers and synthetic crop-protection chemicals. As a consequence, a considerable amount of research has been directed towards evaluating plant extracts or preparations that can be produced locally and used as molluscicides against gastropod species that are secondary pests of rice and/or vectors of trematode parasites in rice-growing areas. Reynaud

Table 7.2. Published reports of Planorbidae and Lymnaeidae as secondary pests of rice.

Species	Country	Host	References
Planorbidae			
Gyraulus convexiusculus (Hutton)	Philippines	*Azolla*[1]	Mochida, 1987
Gyraulus Charpentier sp.	India	BGA[2]	Roger *et al.*, 1985
Indoplanorbis exustus (Deshayes)	India	*Azolla*	Sasmal *et al.*, 1984a
Indoplanorbis exustus (Deshayes)	India	BGA (*Anabaena, Nostoc*)	Sinha *et al.*, 1986a,b
Indoplanorbis exustus (Deshayes)	Philippines	*Azolla*	Mochida, 1987
Indoplanorbis Annandale sp.	India	BGA	Roger *et al.*, 1985
Indoplanorbis Annandale sp.	Philippines	*Azolla*	De Lara, 1986
Physastra hungerfordiana Nevill (now *Glyptophysa*)	Philippines	*Azolla*	De Lara, 1986
Unidentified species	China	*Azolla*	Jin-rong, 1965; Guo-tao *et al.*, 1965 (cited in Lumpkin and Plucknett, 1982)*
Unidentified species	Pakistan	*Azolla*	Ali and Malik, 1987
Unidentified species	Philippines	*Azolla*	Calilung and Lit, 1985
Lymnaeidae			
Lymnaea acuminata de Lamarck	India	*Azolla*	Ittyaverah *et al.*, 1979a,b; Kushari and Taheruzzaman, 1991
Lymnaea acuminata de Lamarck	Nepal	*Azolla*	Mochida, 1987
Lymnaea auricularia (Linnaeus)	Philippines	*Azolla*	Mochida, 1987
Lymnaea auricularia (Linnaeus) race *rufescens* (Gray)	India	*Azolla*	Sasmal *et al.*, 1984a
Lymnaea auricularia (Linnaeus) race *swinhoei* (Adams)	India	*Azolla*	Sasmal and Kulshreshtha, 1984, 1987; Sasmal *et al.*, 1984a,b, 1985
Lymnaea natalensis (Krauss)	Senegal	*Azolla*	Reynaud, 1986
Lymnaea (*Pseudosuccinea*) *luteola* de Lamarck	India	*Azolla*	Chatterjee and Dutta, 1980
Lymnaea (*Pseudosuccinea*) *luteola* de Lamarck form *typica* de Lamarck	India	*Azolla*	Chatterjee, 1982
Lymnaea (*Pseudosuccinea*) *luteola* de Lamarck form *typica* de Lamarck	India	BGA (*Anabaena, Nostoc*)	Sinha *et al.*, 1986a,b
Lymnaea rubiginosa (Michelin)	Philippines	*Azolla*	Mochida, 1987
Lymnaea viridis Quoy & Gaimard	Not stated	BGA	Roger *et al.*, 1987
Lymnaea de Lamarck sp.	India	BGA	Roger *et al.*, 1985
Lymnaea de Lamarck sp.	Pakistan	*Azolla*	Ali and Malik, 1987
Lymnaea de Lamarck sp.	Philippines	*Azolla*	Calilung and Lit, 1985
Radix swinhoei Adams	China	*Azolla*	Chung-Chu, 1984
Radix (*Lymnaea*) *quadrasi* (Moellendorff)	Philippines	*Azolla*	De Lara, 1986
Unidentified species	China	*Azolla*	Jin-rong, 1965; Guo-tao *et al.*, 1965 (cited in Lumpkin and Plucknett, 1982)*

*Lumpkin and Plucknett (1982) list these taxa as *Planorbidus* and *Lymnaeidus*, probably spelling errors for Planorbidae and Lymnaeidae, respectively.
[1] *Azolla* de Lamarck (Azollaceae).
[2] Blue-green algae (Cyanophyceae): *Anabaena* (Bory) Bornet & Flahualt; *Nostoc* (Vaucher) Bornet & Flahualt).

(1986) found powders derived from *Balanites aegyptiaca* (Linnaeus) Deleuil (Zygophyllaceae) and *Detarium heudelotianum* Baillon (= *Detarium senegalense* Gmelin; Caesalpiniaceae) to be effective (100% control at 0.25 g l^{-1} 15 days after application) for controlling the *Azolla* pest *Lymnaea natalensis* (Krauss). *B. aegyptiaca* leaf powder was toxic to *Azolla* at rates in excess of 0.75 g l^{-1}. In the same trials neem cake (from *Azadirachta indica* de Jussieu, Meliaceae) and the synthetic molluscicide metaldehyde were found to be ineffective. Kushari and Taheruzzaman (1991) indicate that neem has a toxic effect on the *Azolla* pest *L. acuminata*. Reviews on the efficacy of plant-derived molluscicides have been provided by Mott (1987) and Singh *et al.* (1996).

There is a large amount of literature available on the control of lymnaeids and planorbids with synthetic molluscicides, primarily reflecting the importance of these gastropods as disease vectors. Studies on the chemical control of planorbids and lymnaeids directed towards suppressing damage to *Azolla* and BGA are, however, much less common. Sasmal and Kulshreshtha (1984, 1987) found that *Lymnaea auricularia* race *swinhoei* (Adams) could be effectively controlled (> 70% mortality 3 days after application) with metaldehyde bait pellets (1 kg a.i. ha^{-1}, broadcast), benzene hexachloride (1 kg a.i. ha^{-1}, on inoculum), and phorate (*O,O*-diethyl *S*-ethylthiomethyl phosphorodithioate) granules (1 kg a.i. ha^{-1}, broadcast), but that at rates between 0.5 and 1 kg a.i. ha^{-1} quinalphos (*O,O*-diethyl *O*-quinoxalin-2-yl phosphorothioate), carbaryl (1-naphthyl methylcarbamate) and carbofuran (2,3-dihydro-2,2-dimethylbenzofuran-7-yl methylcarbamate) provided less than 40% mortality. *Azolla* biomass varied from 250 kg ha^{-1} in the control plots to 10,125 kg ha^{-1} in plots where gastropods were controlled with metaldehyde pellets. Phorate granules applied at 1.25 kg a.i. ha^{-1} had earlier been found to be effective against an unidentified gastropod feeding on *Azolla* (Rajan Asari and Dale, 1978). Simpson *et al.* (1994) found that a gastropod community (primarily species of *Melanoides* Olivier (Thiaridae), but also containing species of *Lymnaea*, *Hippeutis* Charpentier (Planorbidae) and *Gyraulus* Charpentier) (Planorbidae) were not significantly affected by carbofuran applied at 0.5 kg a.i. ha^{-1} or the herbicide butachlor (*N*-butoxymethyl-2-chloro-2′,6′-diethylacetanilide) at 0.35 kg a.i. ha^{-1}. Sinha *et al.* (1986a,b) found thiodicarb (3,7,9,13-tetramethyl-5,11-dioxa-2,8,14-trithia-4,7,9,12-tetra-azapentadeca-3,12-diene-6,10-dione) to be effective (100% mortality in 24 h laboratory bioassays) at 1 mg l^{-1} against a mixed population of gastropods, which included *I. exustus* and *Lymnaea luteola* form *typica* de Lamarck, while chlorpyrifos (*O,O*-diethyl *O*-3,5,6-trichloro-2-pyridyl phosphorothioate), diazinon (*O,O*-diethyl *O*-2-isopropyl-6-methyl-pyrimidin-4-yl phosphorothioate), phorate, carbofuran, carbaryl and endosulfan ((1,4,5,6,7,7-hexachloro-8,9,10-trinorborn-5-en-2,3-ylenebis-methylene) sulphite) were ineffective at rates below 10 mg l^{-1}. Under field conditions thiodicarb applied at 1 mg l^{-1} provided over 50% gastropod mortality 7 days after application.

Growth responses of BGA and *Azolla* to synthetic pesticides are highly variable, and the results obtained in laboratory studies on BGA often do not correlate with those conducted under field conditions (Roger and Kulasooriya, 1980). Pesticides, including some molluscicides commonly used to control both primary and secondary rice pests, may have direct inhibitory or stimulatory effects on BGA and *Azolla* growth and nitrogen assimilation (Roger and Kulasooriya, 1980; Kannaiyan and Nandabalan, 1987a; Pablico and Moody, 1991). BGA are generally more tolerant to pesticides than eukaryotic algae, although susceptibility varies widely between algal strains. Pesticide application often initially depresses nitrogen fixation by BGA, but this may be followed by either an increase or decrease in nitrogen assimilation rates (Roger and Kulasooriya, 1980).

When used at the comparatively high rates necessary to control *P. canaliculata*, fentin acetate (triphenlytin acetate) killed more than 90% of *Azolla pinnata* Brown plants (Pablico and Moody, 1991). Application method can strongly influence the effect of pesticide treatments on *Azolla* growth. Kannaiyan and Nandabalan (1987a) found that carbofuran, monocrotophos (dimethyl (*E*)-1-methyl-2-(methylcarbamoyl)vinyl phosphate), phorate, endosulfan, carbosulfan (2,3-dihydro-2,2-dimethylbenzofuran-7-yl(dibutylaminothio) methylcarbamate) and chlorpyrifos applied to small plots at 0.5 kg a.i. ha^{-1} prior to *Azolla* inoculation all increased both *Azolla* yield and nitrogenase activity. When *Azolla* fronds were treated directly with carbofuran at rates of between 2 and 12% (w/w), reduced yields were recorded.

While control of secondary rice pests with synthetic molluscicides is feasible, it may not always be cost-effective. This is particularly the case in regard to providing longer-term protection of BGA from grazing gastropods, where economic viability may vary regionally according to the composition of gastropod faunas, the abundance of individual species and the nitrogen fixing potential of BGA species present in the fields. Grant *et al.* (1983) found that suppression of gastropod grazers in newly established rice crops had little effect on BGA, while under the same conditions effective control of ostracods resulted in large increases in BGA biomass and nitrogen-fixation rates. In general, *Azolla* has a greater potential for nitrogen fixation than BGA (Roger, 1991) and therefore the short-term protection of *Azolla* with molluscicides prior to its incorporation into the soil may represent a more economically viable proposition.

Cultural control of secondary rice pests

Cultural approaches to gastropod control, such as changes in crop-management practices and the use of herbivore-resistant BGA and *Azolla*, offer opportunities for increasing biological nitrogen fixation in rice-fields without the use of molluscicides. Planting the crop as soon as possible after ploughing rice-fields can alleviate gastropod problems where BGA are being encouraged (Grant *et al.*, 1986). In rice paddy fields flooded from irrigation channels, mesh screens placed over inlets should reduce the rate of colonization by gastropods. Field cultivation during fallow periods can be used to mechanically destroy aestivating animals, while the seasonal drying of fields is effective for controlling grazers of BGA (Roger, 1991), including gastropods.

The use of BGA and *Azolla* species, or selected strains, resistant or tolerant to herbivores is a possible alternative to providing chemical protection. Chung-Chu (1984) and Wen-xiong *et al.* (1987) have reported that *Azolla caroliniana* von Willdenow has a broad spectrum of tolerance to pests and diseases, including gastropods. Mucilaginous strains of some BGA are resistant to invertebrate grazers (Roger, 1991). However, these strains are usually slow-growing and have only low nitrogen-fixation efficiency (Grant *et al.*, 1986). Roger (1991) has suggested that genetic engineering may, in the longer term, provide improved strains of BGA that combine herbivore resistance with rapid growth and high rates of nitrogen assimilation.

Conclusion

In comparison with *Pomacea* and major insect pests such as plant-hoppers, planorbids and lymnaeids are of limited importance as primary pests of rice. They have been reported as damaging rice crops in the field in India, Italy and Australia, and only one Australian species, *I. newcombi*, can currently be regarded as a consistently serious problem for rice production. The potential exists, however, for the significance of planorbids and lymnaeids as primary rice pests to increase in response to changing farming practices, particularly in Asia. The popularity of aerial sowing in Australia has been a major factor in increasing the incidence and severity of crop damage caused by *I. newcombi*. Aerial sowing provides a stable aquatic environment where the germinating rice seed is continuously exposed to attack from a range of pests, including *I. newcombi*, that seldom cause substantial damage to crops sown using other methods. Similarly, a move away from traditional transplantation techniques to broadcast sowing, prompted by increasing labour costs, may lead to planorbids and lymnaeids assuming greater significance as primary pests of rice in Asia.

Planorbids and lymnaeids are of greater importance as secondary pests in countries where rice is grown as a subsistence crop. Although their impact on *Azolla* and BGA can be reduced through the use of synthetic molluscicides, this detracts from the overall economic goal of enhancing biological nitrogen fixation in rice-fields, which is to reduce the dependence of high rice yields on expensive synthetic compounds. Plant-derived molluscicides, manufactured either locally or by rice farmers themselves, have considerable potential to reduce the impact of planorbids and lymnaeids on *Azolla* and BGA production, particularly if they are used in conjunction with cultural methods of gastropod control. Further research in these areas should form a significant part of programmes directed towards optimizing the use of biological nitrogen fixation in subsistence rice production.

Acknowledgements

Susan Boyd, Lucy Borgese (New South Wales Agriculture) and Milagros Zamora (International Rice Research Institute) are thanked for their assistance with literature searches. Phil Colman (Australian Museum, Sydney) is thanked for his advice on nomenclatural matters.

References

Ali, S. and Malik, K.A. (1987) Use of *Azolla* in Pakistan. In: Azolla *Utilization. Proceedings of the Workshop on* Azolla *Use, Fuzhou, Fujian, China, 31 March–5 April 1985.* International Rice Research Institute, Manila, pp. 177–185.

Butani, D.K. and Jotwani, M.G. (1976) Crop pests and their control. 2. Rice. *Pesticides (India)* 10, 29–35.

Calilung, V.J. and Lit, I.L. Jr (1985) Watch out against these *Azolla* pests. *Azolla Notes* 2, 27–29.

Chatterjee, P.B. (1982) Molluscan pests of *Azolla. International Rice Research Newsletter* 7, 16.

Chatterjee, P.B. and Dutta, S. (1980) Snails – a new pest of *Azolla. International Rice Research Newsletter* 5, 24.

Cherian, M.C. and Krishnamurthy, P.S. (1940) Snails as pests of paddy. *Madras Agricultural Journal* 28, 142–143.

Chiappelli, R. (1952) La lotta contro i parassiti. *Riso* 1, 26.

Chung-Chu, L. (1984) Recent advances on *Azolla* research. In: Silver, W.S. and Schröder, E.C. (eds) *Practical Application of* Azolla *for Rice Production. Proceedings of an International Workshop, Mayaguez, Puerto Rico, 17–19 November 1982.* Martinus Nijhoff/Dr W. Junk, Dordrecht, pp. 45–54.

De Lara, A.V. (1986) *Studies of the Effect of Snail on the Growth, Reproduction and Doubling Time of* Azolla *and the Latter's Effect on the Growth and Fecundity of the Snail.* Terminal Report, IBS *Azolla* Project, University of the Philippines at Los Baños, 22 pp.

FAO (2000) FAOSTAT Food and Agriculture Organization of the United Nations Statistical Databases. http://apps.fao.org/default.htm

Foschi, S. (1973) *Research on Copper Molluscicides in Rice Fields.* Final report, INCRA Project 204, International Copper Research Association, New York, 14 pp.

Foschi, S. (1975) Practical aspects of copper treatments on rice. *International Rice Commission Newsletter* 24, 19–22.

Grandori, R. (1952) Danni dei parassiti nel Delta Padano. *Riso* 1, 30.

Grant, I.F., Tirol, A.C., Aziz, T. and Watanabe, I. (1983) Regulation of invertebrate grazers as a means to enhance biomass and nitrogen fixation of Cyanophyceae in wetland rice fields. *Soil Science Society of America Journal* 47, 669–675.

Grant, I.F., Roger, P.A. and Watanabe, I. (1986) Ecosystem manipulation for increasing biological N_2 fixation by blue-green algae (Cyanobacteria) in lowland rice fields. *Biological Agriculture and Horticulture* 3, 299–315.

Grist, D.H. and Lever, R.J.A.W. (1969) *Pests of Rice*. Longmans, London, 520 pp.

Guo-tao, L., Jing-you, D. and Zhen-ru, L. (1965) Studies on the control and eradication of gastropods while growing *Azolla* in summer and autumn in Guangdong. *Zhongguo Nongye Kexue* 11, 24–26 [in Chinese].

Ittyaverah, P.J., Nair, N.R., Thomas, M.J. and John, P.S. (1979a) *Limnaea acuminata* a snail for the biological control of *Salvinia molesta*. *Indian Journal of Weed Science* 11, 76–77.

Ittyaverah, P.J., Nair, N.R. and Thomas, M.J. (1979b) *Limnaea acuminata* Lamarck (Pulmonata: Limnaeidae), a pest on azolla (*Azolla pinnata*). *Agricultural Research Journal of Kerala* 17, 296.

Jin-rong, H. (1965) The skin of white gourd may be used to lure the biggest enemy of red *Azolla*, the Tsueshih snail, to its death. *Fujian Nongye* 9, 44 [in Chinese].

Jones, E.L. (1975) Aquatic snails attack rice. *Farmers' Newsletter (Large Area)* 95, 20–21.

Jones, E.L. (1976) *Annual Report for 1975/76. Pests of Rice – Aquatic Snails*. New South Wales Department of Agriculture, Yanco, 2 pp.

Jones, E.L. (1979) Persistence and change in the status of rice pests in the Riverina. In: *Australian Rice Workshop, Yanco*. New South Wales Department of Agriculture, Yanco, pp. E6–E7.

Jones, E.L. (1984) Insect pests. In: *Rice Growing in New South Wales*. Department of Agriculture New South Wales and the Rice Research Committee, Yanco, 6 pp.

Kannaiyan, S. (1987) Use of *Azolla* in India. In: *Azolla Utilization. Proceedings of the Workshop on* Azolla *Use, Fuzhou, Fujian, China, 31 March–5 April 1985*. International Rice Research Institute, Manila, pp. 109–118.

Kannaiyan, S. and Nandabalan, K. (1987a) Effect of insecticides on the growth and nitrogen fixation in *Azolla*. In: Azolla *Utilization. Proceedings of the Workshop on* Azolla *Use, Fuzhou, Fujian, China, 31 March–5 April 1985*. International Rice Research Institute, Manila, p. 286.

Kannaiyan, S. and Nandabalan, K. (1987b) Influence of neem cake on black rot disease incidence in *Azolla*. In: Azolla *Utilization. Proceedings of the Workshop on* Azolla *Use, Fuzhou, Fujian, China, 31 March–5 April 1985*. International Rice Research Institute, Manila, p. 287.

Kushari, D.P. and Taheruzzaman, Q. (1991) Seasonal effect of leaf extracts of *Azadirachta indica* (neem) and *Albizzia lebbek* (sirish) on the productivity of *Azolla pinnata*. In: Dutta, S.K. and Sloger, C. (eds) *Biological Nitrogen Fixation Associated with Rice Production*. Howard University Press, Washington, DC, pp. 110–116.

Lewis, G.J. (1994) *An Illustrated History of the Riverina Rice Industry*. Ricegrowers' Cooperative, Leeton, 230 pp.

Lumpkin, T.A. (1987) Environmental requirements for successful *Azolla* growth. In: Azolla *Utilization. Proceedings of the Workshop on* Azolla *Use, Fuzhou,*

Fujian, China, 31 March–5 April 1985. International Rice Research Institute, Manila, pp. 89–97.

Lumpkin, T.A. and Plucknett, D.L. (1982) Azolla *as a Green Manure: Use and Management in Crop Production.* Westview Press, Boulder, Colorado, 230 pp.

Mabbayad, B.B. (1987) The *Azolla* program of the Philippines. In: Azolla *Utilization. Proceedings of the Workshop on* Azolla *Use, Fuzhou, Fujian, China, 31 March–5 April 1985.* International Rice Research Institute, Manila, pp. 101–108.

McDonald, D.J. (1994) Temperate rice technology for the 21st century: an Australian example. *Australian Journal of Experimental Agriculture* 34, 877–888.

Madsen, H. (1992) *Interspecific Competition between* Helisoma duryi *(Wetherby 1879) and Intermediate Hosts of Schistosomes (Gastropoda: Planorbidae). An Evaluation of Biological Control of Schistosome Intermediate Hosts by Competitor Snails.* Danish Bilharziasis Laboratory, Charlottenlund, 86 pp.

Malek, E.A. and Cheng, T.C. (1974) *Medical and Economic Malacology.* Academic Press, New York, 398 pp.

Mochida, O. (1987) *Pests of* Azolla *and Control Practices. Training Course for* Azolla *Use, 6–19 June, 1987.* Fujian Academy of Agricultural Science and International Rice Research Institute, Fuzhou, Fujian, China, 60 pp.

Mochida, O., Yoshiyasu, Y. and Dimaano, D. (1987) Insect pests of *Azolla* in the Philippines. In: Azolla *Utilization. Proceedings of the Workshop on* Azolla *Use, Fuzhou, Fujian, China, 31 March–5 April 1985.* International Rice Research Institute, Manila, pp. 207–221.

Mott, K.E. (ed.) (1987) *Plant Molluscicides. Papers Presented at a Meeting of the Scientific Working Group on Plant Molluscicides, UNDP/WORLD BANK/WHO Special Programme for Research and Training in Tropical Diseases, Geneva, Switzerland, 31 January to 2 February 1983.* John Wiley & Sons, New York, 326 pp.

Pablico, P.P. and Moody, K. (1991) Effect of fentin acetate on wet-seeded rice, *Pistia stratiotes* and *Azolla pinnata. Crop Protection* 10, 45–47.

Radostits, O.M., Blood, D.C. and Gay, C.C. (1994) *Veterinary Medicine. A Text-book of the Diseases of Cattle, Sheep, Pigs, Goats and Horses*, 8th edn. Baillière Tindall, London, 1763 pp.

Rajan Asari, P.A. and Dale, D. (1978) A new enemy of *Azolla. International Rice Research Newsletter* 3, 17.

Reynaud, P.A. (1986) Control of the *Azolla* pest *Limnea natalensis* with molluscicides of plant origin. *International Rice Research Newsletter* 11, 27–28.

Roger, P.A. (1991) Reconsidering the utilization of blue-green algae in wetland rice cultivation. In: Dutta, S.K. and Sloger, C. (eds) *Biological Nitrogen Fixation Associated with Rice Production.* Howard University Press, Washington, DC, pp. 119–141.

Roger, P.A. (1996) *Biology and Management of the Floodwater Ecosystem in Ricefields.* International Rice Research Institute, Manila, 250 pp.

Roger, P.A. and Kulasooriya, S.A. (1980) *Blue-green Algae and Rice.* International Rice Research Institute, Manila, 112 pp.

Roger, P.A. and Watanabe, I. (1985) Technologies for utilizing biological nitrogen fixation in wetland rice: potentialities, current usage, and limiting factors. *Fertilizer Research* 9, 39–77.

Roger, P.A., Grant, I.F. and Reddy, P.M. (1985) *Report on the Blue-green Algae Monitoring Tour to India, 7–22 March 1985*. International Rice Research Institute, Manila, 93 pp.

Roger, P.A., Grant, I.F., Reddy, P.M. and Watanabe, I. (1987) The photosynthetic aquatic biomass in wetland rice fields and its effect on nitrogen dynamics. In: *Efficiency of Nitrogen Fertilizers for Rice. Proceedings of the Meeting of the International Network on Soil Fertility and Fertilizer Evaluation for Rice, Griffith, New South Wales, Australia, 10–16 April 1985*. International Rice Research Institute, Manila, pp. 43–68.

Sasmal, S. and Kulshreshtha, J.P. (1984) Chemical control of the snail *Lymnaea auricularia* (L) race *swinhoei* (H.Adams), a pest of *Azolla pinnata*. *Rice Research Newsletter* 5, 2.

Sasmal, S. and Kulshreshtha, J.P. (1987) Biology and control of the snail *Lymnaea auricularia* (L) race *swinhoei* (H.Adams), a pest of *Azolla pinnata*. *Oryza* 24, 177–179.

Sasmal, S., Kulshreshtha, J.P. and Mathur, K.C. (1984a) Gastropod pests on *Azolla*. *Rice Research Newsletter* 5, 3.

Sasmal, S., Kulshreshtha, J.P. and Mathur, K.C. (1984b) A predatory snail on rice brown planthopper. *Rice Research Newsletter* 5, 3.

Sasmal, S., Kulshreshtha, J.P. and Mathur, K.C. (1985) *Lymnaea auricularia* (L) race *swinhoei* (H. Adams), a predatory snail of rice brown planthopper, *Nilaparvata lugens* (Stål). *Oryza* 22, 57.

Simpson, I.C., Roger, P.A., Oficial, R. and Grant, I.F. (1994) Effects of nitrogen fertiliser and pesticide management on floodwater ecology in a wetland ricefield. III. Dynamics of benthic molluscs. *Biology and Fertility of Soils* 18, 219–227.

Singh, A., Singh, D.K., Misra, T.N., Agarwal, R.A. and Singh, A. (1996) Molluscicides of plant origin. *Biological Agriculture and Horticulture* 13, 205–252.

Sinha, P.K., Pal, S. and Triar, S.B. (1986a) An effective molluscicide for grazer snails of blue green algae. *Pesticides (India)* 20, 44–45.

Sinha, P.K., Pal, S., Kumar, K., Triar, S.B. and Singh, R. (1986b) Thiodicarb, an effective molluscicide for grazer snails of blue green algae. *Journal of Entomological Research* 10, 116–118.

Smith, B.J. and Burn, R. (1976) *Glyptophysa cosmeta* (Iredale, 1953) in Victoria (Lymnaeoidea: Planorbidae), with notes on aestivation. *Journal of the Malacological Society of Australia* 3, 175–176.

Stevens, M.M. and Coombes, N.E. (2001) Aestivation in *Isidorella newcombi sens. lat* (Gastropoda: Basommatophora: Planorbidae) and its implications for integrated snail management in Australian rice fields. *Journal of Medical and Applied Malacology* (in press).

Stevens, M.M., Faulder, R.J. and Coombes, N.E. (1996) Microcosm assessment of potential molluscicides for control of the rice snail *Isidorella newcombi* sens. lat. (Gastropoda: Basommatophora: Planorbidae). *Australian Journal of Agricultural Research* 47, 673–680.

Tirumala Rao, V., Ramadoss, P. and Koteswara Rao, M. (1953) Snails damage to paddy in the Northern Circars and their control. *Madras Agricultural Journal* 40, 445–452.

Tomlin, C. (ed.) (1994) *The Pesticide Manual*, 10th edn. British Crop Protection Council and the Royal Society of Chemistry, Farnham, 1341 pp.

Walker, J.C. (1988) Classification of Australian buliniform planorbids (Mollusca: Pulmonata). *Records of the Australian Museum* 40, 61–89.

Wen-xiong, W., Guo-tian, Y., Guo-zhang, Z., Feng-yue, C., Gui-ying, J., Pei-ji, L. and Wei-wen, Z. (1987) Tolerance of *Azolla caroliniana* and its application. In: Azolla *Utilization. Proceedings of the Workshop on* Azolla *Use, Fuzhou, Fujian, China, 31 March–5 April 1985.* International Rice Research Institute, Manila, pp. 283–284.

Woodlands, K., Fowler, J., Lacy, J. and Clampett, W. (1984) Crop establishment. In: *Rice Growing in New South Wales.* Department of Agriculture New South Wales and the Rice Research Committee, Yanco, 12 pp.

8 *Urocyclus flavescens* Kerferstein (Urocyclidae) as a Pest of Banana in South Africa

KAREN DE JAGER AND MIEKE DANEEL

Institute for Tropical and Subtropical Crops, Private Bag X11208, Nelspruit 1200, South Africa

The Banana-cropping Environment

The main production areas for banana (*Musa acuminata* Colla) (Musaceae) in South Africa are Komatipoort, Hazyview, South Kwazulu/Natal, North Kwazulu/Natal, Tzaneen and Levubu (Table 8.1). The total area devoted to banana production is approximately 12,750 ha (BGASA, 1998). Most of the banana plantings in South Africa are commercial plantations, with an average size of 35 ha. Bananas in South Africa are exclusively used for the local market. The bananas grown in South Africa are from the Cavendish group, with the main cultivars currently planted being Grand Nain, Williams, Chinese Cavendish and Dwarf Cavendish.

Climatic conditions in South Africa are subtropical. Production of banana, an essentially tropical species, is therefore severely constrained by suboptimal growth conditions (Robinson, 1993a,b). Rainfall is sporadic from October until March, resulting in a suboptimal average precipitation of 800 mm per annum. In the extremely high temperature conditions of summer, severe moisture stress and wilting of plants is not uncommon. The high radiation intensities can also cause permanent leaf scorch which reduces leaf area and photosynthetic efficiency. Maximum temperatures on summer afternoons can sometimes reach between 40 and 45°C. Furthermore, night air temperatures in winter frequently decline to 12–15°C or even lower, reducing plant growth and leading to various physiological problems which are manifested as yellowing of the foliage and deformation of the developing fruit bunches.

South African banana growers are using several water reticulation systems to reduce moisture and heat stress in the crop, including overhead sprinklers, under-canopy sprinklers, microjet and drip irrigation. The overhead sprinklers are especially useful in very hot areas, such as Komatipoort, to cool down the leaf canopy. None the less, the

Table 8.1. The extent of commercial banana (*Musa acuminata* Colla) (Musaceae) production in different regions of South Africa (BGASA, 1998).

Region	Area (ha)
Komatipoort	4700
Hazyview	2450
South Kwazulu/Natal	2600
North Kwazulu/Natal	800
Tzaneen	1300
Levubu	900

under-canopy irrigation systems are superior to overhead sprinklers due to the lesser effect of wind on drift and evaporation loss. An increasing number of growers are now installing permanent microjet and drip irrigation systems due to these advantages (Robinson, 1993b).

Bananas may be cultivated on soils ranging from fairly sandy soils to heavy clay soils. It has been found that the most suitable soils for banana cultivation in South Africa have a clay content of 30–55%. The physical condition of soils is very important since root development is determined mainly by the degree of aeration of the soil (Smith and Abercromie, 1993).

Emergence of *Urocyclus flavescens* Kerferstein as a Pest

The urocyclid slug *Urocyclus flavescens* Kerferstein has been known for many years in South Africa along the south coast of Kwazulu/Natal, but was first recognized as a serious pest in Mpumalanga during 1967 (de Villiers, 1973). This species is now known in all banana-producing areas of South Africa.

The economic importance of *U. flavescens* arises from its damage to the marketable part of the banana crop, the fruit. Animals infesting the bunches rasp the fruit peel, causing irregular shaped, shallow pits or scars, which eventually take on a brown, corky appearance (Plate 4). The economic impact arises primarily through downgrading or even rejection of the fruit because of the reduced cosmetic appearance. The corky blemish on the fruit caused by *U. flavescens* may be confused with that caused by grasshoppers, although the latter produce scars of rougher appearance due to exposure of the peel fibres; the correct diagnoses are obviously important for the management of pest infestations.

In a severe infestation all the fruit from a bunch can be lost to *U. flavescens* damage. In certain banana-producing areas of South Africa, 10% damage is considered as a relatively low level of loss (Pretorius, 1986).

Pest Ecology

U. flavescens is primarily nocturnal and feeds at night or during overcast, rainy days. It shelters during the day either under debris on the ground or among the leaves and hands of bunches on the banana plants (de Villiers and Jones, 1981). The eggs are deposited during spring under plant material or in crevices in the soil. Hatching occurs after the first spring rains and, if conditions are favourable, the developing animals move up into the banana plants (de Villiers and Jones, 1981). De Villiers and Jones (1981) reported the young *U. flavescens* to be very resistant to drought and able to remain for several months in an aestivatory state in the soil or under plant material on the ground.

Integrated Pest Management

Control of *U. flavescens* is primarily dependent on molluscicidal baits containing metaldehyde or methiocarb. The bulk of the baits are scattered on the ground around the plants, but a few baits may be placed between the fingers of bunches (de Villiers, 1973; de Villiers and Jones, 1981). Where the baits are placed around the plants, growers have found that it is essential to clear all plant debris away from the plants prior to application (de Villiers and Jones, 1981).

In South Africa, bananas are harvested throughout the year. There is a tendency for growers to apply control measures when *U. flavescens* damage is detected during grading of fruit in the packing shed. However, de Villiers (1973) and Jones (1981) noted that much of the damage occurs when the fruit is immature and, if economic damage is to be effectively prevented, control measures need to be applied as soon as *U. flavescens* is seen in the banana plantings (Jones, 1981). Therefore it is of utmost importance to scout regularly for the presence of these gastropods. Molluscicide treatments are most effective in reducing pest abundance and fruit damage when applied after the first spring rains, when young *U. flavescens* are active (Jones, 1981). Adult *U. flavescens* are more difficult to control than the young (de Villiers and Jones, 1981).

As *U. flavescens* feed by a rasping action, bait palatability and thus level of control is best achieved with baits that are firm. Under conditions of high humidity, such as that provided by weedy ground cover and irrigation, baits deteriorate rapidly and applications must be made at frequent intervals to maintain control (Jones, 1981).

Weedy plantations also provide shelter and food conditions favourable to *U. flavescens* population growth (Jones, 1981). Jones identified the surrounding veld as a source of *U. flavescens* infestation in the plantations and recommended the maintenance of a weed-free, molluscicide-baited 5–10 m strip to reduce reinfestation.

Barriers may be placed around the banana stem to prevent *U. flavescens* from infesting the plants. The vein of an old banana leaf, respectively, to *B. similaris* movement into the vines. In experimental situations the physical barriers fixed to the stem provided as good a control as the traditional approach of scattering metaldehyde bait pellets on the ground. 'Bottle traps', made from disposable, plastic, drink bottles (Plate 8), have proved to be a very cheap, highly effective method of controlling *B. similaris*. In field experiments, bottle traps provided 100% reduction in the number of *B. similaris* on the vines after 100 weeks (Chang, 1990). Bottle traps are actively recommended to grape growers (Chang, 1988a, 1990); the proper time for application of the bottle traps is after pruning the grapevines when most of the stems and branches have been removed and *B. similaris* abundance is at its lowest.

References

Chang, C.-P. (1988a) The occurrence and control of *Bradybaena similaris* (Ferussac). *Bulletin of the Taichung District Agricultural Improvement Station, Taiwan Republic of China* 21, 1–2.

Chang, C.-P. (1988b) The investigation on insect and other animal pests on grapevine and their seasonal occurrences in Taiwan. *Chinese Journal of Entomology* 8, 39–49 [in Chinese, with English abstract].

Chang, C.-P. (1990) Evaluation of chemical and exclusion methods for control of *Bradybaena similaris* (Férussac) on grapevine in Taiwan. *Agriculture, Ecosystems and Environment* 31, 85–88.

Chang, C.-P. and Chen, W.-Y. (1989) Morphology and behavior of *Bradybaena similaris* (Férussac) on grape-vine in Taiwan. *Plant Protection Bulletin, Taiwan* 31, 217–224 [in Chinese, with English abstract].

Smith, B.J. (1989) Travelling snails. *Journal of Medical and Applied Malacology* 1, 195–204.

Solem, A. (1959) Systematics of the land and fresh-water Mollusca of the New Hebrides. *Fieldiana, Zoology* 43, 1–238, 34 pls.

Sun, P.M.H. and Hsieh, S.C. (1992) The establishment of a sustainable agricultural system in Taiwan. In: *Sustainable Agriculture for the Asian and Pacific Region. Meeting of the Technical Advisory Committee of the Food and Fertilizer Technology Center for the Asian and Pacific Region, 18–24 May 1992, Suweon, Korea.* Food and Fertilizer Technology Center for the Asian and Pacific Region, Taipei.

van Benthem Jutting, T. (1950) Systematic studies on the non-marine Mollusca of the Indo-Australian Archipelago, II. Critical revision of the Javanese pulmonate land-shells of the families Helicarionidae, Pleurodontidae, Fruticicolidae and Streptaxidae. *Treubia* 20, 381–505.

Yen, T. (1938) Notes on the gastropod fauna of Szechwan Province. *Mitteilungen der Zoologische Museum, Berlin* 23, 438–457.

1

2

3

4

5

6

Plate 1. Mature *Isidorella newcombi* (Adams & Angus). Shell length to 18 mm.

Plate 2. Mature *Glyptophysa* sp. Shell length to 20 mm.

Plate 3. Effect of feeding by *Isidorella newcombi* (Adams & Angus) on rice (*Oryza sativa* Linnaeus) plant development. The two plants on the left exhibit delayed growth as a consequence of root damage during the seedling stage. The plant on the right was taken from an non-infested area of the same crop, and had started to produce tillers.

Plate 4. *Urocyclus flavescens* Kerferstein on banana (*Musa acuminata* Colla), with brown, corky lesions on the fruit caused by feeding by this gastropod pest.

Plate 5. *Bradybaena similaris* (de Férussac).

Plate 6. *Bradybaena similaris* (de Férussac) infesting a bunch of grapes.

7

8

9

10

11

12

Plate 7. During spring, the majority of *Bradybaena similaris* (de Férussac) in Taiwan vineyards occur on the vines.
Plate 8. 'Bottle traps', made from disposable plastic soda bottles, have been shown to be effective in reducing the level of *Bradybaena similaris* (de Férussac) infestation of vines in Taiwan vineyards.
Plate 9. *Cantareus aspersus* (Müller) aggregated on a vine trellis post. As a heat-avoiding strategy, this species often rests during the day or aestivates over longer time periods, on structures above ground level.
Plate 10. *Theba pisana* (Müller) infesting bunches of grapes destined for processing as sultanas. In search of food and as a heat-avoiding strategy, this species readily colonizes vines.
Plate 11. Shells of *Theba pisana* (Müller) and *Cantareus aspersus* (Müller) killed over a 7-day period by metaldehyde bait pellets applied to a 1 m^2 quadrant around a study vine in spring.
Plate 12. Sultana grape drying rack, in New South Wales, Australia, 2 days after application of desiccating oil spray to hasten the drying process. Individual bays are periodically machine shaken, with fruit and any contaminants collected above ground level and spread on ground sheets.

13

14

15

16

17

18

Plate 13. Sultanas spread on ground drying sheets in New South Wales, Australia. The fruit is periodically hand raked and inspected to remove grape stalks, contaminants (gastropod shells) and 'finish' dry the product to <14% moisture content before placement into bins and delivery to the process factories.
Plate 14. Maize (*Zea mays* Linnaeus) seedling severely defoliated by gastropods.
Plate 15. Plant population in a maize (*Zea mays* Linnaeus) crop, sown by conservation tillage in Ohio, severely depleted by gastropod damage to the young seedlings.
Plate 16. Soybean (*Glycine max* (Linnaeus) Merrill) seedlings severely defoliated by gastropods.
Plate 17. Plant population in a soybean (*Glycine max* (Linnaeus) Merrill) crop, sown by conservation tillage in Ohio, severely depleted by gastropod damage to the young seedlings.
Plate 18. Maize damaged by *Bradybaena ravida* (Benson).

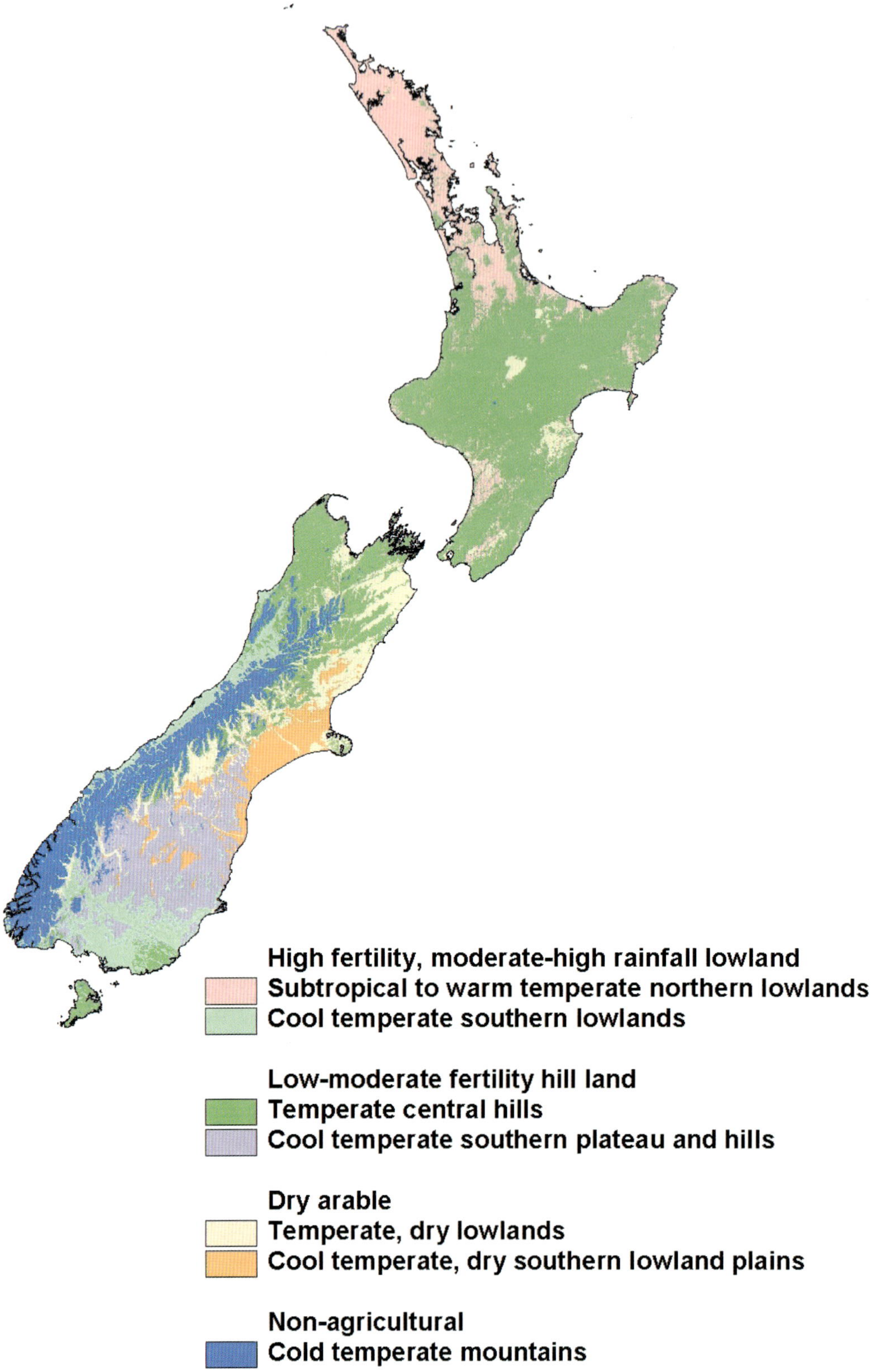

Plate 19. Broad agro-climatic zonation in mainland New Zealand.

10 Agriolimacidae, Arionidae and Milacidae as Pests in West European Sunflower and Maize

GÉRARD HOMMAY

UR Biologie des Interactions Virus Vecteur, INRA, 28 rue de Herrlisheim, 68021 Colmar, Cedex, France

Crop Damage in the Different Countries

Sunflower (*Helianthus annuus* Linnaeus) Asteraceae and maize (*Zea mays* Linnaeus) (Gramineae) are widely cultivated throughout southern Europe (Tables 10.1–10.3). The European Community (EC) ranks third after the USA and China, with maize production varying around 30 million t $year^{-1}$ (AGPM, 1996). The total area of maize in the EC is 7.5 million ha. Of this total, 40% is situated in France, 20% in Germany and 16% in Italy. Half of the annual maize production is used for grain. The other half is grown for forage or silage and occurs more in the northern regions. Sixty per cent of the maize in Europe is used for animal feed and 20% for industry.

In 1996, France was the main producer of maize in the EC, with approximately 40% of the area, i.e. over 1.73 million ha (AGPM, 1996, 1997). Within France, almost 35% of maize for grain was produced in the south-west region. In the same year, maize for silage represented 1.58 million ha in France (AGPM, 1997), with production primarily in the western regions. Italy ranks second as a producer of maize for grain, with 80% of the yield coming from the northern region, particularly in the Pô plain (AGPM, 1996). In Spain, which is the third producer, a decrease of the area devoted to maize has continued since 1988, because of competition from American producers and problems with water resources. Production in Spain is localized in the river-valley systems of Guadalquivir in Andalusia, Ebro in Aragon and Tagus in Estramadura, where irrigation is available. Of approximately 1.5 million ha in maize in Germany, more than 80% is destined for silage (Tables 10.1 and 10.2).

Spain and France are the main producers of sunflower in the EC (Table 10.3). In France, production is concentrated in the central and south-west regions. Sunflower is primarily used for oil and animal feed.

Table 10.1. Area from which maize (*Zea mays* Linnaeus) (Gramineae) was harvested for grain in West Europe (source Eurostat, 2000).

	1000 hectares			
	1997	1998	1999	2000
Austria	161	144	153	164
France	1858	1797	1759	1834
Germany	368	341	371	363
Greece	223	232	210	215
Italy	1039	970	1028	1087
Portugal	186	193	183	172
Spain	486	459	398	425
EC (15 countries)	4358	4179	4151	4307

Table 10.2. Area from which maize (*Zea mays* Linnaeus) (Gramineae) was harvested green in West Europe (source Eurostat, 2000).

	1000 hectares			
	1997	1998	1999	2000
Austria	84	79	76	74
Belgium	185	172	179	166
England	109	103	107	105
France	1477	1461	1399	1403
Germany	1294	1236	1203	1153
The Netherlands	232	219	231	206
Italy	290	282	283	175
Portugal	131	131	131	131
Spain	105	89	84	89
EC (15 countries)	3968	3834	3758	3581

Table 10.3. Area from which sunflower (*Helianthus annuus* Linnaeus) (Asteraceae) was harvested in West Europe (source Eurostat, 2000).

	1000 hectares			
	1997	1998	1999	2000
Austria	20	22	24	22
France	875	782	799	720
Germany	34	34	33	25
Greece	25	35	30	32
Italy	230	233	209	216
Portugal	67	60	55	48
Spain	1004	1048	850	863
EC (15 countries)	2255	2213	2001	1927

Conservation tillage practices, such as no-tillage, have not been widely adapted for maize production in Europe. Maize is sown in spring on fields that are generally bare (fallow) during the preceding winter. In

France, previous crops are mainly maize, wheat (*Triticum aestivum* Linnaeus) (Gramineae) and soybean (*Glycine max* (Linnaeus) Merrill) (Fabaceae) in the south-west with maize for grain and ryegrass (*Lolium* Linnaeus spp.) (Gramineae) with maize for silage. A frequent rotation is maize, wheat and barley (*Hordeum vulgare* Linnaeus) (Gramineae).

For sunflower, conventional tillage is the main cultural practice. Clay soils are cultivated by ploughing in summer/autumn, while for loam soils ploughing generally occurs in spring. The crop is sown in spring. A limited area is under conservation tillage, which includes no-till, where the soil is left undisturbed, and mulch-till, where soil is prepared before sowing with disc openers, chisels or field cultivators. A frequent rotation is oilseed rape (*Brassica napus* Linnaeus var. *oleifera* Linnaeus) (Brassicaceae), wheat and sunflower, and then wheat. In the south-west of France, previous crops are mainly cereals (Jouffret, 1997) and 80% of crops are sown before 1 May.

In countries where the weather conditions in spring are rather dry, gastropods apparently cause little damage. Thus, although large areas of sunflower and maize are cultivated in Spain, gastropods do not seem to be injurious there (Castillejo *et al.*, 1996; Vesperinas and Del Estal, 1996). In countries where the spring is moist, gastropods can be important pests during the crop-establishment phase. Especially in France, various species of Agriolimacidae, Arionidae and Milacidae are collectively considered as the most important pest for sunflower and the second most important pest group for maize (Chabert, 1993).

Gastropod damage in sunflower is mainly observed in western and central France (Hommay and Briard, 1989). Gastropods are considered important pests in localized areas in the west and occasionally in the east (Chabert and Maurin, 1994). In 1994, the area of sunflower treated in France with molluscicide bait pellets was mostly situated in the south-west (Poitou-Charentes, Midi-Pyrénées) and in the south-east (Burgundy, Rhône-Alpes) (Service Central des Enquêtes et Etudes Statistiques, 1996). Sometimes more than two applications of molluscicide are made to a field, the mean number of applications per field varies between 1 and 1.2 according to the region. A survey made in the south-west region (Jouffret, 1997) found 74% of the sunflower crops were treated with molluscicides in 1997 (59% at sowing, 7% after damage, 8% at sowing and after damage).

D'Aguilar and Pacquetian (1963) reported damage by *Milax* Gray (Milacidae) in sunflower fields in central France during the 1962 season, where up to 50% of the crop was destroyed. Low temperatures had reduced growth of seedlings and the crop remained at the susceptible stage over a longer period than usual. Crops on chalky clay soils were generally more severely damaged. Sunflower crops that followed a barley crop were the most severely damaged, while those following winter wheat, oats (*Avena sativa* Linnaeus) (Gramineae), or leguminous plants suffered little. Ballanger and Champolivier (1990, 1996) studied pest populations in sunflower crops in this region from 1986 to 1989. Among

35 fields surveyed, at sites recognized as being prone to gastropod infestation, only in seven was the crop severely damaged. The main gastropod pest species were *Deroceras reticulatum* (Müller) (Agriolimacidae), *Arion hortensis* (de Férussac) and *Arion subfuscus* (Draparnaud) (Arionidae). Sunflower crops did not appear to be an environment favourable for high populations of these gastropod species. Damage arose when the emerged seedlings were defoliated or severed and, particularly the case with juvenile gastropods, through grazing on the hypocotyl. As the seedling crop took several weeks to develop a canopy sufficient to protect the soil from drying, conditions were generally unfavourable for gastropod population recruitment and activity over this spring period. An exception was the wet spring of 1988.

A similar pattern has been observed in the east of France (Hommay, 1994), in a field regularly controlled with 50 cm × 50 cm mat traps. Hommay observed a very high population of *D. reticulatum* in white clover (*Trifolium repens* Linnaeus) (Fabaceae) during autumn 1985 (up to 61 *D. reticulatum* per trap). However, with cultivation and establishment of the sunflower crop in spring 1986, the population declined markedly (a maximum of 2.8 per trap) and no crop damage was observed.

Maize is most frequently damaged by gastropods in crops of the western regions of France (Ritter, 1955; Mallet, 1973; Hommay and Briard, 1989; Chabert, 1994). A survey conducted in 1983 by Mouchart and Drogeat (1984) showed that maize was only occasionally treated with molluscicides, except in the south-west and in most cases only by a small number of farmers. Crops were rarely damaged to the extent that redrilling was required. A further survey conducted in 1993 (Chabert and Maurin, 1994) showed that gastropods were considered important in localized areas in the south-west, occasional pests in the north-west and rarely important in the east regions. In 1994, the greatest area of maize treated in France with molluscicide bait pellets was situated in the south-west (Service Central des Enquêtes et Etudes Statistiques, 1996). The mean number of applications per field varied with region, between 1 and 1.2.

According to the Association Générale des Producteurs de Maïs (B. Naibo, personal communication, 1996), severe infestations and damage on maize are rare. Damage is generally greatest in maize rotations after crops such as oilseed rape and pea (*Pisum sativum* Linnaeus) (Fabaceae), and in no-tillage maize after previous maize crops. In the Landes region (south-west France), where the soils are sandy, crop losses due to gastropods do not occur, despite the use of irrigation.

The extent to which sunflower and maize crops in France were treated with molluscicides increased considerably from 1993 to 1995 (Table 10.4). This was principally due to three successive moist springs that favoured gastropod infestations, and the trend for farmers, under these conditions, to adopt a preventive, prophylactic control strategy. Gastropod populations and molluscicide treatments increased again in 1999 and 2000 because of mild winters.

Table 10.4. Areas of maize (*Zea mays* Linnaeus) and sunflower (*Helianthus annuus* Linnaeus) treated with molluscicide baits in France during the period 1993 to 2000 (source BAYER Society).

	1993	1994	1995	1996	1997	1998	1999	2000
Maize								
Treated area (1000 ha)	270	326	487	446	421	275	491	510
% of area in maize	8	10	15	12	13	8	16	16
Sunflower								
Treated area (1000 ha)	322	581	652	418	379	345	535	496
% of area in sunflower	37	54	66	44	43	44	67	69

Godan (1973) summarized the reports of the Plant Protection Office of the German Federal Republic, from 1962 to 1971. While maize was not listed among the crops most subject to damage by gastropods, the frequency of reports of heavy to very heavy infestations expressed as a percentage of the total number of reports for this crop was 75%. In the Munich region, Stauber (1954) observed infestations by *D. reticulatum* in forage maize with up to ten slugs per plant, which resulted in total loss of plants in a third of the crops and an estimated 50% loss in a further one-third. In summaries of gastropod control for Belgium (Moens, 1980) and England (Port and Port, 1986; South, 1992), maize was not mentioned as a crop frequently damaged by these pests. According to a questionnaire on gastropod problems sent to Swiss organic farmers, damage in maize was slightly more prevalent than in cereals, oilseed rape, beet (*Beta vulgaris* Linnaeus) (Chenopodiaceae) and potato (*Solanum tuberosum* Linnaeus) (Solanaceae), which themselves suffered little damage (Speiser, 1994). In Switzerland, Charles and Calame (2000) observed the development and yield of maize close to an extensive meadow. Gastropods caused severe damage to maize leaves, but did not reduce the number of plants or weaken their development.

Susceptibility of Sunflower and Maize to Injury by Gastropods

Sunflower seeds sown into infested soil are susceptible to attack from gastropod slugs as soon as the seeds imbibe moisture and their outer coating is broken. As the seeds germinate and the seedlings grow, damage may occur to the hypocotyl, cotyledons and first leaves (d'Aguilar and Pacquetian, 1963; Hommay, 1995). While susceptibility declines with age, plants up to the sixth-leaf stage can sustain lethal damage, especially if damage occurs to the meristematic tissues.

In maize, seeds are less frequently attacked than the leaves of the young seedlings, but grain hollowing is generally lethal. Grazing on the coleoptile often leads to plant mortality, while defoliation is rarely sufficiently intensive to cause mortality. However, plants up to the fourth-leaf stage have been shown to be weakened by defoliation, making them more susceptible to other pests or to herbicides, delaying their growth and reducing their yield (Byers and Calvin, 1994; Hommay, 1995). Maize varieties are known to vary in susceptibility as seedlings to damage by *D. reticulatum* when tested in the greenhouse (Hommay, 1994). Varieties with lower concentration of soluble glucosides in the leaves generally suffered less defoliation, but further research is needed to demonstrate a causal relationship between gastropod damage and glucoside content.

Pest Management

Cultivation generally reduces gastropod populations. In a comparison of different cultivation practices in maize, Chabert and Maurin (1994) demonstrated a decline in gastropod damage when the intensity of cultivation (number of times the field was tilled) was increased. Spring cropping systems generally require more ground preparation and consequently, on average, maintain lower gastropod numbers than winter cropping systems. In addition, seed germination is more rapid in spring due to higher soil temperatures, reducing the period when the crop is at the susceptible seedling stage. However, some spring-sown crops, such as sunflower, are highly attractive to gastropods as food plants. Moreover, the vulnerability of these crops is heightened by the low plant population, as yield increases from surviving plants in response to stand thinning are unable to compensate for plant losses.

In spring, populations of *D. reticulatum* and *A. hortensis* comprise mostly juveniles (Hommay and Briard, 1988; Hommay, 1994), whose dispersal and crop-damage capabilities are less important than those of adults. The majority of *D. reticulatum* are recruited into the population by hatching in April–May from overwintered eggs. In *A. hortensis* this juvenile recruitment is predominantly in May–July. Thus it is difficult to forecast crop damage because the period of greatest pest pressure is strongly influenced by the ratio of the two gastropod species. Crops sown in early spring, as is generally the case with maize, are less vulnerable, simply because the crop is established prior to the main period of *D. reticulatum* and *A. hortensis* recruitment. In crops sown later in the spring, such as sunflower, pest pressure can be expected to be high, as sowing often coincides with the main period of gastropod recruitment. However, as experience in the south-west of France indicates, periods of dry weather in later spring often lead to inactivity in the gastropod populations and crop damage is coincident with rainfall events (Ballanger and Champolivier, 1996).

Using enclosed miniplots of winter wheat, oilseed rape and sunflower infested with *D. reticulatum*, Maurin and Chabert (1990) evaluated three strategies for molluscicide application at the time of sowing seed: in the soil as a band with the seed; applied to the soil surface as a band over the seed row; and broadcast over the entire soil surface. Irrespective of the molluscicide-bait product used, the best protection was obtained with broadcast applications. In the field, broadcasting baits in a band over the seed rows was shown to be as effective as broadcasting over the entire crop area (Maurin, 1989). Based on these results, the French Centre Technique Interprofessionnel des Oléagineux Métropolitains (CETIOM, 1995) recommends that, for gastropod control in sunflower, molluscicide baits are best applied to the soil surface at sowing. G. Meyer (BAYER Society, unpublished experimental report for 1994/95) found that best crop protection was afforded by dual application of molluscicide baits, comprising a broadcast application prior to sowing and a banded application with the seed at sowing.

For prevention of gastropod damage in maize in France, the Association Générale des Producteurs de Maïs (B. Naibo, personal communication, 1996) recommends application of molluscicide baits to the soil surface at sowing, either broadcast or as a band over the seed row. If a sustained period of wet weather occurs, the recommendation is for a second treatment after seedling emergence. Moens and Gigot (1988) demonstrated that protection of maize seedlings can be achieved with molluscicide baits applied after sowing.

In France, gastropod management is similar in maize and sunflower and consists of preventive or curative molluscicide applications. For high-value crops, such as maize for seed grain, crops are routinely protected at sowing by molluscicide applications. Molluscicide products authorized for slug control are pellets containing either 5% bensultap (7.5 kg ha^{-1}), 4% thiodicarb (5 kg ha^{-1}), 4% methiocarb (3 kg ha^{-1}) or 5% metaldehyde (5–10 kg ha^{-1}), the last two being the more frequently used. To determine the need for molluscicide treatment, it is recommended that the fields be surveyed prior to sowing, by direct observation at night or by trapping. Farmers are encouraged to use cultural practices that minimize the impact of gastropods in the crops. Such practices as no-tillage are favourable for a build-up of gastropod populations (Hughes and Gaynor, 1984; Willson and Eisley, 1992). Therefore, fields known to support high gastropod numbers should be included in rotation with less favourable crops and/or be subject to cultivation such as ploughing.

Conclusion

Recent changes in agricultural practices, partly due to economic pressures and attention to minimization of expensive inputs, have increased risks of slug damage. Some of these altered practices include greater reliance on incorporation of crop residues (as a result of laws prohibiting burning of straw), conservation cultivation techniques and use of cover crops. Common Agricultural Policy (CAP) reform may have also contributed to an increased incidence of gastropods as pests, since set-aside (5% of the cultivated area in the EC) provides good habitat conditions for these animals. Thus, the use of molluscicide baits has been increasing. One of the main outstanding issues is that of forecasting gastropod damage in crops. Gastropod populations are generally considered difficult to assess. Further, because of variable environmental conditions (weather, previous crops, cultivations, time of sowing, soil structure), there has been little progress in development of prediction tools or economic thresholds for gastropod management. Actually, farmers recognize that baiting with molluscicide is the most cost-effective way of managing gastropod pests, even if it is sometimes necessary to make several applications to a field in a single season. However, there remains a need for effective molluscicide formulations that are less harmful to the environment.

Integrated gastropod control can be achieved. The use of molluscicides can be reduced by improved prediction of damage and, because gastropod infestations tend to be localized, by focusing applications on the areas within fields with the highest pest numbers. Factors such as successive wet summers followed by mild winters, leguminous and cruciferous crops in the rotation and heavy soils all favour high gastropod numbers and are key variables to include in predictions of crop losses. Gastropod damage can be reduced by removal of crop residues from the soil surface immediately after crop harvest, by soil cultivation and by adopting agronomic practices that limit pest access to sown seed and promote high seedling vigour. Further research is needed to augment existing chemical-control techniques with novel approaches, such as coating maize seed with repellent or molluscicide compounds.

References

AGPM (Association Générale des Producteurs de Maïs) (1996) *Le Maïs, tous les chiffres.* Les Dossiers AGPM, 88 pp.

AGPM (Association Générale des Producteurs de Maïs) (1997) *L'Année céréalière et le maïs. Congrès du maïs, Montpellier 1997.* Les dossiers AGPM, 98 pp.

Ballanger, Y. and Champolivier, L. (1990) Dynamique des populations de limace grise (*Deroceras reticulatum* Müller) en relation avec la culture de tournesol. In: *2ème Conférence Internationale sur les Ravageurs en Agriculture, Versailles, 4–6 December 1990.* Association Nationale pour la Protection des Plantes, pp. 143–150.

Ballanger, Y. and Champolivier, L. (1996) Slug damage to sunflower crops in the south-west of France. In: Henderson, I.F. (ed.) *Slug and Snail Pests in Agriculture.* Symposium Proceedings No. 66, pp. 321–326.

Byers, R.A. and Calvin, D.D. (1994) Economic injury levels to field corn from slug (Stylommatophora: Agriolimacidae) feeding. *Journal of Economic Entomology* 87, 1345–1350.

Castillejo, J., Seijas, I. and Villoch, F. (1996) Slug and snail pests in Spanish crops and their economical importance. In: Henderson, I.F. (ed.) *Slug and Snail Pests in Agriculture.* Symposium Proceedings No. 66, British Crop Protection Council, Farnham, pp. 327–332.

CETIOM (Centre Technique Interprofessionnel des Oléagineux Métropolitains) (1995) Tournesol – Limaces: à traiter le plus tôt possible. *Oléoscope* 28, 16–17.

Chabert, A. (1994) Importance relative des ravageurs souterrains en grandes cultures. Résultats de l'enquête d'un groupe de travail. *Phytoma* 461, 20–24.

Chabert, A. and Maurin, G. (1994) Le point sur les limaces. Ne nous laissons pas dépasser. *Phytoma* 467, 35–38.

Charles, R. and Calame, F. (2000) Effets des surfaces de compensation écologique sur les bords des cultures: la part liée aux limaces. *Revue Suisse d'Agriculture* 32, 5–9.

d'Aguilar, J. and Pacquetian, B. (1963) Dégâts de limaces sur Tournesol. *Phytoma* 144, 27–28.

Eurostat (2000) *Agricultural Statistics – Quarterly Bulletin*. European Commission (ed.) Theme 5. Agriculture and Fisheries 7, 20–25.

Godan, D. (1973) Schadwirkung und wirtschaftliche Bedeutung der Schnecken in der Bundesrepublik Deutschland. *Nachrichtenblatt des Deutschen Pflanzenschutzdienstes, Braunschweig* 25, 97–101.

Godan, D. (1983) *Pest Slugs and Snails. Biology and Control.* Springer-Verlag, Berlin, 445 pp.

Hommay, G. (1994) Contribution à la biologie et à l'écologie des limaces (Mollusques Gastéropodes Pulmonés) de grandes cultures. PhD thesis, University of Rennes, France.

Hommay, G. (1995) Les limaces nuisibles aux cultures. *Revue Suisse d'Agriculture* 27, 267–286.

Hommay, G. and Briard, P. (1988) Apport du piégeage dans le suivi des peuplements de limaces en grandes cultures. *Haliotis* 18, 55–74.

Hommay, G. and Briard, P. (1989) A few aspects of slug damage in France. In: Henderson, I.F. (ed.) *Slugs and Snails in World Agriculture.* Monograph No. 41, British Crop Protection Council, Thornton Heath, pp. 379–384.

Hughes, K.A. and Gaynor, D.L. (1984) Comparison of Argentine stem weevil and slug damage in maize direct-drilled into pasture or following winter oats. *New Zealand Journal of Experimental Agriculture* 12, 47–53.

Jouffret, P. (1997) Les conduites du tournesol dans le sud-ouest en 1997. In: *Les Rencontres Annuelles du CETIOM-Tournesol, Paris, 9 December*, pp. 11–15.

Mallet, C. (1973) Les limaces, ennemies des jardins, mais aussi des grandes cultures. *Phytoma* 1973, 10–12.

Maurin, G. (1989) Les limaces – La lutte: quelques éléments de réflexion pratique. *Phytoma* 411, 28–29.

Maurin, G. and Chabert, A. (1990) Efficacité de différentes méthodes d'épandage des produits molluscicides. In: *2ème Conférence Internationale sur les Ravageurs en Agriculture, Versailles, 4–6 December 1990.* Association Nationale pour la Protection des Plantes, pp. 167–173.

Moens, R. (1980) Le problème des limaces dans la protection des végétaux. *Revue de l'Agriculture* 1, 117–132.

Moens, R. and Gigot, J. (1988) Un nouveau granulé anti-limaces à base de thiodicarbe 4%. *Revue de l'Agriculture* 5, 1213–1225.

Mouchart, A. and Drogeat, C. (1984) *Résultats de l'enquête Réalisée en 1983. Limaces en Grandes Cultures.* Association de Coordination Technique Agricole, Paris, 60 pp.

Office Statistique des Communautés Européennes (1997) *Agriculture. Statistical Yearbook*, 256 pp.

Port, C.M. and Port, G.R. (1986) The biology and behaviour of slugs in relation to crop damage and control. *Agricultural Zoology Reviews* 1, 255–299.

Ritter, M. (1955) Les limaces et les escargots. Importance économique et moyens de lutte. *Revue de Zoologie Agricole et Appliquée* 54, 1–10.

Service Central des Enquêtes et Etudes Statistiques (1996) Les pratiques culturales sur grandes cultures en 1994. *Agreste. Données chiffrées – Agriculture* 85, 195 pp.

South, A. (1992) *Terrestrial slugs. Biology, Ecology and Control.* Chapman & Hall, London, 428 pp.

Speiser, B. (1994) Slug problems in Swiss organic agriculture. *Integrated Management of Mollusc Pest Newsletter* 2, 4.

Stauber (1954) Starker Schneckenfrass an Grünmaïs. *Pflanzenschutz, München* 6, 137–138.

Vesperinas, E.S. and Del Estal, P. (1996) Plagas potenciales del girasol (*Helianthus annuus* L.) en España. In: *Proceedings of 14th International Sunflower Conference, Beijing/Shenyang, China, 12–20 June 1996*, pp. 491–493.

Willson, H.R. and Eisley, J.B. (1992) Effects of tillage on the incidence of five key pests on Ohio corn. *Journal of Economic Entomology* 85, 853–859.

11 Helicidae as Pests in Australian and South African Grapevines

GRAEME SANDERSON[1] AND WILLEM SIRGEL[2]

[1]*Agricultural Research and Advisory Station, NSW Agriculture, PO Box 62, Dareton, NSW 2717, Australia;* [2]*Zoology Department, University of Stellenbosch, Private Bag X1, Matieland 7602, South Africa*

Emergence of Gastropods as Viticultural Pests

Grapes (*Vitis vinifera* Linnaeus) (Vitaceae) were introduced to Australia in 1788 and wine production was initiated by the mainly English settlers. Migration associated with the 'gold rush' in the 1860s and subsequent settlement by Europeans provided the knowledge base and grape-growing skills required for production of improved wines. Australia currently contributes only 2% of the world wine production, but none the less is an important New World wine-exporting nation, along with the USA (California), Chile and South Africa. The Australian wine industry is currently in a period of rapid expansion. Dried-grape production began in the Mildura region, Victoria, in the 1880s. This dry-grape industry has recently been in decline due to poor harvest seasons in the late 1990s and competition for sultana grapes from wineries. Currently Australia has 146,000 ha in grape production.

Viticulture in South Africa dates back to 1658, when the first vines were planted in Cape Province. The grape-growing areas of Stellenbosch and Paarl were established in the 1680s and the Orange River development in 1876. Wine production is the main grape enterprise, with export of high-quality table wines an industry focus. The dried-fruit industry in South Africa produces between 25,000 and 40,000 t of dried grapes per year. With the contraction of the Australian industry, both countries now have equivalent-sized drying industries. South Africa has approximately 110,000 ha of land devoted to grape production.

There are approximately 2000 species of terrestrial gastropods native to Australia and 650 in southern Africa (inclusive of South Africa, Namibia, Zimbabwe, southern Mozambique, Swaziland and Lesotho) (van Bruggen, 1978; P. Colman, personal communication, 1995). The number of terrestrial gastropods native to South Africa is around 525–530

(D. Herbert, personal communication, 1995). In addition, various species have been introduced into Australia and South Africa, primarily from Europe. In Australia these introduced terrestrial gastropods number about 50 and in South Africa about 30. Some of these introduced species have attained pest status in agricultural and horticultural farming enterprises.

In the South African viticultural regions, the introduced helicids *Cantareus aspersus* (Müller) and *Theba pisana* (Müller) can reach pest status, although the importance varies among areas where the fruit is grown. Their pest status is twofold. First, juveniles feed during spring on the developing foliar buds and young leaves of the vines. Affected vines display poor, stunted shoot growth with misshapen leaves and a concomitant poor yield. Extended infestation can drastically reduce the vigour and productive life of the vines (Loubser, 1982). In areas where *C. aspersus* infestations are most severe, growers estimate crop losses of up to 25%. Second, during the misty evenings that commonly occur in spring and early summer, the active animals leave mucus trails on the developing grapes, reducing their aesthetic appearance and rendering table grapes unsuitable for export markets. In certain areas, production of grapes cannot be sustained without control of gastropods. The pest status of *C. aspersus* and *T. pisana* in South Africa has evolved over the period since establishment of these species in the country during the latter part of the 19th century, but has become most pronounced over the past 25 years.

C. aspersus and *T. pisana* cause serious damage in the Stellenbosch, Somerset West, Paarl, Worcester and Rawsonville areas of the Western Cape, but are found as far north as Lutzville and Vredendal in the Olifant River irrigation area. Infestations of up to 400 gastropods have been recorded on a single grapevine (Loubser, 1982).

In many sections of the Australian viticultural industry, there is potential for significant damage by gastropods to newly emerged vegetative parts of the vine, which reduce subsequent fruit yield. These infestations of vines primarily occur when gastropod populations on the ground are disturbed by vineyard management practices, such as cultivation or mowing of ground cover to reduce frost likelihood. However, it is the dried vine fruit (sultana grape) industry that has been most affected by introduced terrestrial gastropods. This industry is centred on Mildura in the north-west part of Victoria and has an export emphasis. The main species of concern are the helicids *C. aspersus* and *T. pisana*.

These gastropods are of concern to the Australian industry primarily because of produce contamination. During the late 1980s there was a significant increase in the level of gastropod contamination in dried fruit being delivered to packing sheds for processing. This emergence of gastropods as viticultural pests in Australia was linked to changes in soil-management practices, especially reductions in tillage, improved irrigation and use of ground-cover crops and mulches. There is a trend in Australian viticulture away from furrow or flood irrigation towards overhead spray and under-vine drip and minisprinkler irrigation systems. Furrow irrigation is traditionally applied at 2–4-week intervals. Modern spray and under-vine irrigation systems are run at shorter time intervals and can maintain moist surface conditions, which can promote weed growth in the vineyard. These conditions appear advantageous to gastropod populations.

Cover crops are used in Australian and South African vineyards for soil structural benefits they impart and moisture conservation ability as a mulch. Cover crops include annual cereals (oats, *Avena sativa* Linnaeus; rye, *Secale cereale* Linnaeus; barley, *Hordeum vulgare* Linnaeus) (all Gramineae) and legumes (faba bean, *Vicia faba* Linnaeus; vetch, *Vicia villosa* Roth; field pea, *Pisum sativum* Linnaeus; medic, *Medicago* Linnaeus spp.; clover, *Trifolium* spp.) (all Fabaceae), which are either incorporated into the soil by cultivating prior to vine bud burst or mown and left as a surface mulch. There is also trial use of grass-based, permanent inter-row sod systems in Australia, but this is restricted to higher-rainfall regions or sprinkler-irrigated vineyards. Volunteer inter-row weed growth is also mown or treated with herbicide to provide surface mulch, thus reducing the need for cultivation in vineyard weed control.

The sandy loam soil types that occupy a large part of the Murray Valley viticultural region of Australia are very low in soil organic matter (< 2%). Cover cropping can maintain and improve surface organic-matter levels, soil aggregate stability and water infiltration rates (G. Sanderson, unpublished data). Winter cover cropping is a widely adopted practice in Australian viticulture, particularly in vineyards with full ground-cover irrigation systems such as sprinklers. The green, inter-row plant material is either incorporated into the soil with a cultivator or slashed/mown in late winter/early spring. In South Africa, cover cropping is also a common viticultural practice with greater emphasis on the use of mulching techniques for soil moisture conservation. Cereals are grown as cover crops in the alleys between vine rows and are rolled flat and killed with herbicide to create soil surface mulches.

The dried-fruit industry in Australia has imposed penalties on those growers who deliver contaminated produce, with certain weed seeds, shelled gastropods (snails) and stones being the main contaminants that attract penalty charges. The penalties are imposed to encourage sultana-grape producers to deliver a clean product and to cover the additional costs associated with processing contaminated fruit. The dried fruit is delivered to the process factories in bin quantities of *c.* 500 kg and the penalties operating in 1997 were:

Category A: 1–5 gastropod shells per bin	$10 t^{-1}$
Category B: 6–20 gastropod shells per bin	$50 t^{-1}$
Category C: > 20 gastropod shells per bin	$100 t^{-1}$

With gross returns to growers of A\$1000–1500 t^{-1}, category C penalties thus equate to about 10% loss.

C. aspersus and *T. pisana* form clusters on vine trunks, posts and sometimes within the foliage or grape bunches (Plates 9, 10). Infestation levels of 200–250 *T. pisana* and 50–70 *C. aspersus* per vine have been recorded. Contamination of produce occurs when the gastropods remain in the harvested grape bunches or when gastropods are dislodged from their resting sites and fall to the ground but subsequently gain entry to fruit-picking buckets left overnight in the vineyard.

Studies in a highly infested sultana vineyard have indicated on-vine gastropod biomass (*T. pisana*) in the order of 87 kg for blocks of vines yielding 30 t of fresh fruit (Sanderson, 1995). In crops harvested manually, contamination is primarily restricted to those gastropods residing within the fruit bunches. However, with mechanical harvesting the entire vine canopy is disturbed and there is potential for a high proportion of the gastropods resident in the vines to be dislodged and thus harvested with the fruit.

Contamination of produce is also a problem for some Australian producers of wine grapes.

The Viticultural Regions

The viticultural regions of Australia are concentrated in the southern half of the continent (Fig. 11.1). Viticulture in Australia occurs across many climates and soil types. From the perspectives of temperature and soil types, there are vast areas of Australia suitable for grape production, but the absence of significant rainfall precludes viticulture in many areas without the use of irrigation. Large sections of inland South Australian, inland Victorian and Western Australian viticultural regions are in water deficit in November (i.e. evaporation exceeds rainfall). By January this deficit extends to coastal sectors of Victoria and South Australia, but eastern coastal Australia is typically still in water balance at that time (Dry and Smart, 1986). Availability of irrigation water to supplement rainfall is thus a major limitation to viticultural development. Mildura, Victoria, receives only 275 mm of annual rainfall and the viticultural industry in this region is dependent on irrigation water supplied from the Murray River. A high proportion of Australian viticultural regions have a winter-dominant rainfall pattern and require supplementary summer irrigation.

Viticultural soil types in Australia can range from red duplex soils (two layers) with highly alkaline, clay subsoils in central Victoria to mildly acidic, siliceous sands in Western Australia. Calcareous-earth soil types are very important viticultural soils in Australia (Dry and Smart, 1986). They have brown to red-brown loamy sand, sandy loam or loam surface soils, in which carbonate content is < 10% in the top 0.3 m. With depth the clay and calcareous content increase, while the subsoils range from sandy loams to clay loams. Surface soils are neutral to alkaline and subsoils alkaline to strongly alkaline: typically the upper 50 cm soil has pH ($CaCl_2$) 7.3–8.5. Calcareous-earth soils cover large areas of the irrigated districts along the western section of the Murray River, through Mildura, Victoria, and into the major Riverland viticultural region in South Australia. New vineyard developments are now typically planned on a thorough investigation of soil physical and chemical properties, as well as specific regional climatic data.

The Australian wine-grape industry has undergone rapid expansion in the last decade, with extensive plantings in eastern South Australia, north-west Victoria, and south-central New South Wales. Currently, three-quarters of Australia's grapevines are located in South Australia and Victoria (Fig. 11.1). Wine production is the dominant viticultural industry, with South Australia producing approximately 54% of Australia's red-wine grapes in 2000/01 and 46% of white-wine grapes. Victoria is the major dried-grape producer, with the industry concentrated on the Mildura region. Table grapes are an important crop in the Mildura region and neighbouring parts of north-west Victoria along the Murray River.

Fig. 11.1. Distribution of viticulture in Australia, with size of the symbol relative to the extent of vineyard plantings.

There are at present three regions in which viticulture is practised in South Africa (Fig. 11.2). The first of these occurs along the south-western coastal lowlands in the vicinity of Cape Town (Stellenbosch, Paarl and Somerset West), extending 250 km to the north to Vredendal. The coastal lowlands around Cape Town receive an annual rainfall of 500–750 mm, which is adequate for grape production without irrigation. Soil types in the coastal regions are mainly derived from granite, sandstone or shale. In the Western Cape region, duplex soils are common, with a prominent textural difference between the topsoil and subsoil – typically with sandy topsoil overlying a clay pan, which can be highly acidic. Rainfall in the Western Cape region is winter-dominant with 75% of annual precipitation occurring in winter and 25% in summer. In contrast, the annual precipitation at Vredendal is only 250 mm and must be supplemented by irrigation. Wine is the principal product of the south-western coastal lowlands region, although some table grapes are also produced.

The second South African viticultural region occurs in the shielded inland valley system of the coastal mountain ranges of Cape Province. These encompass the districts of Worcester, Robertson, Ceres and further into the Little Karoo (meaning small dry land) to the districts of Ladismith and Oudtshoorn. The annual rainfall there is 250–375 mm and occurs between the months of May and September, outside the grapevine-

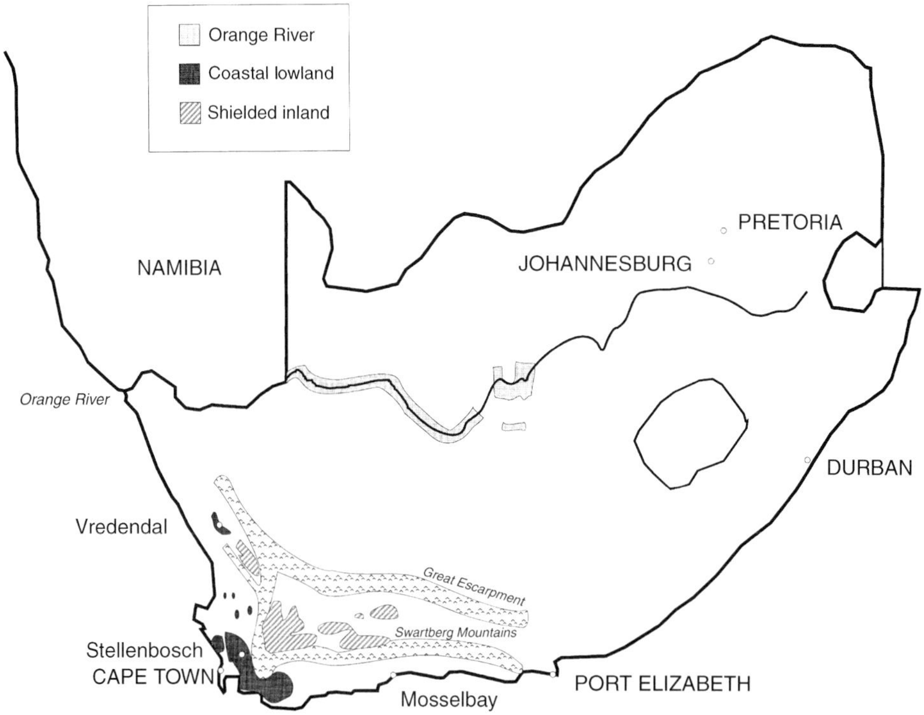

Fig. 11.2. Distribution of viticulture in South Africa.

growing season. Summer irrigation is required for grape production. The soils are predominantly alkaline. Table grapes are produced mainly in the north-west sector, while in the remaining sectors a mix of table grapes, wine and dried fruit are produced. Under-vine drip and minisprinkler irrigation systems are common in Cape Province.

The third South African viticultural region is that extending as a narrow strip along inland parts of the Orange River and sections of its tributaries. With an annual rainfall of less than 63 mm, grape production in this region is mainly focused on dried fruit (95% of South African production) and is dependent on irrigation. The rainfall occurs in summer but crop losses through fruit splitting and fungal-induced decay are of minor concern. The narrow 360 km strip of the Orange River horticultural area is comprised of highly fertile, alluvial soil. Dried-fruit vineyards are predominantly irrigated with water that is gravity-fed from the Orange River via weirs and small dams. Flood irrigations usually occur every 3 weeks till December and then fortnightly until harvest.

Climatic data for Upington (Orange River) indicates higher maximum and lower minimum daily temperatures than Mildura, Victoria. In January Upington has a daily maximum temperature of 40°C and daily minimum temperature of 10°C, whereas in Mildura the corresponding temperature are 32 and 16°C. Mildura is considered a hot, dry-grape-growing region in Australia, but South African conditions of temperature and rainfall are more extreme in the Orange River region.

Distribution and Ecology of Pestiferous Gastropods

Cantareus aspersus

C. aspersus occurs widely in the non-arid regions of Australia, from south-east Queensland to much of New South Wales and Victoria, the south-east of South Australia, the south-west of Western Australia and Tasmania. Thus, the species occurs in the majority of the viticultural regions. *C. aspersus* is also a cosmopolitan pest of urban gardens and a large range of horticultural crops, such as vegetables and citrus. In areas planted to vines, most reports of its occurrence as a pest emanate from inland irrigated situations. The soils in the areas where gastropods have pest status are typically pH-neutral to alkaline and sandy loam to loam in texture. The biology in vineyards has not been studied in Australia, but studies of the captive reproductive behaviour of *C. aspersus* for human consumption showed that most individuals required two seasons to reach reproductive maturity (Simpson, 1993). Reproductive performance declined in *C. aspersus* > 2 years old and those hatched in autumn grew poorly when compared with summer hatchlings. Observation on populations in southern New South Wales vineyards indicate reproductive activity and oviposition in *C. aspersus* during spring, summer and autumn, usually in association with rainfall events.

C. aspersus occurs throughout South Africa wherever gardening is practised. It is apparently dependent on habitats occupied by cultivated and adventive plant species, as populations do not occur in natural habitats. With respect to viticulture, *C. aspersus* occurs in vineyards with acid and alkaline soil conditions but abundance is apparently closely linked to moisture. Thus, while *C. aspersus* occurs throughout the coastal lowland region, populations are highest in the elevated, more humid sectors forming the inland margins of the region. In the inland, shielded valleys, high abundance is favoured by irrigation. In the absence of irrigation in the arid areas, *C. aspersus* may persist at low numbers but become particularly abundant in seasons of unusually high rainfall. Along the Orange River, *C. aspersus* is rarely abundant, despite the wide use of irrigation.

The phenology of *C. aspersus* populations in South Africa is somewhat different from that which occurs in Australia. In the Mediterranean-like climate of the South African viticultural regions, especially inland, *C. aspersus* has an aestivatory period during summer (December–March). After the first rains in autumn, the animals become active and reproductive, with oviposition in April and May. During the coldest period of June and the early part of July, *C. aspersus* adults usually hibernate in the ground or under dead plant material. Hatch from the autumn-laid eggs occurs during July and the juvenile animals feed on green vegetation. When grapevines leave winter dormancy, *C. aspersus* often feed on developing foliar buds. *C. aspersus* grow rapidly over the later winter and spring period to reach maturity by October or November. Over the same period, some adult *C. aspersus* that emerge from hibernation sites oviposit further eggs, but apparently this activity contributes little to population recruitment, as the survival rate of the progeny is low over the ensuing summer.

Theba pisana

On the Australian continent, *T. pisana* was first recorded in Western Australia early in the 20th century. It has since spread across south-west Western Australia, to reach Port Adelaide, South Australia, by 1928 (Baker, 1986) and along sections of the coastal zones of South Australia, Victoria and New South Wales by the 1990s (Baker, Chapter 6, this volume; P. Colman, personal communcation, 1995). *T. pisana* emerged as a pest in dried-fruit viticulture in the late 1980s, coincident with its spread into viticultural areas and changes in vineyard management practices, such as ground-cover cropping and mulch retention.

Studies on the biology of *T. pisana* in vineyards were undertaken in the Coomealla Irrigation District of New South Wales during the period 1992–1994. In this environment, *T. pisana* was found to move from the inter-row cover crop and adjacent weedy areas on to the vines in late spring and early summer as a heat-avoidance strategy – elevated and shaded aestivating sites are cool and more humid than ground conditions. Mowing or cultivation of these ground home sites hastens their movement on to grapevines. The gastropods aestivate on the trunk, in the canopy and within the fruit bunches. They may become active during periods of summer rainfall and may descend from the vines to feed at ground level. A small mark (with shell painting)-and-recapture (by baiting) study in summer indicated that 74–75% of animals aestivating in the vines were activated by rainfall events and immigrated out of the vines to feed. Mating was observed to occur in early autumn and oviposition periodically in winter and early spring in association with rainfall events.

Detailed studies have been undertaken on *T. pisana* biology in pasture and cereal land in Australia, principally in South Australia (reviewed by Baker, Chapter 6, this volume). The rapid movement of *T. pisana* from pasture to aestivate on roadside trees and other vegetation is mirrored in *T. pisana* movement into vineyards. Baker (Chapter 6, this volume) found that *T. pisana* was capable of moving > 55 m in 1 month. This would explain the speed of infestation that occurs on vineyards bordering weedy, *T. pisana*-infested drainage channels and road verges. The life cycle of *T. pisana* in broad-acre agriculture in South Australia is either annual or biennial, depending on the nature of the vegetation cover.

T. pisana was introduced to South Africa towards the latter part of the 19th century (Connolly, 1916). At present, it is primarily distributed in a 3–10 km wide zone along the coastline from Durban in the east and extending into Namibia in the west. Abundance is particularly high in sections of the coast about 300 km north of Cape Town, with counts of up to 4000 individuals m^{-2} having been recorded (W. Sirgel, unpublished data). *T. pisana* infestations in vineyards are restricted to those on calcareous, sandy soils in the vicinity of Vredendal and Cape Town. A rather common practice is the transport of calcareous sand into areas with acidic clay soils as a base in road construction and in house construction. Often this sand is contaminated with *T. pisana* and the species often persists in the new areas for several years, associated with small quantities of the sand that have been discarded. These temporary colonies may inflict some damage in nearby vineyards.

T. pisana is highly invasive in the Fynbos Biome and during the last 20 years has become established on soils within the shielded inland valley region in both natural and modified landscapes (Macdonald and Jarman, 1984; Robertson *et al.*, 2000). It has rapidly attained pest status in vineyards there. The mode of introduction is believed to be as contaminates in lime transported from western coastal regions.

In South Africa, Barnes and Swart (1979) found that *T. pisana* became full-grown in about 1 year and under favourable conditions can live for up to 4 years.

and is promoted as detering gastropod movement on to the vines for up to 60 days.

The South African approach to overcoming the inefficacy of ground-applied molluscicide baits has been to develop a paste form of bait that can be applied to the vine. A common practice is to apply a paste made from bran and metaldehyde to the trunk of the vines, primarily to prevent movement of the gastropods into the vines (Schwartz and Siebert, 1987). Metaldehyde paste was shown to kill *C. aspersus* for up to 52 days following application to vine cordons (Schwartz and Capatos, 1990). In some areas, tobacco (*Nicotiana tabacum* Linnaeus; Solanaceae) dust is spread at the base of the vines as a deterrent to the gastropods. Animals crawling over the dust become irritated and retract into their shells. Subsequent mortality in animals exposed in this way can be high. However, the tobacco dust tends to form a hard crust on exposure to heavy dew or rain and rapidly becomes ineffective.

The viticultural industries in both Australia and South Africa are keen to find sustainable, biologically based solutions to helicid gastropod problems in the vineyards. However, to date, biological control has been actively pursued only in South Africa. A novel suppression strategy in South Africa is the deployment of ducks (*Anas* Linnaeus and *Cairina* Fleming spp.; Anatidae) in vineyards as biological control agents. Experience with this system indicates that the stocking rate must be such that the ducks are not overfed and are thus eager to search out the gastropods. Efforts to establish biological control through the introduction of predatory gastropods of the genera *Gonaxis* Taylor (Streptaxidae), from Kenya, and *Euglandina* Crosse & Fischer (Oleacinidae), from the USA, have failed (S. Walters, personal communication, 1999).

The rapid emergence of gastropods as viticultural pests in Australia during the late 1980s saw the sultana industry unprepared. Basic information on biology and control strategies was lacking. Gastropod control is now part of the pest-management portfolio, and general vineyard cultural practices aim to minimize the impacts of these pests. Experience in the sultana industry indicates that preventive practices are more cost effective than remedial ones, as gastropod populations at pest levels may require substantive changes in vineyard management and several years to bring under control. Essential components of integrated gastropod management include the following.

1. Routine surveillance of the vineyards to detect increases in gastropod abundance.
2. Minimization of the extent of uncultivated or weedy areas surrounding the vineyards as sources of reinfestation. Where reduction in the extent of these potential source areas is not possible, because of terrain or land tenure, a buffer zone around the vineyard may be established by periodic application of molluscicidal baits.
3. Legume cover crops and minimum tillage are preferred vineyards practices, but are recognized as being conducive to gastropod population

increase. Where gastropod abundance within the vineyard approaches levels that compromise crop yields or quality, then gastropod control practices must be implemented. These may include the application of molluscicides, adoption of soil cultivation and selection of cereals as the winter cover-crop species.

The instigation of an integrated approach to control in a 19 ha dried-fruit vineyard in New South Wales, as an on-farm demonstration, reduced gastropod contamination from an average of 1246 per rack in 1993 to 88 per rack in 1994 (93% control) (Sanderson and Davidson, 1994).

Control strategies outlined by Loubser (1982) for South African vineyards are similar to the recommendations made in the 1990s for Australian vineyards. He suggests the removal of weed growth to reduce breeding areas, baiting at the base of vines prior to cover-crop cultivation, bait replacement every 2–3 weeks and maintaining a clean, cultivated vineyard if the gastropod population is high. Regular shallow cultivation is also recommended to expose gastropod eggs to the desiccating effect of sunlight. Several differences from Australian control measures include the use of ducks to suppress the gastropod populations and hand collection by labourers if infestations occur in localized patches.

Postharvest Decontamination

In the Australian industry, sultana-grape bunches are predominantly harvested by hand. The labourers are paid on quantities harvested and there is little incentive for them to remove gastropod contaminants while harvesting bunches from the vines. Therefore, in infested vineyards, gastropod removal from the harvested fruit begins when bunches are spread on drying racks and continues by periodic search as the fruit dries. When drying racks are full of grape bunches a vegetable-based drying oil is sprayed to hasten the drying process. This oil also has the effect of desiccating a high proportion of the gastropods, resulting in the shell remnants becoming the contaminant of concern. Typically, drying racks are 50 m long and hold 15 t of fresh fruit, which dries down to about 3.5 t of dried sultanas (Plate 12).

Controlled studies at a New South Wales vineyard adjacent to Mildura, Victoria, illustrate the magnitude of the task of decontamination of *T. pisana*-infested fruit on the drying racks. The higher the contamination level, the more time the growers tended to spend searching for gastropods during the bunch-spreading operation, resulting in a higher proportion of contaminating animals being removed compared with lightly contaminated harvests (Table 11.1). None the less, significant levels of contamination can remain despite intensive searching at the time of spreading the bunches on the racks and, in this example, produce from racks A–F would all have incurred penalty charges had not further decontamination been undertaken. *T. pisana* are difficult to detect during

Table 11.1. Success of postharvest disinfestation activities in a New South Wales dried-fruit vineyard, estimated by the percentage of *Theba pisana* (Müller) (Helicidae) removed by labourers from drying racks* loaded with fresh fruit varying in gastropod contamination level.

Rack	Fruit contamination level: gastropods per rack	Per cent of gastropods removed		
		At spreading	During drying	At raking
B	Low < 250	2	45	53
C	Medium 500–600	7	84	9
A	Medium 500–600	7	80	13
E	High > 2500	27	–	73
F	High > 2500	34	–	66

*Approximately 15 t fruit fresh weight per rack.

the early stages of fruit drying, but as the fruit darken and shrink the gastropod shells become more visible. However, even with an intensive search effort at spreading and periodically during drying, a significant level of contamination may remain unless a final search effort is made prior to binning the dried fruit (Plate 13). The average time over all racks to remove *T. pisana* at spreading, once from the rack before shaking and at raking before binning was 6 person-hours per rack. This equated to a cost of approximately \$20 t^{-1} of dried fruit, which at a market value of A\$1000–1500 t^{-1} is a minor monetary investment. At the time of this study (1993), the maximum penalty for gastropod contamination was \$200 t^{-1} (currently \$100 t^{-1}).

Conclusion

Exotic helicid snails have had a detrimental effect on viticultural production in Australia and South Africa. The emergence of these terrestrial gastropods as significant viticultural pests has occurred over the past 25 years in South Africa and more recently in Australia. Gastropods are rated among the most important viticultural pests in both South Africa and Australia.

Gastropod feeding on grapevine buds and young foliage can reduce the yield potential of vines and, as a contaminant, downgrade dried-grape production. Wine industries in both countries are implementing quality-assurance schemes and integrated pest-management and soil/water-management programmes for grape production. The aim is to ensure environmental sustainability and improve product quality while maintaining or improving profitability. Gastropod infestations in modern, mechanized vineyards could have an impact on perceived wine quality.

The Australian and South African grape industries have recommendations on control strategies for terrestrial gastropods based on crop cultural practices and molluscicide baits. Gastropod-control research has not been a high priority in Australia, and research in South Africa occurred primarily in the early 1980s. There is a need to gain a better understanding of gastropod ecology in vineyards in order to develop improved bait treatments and strategies for their use in irrigated viticulture. Biological control agents also require evaluation in both Australian and South African conditions, although it is well recognized that proper assessment of the environmental effects of potential biological agents must accompany assessments of efficacy in control of pest species.

References

Baker, G.H. (1986) *The Biology and Control of White Snails (Mollusca: Helicidae), Introduced Pests in Australia.* Division of Entomology Technical Paper No. 25, Commonwealth Scientific and Industrial Research Organization, 31 pp.

Barnes, B.N. and Swart, P.L. (1979) *Snails on Table Grapes in the Winter Rainfall Area.* Table Grapes: Winter Rainfall E.15, Farming in South Africa, Stellenbosch, South Africa.

Connolly, M. (1916) Notes on South African non-marine Mollusca. V. On the introduced land-molluscan fauna of South Africa. *Annals of the South African Museum* 13, 181–190.

Davidson, R., Sanderson, G. and Schache, M. (1993) Snail control studies in vines. *Australian Dried Fruits News* 20(3), 12–14.

Dry, P.R. and Smart, R.E. (1986) The grape growing regions of Australia. In: Coombe, B.G. and Dry, P.R. (eds) *Resources in Australia. Viticulture,* Vol. 1. Australian Industrial Publishers, Adelaide, pp. 37–60.

Godon, D. (1983) *Pest Slugs and Snails.* Springer-Verlag, Berlin, 445 pp.

Loubser, J.T. (1982) *The Control of Snails in Vineyards.* Oenology and Viticulture F.19, Farming in South Africa, Stellenbosch, South Africa.

Macdonald, I.A.W. and Jarman, M.L. (1984) Invasive alien organisms in the terrestrial ecosystems of the fynbos biome, South Africa. *South African National Scientific Programmes Report* No. 85.

Robertson, M., Herbert, D. and Villet, M. (2000) Predictive modelling of invasive snail distributions in South Africa – *Theba pisana* (Müller, 1774) (Helicidae). In: Ponder, W. (ed.) *Molluscs 2000. Understanding Molluscan Biodiversity in Our Region into the 21st Century,* Malacological Society of Australasia, Sydney, pp. 72–73.

Sanderson, G. (1995) Snails in viticulture. *Australian Grapegrower and Winemaker* 378A, 115–118.

Sanderson, G. and Davidson, R. (1994) The cost of snail control. *Australian Dried Fruits News* 21(5), 15–18.

Schwartz, A. and Capatos, D. (1990) An evaluation of chemicals for the toxicity to brown snail (*Helix aspersa* Müller) on grapevines. *South African Journal of Enology and Viticulture* 11, 55–58.

Schwartz, A. and Siebert, M.W. (1987) Field trials with toxic bait for the control of the brown garden snail (*Helix aspersa* Müller). *South African Journal of Enology and Viticulture* 8, 80–81.

Simpson, R.D. (1993) The reproductive biology of *Helix aspersa* (Müller) in relation to its production in cool temperate conditions in Australia. In: *Proceedings of the Third International Congress of Medical and Applied Malacology.* Camden, Sydney, Australia.

van Bruggen, A.C. (1978) Land molluscs. In: Werger, M.J. (ed.) *Biology and Ecology of Southern Africa*. W. Junk, The Hague, pp. 877–923.

12 Agriolimacidae, Arionidae and Milacidae as Pests in West European Cereals

DAVID M. GLEN[1] AND ROBERT MOENS[2]

[1]IACR – Long Ashton Research Station, Department of Agricultural Sciences, University of Bristol, Long Ashton, Bristol BS41 9AF, UK; [2]Département Lutte biologique et Resources phytogénétiques, CRA Chemin de Liroux, 2, B-5030 Gembloux, Belgium

Introduction

Small-grain cereals (mainly wheat (*Triticum aestivum* Linnaeus) and barley (*Hordeum vulgare* Linnaeus)) (Gramineae) are the most important arable crops grown in western Europe. For example, in the European Union in 1993/94, production of wheat and barley (74 and 43 million t, respectively) was substantially greater than that of maize (*Zea mays* Linnaeus) (Gramineae) (29 million t) (Renshaw, 1994). In the UK in 1994 and 1996 (Table 12.1), wheat alone occupied 38–41% of the total area of arable land and barley occupied 23–26%. Smaller areas of oats (*Avena sativa* Linnaeus) (Gramineae), rye (*Secale cereale* Linnaeus) (Gramineae) and triticale (the allohexaploid between diploid rye and tetraploid wheat) are also grown. Wheat, barley, oats and rye are grown for human consumption (wheat as bread, pasta, breakfast cereals, etc., barley mainly for malting, oats mainly for breakfast cereals and rye mainly for bread) and animal feed. Triticale is used mainly for animal feed.

Wheat is undoubtedly the most important crop damaged by gastropods in western Europe, in terms of the area at risk and the area requiring treatment with molluscicides. For example, 27% of the wheat area in the UK was treated with molluscicides in 1994, representing 61% of the total area treated in that year (Table 12.1). Barley, oats, rye and triticale are also susceptible to damage. Surveys in Great Britain in the 1960s and 1970s (Strickland, 1965; Hunter, 1969; Stephenson and Bardner, 1977) indicated that 0.2–2.2% of the wheat crop was lost to gastropods. Port and Port (1986) pointed out that more recent estimates are not available for the extent of losses in cereal crops, but the increase of molluscicide use from the 1960s to 1982 clearly demonstrated that farmer perception of gastropods in these crops had increased considerably. Subsequent surveys indicate a 67-fold increase in molluscicide usage in

Table 12.1. Total areas of cereals and other arable crops grown and areas treated with molluscicides in Great Britain, 1994 (from Garthwaite *et al.*, 1995) and 1996 (from Thomas *et al.*, 1997).

Crop	Total area (ha) grown	Area (ha) treated with molluscicides	Area treated as % of total crop area	Area treated as % of total area treated
Wheat (*Triticum aestivum* Linnaeus) (Gramineae)				
1994	1,802,190	485,950	27.0	60.5
1996	1,967,270	266,290	13.5	56.8
Spring barley (*Hordeum vulgare* Linnaeus) (Gramineae)				
1994	450,600	280	0.1	0.1
1996	491,210	2,730	0.6	0.6
Winter barley (*Hordeum vulgare* Linnaeus) (Gramineae)				
1994	620,130	73,150	11.0	9.1
1996	740,880	34,180	4.6	7.3
Oats (*Avena sativa* Linnaeus) (Gramineae)				
1994	105,950	7,840	7.4	1.0
1996	93,450	1,200	1.2	0.2
Rye (*Secale cereale* Linnaeus) (Gramineae)				
1994	7,350	0	0	0
1996	8,220	0	0	0
Triticale (*Secale cereale* Linnaeus × *Triticum aestivum* Linnaeus) (Gramineae)				
1994	5,660	1,220	21.5	0.2
1996	7,100	0	0	0
Oilseed rape (*Brassica napus* Linnaeus var. *oleifera* Linnaeus) (Brassicaceae)				
1994	403,470	120,830	29.9	15.0
1996	355,850	72,390	20.3	15.5
Other arable				
1994	634,840	86,960	13.7	10.8
1996	610,270	83,700	13.7	17.9
Set-aside				
1994	725,930	25,870	3.6	3.2
1996	506,220	7,940	1.6	1.7
All arable crops				
1994	4,756,120	802,100	16.9	–
1996	4,780,470	468,430	9.8	–

Great Britain between the early 1970s and 1994/95 (Garthwaite and Thomas, 1996). Autumn-sown cereals are at greater risk than spring-sown cereals, so the trend in recent decades towards autumn sowing has contributed greatly to the increase in the use of molluscicides in cereal crops. Other agronomic changes have also contributed, as described in this chapter. However, it is important to note that molluscicide use on arable crops (mainly cereals) over the period from 1980 to 1995 (Garthwaite and Thomas, 1996) and into 1996 (Thomas *et al.*, 1997) shows considerable year-to-year fluctuations with no general upward trend.

Gastropods are most important during the crop-establishment phase with damage to seeds and seedlings (Moens, 1980, 1989; Martin and Kelly, 1986; Port and Port, 1986; Glen, 1989; Gratwick, 1992). Gastropods kill wheat seeds by eating the embryo, with destruction of part or all of the endosperm. The extent of damage to individual wheat seeds varies greatly, but, because the gastropods always destroy critical meristem tissues, the seed is always killed, irrespective of the amount of the tissue consumed. Gastropods also kill young seedlings after germination by destroying critical tissues, such as the meristem at the base of the shoot (Gair *et al.*, 1987; Gratwick, 1992). These animals may also graze on and destroy the leaves of seedlings after emergence, but, once the plants have reached the tillering stage, this leaf damage is generally not considered to be of any great importance. In wet summers, however, gastropods can damage the flag leaves just below the developing wheat seed heads (Kemp and Newell, 1987; Gratwick, 1992), which may lead to a reduction in yield (Kemp and Newell, 1987).

Barley and oats suffer similar damage to wheat, but these crops are considered to be at lesser risk. This is because gastropods prefer wheat (Duthoit, 1964), partly due to the absence of the natural seed coating found in barley and oats and also because of agronomic practices: barley and oats are seldom grown immediately after oilseed rape (*Brassica napus* Linnaeus var. *oleifera* Linnaeus) (Brassicaceae) in the rotation (Glen, 1989), barley is normally drilled earlier in the autumn than winter wheat (Port and Port, 1986) and barley is not usually grown on the clay soils where gastropods are most troublesome (Brown, 1955). The role of oilseed rape is explained below.

Surveys of arable farmers and crop consultants in England and Wales in the mid-1980s amply demonstrated industry concern about gastropod damage to cereals (Glen, 1989). In one questionnaire, farmers and crop consultants belonging to the Long Ashton Members' Association (LAMA) were invited to identify the pest causing them greatest concern in a range of crops. The replies showed (Fig. 12.1) that gastropods were the pests causing most concern to wheat growers and these came second only to aphids as a source of anxiety among producers of barley crops.

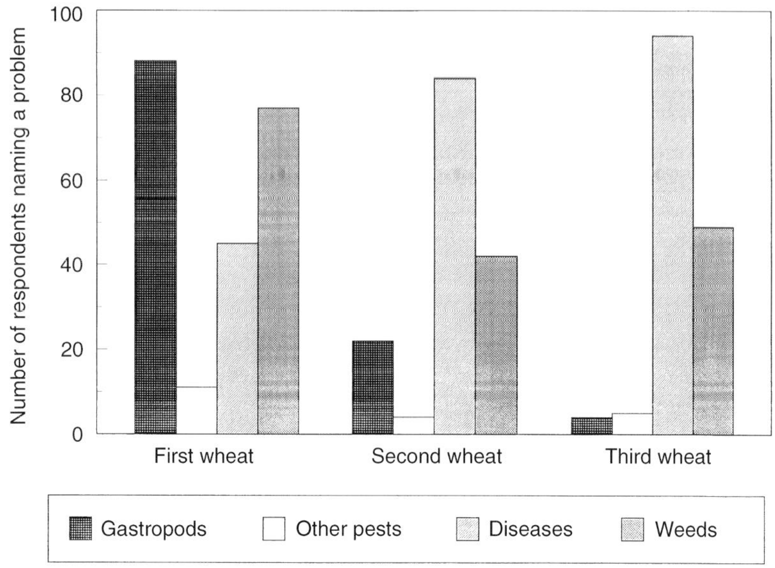

Fig. 12.1. Perceptions by farmers and consultants in England and Wales of the importance of gastropods, together with other pests, diseases and weeds, as problems in first, second and third wheat (*Triticum aestivum* Linnaeus) (Gramineae) crops after oilseed rape (*Brassica napus* Linnaeus var. *oleifera* Linnaeus) (Brassicaceae) (from Glen, 1989).

Gastropod Species Responsible for Damage

The agriolimacid *Deroceras reticulatum* (Müller) is considered to be the most common gastropod pest species in cereal crops and cereal-dominated rotations in western Europe (Runham and Hunter, 1970; Glen and Wiltshire, 1988; Moens, 1989). However, *D. reticulatum* usually occurs together with other gastropod species with the slug body form, particularly members of the families Arionidae and Milacidae (Brown, 1955; Gould, 1961; Duthoit, 1964; Glen *et al.*, 1984, 1989, 1992b; Kemp and Newell, 1987; Glen and Wiltshire, 1988; Hommay *et al.*, 1991; Hommay, 1995). Shelled gastropods (snails) are relatively rare in cereal fields in western Europe and are not recognized as pests in these environments, in contrast to southern Australia (see Baker, Chapter 6, this volume). Duthoit (1964) showed that all species of gastropod that she found in UK cereal fields were capable of damaging cereal seeds and seedlings in the laboratory. While *D. reticulatum* consumed more seeds of wheat than barley, *Arion hortensis* agg. (Arionidae) ate about equal numbers of each and *Tandonia budapestensis* (Hazay) (Milacidae) ate more barley than wheat seeds. All species tended to eat both the embryo and the endosperm of wheat seeds, but only the embryo of barley seeds. Oat seeds were virtually undamaged by the gastropods under these conditions. When given the choice of seeds or seedlings, *D. reticulatum*, *Arion ater* (Linnaeus) and *Arion fasciatus* agg. all caused equal damage to both, whereas *A. hortensis* agg. and *T. budapestensis* were more likely to damage seeds than seedlings. On balance, Duthoit (1964) concluded that *D. reticulatum* and *A. ater* were potentially the most damaging species present in cereal crops, mainly because of their greater appetite compared with other species, but also because of their feeding preferences. Given that *D. reticulatum* is much more prevalent than *A. ater* in arable fields with cereal-dominated crop rotations (Glen and Wiltshire, 1988), the former species is generally considered to be the most important pest species. However, it should be noted that, if gastropod biomass is more important than numbers in determining the severity of damage, as suggested by Glen *et al.* (1989), other gastropod species may be of more importance than indicated by their abundance. For example, although Glen *et al.* (1989) found that *D. reticulatum* (96 m^{-2}) was about six times as abundant as *Arion distinctus* (Mabille) and *Arion subfuscus* (Draparnaud) (15 m^{-2} and 13 m^{-2}, respectively) in a wheat seed-bed, the biomass of all three species was similar (1.3–1.5 g m^{-2}).

Factors Affecting Gastropod Damage to Cereals

Damage to cereals depends on gastropod abundance, feeding rate per individual and the vulnerability of the crop to damage. Each of these factors is considered below.

Gastropod abundance in cereal crops

The difficulty of estimating gastropod abundance in arable fields has greatly limited our understanding of their pest status in cereals. Gastropod slugs live in the soil as well as on the soil surface and methods of extracting these animals from soil are generally considered slow and laborious. The soil-flooding process devised by South (1964) for grassland and modified by Hunter (1968) for arable crops, has been refined at Long Ashton and used extensively to study populations in the upper 100 mm of soil in cereal fields (Glen *et al.*, 1984, 1988, 1989, 1990, 1992a,b, 1994a,b, 1996; Glen and Wiltshire, 1986; Wiltshire and Glen, 1989; Wilson *et al.*, 1994a; Bohan *et al.*, 1997, 2000a,b). It is widely recognized that gastropod populations in cereals are greatly influenced by agricultural practices. Furthermore, their populations in cereal crops exhibit considerable fluctuation both within and between years, even when agricultural practices do not appreciably change. These gastropods are strongly dependent on moisture for feeding, reproduction and survival, so that their numbers and distribution are greatly influenced by soil moisture, as well as by temperature. However, populations in cereal crops have been observed to decline in certain years when soil moisture and temperature were apparently favourable (Glen *et al.*, 1988, 1996), when there were no obvious changes in crop husbandry and when populations remained high or increased at other sites. This pattern suggests that natural enemies may be responsible for marked variation in population size, but our lack of understanding of the system precludes the prediction of abundance so vital to the management of the pests.

Spatial pattern in cereal fields

Hunter (1966) found that *D. reticulatum*, *A. hortensis* agg. and *T. budapestensis* moved to greater depth in arable soil during a dry period, as they also did in cold weather in winter. In a winter wheat crop, Glen *et al.* (1984, 1992b) found that *D. reticulatum*, *Arion intermedius* Normand and *Arion silvaticus* Lohmander were virtually absent from the upper 10 cm of soil during the dry summers of 1983 and 1984, but numbers rapidly recovered when the soil became moist again in autumn. This suggests that these gastropods survived dry conditions by moving deep into the soil, possibly using cracks that opened in the clay soil as it dried out. The ability of gastropod slugs to move to sources of moisture at depth in clay and silt soils may, in part, explain why they are more troublesome pests of cereals grown on such soils, but a higher moisture-retention capacity in such soils also contributes to better survival. In shallow soils and those of a sandy nature and thus of low moisture retention, gastropod slugs are unable to survive dry weather conditions in this way.

Gastropods are known to have underdispersed (aggregated) dispersion patterns and this has been confirmed for species resident in agricultural fields (South, 1965; Hunter, 1966; Airey, 1984). Recent studies of the distribution patterns of *D. reticulatum* and *A. intermedius* in a cereal field (Bohan *et al.*, 1997, 2000a,b; Shirley *et al.*, 1998) have revealed spatial dynamics not previously appreciated. Hot spots of abundance of *D. reticulatum* were found distributed at random throughout the cereal field (Bohan *et al.*, 2000b) in a spatial pattern that was consistent with predictions from a model of the movement and survival of individuals of this species (Shirley *et al.*, 1998). However, *A. intermedius* was found in a stable patch, within an area about 40 m in diameter (Bohan *et al.*, 2000b). This difference in distribution patterns is consistent with the contrasting biology of these two species, *D. reticulatum* being more surface-active, breeding at any time of year and reaching maturity faster than the less active, strictly annual, autumn-breeding *A. intermedius*. The action of natural enemies can be expected to operate in a non-random way but their contribution to the spatial dynamics of species like *D. reticulatum* and *A. intermedius* is only beginning to be understood, as explained below. The available evidence suggests that gastropod individuals move only relatively small net distances in cereal fields (South, 1965; Pinder, 1969; Fleming, 1989; Glen *et al.*, 1991) and populations in cereals are thought not to be greatly influenced by migration from adjacent field margins, such as wild-flower strips (Frank, 1998). This conclusion is supported by observations that gastropod damage is typically more severe in the middle of cereal fields than at the edges (Gould, 1961), which is the opposite of the pattern expected if migration from field margins were important.

Effects of cultural practices on gastropod populations

CROP ROTATION. Gould (1961) showed, in surveys of crops in East Anglia, that winter-wheat crops following in rotation such dense, leafy crops as pea (*Pisum sativum* Linnaeus) (Fabaceae) were at greater risk from gastropod damage than wheat following either fallow or crops such as potato (*Solanum tuberosum* Linnaeus) (Solanaceae), which leave relatively more bare soil between vegetated rows. The incidence of gastropod damage in wheat has increased greatly since Gould's survey because of a large expansion since the 1970s in the area planted to oilseed rape and its prevalence in rotation with wheat (Stephenson and Bardner, 1977; Martin and Kelly, 1986; Port and Port, 1986). In a survey in 1986/87 (Glen, 1989), farmers and consultants in LAMA were invited to name one pest, disease and weed (only one of each) that they had most encountered in first, second and third wheat crops after a break crop of oilseed rape. If they had not encountered a pest, disease or weed problem in these crops, then no reply was given. It is clear (Fig. 12.1) from the respondents that gastropods were not only considered to be the most important pest invertebrate group, but also the most pressing crop-protection issue in the first wheat crops to follow rape in the rotation. Gastropods were considered important compared with other pest invertebrates in second and third wheat crops after rape, but invertebrates were considered to be relatively unimportant in comparison with disease and weed problems in these crops (Fig. 12.1).

Further evidence of the greater risk of gastropod damage to winter-wheat crops following rape, compared with those following cereals, was provided by a survey throughout the UK from 1987 to 1990 (Glen *et al.*, 1993). This increased risk has been attributed to a higher abundance of gastropods, but the evidence for this was not conclusive. It is important to note here that populations do not inevitably increase within oilseed-rape crops: Glen *et al.* (1996), for example, reported no increase in gastropod populations in an oilseed-rape crop that followed 3 years of cereals that had supported high gastropod numbers.

CROP-RESIDUE DISPOSAL. In the 1970s and 1980s in the UK, it was common practice for farmers to dispose of unwanted cereal straw by burning *in situ* (Prew and Lord, 1988). Studies during the period 1982–1988 showed that gastropod populations tended to be greater in plots where straw or stubble was not burned than where straw was burned (Glen *et al.*, 1984, 1988). Thus, straw burning tended to depress gastropod populations, although gastropods did not appear to be affected directly by heat, but rather by removal of food and shelter (Glen *et al.*, 1988). However, atmospheric pollution problems associated with straw burning led initially to restrictions in straw burning in the 1980s (Prew and Lord, 1988), followed by a complete ban on the practice in the UK in 1993.

Farmers currently dispose of cereal straw either by baling and transport from the fields, thus leaving only the stubble behind, or they chop and spread the straw on the soil surface for later incorporation. On a clay soil in Oxfordshire, no consistent difference was found between gastropod populations residing in plots subjected to these two practices (Glen *et al.*, 1984, 1988). However, at this site, severe reduction in seedling numbers due to gastropod damage and the resulting poor yield of wheat crops (Christian *et al.*, 1999) probably restricted the amount of chopped straw returned and thus may not have been typical of normal farm-practice conditions. In a more recent long-term study, from 1988 onwards, gastropod populations have been consistently greater where straw was chopped, spread on the soil surface and subsequently incorporated by cultivation than where the straw was baled and removed (Glen *et al.*, 1994b; Kendall *et al.*, 1995; Symondson *et al.*, 1996).

Incorporation of crop residues improves soil structure and returns nutrients to the soil for recycling. For these reasons, in farming systems designed to combine profitable farming with environment protection, incorporation is the preferred method of disposing of crop residues (Jordan and Hutcheon, 1996), despite the resulting increase in gastropod numbers, at least in the initial years where such systems are adopted (Glen *et al.*, 1996).

CULTIVATION. It is generally accepted that gastropod populations are favoured by conservation tillage relative to traditional cultivation, which reduces numbers (see reviews by Martin and Kelly, 1986; Port and Port, 1986). Hunter (1967) observed that gastropod numbers were reduced by cultivation of an arable loam soil and considered much of this to have been caused by mechanical injury to the gastropods. Exposure of gastropods to high radiant temperatures and predation on the soil surface could also contribute to reductions in numbers following cultivation (Martin and Kelly, 1986). Ploughing and subsequent cultivations to produce a seed-bed for drilling winter cereals often result in substantial reductions in gastropod populations (Glen *et al.*, 1988, 1990; Kendall *et al.*, 1995) compared with uncultivated plots, but effects are variable in different years and sites (Glen *et al.*, 1988). Non-inversion methods of tillage, such as minimum tillage with tines or disc, or single-pass cultivation systems, such as that with the Dutzi cultivator, generally have less effect on gastropod populations than ploughing combined with subsequent cultivations to produce a seed-bed for cereals (Glen *et al.*, 1988, 1994b, 1996; Kendall *et al.*, 1995). In long-term experiments, the numbers of gastropods in plots under non-inversion tillage regimes were generally intermediate between the high numbers in zero-tillage plots and low numbers in ploughed plots (Kendall *et al.*, 1995; Symondson *et al.*, 1996).

The conclusion that emerges from these studies is that the greater the number of cultivation operations and the more intensive the method of cultivation, the more likely it is that gastropod numbers will be reduced substantially. However, even where tillage results in substantial reductions in populations, sufficient gastropods may still survive to cause severe damage to cereal seeds and seedlings, as shown, for example, by Glen *et al.* (1990).

DRAINAGE. Since gastropods are moisture-dependent animals, Martin and Kelly (1986) concluded that good drainage may reduce the risk of gastropod damage to winter cereals. However, experimental evidence suggests that drainage of soils to prevent waterlogging can improve survival and reproductive success. Carrick (1942) found that *D. reticulatum* laid few eggs in soil that was 100% saturated with water and those few eggs that were laid failed to hatch. Moreover, Stephenson and Bardner (1977) reported that, when soil was flooded for 34 days in winter, substantial numbers of *D. reticulatum*, *A. ater* and *A. hortensis* agg. were killed.

Effects of natural enemies on gastropod populations in cereals

There is little published information on the impact of natural enemies on gastropod populations in cereals in western Europe. However, as noted earlier, substantial population declines, during periods of apparently favourable weather and following 3 years or more with relatively high gastropod populations (Glen *et al.*, 1988, 1996), suggest that natural enemies may play a role in regulating numbers below the carrying capacity of the environment. The identity of the natural enemies that may be involved is not currently known.

A wide range of vertebrate and invertebrate predators are known to feed on gastropods (see Port and Port, 1986; Barker, 2002). Polyphagous predatory carabid beetles are the group considered most likely to have a substantial impact on gastropod populations in cereal fields (Symondson, 2002). Burn (1988) recorded more gastropods (mainly *D. reticulatum*) in traps in areas of cereal crops that had been surrounded by barriers to exclude polyphagous predatory beetles than in areas where the numbers of these predators had not been manipulated. Evidence is accumulating to show that one species of carabid beetle that is widespread and common in cereal fields, *Pterostichus melanarius* (Illiger), is an important predator of gastropods. Stephenson (1965) showed that *P. melanarius* could eliminate *D. reticulatum* from outdoor enclosures. Symondson *et al.* (1996) showed that gastropods (mainly *D. reticulatum* and *A. intermedius*) were important prey of *P. melanarius* in an arable field during the period from July to September, immediately before a crop of winter wheat was sown. Polyclonal antibody analysis demonstrated that, on average, 80% of *P. melanarius* adults had fed on gastropod tissue. Moreover, the numbers of beetles recorded in pitfall traps and the amount of gastropod material in the guts of these predators was positively correlated with the biomass of gastropods m^{-2} in soil samples. At the same experimental site, trends in the numbers and nutritional status of *P. melanarius* over a 5-year period from 1992 to 1996 were consistent with *D. reticulatum* and *A. intermedius* being important prey. When numbers of these gastropods were low, this predator was unable to maintain its numbers or nutritional status on alternative prey (Symondson *et al.*, 2002). Recent studies (Bohan *et al.*, 2000a) indicate that predation by *P. melanarius* can reduce the rate of growth of populations of *D. reticulatum* and *A. intermedius* in a spatially density-dependent manner in cereal fields during the summer months, thus possibly contributing to population regulation and reducing the risk of damage to autumn-sown cereal crops.

Gastropod feeding rate

Availability and accessibility of cereal seeds and seedlings

Gastropod activity on the soil surface in cereal fields is dependent on temperature and soil surface moisture (Young and Port, 1989, 1991; Yang *et al.*, 1991, 1993; Chabert, 1999). Furthermore, their activity is closely correlated with the severity of grazing damage to wheat seedling leaves (Glen *et al.*, 1993). However, damage to cereal seeds and seedlings before emergence, the most important damage, is not correlated with gastropod surface activity, estimated by bait-trapping, between sowing and emergence (Glen *et al.*, 1993). This latter damage is more dependent on the biomass of gastropods in the soil, together with the availability and accessibility of seeds.

Wheat is especially vulnerable to damage at establishment if seed-bed conditions enable the gastropods to move through the soil and locate the seeds and young seedlings. Field surveys have shown that winter wheat sown in a cloddy seed-bed in soil with a high clay or silt content is especially susceptible to damage (Gould, 1961; Moens, 1980), because of the availability of air spaces between soil aggregates. Moens (1989) emphasized the importance of seed cover in determining the severity of damage to wheat seeds. Consolidation of cloddy seed-beds often reduces the severity of damage. Gastropod damage to cereals is characteristically less severe around field edges (headlands) than in the middle of the field, because machinery turning at the edge of the field consolidates the soil (Gould, 1961). The influences of seed-bed conditions on the accessibility of wheat seeds to *D. reticulatum* (Stephenson, 1975; Moens, 1983, 1986; Davies, 1989), *A. hortensis* agg. and *T. budapestensis* (Davies, 1989) have been demonstrated in the laboratory.

Stephenson (1975) noted that damage by *D. reticulatum* to wheat seeds sown at 20–38 mm depth in trays was considerably less than damage to shallower-sown seeds (13 mm). However, he suggested that deeper sowing would be unacceptable in practice as a means of reducing damage, because it would result in delayed emergence, which, in turn, would prevent wheat plants from becoming established before winter.

It has also been considered (Gair *et al.*, 1987) that deeper sowing, by increasing the time to emergence, would make the seeds and seedling vulnerable to damage for a longer period. However, a series of three field experiments (Glen *et al.*, 1989, 1990, 1994b) has shown clearly that sowing at 40–50 mm depth in coarse seed-beds during autumn resulted in substantial reductions in gastropod damage compared with shallow sowing (20–25 mm). No appreciable delay in emergence was associated with the 40–50 mm sowing depth in these experiments, and a yield benefit from the increased sowing depth were observed in 1 year (Fig. 12.2).

Glen *et al.* (1989) showed that the percentage of wheat seeds killed by a mixed-species gastropod community in different seed-beds in a clay soil was directly related to the biomass of gastropods living in the upper 100 mm of soil (Fig. 12.3A) (damage was less well correlated with numbers). Furthermore, per cent kill was also inversely related to the depth of seed placement (Fig. 12.3B) and the percentage of fine soil aggregates (< 6 mm) in the seed-bed (Fig. 12.3C). It thus seemed that gastropod biomass in the soil (*S*) provided a measure of the potential for kill of wheat seeds, but the potential was diminished with greater sowing depth (*D*) and a greater percentage of fine soil aggregates (*F*) in the seed-bed. The influence of these three factors on the percentage of wheat seeds killed by gastropods (*P*) could be described in a simple model:

$$P = \frac{a\sqrt{S}}{DF}$$

where *a* is a constant. In the experiment of Glen *et al.* (1989) this model accounted for 94% of the variance in seed kill among nine combinations of seed-bed tilth and consolidation (Fig. 12.3D). This model indicates that, at the median gastropod biomass recorded in this experiment (1.6 g m^{-2}), the combined effects of seed-bed tilth and sowing depth together would have resulted in the percentage of seeds killed by gastropods ranging from as little as 5% to as much as 24%. Although, as described above, other work has established that seed-bed consolidation can reduce seed kill by gastropods, Glen *et al.* (1989) noted higher seed kill in seed-beds that had been consolidated before sowing than in unconsolidated seed-beds. This was because consolidation before sowing resulted in seeds being sown at shallower depth than in looser seed-beds, where drill penetration was greater. Glen *et al.* (1989) also noted that consolidation by rolling after drilling had no significant effect on the percentage of seeds killed, because rolling the soil failed to break down soil aggregates, a possibility also noted previously by Stephenson (1975) in the laboratory.

Palatability

Differences in susceptibility of wheat, barley and oat seeds are due, at least in part, to the lack of an outer seed coating on wheat seeds (most susceptible), compared with barley and oat seeds (least susceptible). In

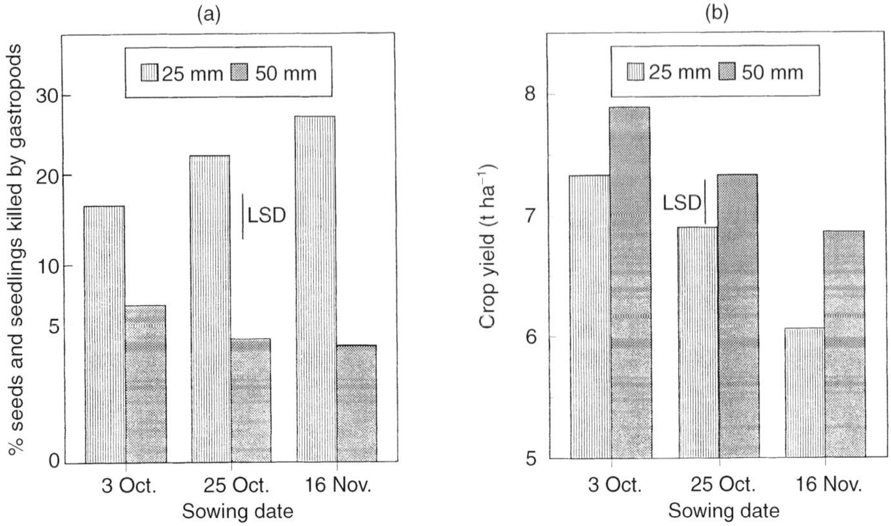

Fig. 12.2. Relationship between sowing depth and (a) percentage of seeds and seedlings killed by gastropods (square-root scale), (b) crop yield for wheat (*Triticum aestivum* Linnaeus) (Gramineae) sown at three dates in autumn 1989 (adapted from Glen *et al.*, 1994b). LSD, least significant difference ($P = 0.05$).

laboratory studies, Spaull and Eldon (1990) and Evans and Spaull (1996) found differences in the degree of grain hollowing in different wheat cultivars by *D. reticulatum*. However, Cook *et al.* (1996) failed to find significant differences between wheat cultivars in laboratory experiments that included the most and least preferred cultivars in the above experiments, and concluded that cultivars possess no inherent differences in palatability to *D. reticulatum*.

Availability of alternative food

At the time when crops are most susceptible and vulnerable to gastropod damage, residues of the previous crop are generally the most abundant alternative food. However, gastropod numbers tend to increase and their damage is often severe where large amounts of crop residues are returned to the soil. These observations indicate that the availability of food in the form of crop residues does not greatly influence the severity of damage to establishing wheat crops. This may be simply because high gastropod numbers outweigh any tendency for individuals to feed on alternative food. However, the known preference of *D. reticulatum* for novel foods (Frain and Newell, 1982), in this case wheat seeds and seedlings, may also be important. Recent laboratory and field studies (Cook *et al.*, 1996, 1997) do indicate, however, that certain weeds common in arable fields, such as dandelion (*Taraxacum officinale* Weber) (Asteraceae), are highly preferred food for *D. reticulatum*, in comparison with wheat seedlings, and their presence can reduce damage to wheat crops.

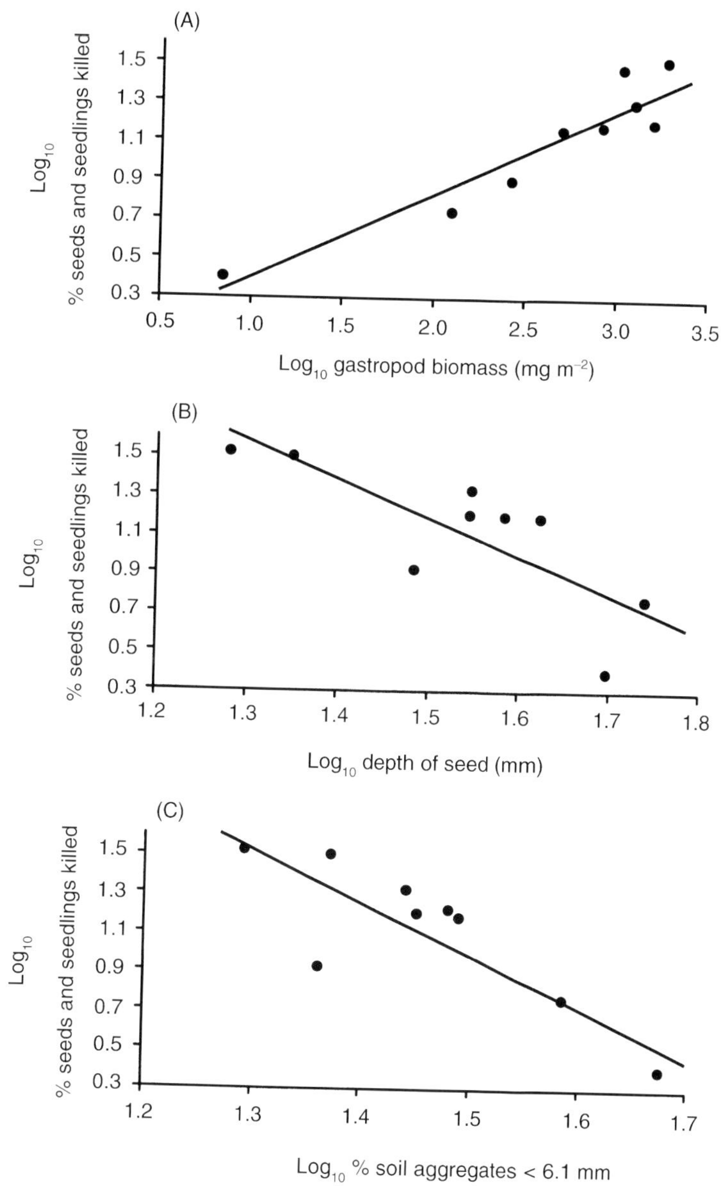

Vulnerability of crop to damage

Specific vulnerability and duration of vulnerable stages

Wheat is most vulnerable to gastropod damage shortly after drilling (Moens, 1983, 1986, 1989). As soon as the seed has imbibed water (Zadok's growth stages (GS) 02–05), gastropods gain entry to and are able to hollow the developing embryo (Fig. 12.4A). As noted above, gastropods

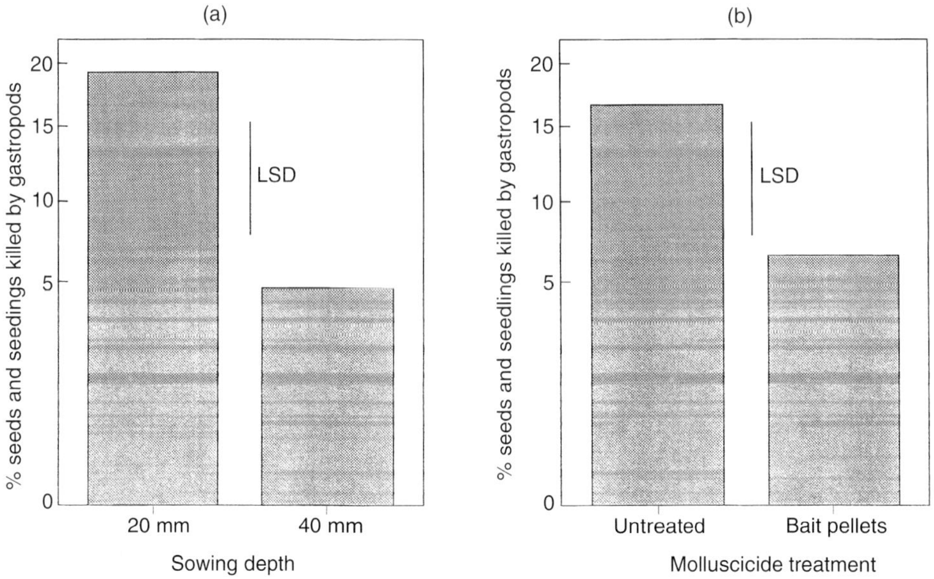

Fig. 12.5. Percentage of wheat (*Triticum aestivum* Linnaeus) (Graminaceae) seeds and seedlings killed by gastropods (square-root scale) as influenced by (a) sowing depth and (b) molluscicidal bait pellets applied as a broadcast treatment on the soil surface immediately after sowing (after Glen *et al.*, 1990). LSD, least significant difference ($P = 0.05$).

Although the presence of certain weed species has been shown to reduce damage to wheat seedlings, it is unlikely that farmers will ever deliberately sow weeds in order to reduce gastropod damage. However, it may be possible to grow companion crops instead of weeds to reduce the severity of gastropod damage. For example, George *et al.* (1995) demonstrated in the laboratory that damage to wheat seedlings by *D. reticulatum* was reduced when wheat was grown together with white clover (*Trifolium repens* Linnaeus) (Fabaceae). This approach has yet to be fully developed and used by farmers as a gastropod control strategy.

Biological control

The nematode *Phasmarhabditis hermaphrodita* (Schneider) (Rhabditidae) is a parasite capable of killing Agriolimacidae, Arionidae, Milacidae and other gastropods (Wilson *et al.*, 1993). It has been developed as a commercial biological control agent and has been used successfully to protect a range of crops from gastropod damage (Glen and Wilson, 1997), including winter wheat (Glen *et al.*, 1994a; Wilson *et al.*, 1994a,b, 1996; Hass *et al.*, 1999). The effectiveness of the nematode is dose-dependent. In winter wheat, a dose of 3×10^9 ha^{-1} was found to give protection to seeds and seedlings similar to that provided by a broadcast application of methiocarb bait pellets applied immediately after drilling (Wilson *et al.*,

1994a). In dry soil conditions, shallow cultivation by tines after nematode application was found to result in improved efficacy, probably as a result of protection of the nematodes from desiccation (Wilson *et al.*, 1996). Hass *et al.* (1999) noted that the machinery used for shallow incorporation of nematodes can have a profound influence on nematode efficacy. Thus, field experiments have demonstrated that it is technically feasible to use *P. hermaphrodita* for control of gastropod damage in cereals. It is compatible with the cultural control measures outlined above and it has two important advantages over chemical control. First, it does not affect non-target invertebrates or vertebrates (other than non-pest gastropod species) (Glen and Wilson, 1997) and, secondly, it is not adversely affected by heavy rain even in conditions where molluscicidal bait pellets are rendered ineffective (Hass *et al.*, 1999). However, *P. hermaphrodita* is not currently used for gastropod control in cereals because of its high cost, rendering it non-competitive with chemical molluscicides, and the limited storage life of this biocontrol agent, even under refrigerated conditions. For these reasons, its use is likely to be restricted to the home-garden market and high-value horticultural crops for the foreseeable future (Glen and Wilson, 1997).

Chemical control

Control of gastropod damage in west European cereals relies mainly on the application of molluscicidal bait pellets containing metaldehyde, methiocarb or thiodicarb as the active ingredient. Pellets are normally applied shortly before or after drilling as broadcast treatments on the soil surface, or the pellets are applied as an admixture with the seeds at drilling. Laboratory studies have shown that, following an encounter, *D. reticulatum* and *A. distinctus* are more likely to feed on molluscicidal pellets than on wheat seeds (Bourne *et al.*, 1988; Bailey and Wedgwood, 1991). However, the amount of bait pellet eaten is reduced in the presence of wheat seeds (Bourne *et al.*, 1988). This suggests that molluscicidal bait pellets should be applied before wheat seeds are available or that, if they are applied when wheat seeds are sown, it is important to maximize the availability of pellets and minimize the availability of wheat seeds, as outlined below.

Laboratory and field studies have demonstrated that it is feasible to protect cereal seeds and seedlings from gastropod damage by applying molluscicidal or repellent chemicals to seeds (Gould, 1962; Scott *et al.*, 1984; Ester and Nijënstein, 1995; Ester *et al.*, 1996; Watkins *et al.*, 1996; Nijënstein and Ester, 1998). Certain fungicidal seed dressings also protect seeds from gastropod damage (Moens *et al.*, 1992). However, no molluscicidal seed treatments have yet been made available commercially. Reasons for this are not always clear but considerations such as potential toxicity of dressed seeds to seed-eating birds are important in some cases.

Timing and placement of molluscicide bait applications

The most reliable control of gastropod damage to winter wheat is generally achieved by applying molluscicide bait pellets shortly before or after sowing (Glen *et al.*, 1992a; Gratwick, 1992; Port *et al.*, 1992). Gratwick (1992) states that the best results are generally obtained by applying bait pellets before sowing, when gastropods are active on the soil surface, and then avoiding further tillage for at least 3 days after treatment. This advice is consistent with the recommendation by Bourne *et al.* (1988) described above. However, because of the importance of timely sowing in maximizing crop establishment, growers are not recommended by Gratwick (1992) to delay sowing simply to apply bait pellets before drilling. In most situations the best practical option is to broadcast bait pellets on the soil surface at or immediately after drilling (Gratwick, 1992). Experimental applications of bait pellets to the stubble of a previous crop have given relatively poor results compared with applications closer to the time of seed sowing, which are more effective probably because they are targeted on residual populations already reduced by tillage (Port *et al.*, 1992). Typically, molluscicide applications to wheat crops kill only about 50% of the gastropod population that is resident in the upper 10 cm of soil at the time of application, with a slightly greater reduction in biomass of *c.* 60% (Glen and Wiltshire, 1986; Wiltshire and Glen, 1989; Glen *et al.*, 1991). In addition, gastropod eggs in soil are unaffected by molluscicides. Thus, gastropod populations often have sufficient time to recover from treatments applied to stubble before a wheat crop is sown. Nevertheless, there appear to be differences in the resilience of gastropod species. Glen *et al.* (1992b) reported recovery of *D. reticulatum* and *A. intermedius* populations within a few months of autumn molluscicide treatment, but year-long reductions in *A. silvaticus* and *A. subfuscus*.

Application of molluscicidal bait pellets to the soil surface at seed sowing protects seedlings from grazing damage after plant emergence, in addition to protecting seeds and seedlings before emergence (Glen *et al.*, 1990, 1992a). If gastropod grazing damage to emergent seedlings causes concern, it may be worthwhile to apply pellets after crop emergence to reduce plant losses, particularly if the crop has been thinned by seeds being killed before emergence (Gratwick, 1992). However, by the time that damage becomes evident after emergence, it may be too late to take the most effective action (e.g. Glen *et al.*, 1992a). This emphasizes the importance of reliable prediction of the risk of gastropod damage and the need for control measures.

Molluscicidal bait pellets can either be applied to the soil surface or drilled with wheat seed. The latter option appears attractive, because the bait pellets are close to the seeds, which are the most vulnerable stage. In addition, bait pellets drilled with seeds are less likely to kill non-target invertebrates and vertebrates than pellets broadcast on the soil surface (Kennedy, 1990; Johnson *et al.*, 1991). However, bait pellets drilled with wheat seeds may be unavailable to gastropods (Glen *et al.*, 1992a), especially if the recommended seed-cover measures are adopted to protect seeds from feeding gastropods. Shallow-sown seeds in a coarse seed-bed, which are vulnerable to gastropod damage, can be protected by bait pellets drilled with them, but this protection may be no better than would be achieved by drilling seed a little deeper in a coarse seed-bed (Glen *et al.*, 1992a). In order to improve the protection from seed kill obtained by deeper drilling alone, bait pellets should be broadcast on the soil surface after drilling a little deeper than normal (Glen *et al.*, 1990, 1992a).

Damage Forecasting and Risk Assessment

Because of the importance of taking appropriate action before or shortly after cereal crops are sown in order to provide the best protection from gastropod damage (Glen *et al.*, 1992a; Gratwick, 1992; Port *et al.*, 1992), a reliable system of forecasting gastropod damage to cereals would be extremely valuable to farmers and consultants. Experience clearly indicates that, by the time that potentially severe damage becomes evident, either as gaps in the rows of an emerging wheat crop or as grazing damage to seedlings, it is already too late to take the most effective action. The relationships in Fig. 12.3 clearly indicate the importance of gastropod biomass in soil, seed-bed tilth and sowing depth in assessing the risk of damage. However, because direct methods of estimating gastropod biomass are too slow and labour-intensive for commercial use, consultants and growers currently rely on refuge traps, whose catch provides a composite index of gastropod abundance and the degree of surface activity. Whilst gastropod activity on the soil surface can be predicted on the basis of temperature and surface moisture (Young and Port, 1989, 1991; Yang *et al.*, 1991, 1993; Chabert, 1999), surface activity is almost certainly considerably disrupted by the cultivations involved in seed-bed preparation and, as pointed out earlier, gastropod activity on the soil surface at drilling and between drilling and emergence is poorly correlated with the severity of seed hollowing by these animals (Glen *et al.*, 1993).

In a study throughout the UK, Glen *et al.* (1993) found that the best predictor of damage was simply the peak number of gastropods trapped at times when the soil surface was visibly moist, during the period from July until the soil was disturbed by cultivation. However, even this predictor was imprecise, accounting for only 26% of the variance in damage, and the threshold trap catch for molluscicide treatment based on this predictor inevitably includes a large safety margin. Thus, even if farmers use this prediction method, it is likely that many fields treated with molluscicide do not have sufficiently high gastropod numbers to justify such treatment. Thus, there is considerable scope for improvement in damage forecasting. Current effort to improve forecasting is focused on improved understanding of the spatial and temporal dynamics of gastropod populations in arable crops (Bohan *et al.*, 1997, 2000a,b; Shirley *et al.*, 1997, 1998, 2001), including the influence of weather (Chabert, 1999). However, improved practical methods of assessing gastropod biomass in soil are also needed in order to give better predictions of the severity of damage and the need for control measures. It may be possible to assess gastropod biomass in soil by using traps, such as those described by Hommay and Briard (1988) or Young *et al.* (1996), to assess surface activity, in combination with a simulation model of population dynamics, such as that for *D. reticulatum* described by Shirley *et al.* (2001). Young *et al.* (1996) have stressed the desirability of using non-toxic baits rather than molluscicide baits (as used, for example, by Glen *et al.*, 1993) in traps.

Because gastropod surface activity is strongly correlated with grazing damage to seedlings (Glen *et al.*, 1993), weather information (Young and Port, 1989, 1991; Chabert, 1999) is of considerable value in helping to decide on the need for and likely success of control measures after crop emergence.

Conclusions

Gastropod damage to cereals is influenced by several interacting factors and understanding of these has improved steadily in recent years. Because of this and improvements in cultural and chemical control measures, it might be expected that problems would be considered to be less severe now than in the past. Survey data from 1982 to 1996 indicate no trend in molluscicide use in cereals (Thomas *et al.*, 1997). However, farmers' perception of the severity of gastropod problems in cereals had, if anything, increased in the late 1990s. Reasons for this are not fully clear, but a number of plausible contributory factors can be proposed.

First, as stressed by Martin and Kelly (1986), farmers are well aware of the need to produce an adequate plant stand in order to achieve sufficient yield to provide profit, and they are alert to the potential for gastropods to prevent such a stand being achieved. It is sometimes suggested that farmers could simply increase the seeding rate in order to compensate for anticipated losses from gastropods. However, this is not a sensible alternative to molluscicide use, because: (i) if damage is less severe than anticipated, then an overly dense stand could lead to yield loss; and (ii) in severe damage a greater initial plant density could not compensate for losses.

Secondly, cereal growers and consultants frequently express the opinion that gastropods have become more abundant in recent years. Much evidence has been presented in this chapter to show that modern agronomic practices could be responsible for increased gastropod abundance. Similarly, trends in weather patterns (mild winters and wet summers) in many recent years have been favourable to gastropods (Chabert, 1999). Reductions in the impact of natural enemies and other biotic constraints on gastropod population growth could also be responsible, perhaps mediated by changes in agricultural practices. This 'natural enemy' hypothesis has not yet been properly explored. However, evidence is steadily accumulating that natural enemies, such as carabid beetles, can reduce the growth rate of gastropod populations (Bohan *et al.*, 2000a; Buckland and Grime, 2000; McKemey, 2000; Symondson *et al.*, 2002) and further research is warranted.

Thirdly, while the widespread adoption of cultural measures to reduce kill of seeds and seedlings has improved crop stand at emergence, it seems that, in fields with an abundance of gastropods, this has sometimes resulted in attack being only delayed until after emergence, with severe grazing and consequent loss of plant stand.

Fourthly, although molluscicide bait pellets are normally highly effective in reducing the severity of damage, many farmers and consultants continue to report inadequate control. The reasons for this are not known and must be identified. For example, because molluscicides typically kill only about 50% of the population in the soil, then, where gastropods are abundant, enough may survive treatment to inflict severe damage. Gastropods hatching from eggs or resuming normal surface-activity patterns some time after the disruption caused by seed-bed preparation and after molluscicides have ceased to be effective may also be responsible. It is also possible that, as suggested by Hass *et al.* (1999), molluscicidal bait pellets are rendered ineffective when heavy rain falls shortly after bait pellets are applied, with pellets becoming covered in mud splash or washed into soil crevices and hidden from gastropods. If this is so, then in some years it would be possible to adjust timing of bait-pellet application in order to avoid exposure of bait pellets to such rain. In very wet autumns this may not be achievable, but nevertheless knowledge of the effects of rainfall would indicate which molluscicide applications are likely to be ineffective and where further treatment may be necessary. Unfortunately, biological control using the nematode *P. hermaphrodita*, which Hass *et al.* (1999) found to be highly effective in wet conditions, is not economically viable as a practical control measure in cereals.

It must be emphasized that the above suggestions are merely pointers to future research needed to provide the improved understanding of gastropod population ecology and behaviour and the factors that influence molluscicide efficacy in cereal crops. This new understanding will not be easily achieved. However, the potential benefits of better prediction and control of gastropod damage make such studies well worthwhile. Precise estimation of risk will lead to a reduction in unnecessary molluscicide usage, which will result in less widespread environmental side-effects as well as lower cost to the farmer. Growers will also be able to concentrate their efforts in fields where the real risk is high and adequate control is difficult to achieve. More reliable and economic control would have obvious benefits for cereal growers. In addition to investigations into ways to achieve better use of existing control methods, new methods, such as seed treatment, should continue to be explored.

Acknowledgements

IACR receives grant-aided support from the Biotechnology and Biological Sciences Research Council of the UK.

References

Airey, W.J. (1984) The distribution of slug damage in a potato crop. *Journal of Molluscan Studies* 50, 239–240.

Bailey, S.E.R. and Wedgwood, M.A. (1991) Complementary video and acoustic recordings of foraging by two pest species of slugs on non-toxic and molluscicidal baits. *Annals of Applied Biology* 119, 163–176.

Barker, G.M. (ed.) (2002) *Natural Enemies of Terrestrial Molluscs.* CAB International, Wallingford.

Bohan, D., Brain, P., Glen, D.M., Wiltshire, C.W., Hughes, L., Shirley, M. and Port, G. (1997) Decision making in slug pest control: slug population dynamics in space and time. In: *Optimising Cereal Inputs: its Scientific Basis.* Aspects of Applied Biology 50, Association of Applied Biologists, Wellesbourne, Warwick, pp. 323–332.

Bohan, D.A., Bohan, A.C., Glen, D.M., Symondson, W.O.C., Wiltshire, C.W. and Hughes, L. (2000a) Spatial dynamics of predation by carabid beetles on slugs. *Journal of Animal Ecology* 69, 367–379.

Bohan, D.A., Glen, D.M., Wiltshire, C.W.W. and Hughes, L. (2000b) Parametric intensity and the spatial arrangement of the terrestrial mollusc herbivores *Deroceras reticulatum* and *Arion intermedius. Journal of Animal Ecology* 69, 1031–1046.

Bourne, N.B., Jones, G.W. and Bowen, I.D. (1988) Slug feeding behaviour in relation to control with molluscicidal baits. *Journal of Molluscan Studies* 54, 327–338.

Brown, E.B. (1955) Some current British soil pest problems. In: Kevan, D.K.McE. (ed.) *Soil Zoology.* Butterworths, London, pp. 256–258.

Buckland, S.M. and Grime, J.P. (2000) The effects of trophic structure and soil fertility on the assembly of plant communities: a microcosm experiment. *Oikos* 91, 336–352.

Burn, A.J. (1988) Assessment of the impact of pesticides on invertebrate predation in cereal crops. *Environmental Aspects of Applied Biology,* Part 2. Aspects of Applied Biology 17, Association of Applied Biologists, Wellesbourne, Warwick, pp. 173–179.

Carrick, R. (1942) The grey field slug *Agriolimax agrestis* L. and its environment. *Annals of Applied Biology* 29, 43–55.

Chabert, A. (1999) *Prévision des attaques de limaces: mise au point d'un systeme d'aide à la décision par la definition des zones à risques et la modelisation de la mise en activité des limaces.* Dossier 95/02-4, Enveloppe Recherche de l'ACTA, 71 pp.

Christian, D.G., Bacon, E.T.G., Brockie, D., Glen, D., Gutteridge, R.J. and Jenkyn, J.F. (1999) Interactions of straw disposal methods and direct drilling or cultivations on winter wheat (*Triticum aestivum*) grown on a clay soil. *Journal of Agricultural Engineering Research* 73, 297–309.

Cook, R.T., Bailey, S.E.R. and McCrohan, C.R. (1996) Slug preferences for winter wheat cultivars and common agricultural weeds. *Journal of Applied Ecology* 33, 866–872.

Cook, R.T., Bailey, S.E.R. and McCrohan, C.R. (1997) The potential for common weeds to reduce slug damage to winter wheat: laboratory and field studies. *Journal of Applied Ecology* 34, 79–87.

Davies, A.J. (1989) Effects of soil compaction on damage to wheat seeds by three pest species of slug. *Crop Protection* 8, 118–121.

Duthoit, C.M.G. (1964) Slugs and food preferences. *Plant Pathology* 13, 73–78.

Ester, A. and Nijënstein, J.H. (1995) Control of the field slug *Deroceras reticulatum* (Müller) (Pulmonata: Limacidae) by pesticides applied to winter wheat seed. *Crop Protection* 5, 409–413.

Ester, A., Darwinkel, A., Floot, H.W.G. and Nijënstein, J.H. (1996) Control of field slugs (*Deroceras reticulatum*) in winter wheat using seeds treated with pesticides. In: Henderson, I.F. (ed.) *Slug and Snail Pests in Agriculture.* Symposium Proceedings No. 66, British Crop Protection Council, Farnham, pp. 165–172.

Evans, K.A. and Spaull, A.M. (1996) Differential tolerance of winter wheat cultivars to grain hollowing by the grey field slug (*Deroceras reticulatum*): a laboratory study. In: Henderson, I.F. (ed.) *Slug and Snail Pests in Agriculture.* Symposium Proceedings No. 66, British Crop Protection Council, Farnham, pp. 433–438.

Fleming, C.C. (1989) Population structure of *Deroceras reticulatum.* In: Henderson, I.F. (ed.) *Slugs and Snails in World Agriculture.* Monograph No. 41, British Crop Protection Council, Thornton Heath, pp. 413–415.

Frain, J.M. and Newell, P.F. (1982) Meal size and a feeding assay for *Deroceras reticulatum* (Müll.). *Journal of Molluscan Studies* 48, 98–99.

Frank, T. (1998) Slug damage and number of slugs (Gastropoda: Pulmonata) in winter wheat in fields with sown wildflower strips. *Journal of Molluscan Studies* 64, 319–328.

Gair, R., Jenkins, J.E.E. and Lester, E. (1987) *Cereal Pest and Diseases*, 4th edn. Farming Press, Ipswich, 268 pp.

Garthwaite, D.G. and Thomas, M.R. (1996) The use of molluscicides in agriculture and horticulture in Great Britain over the last 30 years. In: Henderson, I.F. (ed.) *Slug and Snail Pests in Agriculture.* Symposium Proceedings No. 66, British Crop Protection Council, Farnham, UK, pp. 39–46.

Garthwaite, D.G., Thomas, M.R. and Hart, M. (1995) *Arable Farm Crops in Great Britain 1994.* Pesticide Usage Survey Report 127, Ministry of Agriculture, Fisheries and Food, London, 97 pp.

George, S.K., Kendall, D.A., Clements, R.O. and Asteraki, E.J. (1995) Slug damage to clover and wheat grown singly and in mixtures. In: McKinlay, R.D. and Atkinson, D. (eds) *Integrated Crop Protection: Towards Sustainability?*

Symposium Proceedings No. 63, British Crop Protection Council, Farnham, pp. 139–142.
Glen, D.M. (1989) Understanding and predicting slug problems in cereals. In: Henderson, I.F. (ed.) *Slugs and Snails in World Agriculture.* Monograph No. 41, British Crop Protection Council, Thornton Heath, pp. 253–262.
Glen, D.M. and Wilson, M.J. (1997) Slug-parasitic nematodes as biocontrol agents for slugs. *Agro-Food-Industry Hi-Tech* 8(2), 23–27.
Glen, D.M. and Wiltshire, C.W. (1986) Estimating slug populations from bait-trap catches. In: *Proceedings of the 1986 British Crop Protection Conference – Pests and Diseases*, Vol. 3, pp. 1151–1158.
Glen, D.M. and Wiltshire, C.W. (1988) Distribution of slug species. In: *IACR Long Ashton Research Station Annual Report 1987*, p. 28.
Glen, D.M., Wiltshire, C.W. and Milsom, N.F. (1984) Slugs and straw disposal in winter wheat. In: *Proceedings of the 1984 British Crop Protection Conference – Pests and Diseases*, Vol. 1, pp. 139–144.
Glen, D.M., Wiltshire, C.W. and Milsom, N.F. (1988) Effects of straw disposal on slug problems in cereals. In: *Environmental Aspects of Applied Biology*, Vol. 2. Aspects of Applied Biology 17, Association of Applied Biologists, Wellesbourne, pp. 173–179.
Glen, D.M., Milsom, N.F. and Wiltshire, C.W. (1989) Effects of seed-bed conditions on slug numbers and damage to winter wheat in a clay soil. *Annals of Applied Biology* 115, 177–190.
Glen, D.M., Milsom, N.F. and Wiltshire, C.W. (1990) Effect of seed depth on slug damage to winter wheat. *Annals of Applied Biology* 117, 693–701.
Glen, D.M., Wiltshire, C.W. and Butler, R.C. (1991) Slug population changes following molluscicide treatment in relation to distance from edge of treated area. *Crop Protection* 10, 408–412.
Glen, D.M., Wiltshire, C.W. and Langdon, C.J. (1992a) Influence of seed depth and molluscicide pellet placement and timing on slug damage, activity and survival in winter wheat. *Crop Protection* 11, 555–560.
Glen, D.M., Wiltshire, C.W. and Milsom, N.F. (1992b) Some aspects of forecasting slug damage in arable crops. *Journal of Medical and Applied Malacology* 4, 147–152.
Glen, D.M., Spaull, A.M., Mowat, D.J., Green, D.B. and Jackson, A.W. (1993) Crop monitoring to assess the risk of slug damage to winter wheat in the UK. *Annals of Applied Biology* 122, 161–172.
Glen, D.M., Wilson, M.J., Pearce, J.D. and Rodgers, P.B. (1994a) Discovery and investigation of a novel nematode parasite for biological control of slugs. In: *Proceedings of the Brighton Crop Protection Conference – Pests and Diseases – 1994*, Vol. 2, pp. 617–624.
Glen, D.M., Wiltshire, C.W., Wilson M.J., Kendall, D.A. and Symondson, W.O.C. (1994b) Slugs in arable crops: key pests under CAP reform? In: *Arable Farming under CAP Reform.* Aspects of Applied Biology 40, Association of Applied Biologists, Wellesbourne, pp. 199–206.
Glen, D.M., Wiltshire, C.W., Walker, A.J., Wilson, M.J. and Shewry, P.R. (1996) Slug problems and control strategies in relation to crop rotations. In: *Rotations and Cropping Systems.* Aspects of Applied Biology 47, Association of Applied Biologists, Wellesbourne, pp. 153–160.
Gould, H.J. (1961) Observations on slug damage to winter wheat in East Anglia 1957–1959. *Plant Pathology* 10, 142–147.

Gould, H.J. (1962) Tests with seed dressings to control seed hollowing of winter wheat by slugs. *Plant Pathology* 11, 147–152.

Gratwick, M. (ed.) (1992) *Crop Pests in the UK. Collected Edition of MAFF Leaflets.* Chapman & Hall, London, 490 pp.

Hass, B., Hughes, L.A. and Glen, D.M. (1999) Overall versus band application of the nematode *Phasmarhabditis hermaphrodita* with and without incorporation into soil, for biological control of slugs in winter wheat. *Biocontrol Science and Technology* 9, 579–586.

Hommay, G. (1995) Les limaces nuisible aux cultures. *Revue Suisse d'Agriculture* 27, 267–286.

Hommay, G. and Briard, P. (1988) Apport du piégeage dans le suivi des peuplements de limaces en grande culture. *Haliotis* 18, 55–74.

Hommay, G., Peureux, D. and Verbeke, D. (1991) Un réseau d'observations sur les limaces dans l'est de la France. *Phytoma* 431, 14–19.

Hunter, P.J. (1966) The distribution and abundance of slugs on an arable plot in Northumberland. *Journal of Animal Ecology* 35, 543–557.

Hunter, P.J. (1967) The effect of cultivations on slugs of arable land. *Plant Pathology* 16, 153–156.

Hunter, P.J. (1968) Studies on slugs of arable ground: I. Sampling methods. *Malacologia* 6, 369–377.

Hunter, P.J. (1969) An estimate of the extent of slug damage to wheat and potatoes in England and Wales. *National Agricultural Advisory Service Quarterly Review* 85, 31–36.

Jessop, N.H. (1969) The effects of simulated slug damage on the yield of winter wheat. *Plant Pathology* 18, 172–175.

Johnson, I.P., Flowerdew, J.R. and Hare, R. (1991) Effects of broadcasting and of drilling methiocarb molluscicide pellets on field populations of wood mice *Apodemus sylvaticus*. *Bulletin of Environmental Contamination and Toxicology* 46, 84–91.

Jordan, V.W.L. and Hutcheon, J.A. (1996) Multifunctional crop rotation: the contribution and interactions for integrated crop protection and nutrient management in sustainable cropping systems. In: *Rotations and Cropping Systems.* Aspects of Applied Biology 47, Association of Applied Biologists, Wellesbourne, pp. 301–308.

Kemp, N.J. and Newell, P.F. (1987) Slug damage to the flag leaves of winter wheat. *Journal of Molluscan Studies* 53, 109–111.

Kendall, D.A., Chinn, N.E., Glen, D.M., Wiltshire, C.W., Winstone, L. and Tidboald, C. (1995) Effects of soil management on cereal pests and their natural enemies. In: Glen, D.M., Greaves, M.P. and Anderson, H.M. (eds) *Ecology and Integrated Arable Farming Systems.* John Wiley & Sons, London, pp. 83–102.

Kennedy, P.J. (1990) The effects of molluscicides on the abundance and distribution of ground beetles (Coleoptera, Carabidae) and other invertebrates. PhD thesis, University of Bristol.

McKemey, A.R. (2000) Integrating behavioural aspects of a carabid–slug interaction using immunological data on predator ecology. PhD thesis, Cardiff University.

Martin, T.J. and Kelly, J.R. (1986) The effects of changing agriculture on slugs as pests of cereals. In: *Proceedings of the 1986 British Crop Protection Conference – Pests and Diseases*, Vol. 2, pp. 411–424.

Moens, R. (1980) Le problème des limaces dans la protection des végétaux. *Revue d'Agriculture (Bruxelles)* 33, 117–132.
Moens, R. (1983) Essais sur le protection des grains de froment contre l'attaque des limaces. *Revue d'Agriculture (Bruxelles)* 36, 1303–1317.
Moens, R. (1986) Observations sur l'attaque de *Deroceras reticulatum* (Müller) aux emblavures de froment. *Agricontact* 165, 1–5.
Moens, R. (1989) Factors affecting slug damage and control measure decisions. In: Henderson, I.F. (ed.) *Slugs and Snails in World Agriculture*. Monograph No. 41, British Crop Protection Council, Thornton Heath, pp. 227–236.
Moens, R., Latteur, G. and Fayt, E. (1992) Contribution to an integrated slug control. *Parasitica* 48, 83–105.
Nijënstein, J.H. and Ester, A. (1998) Phytotoxicity and control of the field slug *Deroceras reticulatum* by seed applied pesticides in wheat, barley and perennial ryegrass. *Seed Science and Technology* 26, 501–513.
Pinder, L.C.V. (1969) The biology and behaviour of some slugs of economic importance, *Agriolimax reticulatus*, *Arion hortensis* and *Milax budapestensis*. PhD thesis, University of Newcastle upon Tyne.
Port, C.M. and Port, G.R. (1986) The biology and behaviour of slugs in relation to crop damage and control. *Agricultural Zoology Reviews* 1, 255–299.
Port, G.R., Hogan, J.M. and Port, C.M. (1992) Factors affecting the time of slug control in winter wheat. In: *Proceedings of the Ninth International Malacological Congress, 1986*. Unitas Malacologica, Leiden, the Netherlands, pp. 257–261.
Prew, R.D. and Lord, E.I. (1988) The straw incorporation problem. In: *Environmental Aspects of Applied Biology*, Vol. 2. Aspects of Applied Biology 17, Association of Applied Biologists, Wellesbourne, pp. 163–171.
Renshaw, D. (1994) CAP reform from the EC viewpoint. In: *Arable Farming under CAP Reform*. Aspects of Applied Biology 40, Association of Applied Biologists, Wellesbourne, pp. 1–4.
Runham, N.W. and Hunter, P.J. (1970) *Terrestrial Slugs*. Hutchinson University Library, London, 184 pp.
Scott, G.C., Pickett, J.A., Smith, M.C. and Woodcock, C.M. (1984) Seed treatments for controlling slugs in winter wheat. In: *Proceedings of the 1984 British Crop Protection Conference – Pests and Diseases*, Vol. 1, pp. 133–138.
Shirley, M.D., Rushton, S.P., Port, G.R. and Young, A.G. (1997) Decision making in slug pest control: simulation models of slug populations. In: *Optimising Cereal Inputs: its Scientific Basis*. Aspects of Applied Biology 50, Association of Applied Biologists, Wellesbourne, pp. 333–340.
Shirley, M.D., Rushton, S.P., Port, G.R., Young, A.G., Bohan, D.A. and Glen, D.M. (1998) Spatial modelling of slug populations in arable crops. In: *Proceedings of the 1998 Brighton Crop Protection Conference – Pest and Diseases*, Vol. 3, pp. 1095–1102.
Shirley, M.D., Rushton, S.P., Young, A.G. and Port, G.R. (2001) Simulating the long-term dynamics of slug populations: a process-based modelling approach for pest control. *Journal of Applied Ecology* 38, 401–411.
South, A. (1964) Estimation of slug populations. *Annals of Applied Biology* 53, 251–258.
South, A. (1965) Biology and ecology of *Agriolimax reticulatus* (Müll.) and other slugs: spatial distribution. *Journal of Animal Ecology* 34, 403–417.

Spaull, A.M. and Eldon, S. (1990) Is it possible to limit slug damage using choice of winter wheat cultivars? In: *Proceedings of the Brighton Crop Protection Conference – Pests and Diseases – 1990*, Vol. 2, pp. 703–708.

Stephenson, J.W. (1965) Slug parasites and predators. In: *Rothamsted Experimental Station Report for 1964*, Part 1, pp. 187–188.

Stephenson, J.W. (1975) Laboratory observations on the effects of soil consolidation on slug damage to winter wheat. *Plant Pathology* 24, 9–11.

Stephenson, J.W. and Bardner, R. (1977) Slugs in agriculture. In: *Rothamsted Experimental Station Report for 1976*, Part 2, pp. 169–187.

Strickland, A.H. (1965) Pest control and productivity in British agriculture. *Journal of the Royal Society of the Arts* 113, 62–81.

Symondson, W.O.C. (2002) Coleoptera (Carabidae, Staphylinidae, Lampyridae, Drilidae and Silphidae) as predators of terrestrial gastropods. In: Barker, G.M. (ed.) *Natural Enemies of Terrestrial Molluscs.* CAB International, Wallingford, UK.

Symondson, W.O.C., Glen, D.M., Wiltshire, C.W., Langdon, C.J. and Liddell, J.E. (1996) The effects of cultivation techniques and methods of straw disposal on predation by *Pterostichus melanarius* (Coleoptera: Carabidae) upon slugs (Gastropoda: Pulmonata). *Journal of Applied Ecology* 33, 741–753.

Symondson, W.O.C., Glen, D.M., Ives, A.R., Langdon, C.J. and Wiltshire, C.W.W. (2002) Dynamics of the relationship between a polyphagous predator and slugs over five years. *Ecology* (in press).

Thomas, M.R., Garthwaite, D.G. and Banham, A.R. (1997) *Arable Farm Crops in Great Britain 1996.* Pesticide Usage Survey Report 141, Ministry of Agriculture, Fisheries and Food, London, 97 pp.

Watkins, R.W., Mosson, H.J., Gurney, J.E., Cowan, D.P. and Edwards, J.P. (1996) Cinnamic acid derivatives: novel repellent seed dressings for the protection of wheat seeds against damage by the field slug, *Deroceras reticulatum. Crop Protection* 15, 77–83.

Wibberley, E.J. (1989) *Cereal Husbandry.* Farming Press, Ipswich, UK, 258 pp.

Wilson, M.J., Glen, D.M. and George, S.K. (1993) The rhabditid nematode *Phasmarhabditis hermaphrodita* as a potential biological control agent for slugs. *Biocontrol Science and Technology* 3, 503–511.

Wilson, M.J., Glen, D.M., George, S.K., Pearce, J.D. and Wiltshire, C.W. (1994a) Biological control of slugs in winter wheat using the rhabditid nematode *Phasmarhabditis hermaphrodita. Annals of Applied Biology* 125, 377–390.

Wilson, M.J., Glen, D.M., Wiltshire, C.W. and George, S.K. (1994b) Mini-plot field experiments using the rhabditid nematode *Phasmarhabditis hermaphrodita* for biocontrol of slugs. *Biocontrol Science and Technology* 4, 103–113.

Wilson, M.J., Hughes, L.A., Hamacher, G.M., Barahona, L.D. and Glen, D.M. (1996) Effects of soil incorporation on the efficacy of the rhabditid nematode, *Phasmarhabditis hermaphrodita*, as a biological control agent for slugs. *Annals of Applied Biology* 128, 117–126.

Wiltshire, C.W. and Glen, D.M. (1989) Effects of molluscicides on slugs and soil arthropods in winter wheat crops. In: Henderson, I.F. (ed.) *Slugs and Snails in World Agriculture.* Monograph No. 41, British Crop Protection Council, Thornton Heath, pp. 399–406.

Young, A.G. and Port, G.R. (1989) The effect of microclimate on slug activity in the field. In: Henderson, I.F. (ed.) *Slugs and Snails in World Agriculture.*

Monograph No. 41, British Crop Protection Council, Thornton Heath, pp. 263–269.

Young, A.G. and Port, G.R. (1991) The influence of soil moisture on the activity of *Deroceras reticulatum* (Müller). *Journal of Molluscan Studies* 57, 138–140.

Young, A.G., Port, G.R. and Green, D.I. (1991) Development of a forecast of slug activity: models to relate slug activity to meteorological conditions. *Crop Protection* 10, 413–415.

Young, A.G., Port, G.R. and Green, D.B. (1993) Development of a forecast of slug activity: validation of models to predict slug activity from meteorological conditions. *Crop Protection* 12, 232–236.

Young, A.G., Port, G.R., Craig, A.D., James, D.A. and Green, T. (1996) The use of refuge traps in assessing risk of slug damage: a comparison of trap material and bait. In: *Slug and Snail Pests in Agriculture.* British Crop Protection Council, Farnham, pp. 133–140.

13 Agriolimacidae and Arionidae as Pests in Conservation-tillage Soybean and Maize Cropping in North America

Ronald B. Hammond[1] and Robert A. Byers[2]

[1]*Department of Entomology, Ohio Agricultural Research and Development Center, Ohio State University, Wooster, OH 44691, USA;* [2]*USDA-ARS Pasture Systems and Watershed Management Research Unit, US Regional Pasture Research Laboratory Building, University Park, PA 16802, USA*

Growers have been adopting systems that increase the amount of crop residue left on the soil surface to provide benefits for themselves, their fields and the environment. In much of North America, especially the USA, these practices are becoming known as crop-residue management (CRM) systems (CTIC, 1995), and include both conservation tillage and reduced tillage. As defined by the Conservation Technology Information Center (CTIC, 1995), a leading proponent of these practices in the USA, conservation tillage is any cropping system that allows for 30% or more of the soil surface to remain covered by residues following planting. It includes: (i) no-till, where the soil is left undisturbed with planting accomplished in a narrow seed-bed or slot and weed control being achieved primarily with herbicides; (ii) ridge-till, where planting is into a ridged seed-bed with weed control achieved by use of herbicides and/or tillage; and (iii) mulch-till, where soil is disturbed prior to planting with implements that retain at least 30% soil cover with residues, with weed control achieved through herbicides and/or tillage. Reduced tillage is defined as any cropping system that retains 15–30% residue cover on the soil surface following crop planting. As a comparison, conventional tillage is a system where there is less than 15% cover after planting, and usually involves ploughing or other forms of intensive tillage.

Benefits of CRM for growers include reduced labour requirements, reduced machinery wear and savings in time and fuel. Benefits for their crop fields typically include higher soil moisture, reduced soil erosion, improved water infiltration, increased organic matter, decreased soil compaction and improved soil tilth. Finally, benefits for the environment may include improved water quality, more wildlife, reduced release of carbon gases and reduced air pollution.

CRM systems have been gaining acceptance by growers throughout North America. There has been a large increase in areas on which CRM principles were applied since the 1970s (CTIC, 1995). Currently, the areas of North America where no-till is most extensively practised are the USA midwestern states of Illinois (2.4 million ha), Indiana (1.8 million ha), Iowa (1.7 million ha) and Ohio (1.6 million ha). However, on percentage of total hectares in arable crops, the states of Kentucky (47%), Maryland (43%), Tennessee (42%), Delaware (39%) and Ohio (38%) are the most avid users of no-till. The major field crops in the USA planted in no-till hectares are soybean (*Glycine max* (Linnaeus) Merrill; Fabaceae) (6.5 million ha), maize (*Zea mays* Linnaeus; Gramineae) (5.3 million ha), small grains (730,000 ha) and cotton (*Gossypium hirsutum* Linnaeus; Malvaceae) (225,000 ha).

Early Concerns with Gastropods on Maize and Soybean in North America

Gastropods were first recognized as pests of maize in the 1950s, when crops planted into ploughed *lucerne* (*Medicago sativa* Linnaeus) (Fabaceae) sod or heavily manured arable fields were damaged by the introduced agriolimacid *Deroceras reticulatum* (Müller) (Neiswander, 1959). At that time, intensive tillage was the normal practice for seed-bed preparation and usually effected good gastropod control. Although the molluscicide metaldehyde was mentioned as a control option, its use in maize was considered unnecessary unless the gastropod population was unusually high. In a report on the evaluation of molluscicidal chemicals, Barry (1969) reported widespread economic damage by *D. reticulatum* in maize planted in conservation-tilled fields in Ohio during 1968. Rollo (1974) conducted studies on the ecology of various agriolimacid and arionid species in maize fields of Ontario, Canada, in the early 1970s and reported on sampling procedures from no-till maize (Rollo and Ellis, 1974). Gregory and Musick (1976) considered *D. reticulatum* and other gastropods the most serious non-insect problem encountered in conservation-tillage crops in Ohio and other north-eastern states of the USA. They recognized the demand by growers for effective gastropod control options, thus indicating that these animals were indeed significant constraints to conservation-tillage production of maize. Numerous reports of gastropod infestations on maize were reported in the early to mid-1980s from New York (Ramsey, 1984) and Ohio (Hammond and Stinner, 1987).

The association of pestiferous gastropods and soybean did not eventuate until conservation-tillage methods were used for the crop in the early 1980s (Hammond, 1985). Prior to this time, the lack of persistent herbicides, especially post-emergent herbicides, had limited soybean to intensive tillage systems (AAVIM, 1983). Hammond noted that *D. reticulatum* was the primary gastropod causing injury to soybean planted by no-till or reduced-till into maize residue in Ohio. A search of pertinent literature associated with soybean pests revealed no prior mention of gastropods. Hammond and Stinner (1987) later reported that the amount of surface residue influenced gastropod populations in both maize and soybean, with the largest numbers in no-till systems. They also found gastropods to be more prevalent when the previous crop was soybean and suggested that soybean residue might provide a highly favourable habitat. They concluded that, as more soybean was grown in conservation-tillage systems in rotation with maize, the incidence of gastropod pest infestations in both crops would increase.

During the 1980s, in recognition of the importance of gastropods as pests in Ohio, growers were able to obtain a special state registration (known as a 24C) for use of a molluscicidal bait containing methiocarb to control agriolimacids and arionids on maize. The bait was prepared by mixing methiocarb with cracked maize seed moistened with water, beer and molasses. Applied at 22.4 kg ha^{-1} (equivalent to 1.12 kg methiocarb ha^{-1}), the bait proved very effective against these pests (Calvin, 1988; R.B. Hammond and R.A. McCartney, unpublished data). At that time gastropods were just beginning to be recognized as pests on soybean, and submissions were made for extension of the methiocarb registration to soybean. Although such registration attempts failed in Ohio, some US states did grant a similar 24C registration for a grower-prepared, cracked-maize bait with thiodicarb (Larvin®) as the molluscicide active ingredient (Hammond, 1989b). Thiodicarb had already been registered for insect control on soybean, and efficacy trials in 1985 (R.B. Hammond, unpublished data) had showed good activity against agriolimacid and arionid species. Calvin (1988) and McCartney *et al.* (1990) also reported on good efficacy of thiodicarb baits, albeit the work was done in experimental maize plots. However, in the late 1980s the 24C registrations pertaining to both methiocarb and thiodicarb were cancelled by the respective companies, because apparent misuse of the products had led to bird kills.

The majority of gastropod problems in maize and soybean in North America are in the eastern half of the midwest, much of the north-east and the eastern shore region of the USA. Gastropod infestations have increased in both area and intensity since the review by Gregory and Musick (1976) and now occur west of Ohio. In regions where the problem is most intensive, growers consider gastropods a major impediment to profitable crop production, and growers are questioning their ability to continue the use of conservation-tillage practices (Willson and Eisley, 1992). To address these concerns, research on agriolimacid and arionid species in field crops has increased (e.g. Byers and Bierlein, 1984; Hammond and Stinner, 1987; Calvin, 1988; Byers *et al.*, 1989; Byers and

Calvin, 1994; Hammond *et al.*, 1996). Currently, there is limited published information available.

Gastropod Species in Maize and Soybean

Although most early publications mention only *D. reticulatum* as the damaging species, recent research has shown additional species to be present in maize and soybean crops. These additional species are the native agriolimacid *Deroceras laeve* (Müller) and the introduced arionids, *Arion fasciatus* (Nilsson) and *Arion subfuscus* (Draparnaud). Whether the absence of mention of these species in earlier publications accurately reflected their absence or their not having been correctly identified is unknown.

D. reticulatum, known in North America as the gray garden slug, has been confirmed as the most widespread and damaging species in maize and soybean in North America. *D. laeve*, the marsh slug, is rated second in population size and damage potential to maize and soybean (Byers and Calvin, 1994; R.B. Hammond, unpublished data). Much of the earlier literature associated with either crop does not mention this native and widespread species. An important consideration with these two agriolimacid species is that juveniles are quite similar in appearance and are often difficult to differentiate (Chichester and Getz, 1973). We believe that some earlier research reports undoubtedly involved both species, albeit that *D. reticulatum* would have been the more numerous.

A. fasciatus, known as the banded slug, is currently not as common or as damaging as either *Deroceras* species. It is considered to be more of a pest in the north-east and eastern portions of North America. *A. fasciatus* has been found only in a few fields in the eastern maize belt (portions of Ohio) and then only in very low numbers. This species is less likely to climb up on to the plant than other species and thus is more likely to feed at the soil surface and in open seed furrows (Goh *et al.*, 1988c).

A. subfuscus, the dusky slug, has been found in maize and soybean mostly in the eastern maize belt. It is not considered a major pest of field crops in the USA east of Ohio, being found infrequently (R.A. Byers, unpublished data). However, in Ohio *A. subfuscus* has been observed as the predominant species in some fields, with numbers in baited-beer traps often exceeding 50 animals caught within a 24 h period (R.B. Hammond, unpublished data). *A. subfuscus* often causes growers greater concern than other gastropod species, because of its larger size. Because references to its pest potential are infrequent and research on this species in field crops has been limited, the damage potential of *A. subfuscus* is unclear. However in Ohio, *A. subfuscus* appears capable of causing significant injury to young, germinating plants, especially when crops are planted in late spring and germination coincides with activity of the juveniles (R.B. Hammond, unpublished data). The limited ecological literature suggests that *A. subfuscus* is mostly a woodland species and fungivorous (Chichester and Getz, 1969, 1973).

Sampling Gastropods in Maize and Soybean

Practical sampling methods for agriolimacids and arionids are needed by growers and crop consultants. One of the few studies in maize in North America was that by Rollo and Ellis (1974), who described a mechanical process for sampling gastropods and their eggs from fields in Ontario, Canada; however, their method is more suited to research than for growers or consultants. Our experiences in the midwest USA indicate that sampling procedures are not widely adopted by the industry unless they are quick, easy and low-cost. Absolute sampling methods, such as hand-sorting, soil washing and flotation, and slow flooding of soil samples have had limited use in pest-management programmes because they are too costly in terms of time and effort.

The primary means of sampling for gastropods in Ohio in maize and soybean include searching underneath crop residue for the various life stages, use of shingle traps, with and without the use of a bait and/or an attractant such as beer, and *in situ* searching for gastropods on plants after crop seedling emergence. For pest-management purposes, Byers and Calvin (1994) have shown that assessing the degree of plant injury by gastropods offers a quick and practical monitoring method.

Currently in Ohio, we recommend that growers begin evaluating the potential for infestations prior to crop establishment, by searching under the crop residue for gastropods and their eggs (R.B. Hammond, unpublished data). However, Ferguson and Hanks (1990) found that searching the soil surface was not an accurate method in estimating gastropod abundance, with marked underestimation of small (< 20 mm) juvenile stages of species like *D. reticulatum*. In the cropping situation, this inability to detect the smaller individuals in a population and the eggs can lead growers and consultants to underestimate the damage potential of the resident gastropod population.

As a follow-up to sampling in the residues of the previous crop, we recommend the use of traps, with and without an attractant and/or bait, to monitor changes in gastropod populations during the early stages of crop development. There has been considerable research in the USA with these types of traps, albeit mostly for forage situations. Schrim and Byers (1980) compared six different refuge traps in sod-seeded legumes and found that an asphalt roofing shingle, covered with aluminium foil, placed on the soil surface was the best sampling trap; Hammond and Stinner (1987) used similar traps in reduced-tillage and no-till maize and soybean. These traps are considered extremely useful because their flat configuration makes for ease of storage, their low weight aids transportation into the fields and the aluminium foil provides for good visibility and thus ease of relocation in vegetated fields. Other researchers have used the shingle placed over a cylindrical 10 cm deep hole in the soil (Byers and Bierlein, 1984; Goh *et al.*, 1988a,b,c; Byers *et al.*, 1989). The addition of an attractant can increase numbers of trapped gastropods (e.g. Grant *et al.*, 1982), while a small quantity of molluscicide placed under the shingle can provide for killing of the gastropods and thus allows them to be counted (see Glen and Wiltshire, 1986). R.B. Hammond (unpublished data) examined numerous types of traps in no-till maize and soybean with and without different attractants and molluscicidal baits, and concluded that aluminium-covered roofing shingles over a cup of beer provided for the highest catches of agriolimacids and arionids. The addition of beer to the traps, as an attractant, only slightly increases the time required to place the traps in a field compared with placing shingles over a hole, as done by Byers and Bierlein (1984) and Byers *et al.* (1989). None of these traps provide an absolute estimate of the population size; they sample from an unknown area. However, they give growers and consultants an idea of the species that are present and their relative abundance, which is useful information when making pest-management decisions.

Once the crop has emerged, *in situ* sampling of gastropods on the plants also provides an estimate of the species and their abundance. Barry (1969) used this method when determining chemical efficacy, finding juvenile *D. reticulatum* numbers ranging from 0 to 30 per maize plant. R.B. Hammond (unpublished data) has used this method in both maize and soybean in numerous investigations, including population-dynamic studies and molluscicide-efficacy trials. The greatest benefit is that the numbers of gastropods can be expressed on an individual plant basis and thus provide a relative measure of pest burden. In maize, plants can be individually examined for gastropods, as the plant becomes the sampling unit. Our approach in soybean is to count the number of plants and number of gastropods active on them per unit area (for example, per 1 m^2) and then calculate the number of pests per individual soybean plant. No-till soybean is usually planted in narrow inter-row spaces (typically 18 cm between rows), and it is much easier to count gastropods per unit area than per individual plant. This *in situ* sampling requires the counts to be made after dusk, when the gastropods have emerged from underneath the residue and climbed on to the plants. This practice requires late evening or night visits to the fields, which might prove unacceptable.

Gastropod Damage and Thresholds

In any pest-management system, a primary goal is to maintain the pest population below an economic level and to apply a control, such as a pesticide, only when needed to achieve that goal and when benefits exceed costs. We are just now beginning to understand the damage potential of agriolimacids and arionids in maize and soybean. However, little information is currently available on the relationship between the numbers of these gastropods and yield-loss potential. Both maize and soybean can be heavily defoliated by gastropods, which can lead to significant yield reductions. Gastropods can cause reductions in plant populations in both crops, by feeding on germinating seed. Damage to seed is most severe when the seed is inadequately covered with soil within drill furrows. Most of the damage is from juvenile *D. reticulatum*, with populations on maize reported as high as ten to 30 per plant (Barry, 1969) and averaging ten juveniles per plant (R.B. Hammond, unpublished data). Eggs that have overwintered or have been laid by overwintered adults produce the juveniles that cause the damage. The stage that overwinters appears to vary according to the location and environmental conditions. While Rollo (1974) observed eggs to be the overwintering stage in Ontario, Canada, we have seen both adults and eggs overwintering in areas further south (R.B. Hammond, unpublished data).

Gastropod injury to seedling maize in the spring is different from that of other early-season pests, such as cutworms and stalkborers, in that it rarely causes rapid mortality of the seedlings (Byers and Calvin, 1994). Maize damaged by gastropods is characterized by tissue loss between the veins of the leaves, but with the lower epidermis intact. The extent of the defoliation depends on the population size of the gastropods, their feeding activity and the rate of growth of the maize. In severe infestation, the maize seedlings can be completely defoliated (Plates 14 and 15). These plants often survive, but contribute little to crop grain yield.

Byers and Calvin (1994) are the only researchers to date who have investigated the economic impact of gastropods on maize in North America. They estimated feeding injury caused by different infestation levels of *D. reticulatum* and then related this to individual maize-plant yield losses. Their data indicated that the crop was most vulnerable/sensitive to injury 3–4 weeks after planting, when seedlings were at the three- to four-leaf stage. Economic injury levels were established that ranged from 2 to 20% defoliation in a warm, wet season and from 39 to 59% in seasons characterized as either cool and dry or warm and dry.

Because Byers and Calvin (1994) were interested in sublethal effects of *D. reticulatum* feeding, they addressed infestations that produced defoliation, not injury that kills the plants and thus causes reductions in plant populations (although they had to reduce the level of infestations after the first year of the study because plant mortality did occur). Data on gastropod-induced stand reduction are currently not available. When plant numbers are severely reduced, growers must determine whether maize can be replanted or if another crop must be planted. It has been our experience that these fields often require molluscicide treatment before replanting maize (or any other susceptible crop) because the gastropod infestation causing the failure of the first crop remains and, indeed, often increases through growth of individuals.

Gastropod injury to soybeans can also be severe, resulting in both plant-population loss (Plates 16 and 17) and reduced plant growth. In the midwest USA, soybean is often planted late in the spring relative to maize, and seed germination and crop emergence are more coincident with the period of most intensive activity of *D. reticulatum* and other gastropods. Gastropods will feed on germinating seeds and the stems and cotyledons of young seedlings while beneath the residue of the previous crop, resulting in plant death and stand reductions. This type of injury occurs soon after planting, before many growers realize that gastropod infestations are present. Stand reductions at this stage of crop development can be significant, resulting in large areas of total loss, but often the damage is detected too late to take effective action.

Gastropods will also defoliate soybeans, often commencing with the unifoliate leaves and then progressing to the trifoliate leaves as the plant grows. This defoliation is often similar in appearance to that of early-soybean insect pests. We have also observed that gastropods will feed on the tip of the leaves before it opens and expands, resulting in the distal half of the leaf being missing; little further defoliation to the remaining leaf is seen. In severe infestations, plants can be completely defoliated, resulting in plant deaths and stand reductions.

Most of the information on early-season injury and soybean yield–loss relationships comes from simulated defoliation studies (Hammond, 1989a; Hunt *et al.*, 1994). No yield-loss models based on actual or simulated gastropod injury to soybean have been developed. Hammond (1989a) simulated insect injury by damaging the unifoliate leaves at various intensities on a single day and found no consistent impact of defoliation on yields. However, Hunt *et al.* (1994) found that simulating bean-leaf beetle (*Cerotoma trifurcata* Forster; Chrysomelidae) injury to seedling soybean during stages VC (cotyledons present) to V3 (two trifoliates present) reduced yields by up to 12% at a defoliation level of 68%. Yield losses were mainly attributed to the delay in plant canopy reaching the critical leaf-area index (when 90% of the light is intercepted). Based on the data of Hunt *et al.* (1994), seedling defoliation by gastropods may be expected to reduce plant fitness by delaying canopy development and plant height.

While the aforementioned simulated pest studies relate to soybean grown in wide-row (usually 76 cm) planting situations, most gastropod injury is to soybean grown in narrow rows (18 cm), which is the more common practice with no-till soybean. The between- and within-row plant-spacing conditions would undoubtedly influence the competitive interaction among plants and thus alter the response to defoliation. Published agronomic recommendations clearly indicate the plant populations required to maintain yield and thus provide useful guides to expected yield losses resulting from pest-induced stand reductions. However, because pest-induced stand losses are often not uniform throughout a field, the grower must determine the areas that require replanting. Furthermore, this agronomic literature does not provide for estimation of yield losses due to reduced plant vigour or later defoliation. Currently, there is no information available on the relationship between early-season defoliation, stand loss and the use of narrow row spacings.

Gastropod Management

Cultural control

Experience has shown that in both maize and soybean the most effective gastropod-management approach is that of prevention, rather than combating infestations after some level of crop injury has occurred. Tillage has been the primary means of managing Agriolimacidae and Arionidae in field crops (Goh *et al.*, 1988a,b,c). Removing the crop residue or ground cover is the first line of defence because it is the presence of the residue that increases the likelihood of gastropod infestations. The problem with this approach is that it is contrary to the philosophies of conservation tillage. Nevertheless, where erosion is less problematic, light to moderate tillage is recommended as a strategy for combating gastropod infestations that persist from one year to the next and where other control strategies have not proved cost-effective.

Other cultural practices have been suggested for gastropod management. The use of equipment in front of the planter ('row cleaners') that removes residue from the seed-row zone is viewed by growers as effective in reducing gastropod injury to the crops. Row cleaners have been shown to enhance germination and provide for quicker, more vigorous plant growth in the spring, because of drier and slightly warmer soils. However, information available at this time is ambiguous as to the value of row cleaners as a strategy for reducing gastropod infestations on plants. We have seen numerous fields where row cleaners in maize did not prevent gastropod injury, which was often severe (R.B. Hammond, unpublished data). Although row cleaners are gaining use in maize production, their utilization is limited in soybean because of the narrow row spacing used in production of this crop.

Rapid plant growth allows the crop to pass through the seedling stage quickly and avoid prolonged exposure to injurious gastropods (G. Dively, unpublished data). Vigorously growing plants with high leaf area are well able to tolerate injury. Thus, any practice that allows for quicker germination and growth or the presence of greater leaf area will theoretically reduce the significance of gastropods as pests. Along with row cleaners in maize, the application of a 'starter' fertilizer with the seed at planting has been suggested as a strategy for rapidly advancing the crop through the vulnerable seedling stages. In areas where gastropod infestations occur primarily in late spring, earlier planting dates may be adopted to allow sufficient time for the plants to produce adequate leaf area to withstand gastropod injury. Late-season plantings are suggested for areas in which

Rollo, C.D. and Ellis, C.R. (1974) Sampling methods for the slugs, *Deroceras reticulatum* (Müller), *D. laeve* (Müller), and *Arion fasciatus* Nilsson in Ontario corn fields. *Proceedings of the Entomological Society of Ontario* 105, 89–95.

Schrim, M. and Byers, R.A. (1980) A method for sampling three slug species attacking sod-seeded legumes. *Melsheimer Entomological Series* 29, 9–11.

Willson, H.R. and Eisley, B.R. (1992) Effects of tillage and prior crop on the incidence of five key pests on Ohio corn. *Journal of Economic Entomology* 85, 853–859.

14 *Bradybaena ravida* (Benson) (Bradybaenidae) in Cereal–Cotton Rotations of Jingyang County, Shaanxi Province, China

Chen De-niu,[1] Zhang Guo-qing,[1] Xu Wenxian,[2] Wang Man-sheng,[3] Liu Yanhong,[2] Cheng Xingmin[3] and Wu Jiqing[3]

[1]*Institute of Zoology, Academia Sinica, Beijing 100080, China;* [2]*Shaanxi Plant Protection Station-in-Chief, Xian, Shaanxi Province, China;* [3]*Shaanxi Plant Protection Institute, Yanglin, Shaanxi Province, China*

Emergence of *Bradybaena ravida* as a Pest in Jingyang County

The terrestrial gastropod *Bradybaena ravida* (Benson) (Bradybaenidae) is widely distributed in China as an indigenous species, having been reported from the majority of the eastern provinces (De-niu *et al.*, 1995; Fig. 14.1). Outside China the species is found in the coastal and Amur River basin districts of south-eastern Russia (Likharev and Rammel'meier, 1952; Schileyko, 1978). The natural habitat of *B. ravida* is primarily broad-leaved and mixed broad-leaved-conifer forests (Likharev and Rammel'meier, 1952). In 1986 *B. ravida* was considered a pest in China only along certain sections of the Yangtze River valley. At that time a minor infestation was reported on 3 ha of irrigated land in Jingyang County, Shaanxi Province. Over the following 3 years the area of infestation in Jingyang County expanded to 6, 60 and 600 ha, respectively (Yanhong and Wenxian, 1990), and by 1998 some 7000 ha of farmland was severely threatened. The annual precipitation in Jingyang County is typically modest, at around 500 mm. The emergence of *B. ravida* as a pest coincided with 5–6 years of above-average rainfall and followed changes in agricultural practices, which placed heavy reliance on wheat (*Triticum aestivum* Linnaeus; Gramineae) and maize (*Zea mays* Linnaeus; Gramineae) in rotation with minimal soil cultivation and extensive use of

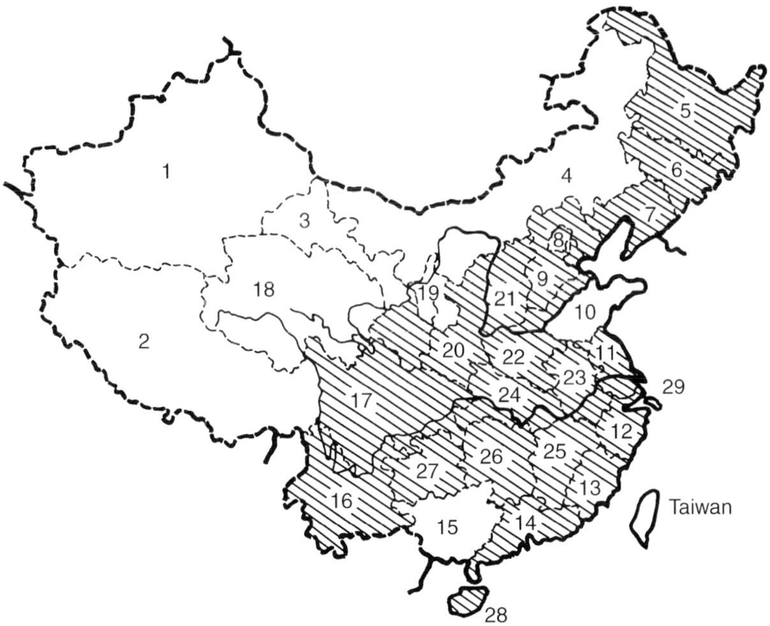

Fig. 14.1. Map of China, showing natural distribution of *Bradybaena ravida* (Benson) (Bradybaenidae) (shaded). Provincial names: 1. Xinjiang Uygur Zizhiqu; 2. Xizang Zizhiqu; 3. Gansu; 4. Nei Mongol Zizhiqu; 5. Heilongjiang; 6. Jilin; 7. Liaoning; 8. Beijing; 9. Hebei; 10. Shandong; 11. Jiangsu; 12. Zhejiang; 13. Fujian; 14. Guangdong; 15. Guangxi; 16. Yunnan; 17. Sichuan; 18. Qinghai; 19. Ningxia Huizu Zizhiqu; 20. Shaanxi; 21. Shanxi; 22. Henan; 23. Anhui; 24. Hubei; 25. Jiangxi; 26. Hunan; 27. Guizhou; 28. Hainan; 29. Shanghai. *B. ravida* is recognized as an agricultural pest primarily in Shaanxi, Hubei and Jiangsu provinces.

pesticides and fertilizers. Due to the decline in cotton (*Gossypium hirsutum* Linnaeus; Malvaceae), increased areas have been devoted to arable production, where maize is sown by transplanting seedlings before the early summer harvest of the wheat and wheat is sown by minimum tillage immediately after the autumn harvest of maize. The absence of soil tillage and, in the areas of Jingyang County most adversely affected by *B. ravida*, the irrigation of crops may have contributed to the emergence of this gastropod as a serious pest.

In a survey of 16 counties in Shaanxi Province (De-niu *et al.*, 1995), 7797 samples were collected to determine the community assemblage of gastropods. From these collections 68 species or subspecies pertaining to 31 genera and 15 families were identified, ten of which were species new to science. The dominant species, often locally abundant, were *B. ravida*, *Bradybaena kiangsiensis* (von Martens), *Cathaica fasciola* (Draparnaud), *Cathaica pulveraticula* (von Martens) (Bradybaenidae), *Opeas striatissimum* (Gredler) (Subulinidae), *Deroceras agreste* (Linnaeus) (Agriolimacidae) and *Meghimatium bilineatum* (Benson) (Philomycidae). Many of these species assume pest status in crops in the central Chinese provinces (De-niu, 1996). Despite widespread occurrence in Shaanxi Province, *B. ravida* was found to be of pest status primarily in Jingyang County, and even there high abundance and extensive damage were not uniformly distributed but were confined to areas of evidently favourable climatic and land-use conditions.

At present in China *B. ravida* is a recognized pest primarily in the Hanzhong Zone in the southern sector of Shaanxi Province, the Jinjiang Plain sector of Hubei Province and certain sectors of the Jiangsu Province (Fig. 14.1).

Throughout Shaanxi Province there has been an increasing awareness of the importance of *B. ravida* as a pest, and intensive field research has been initiated to develop strategies for control. This species is known locally as the 'grey *Bradybaena* snail' or 'huiba snail'.

Biology

B. ravida has one generation a year in central Shaanxi. Hibernation takes place in November, when the temperature is lower than 10°C and the relative humidity is lower than 76%. The adult and juvenile stages overwinter in soil crevices, 20–40 mm deep in the loose soil of tilled ridges and among crop residues. Very few eggs are present during winter. The animals emerge from hibernation and resume feeding in March, when the ambient temperature rises above 8°C and when moisture becomes plentiful. Feeding activity is maximal from April till June, and copulation and oviposition take place during May. The mating process typically takes 5–6 h, but the longest recorded is 16 h. Oviposition takes place 5–7 days after mating, with clutches containing 20–300 eggs deposited 20–40 mm deep in the soil. Observations under controlled conditions indicate that each individual is capable of producing several clutches of eggs as a result of a single mating event (Table 14.1), with total fecundity

Table 14.1. Observations on the reproductive activity of *Bradybaena ravida* (Benson) (Bradybaenidae) collected from farmland in Jingyang Country, Shaanxi Province, and maintained under controlled laboratory conditions. Following their first mating, the *B. ravida* were maintained individually for their entire reproductive season and various fecundity parameters recorded. Br1 and Br2 represent individuals from a mating pair of *B. ravida*.

Date at which mating occurred	Days between mating and first oviposition		Days between first and last oviposition		Frequency of oviposition (number of clutches)		Maximum number of eggs in a clutch		Total number of eggs produced	
	Br1	Br2	Br1	Br2	Br1	Br2	Br1	Br2	Br1	Br2
28 March	20	40	19	12	3	5	27	53	61	140
28 March	20	30	0	28	1	8	24	68	24	371
28 March	33	36	21	22	5	6	62	58	238	260
5 April	30	39	23	20	5	6	101	69	187	336
27 April	4	4	29	37	7	7	122	143	180	212
10 May	5	8	52	49	7	7	140	85	638	346
Mean	22.4		26.0		5.8		79.3		249.4	

within a single season of up to 638 eggs. The spherical eggs, 1.5 mm in diameter and with a calcareous shell, are milky white in colour when oviposited but gradually attain a light yellow colour. The incubation period of *B. ravida* in Shaanxi is 14–20 days.

Aestivation typically occurs through late June to mid-August, when the mean temperature is around 25°C and the relative humidity is lower than 65%. *B. ravida* usually passes the summer in aestivation in loose soil or crop stubble. A second peak of feeding and reproductive activity occurs from mid-August through to late September, when the temperature declines to 20–22°C. During the periods of hibernation and aestivation, *B. ravida* seal the shell aperture with an epiphragm. Observations indicate that *B. ravida* takes 1 full year to reach reproductive maturity, and the longevity of the adults is in the order of 1.5 to 2 years. Individuals thus potentially contribute to two or more periods of population recruitment during their lifespan. There is great variation in the size of shells, but adult animals generally attain a shell size of 19 mm diameter.

Observations under controlled conditions indicate that *B. ravida* activity is confined within the temperature extremes of 10 and 35°C, with an optimum in the range 16–22°C. In respect of high temperatures, experiments under controlled conditions indicated that animals had a survival rate of 73% following a 1-day exposure to 44°C, but survival was nil at temperatures of 46°C or higher. Observations of animals under winter conditions in Shaanxi (Table 14.2) demonstrated that *B. ravida* populations are well able to persist where seasonal minimum temperatures are as low as –29°C. Experiments none the less demonstrate that the ability of *B. ravida* to withstand desiccation is weak: no activity could be observed when the relative humidity was below 68%. Field observations indicated that *B. ravida* has difficulty in ovipositing and the survival of eggs is low when the soil water content is below 15%. Laboratory studies (De-niu *et al.*, 1996a) confirm the importance of soil moisture, with a soil moisture content of *c.* 20% needed for oviposition activity and embryonic development.

The population dynamics of *B. ravida* has been extensively studied over the period from 1989 to 1991 in farmland of the Xichencun and Zhoujiadao villages, where severe infestations occurred (De-niu *et al.*, 1996a). In arable fields in which wheat and maize were grown in rotation,

Table 14.2. Overwintering mortality in *Bradybaena ravida* (Benson) (Bradybaenidae) in farmland at three localities in Shaanxi Province, China. Mortality was recorded for samples of *B. ravida* marked in November 1991 and recovered from field sites in February 1992.

Localities	Minimum temperature during winter (°C)	Sample size	Per cent mortality
Tabai County	– 29.0°C	200	67
Jingyang County	– 18.5°C	286	5
Yangling County	– 16.5°C	70	20

B. ravida abundance was maximal in the spring in wheat and maximal in late summer and autumn in maize, albeit with considerable variation in abundance from one year to the next (Fig. 14.2). Eggs are to be found throughout the growing season in arable fields, but egg abundance tends to be bimodal, with peaks in spring and summer. In cotton (Fig. 14.3), *B. ravida* abundance is generally lower than in arable fields, with numbers low after the winter hibernation period but tending to increase over the growing season of the crop. When wheat is harvested, *B. ravida*

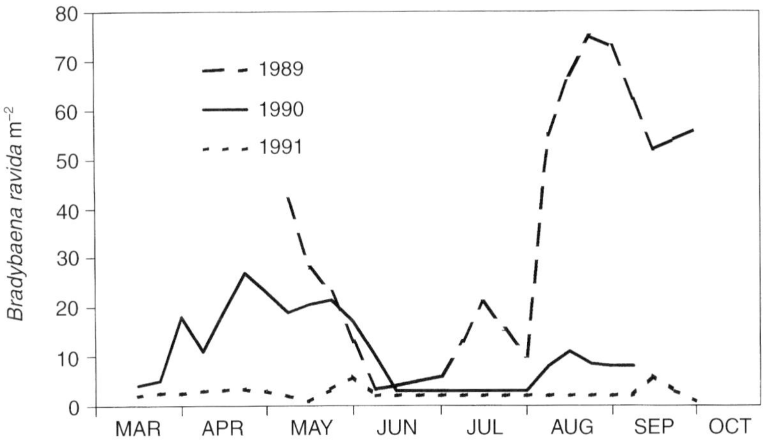

Fig. 14.2. Abundance of *Bradybaena ravida* (Benson) (Bradybaenidae) in an arable field in Xichencun Village, Jingyang County, Shaanxi Province, China, over the period May 1989 to September 1991 (from De-niu *et al.*, 1996).

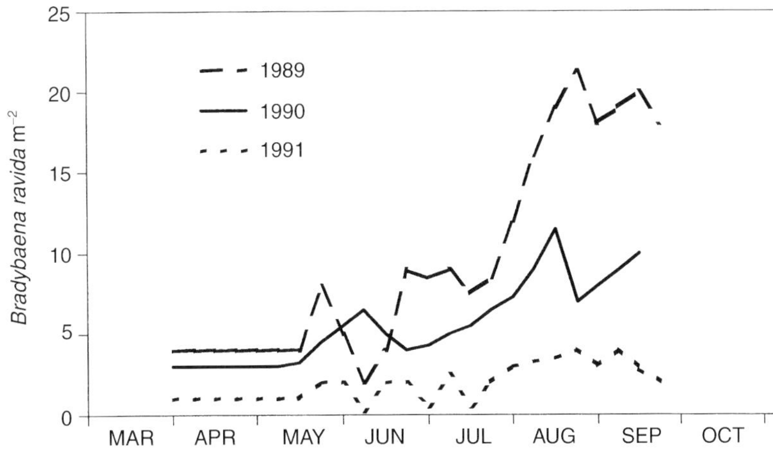

Fig. 14.3. Abundance of *Bradybaena ravida* (Benson) (Bradybaenidae) in a cotton (*Gossypium hirsutum* Linnaeus) (Malvaceae) field in Xichencun Village, Jingyang County, Shaanxi Province, China, over the period May 1989 to September 1991. Note the low abundance in spring and the general increase in numbers over the growing season. (From De-niu *et al.*, 1996.)

Furthermore, *B. ravida* eggs are rapidly destroyed in direct sunlight and therefore tillage practices that expose eggs to the soil surface will contribute to control. Studies by Jiqing and Wenxian (1992) demonstrated that disturbance of the upper 20–40 mm soil layers by harrowing and raking effected egg mortalities as high as 50%.

On the basis of these results, De-niu *et al.* (1996b) recommended that ploughing, with burial of *B. ravida* to a depth of 200 mm, be adopted as a control strategy, with tillage imposed in autumn for fallow fields and in spring ahead of planting the maize. This should be supplemented with shallow inter-row tillage during the period of *B. ravida* oviposition.

In established crops, manual collection and removal of *B. ravida* provide a useful adjunct to cultural and chemical control methods. De-niu *et al.* (1996b) report one example where 1552 kg of *B. ravida* were collected from 25.3 ha of wheat over a 3-day period in May, resulting in a mean reduction in infestation levels from 29.4 m^{-2} to 16.2 m^{-2}.

Ammonium bicarbonate had for many years been the principal nitrogenous fertilizer used on farmland in Jingyang County. Application of ammonium bicarbonate also had the advantage of being molluscicidal. However, the use of ammonium bicarbonate has declined with the advent of urea and diammonium phosphate fertilizers, and field observations indicate that these fertilizers have little impact on *B. ravida* numbers. In farm demonstration plots, where urea was applied at the rate of 375 kg ha^{-1} during autumn sowing of wheat, Jiqing and Wenxian (1992) found that *B. ravida* numbers had declined by only 23% when sampled in the following spring. This contrasts with a 94% decline in *B. ravida* numbers over the same period in plots that had received 750 kg ha^{-1} of ammonium bicarbonate. These contrasting efficacies were confirmed under controlled experimental conditions (De-niu *et al.*, 1996a). The molluscicidal activity of ammonium bicarbonate was shown to be maximal if the material was applied in the evening or early morning, when it came in direct contact with active *B. ravida*. Application made around noon effected low rates of mortality because the *B. ravida* were inactive and were afforded protection by virtue of being retracted into their shell. Experimental applications of ammonium chloride at 75 to 450 kg ha^{-1} resulted in 72–78% mortality in *B. ravida*, indicating the potential of this fertilizer as part of an integrated control programme.

Many organic pesticides have been screened for activity against *B. ravida*, including metaldehyde, phoxim, carbofuran, 1605, monocrotophos, lindane, carbaryl, 2,2-dichlorovinyldimethylphosphate (DDVP), trichlorphon, NRDC-161, acephate, coumachlor and temephos. Only metaldehyde exhibited significant molluscicidal activity. Jiqing and Wenxian (1992) and De-niu *et al.* (1996b) demonstrated that inexpensive and effective baits, with metaldehyde as the active ingredient, may be formulated from organic materials readily available to farmers, such as maize flour, cracked maize seed, wheat bran, cottonseed cake and fresh grass. Bait comprising moistened maize flour and 4% metaldehyde, applied at 21 kg ha^{-1} as small hills in the infested fields, is recommended for *B. ravida* control. Results from evaluations in field plots indicate that 90% reductions in *B. ravida* numbers can be achieved with this bait formulation. Experimental work indicates that metaldehyde mixed with soil at 3% by weight also affords good control of *B. ravida*.

Demonstration of integrated pest management

The integration of control strategies for management of *B. ravida* has been evaluated on a demonstration farm in Zhongzhang. Cotton and arable fields (wheat–maize cropping) under conventional farm practice were contrasted with those in which *B. ravida* populations were managed by deep cultivation, application of ammonium carbonate, application of metaldehyde, hand-collection of gastropods and hoeing or raking of the soil to expose the eggs. Both in cotton and in the arable fields, a substantial reduction in *B. ravida* infestations was achieved in a single year by the integrated approach (Table 14.4).

Conclusions

B. ravida has assumed great significance as a crop pest in Jingyang County, Shaanxi Province, over the past 15 years. It is apparent that *B. ravida* has the capacity for rapid population increase and spread under favourable farmland conditions and can inflict severe damage on most of the principal agricultural crop species. Natural enemies are apparently not effective in regulating populations of this pest. De-niu *et al.* (1996a) suggested that the disturbed ecological conditions on farmland, due to a long history of pesticide and fertilizer application, combined with adoption of minimum tillage and intercropping practices, irrigation and

Table 14.4. Demonstration of the value of an integrated approach to management of *Bradybaena ravida* (Benson) (Bradybaenidae) infestations on farmland at Zhongzhang, Jingyang County, Shaanxi Province, China. Changes in *B. ravida* numbers in cotton (*Gossypium hirsutum* Linnaeus) (Malvaceae) and arable fields under conventional farming practices were compared with those in fields under an integrated pest management regime. *B. ravida* numbers were estimated in each of five replicate plots at the commencement (September 1990) and at the end of the demonstration period (September 1991).

	Conventional farming			Farming with integrated pest management		
	Mean numbers of *B. ravida* m^{-2}		Per cent change in *B. ravida* numbers	Mean numbers of *B. ravida* m^{-2}		Per cent change in *B. ravida* numbers
	1990	1991		1990	1991	
Cotton fields	37.4	34.0	− 9.1	49.8	5.8	− 88.4
Arable fields	29.2	23.8	− 18.5	41.8	10.0	− 76.0

changes in weather patterns may all have contributed to the emergence of *B. ravida* as a pest.

Since its emergence as an agricultural pest, the biology of *B. ravida* has been extensively studied in Jingyang County and a range of control options have been successfully developed. An integrated approach, combining various cultural and chemical methods, is essential for management of *B. ravida*, and the demonstration on farms of the utility of the control options has been important in the adoption of effective control strategies by the farmers of Jingyang County.

References

De-niu, C. (1996) Investigation on terrestrial Mollusca from Wuling Mountain and its neighbourhood, China. *The Papustyla* 10(6), 1–12.

De-niu, C., Wenxian, X., and Yanhong, L. (1995) Zoogeographical analysis of the land molluscs from Shaanxi Province, China, with description of two new species (Gastropoda: Pulmonata: Stylommatophora). *Acta Zootaxonomica Sinica* 20, 398–410 [in Chinese].

De-niu, C., Zhang, G.Q., Wenxian, X., Yanhong, L., Man-sheng, W. and Yonggan, Y. (1996a) A study of outbreaks of *Bradybaena ravida* in China. In: Henderson, I.F. (ed.) *Slug and Snail Pests in Agriculture*. Symposium Proceedings No. 66, British Crop Protection Council, Farnham, pp. 405–410.

De-niu, C., Guo-qing, Z., Wenxian, X., Yanhong, L., Xingmin, C. and Jiqing, W. (1996b) A study of complementary techniques for snail control. In: Henderson, I.F. (ed.) *Slug and Snail Pests in Agriculture*. Symposium Proceedings No. 66, British Crop Protection Council, Farnham, pp. 425–432.

Jiqing, W. and Xinming, C. (1992) Comprehensive control of the grey *Bradybaena ravida*. *Shaanxi Journal of Agriculture* 3, 29–30 [in Chinese].

Likharev, I.M. and Rammel'meier, E.S. (1952) Nazemnye mollyuski fauny SSSR. *Akademiya Nauk SSSR Zoologicheskii Institut* 43, 1–511 [English translation from the Russian: Israel Program for Scientific Translations, Jerusalem, 1962, 574 pp.].

Schileyko, A.A. (1978) Land molluscs of the superfamily Helicoidea. *Fauna SSSR, Molljuski, III* 117, 1–384 [in Russian].

Yanhong, L. and Wenxian, X. (1990) Severe snail attack on crops in Shaanxi Province, China. *Plant Protection* 4, 55 [in Chinese].

15 Agriolimacidae and Arionidae as Pests in Lucerne and Other Legumes in Forage Systems of North-eastern North America

ROBERT A. BYERS

USDA-ARS, Pasture Systems and Watershed Management Research Unit, University Park, PA 16802, USA

Ecology of Forage-legume Systems

Forages are herbaceous plant species grown for mechanical harvest to be fed fresh to livestock or ensiled or dried and later fed to these animals. Most forage species are grasses and legumes, grown either separately or as mixtures. Legumes are highly valued components of forage systems, because of their protein content and thus high nutritional value relative to grasses, and they contribute to the nitrogen economy of the soil through fixation of atmospheric nitrogen. About 85% of N_2 fixation in agricultural soils derives from pulse and forage legumes (Vance *et al.*, 1988).

A large portion of the agricultural land in the north-eastern region of North America has been devoted to forage production since colonial times. Indeed, most present land devoted to forage production in the north-eastern region was forest before European settlement in the 17th, 18th and 19th centuries. By the first agricultural census in 1839, the north-east region was the main hay-producing area in the USA (Washko, 1974). The dairy industry, which relies heavily on forages, developed in the north-east because of its close proximity to heavily populated eastern cities: fluid milk cannot be transported long distances cheaply and must be produced close to the market (Garber *et al.*, 1946). The legumes most commonly grown in the north-east forage systems include lucerne (*Medicago sativa* Linnaeus), red clover (*Trifolium pratense* Linnaeus), bird's-foot trefoil (*Lotus corniculatus* Linnaeus) and white clover (*Trifolium repens* Linnaeus) (all Fabaceae). All four are species introduced from Europe.

Lucerne was brought to the colonies in 1736 but failed to survive except on well-drained calcareous soils. Between 1858 and 1910 three winter-hardy germplasm stocks from Europe and Russia were brought to the upper midwestern USA and eastern Canada. Two intermediate winter-hardy introductions came from Asia Minor between 1898 and 1925 and one other from France in 1947 (Barnes *et al.*, 1988). About 1 million ha of lucerne is currently grown in the north-east USA (Barnes and Sheaffer, 1995).

The spread of clovers was rapid in colonial times. Red clover was grown as early as 1663 in the eastern USA. White clover was introduced into the Ohio River Valley and sown with Kentucky bluegrass (*Poa pratensis* Linnaeus; Gramineae) by French fur traders and missionaries, prior to settlement of the area by the English. With a few exceptions in western North America, the clover species native to the Americas have never contributed much to agriculture (Taylor, 1985).

Bird's-foot trefoil is native to Europe, North Africa and parts of Asia. It is not certain how bird's-foot trefoil was introduced into the USA. A naturalized stand was discovered growing in Albany County, New York, in 1934. This ecotype was named Empire and provided the stock from which other cultivars were developed, characterized by finned stems and a prostrate growth habit and flowering later in the growing season and more winter-hardy than European types. European-type cultivars, such as 'Cascade', 'Viking' and 'Leo', were subsequently developed from European introductions. Extensive areas of bird's-foot trefoil now occur in eastern Canada (200,000 ha) and midwestern and north-eastern regions of the USA (1 million ha) (Beuselinck and Grant, 1995).

Forage legumes in humid regions are often short-lived and periodic resowing is usually necessary to maintain them as high-yielding monocultures or as productive components of mixed swards with forage grasses. Because forage crops are in fields that tend to occur on steep slopes and have shallow erosive soils and/or poor drainage, conservation tillage is generally preferred over conventional-tillage methods for establishment of new or renovating old forage crops. Conservation tillage as a forage establishment method was developed using herbicides to suppress or eliminate competition from the resident plant community (usually grass-dominant) in order to establish desirable legume and grass cultivars. In most cases this conservation tillage takes the form of direct seeding into the sod (turf) without cultivation and thus with minimal disturbance of the soil. As such, this form of conservation tillage is often referred to as no-till. Morris (1969) suggested that the insulating layer of vegetation in grassland produces a more stable microclimate than does open ground and consequently populations of invertebrates, including gastropods, are generally higher in these systems. This insulating layer is retained in conservation tillage systems, albeit comprising mostly of herbicide-killed plant material, and experience globally indicates persistence and activity in gastropod populations is favoured by this crop-establishment method. Furthermore, the slot or groove in the sod made by conservation-tillage seed-drills provides an ideal microhabitat for gastropods because of the high humidity and low light conditions, while the seedlings arising from the sown seed provide a readily available food source. Gastropods and other herbivorous invertebrates resident in fields thus often become pests in legumes established by conservation-tillage methods. Gastropods tend to preferentially feed on, and thus greatly reduce the survival of, legume seedlings growing in grass swards. This can lead to encroachment of weeds and grasses, less preferred by gastropods, into the spaces left open after the legumes are consumed, and thus result in a suboptimal content of legumes in the forage. In cases of severe infestation, the extent of seedling losses may require further sowing of legumes in an effort to achieve a desirable representation of legumes in the forage. Furthermore, gastropods and other pests can contribute to the decline of legumes in established forage swards through their selective herbivory.

Establishment of Legumes by Conservation-tillage Systems and the Emergence of Gastropods as Forage Pests

Legumes such as red clover, lucerne and bird's-foot trefoil were the species preferred for introduction into swards for renovation of fields that had become grass-dominant (e.g. Decker *et al.*, 1969; Taylor *et al.*, 1969). However, establishment of the legumes in these grass-dominant sods often failed because of pests feeding on emerging seedlings. Faix (1980) found seedling losses, attributable to feeding by Agriolimacidae, to be greater with lucerne and red-clover seedlings than with bird's-foot trefoil when sown in *Festuca* Linnaeus sod in Illinois. Grant *et al.* (1982) showed that the agriolimacid *Deroceras reticulatum* (Müller), as well as crickets (*Allonemobius* spp.; Gryllidae), was a major defoliator of lucerne seedlings following seeding by conservation tillage (sod-seeding) in Kentucky bluegrass-dominated sod in Kentucky. In their field experiments, lucerne seedling growth was significantly reduced and the rate of seedling loss due to pests was significantly greater in conservation-tillage plots as compared with conventional-tillage plots.

In some cases, legume establishment in sods was enhanced using pesticides, further validating the importance of gastropods as pests. Dowling and Linscott (1983) found that forage establishment by conservation-tillage in New York was unreliable, with reduced, uneven and patchy emergence of legumes. Many seedlings showed signs of poor vigour and varying degrees of defoliation. Based on responses to treatments with the pesticide methiocarb, it was concluded that feeding by gastropods was probably the main factor that limited establishment. The symptoms observed were more apparent on lucerne than on bird's-foot trefoil and red clover. In a subsequent field experiment, Dowling and Linscott (1985) seeded lucerne during late summer into four grass-dominant sod types, namely hay sod (species not mentioned), annual weeds (species not mentioned), couch grass (*Agropyron repens* (Linnaeus) de Beauvois; Gramineae) sod and rye (*Secale cereale* Linnaeus; Gramineae) stubble. Methiocarb treatment reduced damage from gastropods and increased survival of legume seedlings at most sites. *D. reticulatum* was found in all fields and *Deroceras laeve* (Müller) in low abundance at several sites, while *Arion fasciatus* Nilsson (Arionidae) was largely confined to the rye stubble site.

Byers *et al.* (1985) used conservation tillage to sow three legumes – red clover, lucerne and bird's-foot trefoil – in eight grass fields in Pennsylvania, three in early spring, two in late spring and three in late summer. The grasses were predominantly cocksfoot grass (*Dactylis glomerata* Linnaeus) (Gramineae) and Kentucky bluegrass, with one site smooth brome (*Bromus inermis* Linnaeus) (Gramineae). Methiocarb bait alone or in combination with carbofuran granules or spray applied at sowing provided some reductions in populations of the gastropods *D. reticulatum*, *D. laeve* and *A. fasciatus* and improved the establishment of red clover and lucerne. The establishment of bird's-foot trefoil was not improved. Byers *et al.* (1985) were able to demonstrate that increased yields of total, *in vitro*-digestible, crude protein and legume dry matter resulting from pesticide protection of the seedlings was greater for late-spring sowings than at other times. This suggests that early-season sowing of legumes may avoid injury from gastropods in conservation tillage conditions.

In a multiple-treatment field experiment, Byers and Templeton (1988) studied the effect of various environmental factors on the establishment of lucerne when sown by conservation tillage into cocksfoot grass sod in Pennsylvania in each of 2 years. Their treatments comprised two sowing times (early and late spring) to produce differences in temperature, light and soil moisture; two vegetation-suppression methods (mowing and broadcast spray applications of the herbicide glyphosate) to produce varying degrees of plant competition; and two pesticide treatments (methiocarb bait combined with carbofuran granules vs. no pesticide) to establish different levels of invertebrate feeding. Lucerne dry matter was affected very little by sowing date in either year. Highest dry-matter yields of lucerne were achieved when plant competition was suppressed by broadcast spraying of herbicide. The combined molluscicide–insecticide treatment yielded more lucerne dry matter during the year of establishment. Byers and Templeton (1988) concluded that the gastropods *D. reticulatum* and *D. laeve* were among the pests controlled and may have been partially responsible for the observed effects on dry-matter yields.

Byers and Bierlein (1984) seeded lucerne by conservation tillage into old, spent stands of lucerne in Pennsylvania. Pesticides in 12 treatment combinations were applied at planting time to control plant competition and invertebrates. A pesticide combination treatment of paraquat or glyphosate + carbofuran granules + methiocarb bait significantly increased lucerne seedling numbers, and increased total dry matter, digestible dry matter and protein yields in one field. In the second field the pesticide combinations of glyphosate + carbofuran or carbofuran + methiocarb bait were the only treatments to effect increases in dry-matter and digestible dry-matter yields relative to no-pesticide controls. Invertebrates negatively associated with numbers of lucerne seedling established were several species of Homoptera (Cicadellidae), the isopod

Trachelipus rathkei Brandt (Oniscidae) and the gastropods *D. laeve*, *D. reticulatum* and *A. fasciatus*. The numbers of gastropods (1.3–3.2 per shingle trap) accounted for 28–44% of the variation in seedling numbers, with *D. laeve* comprising the largest principal component at both sites. Byers and Bierlein (1984) suggested that conservation-tillage reintroduction of lucerne into old lucerne fields is not to be recommended, because diseases, weeds and invertebrates (especially gastropods) severely limited establishment success.

In the north-east USA and south-eastern Canada, legumes are often established during late summer by conservation tillage in the stubble residue of small-grain crops. Agriolimacid and arionid species have been shown to be important pests of lucerne seedlings sown into both wheat (*Triticum aestivum* Linnaeus) (Gramineae) and rye stubble in New York (Dowling and Linscott, 1983, 1985) and oat (*Avena sativa* Linnaeus) (Gramineae) stubble in Pennsylvania (Stout *et al.*, 1992). Dowling and Linscott (1985), for example, found that methiocarb treatment reduced damage from *D. reticulatum* and *D. laeve* and increased survival of seedlings when lucerne was seeded during late summer into rye stubble in New York.

Following harvest, maize (*Zea mays* Linnaeus) (Gramineae) fields often provide ideal environments for conservation tillage of legumes, not least because gastropod populations are usually low. Byers *et al.* (1994) planted lucerne in 3 consecutive years into either a conventionally prepared seed-bed (maize harvested for grain in autumn and the following year the residue was ploughed under in late March) or by conservation tillage into residues of maize harvested for grain, silage and silage plus rye used as a winter cover crop. There were three planting dates each spring and subplot treatments comprising methiocarb bait + carbofuran granules and an untreated control. Although the pesticide treatment was effective in reducing the numbers of gastropods (as indicated by catch per shingle trap) and reducing feeding injury to seedlings, lucerne plant numbers and subsequent yield were not affected. Byers *et al.* (1994) concluded that the gastropod populations in these maize residues were too low (0.31–1.74 per shingle trap) to adversely affect lucerne establishment.

Plant-species and Age Effects on Gastropod Herbivory

Clearly seedlings are less tolerant of herbivory than are more mature plants. Evidence suggests that this low tolerance is accentuated in forage legumes because legume seedlings are highly preferred as food for pests like gastropods and, in mixed-species swards, are slower-growing than grasses and thus at a competitive disadvantage. Byers and Bierlein (1982) showed that when given a choice between lucerne seedlings of different ages, *D. reticulatum*, *D. laeve* and *A. fasciatus* preferred the 3-day-old seedlings over those 5 or 7 days old. *D. laeve* and *D. reticulatum*, but not *A. fasciatus*, exhibit a similar preference when presented with red-clover seedlings. *D. reticulatum* and *A. fasciatus* preferred the youngest, while *D. laeve* showed no preference for different-aged bird's-foot-trefoil seedlings. Such preference for young seedlings has also been demonstrated for gastropods feeding on non-forage species (e.g. Hanley *et al.*, 1995a).

Alleochemicals, such as cyanogenic glucosides in white clover, may confer some protection from gastropods. Burgess and Ennos (1987) and Dirzo and Harper (1982a,b) showed that gastropods preferred acyanogenic white clover. Horrill and Richards (1986) showed that *Arion hortensis* de Férussac selectively grazed and killed young seedlings of both cyanogenic and acyanogenic forms of white clover. In two experiments with 5-day-old seedlings and in two of three experiments with 11-day-old seedlings, equal proportions of acyanogenic and cyanogenic seedlings survived or were killed by grazing. In all other experiments with seedlings aged 16 days, 23 days and 35 days, more acyanogenic seedlings were killed by grazing than were cyanogenic seedlings (Horrill and Richards, 1986), indicating increased cyanogenic activity and reduced susceptibility as the seedlings aged.

Selective herbivory among gastropods feeding in plant communities is well known. However, there is relatively little information of the relative susceptibility of forage species to herbivory by gastropods, particularly as it relates to north-eastern North America. As indicated above, seedlings of forage legumes are highly susceptible to agriolimacid and arionid species that occur in forage fields. Byers and Bierlein (1982) found, however, that *D. reticulatum*, *D. laeve* and *A. fasciatus* prefer red-clover and lucerne seedlings over those of bird's-foot trefoil in laboratory experiments. These results generally concur with data from field experiments (e.g. Faix, 1980; Dowling and Linscott, 1983; Byers *et al.*, 1985), which indicate highest susceptibility in lucerne. Ferguson *et al.* (1989) detected cultivar differences in gastropod damage in white clover, while Barker (1989) showed that, in the seedling stage, the glandular-haired *Medicago disciformis* de Candolle is less susceptible to feeding by *D. reticulatum* than are seedlings of various lucerne cultivars.

Damage from Gastropods in Established Legume Crops

The impact of gastropods on established forage legumes in the north-east is not well documented. Arionid and agriolimacid species commonly feed on the lower leaves of mature lucerne, but the consequences for crop yield have not been measured. Cottam (1985, 1986) showed that agriolimacids preferred 'ladino' white clover to cocksfoot grass in the laboratory; in the field preferential feeding on clover resulted in the mixed swards becoming increasingly dominated by cocksfoot grass. That gastropods can significantly affect the productivity and persistence of forage legumes is supported by observations and experimental results in grasslands elsewhere (Ferguson *et al.*, 1989; Clements and Murray, 1991; Barker and Addison, 1992; Hanley *et al.*, 1995b).

Management of Gastropods in Forage Crops

Godan (1983) defined the term 'critical number' as the number of gastropods which a plant can carry, or the maximal intensity of infestation at which no economic losses are incurred. Barratt *et al.* (1994) quantified lucerne losses from *D. reticulatum* using field cages in Pennsylvania. They found that infestations of 12 and 24 *D. reticulatum* m^{-2} significantly reduced lucerne plant populations by 41% and 53% and dry-matter yield by 49% and 61%, respectively, when lucerne was sown by conservation tillage into maize stubble. When sown into oat stubble 12 and 24 *D. reticulatum* m^{-2} significantly reduced seedling populations by 37% and 52%, respectively, but only at the highest gastropod infestation level was dry-matter yield reduced (52% reduction). Despite this information, economic thresholds for losses in forage due to gastropods have not been established and thus do not at present feature in pest-management decisions. While shingle traps have been shown to be useful tools for monitoring gastropod infestations under experimental conditions, they have yet to be used extensively by growers or consultants.

Metaldehyde and methiocarb in many different formulations are widely used for crop protection, but not at present for forage legumes in north-east North America. No metaldehyde-based products are currently registered for use in forages, and methiocarb has been withdrawn from the entire US market because of toxicity to non-target animals, such as birds. As a result of research into alternative molluscicidal materials, the glucoside 6-HT-β-C-3-COOH was isolated from couch grass and demonstrated to have activity against *Arion subfuscus* (Drapamaud) in a bait formulation (Hagin and Bobnick, 1991). However, the high cost of registration and manufacture precludes development of the product through extraction or synthesis, because forages represent a small market (John Pickle, HACCO Inc., personal communication, 1999). If a less expensive production method is developed, this glucoside compound could have a useful role in gastropod management, because of its lower toxicity to vertebrates than conventional molluscicides (Hagin and Bobnick, 1991). There remains a great need for environmentally safe molluscicide formulations, with good efficacy, that could be used in forages.

Gastropod management thus currently relies on cultural practices, such as crop rotation, reduction of crop residues before planting and early planting (Byers and Templeton, 1988). Manipulating the amount of vegetative cover by mowing or herbicide treatment can lead to conditions that are unfavourable for gastropods and thus provide an effective method of pest control in advance of sowing. Welty *et al.* (1981) overcame difficulties in establishing legumes sown into grass sod in Montana by imposing 28 days between treatment of the sod with glyphosate herbicide and sowing by conservation tillage. This delay allowed sufficient time for the grass residue to dry out and to permit sunlight to penetrate the canopy, resulting in desiccation of the surface-dwelling gastropods.

Clements, R.O. and Murray, P.J. (1991) Incidence and severity of pest damage to white clover. *Aspects of Applied Biology* 21, 369–372.

Cottam, D.A. (1985) Frequency-dependent grazing by slugs and grasshoppers. *Journal of Ecology* 73, 925–933.

Cottam, D.A. (1986) The effects of slug grazing on *Trifolium repens* and *Dactylis glomerata* in monoculture and mixed swards. *Oikos* 47, 275–279.

Decker, A.M., Retzer, H.J., Sarna, M.L. and Kerr, H.D. (1969) Permanent pastures improved with sod-seeding and fertilization. *Agronomy Journal* 61, 243–247.

Dirzo, R. and Harper, J.L. (1982a) Experimental studies on slug–plant interactions. III. Differences in the acceptability of individual plants of *Trifolium repens* to slugs and snails. *Journal of Ecology* 70, 101–117.

Dirzo, R. and Harper, J.L. (1982b) Experimental studies on slug–plant interactions. IV. The performance of cyanogenic and acyanogenic morphs of *Trifolium repens* in the field. *Journal of Ecology* 70, 119–138.

Dowling, P.M. and Linscott, D.L. (1983) Use of pesticides to determine relative importance of pest and disease factors limiting establishment of sod-seeded lucerne. *Grass and Forage Science* 38, 179–185.

Dowling, P.M. and Linscott, D.L. (1985) Slugs as primary limitation to establishment of sod-seeded lucerne. *Crop Protection* 4, 394–402.

Faix, J.J. (1980) Evaluation of pesticides for improving alfalfa establishment in conventional and no-till sod planting. *Dixon Spring Agricultural Center* 8, 104–109.

Ferguson, C.M., Lewis, G.C., Hanks, C.B., Parsons, D.M.J. and Asteraki, E.J. (1989) Incidence and severity of damage by slugs and snails to leaves of twelve clover cultivars: tests of agrochemicals and cultivars. *Annals of Applied Biology* 114 (suppl.) (10), 138–139.

Garber, R.J., Myers, W.M. and Sprague, V.G. (1946) *Pasture and Pasture Problems in Northeastern States*. School of Agriculture Bulletin 485, Pennsylvania State College, 44 pp.

Godan, D. (1983) *Pest Slugs and Snails, Biology and Control*. Springer-Verlag, Berlin, 445 pp.

Grant, J.F., Yeargan, K.V., Pass, B.C. and Parr, J.C. (1982) Invertebrate organisms associated with alfalfa seedling loss in complete-tillage and no-tillage plantings. *Journal of Economic Entomology* 75, 822–826.

Hagin, R.D. and Bobnick, S.J. (1991) Isolation and identification of a slug-specific molluscicide from quackgrass (*Agropyron repens*, L. Beauv.). *Journal of Agricultural and Food Chemistry* 39, 192–196.

Hanley, M.E., Fenner, M. and Edwards, P.J. (1995a) The effect of seedling age on the likelihood of herbivory by the slug *Deroceras reticulatum*. *Functional Ecology* 9, 754–759.

Hanley, M.E., Fenner, M. and Edwards, P.J. (1995b) An experimental study of the effects of molluscan grazing on seedling recruitment and survival in grassland. *Journal of Ecology* 83, 621–627.

Horrill, J.C. and Richards, A.J. (1986) Differential grazing by the mollusc *Arion hortensis* Fér. on cyanogenic and acyanogenic seedlings of the white clover, *Trifolium repens* L. *Heredity* 56, 277–281.

Morris, M.G. (1969) Populations of invertebrate animals and the management of chalk grassland in Britain. *Biological Conservation* 1, 225–233.

Rathcke, B. (1985) Slugs as generalist herbivores: tests of three hypotheses on plant choices. *Ecology* 66, 828–836.

Rollo, C.D. (1974) Ecology of the slugs, *Deroceras reticulatum*, *D. laeve*, and *Arion fasciatus* in Ontario corn fields. MSc, University Guelph, Ontario, Canada.

Rollo, C.D. and Ellis, C.R. (1974) Sampling methods for the slugs, *Deroceras reticulatum* (Müller), *D. laeve* (Müller), and *Arion fasciatus* Nilsson in Ontario corn fields. *Proceedings of the Entomological Society of Ontario* 105, 89–95.

Sheaffer, C.C. (1989) Legume establishment and harvest management in the U.S.A. In: Marten, G.C., Matches, A.G., Barnes, R.F., Brougham, R.W., Clements, R.J. and Sheath, G.W. (eds) *Persistence of Forage Legumes. Proceedings of an Australia–New Zealand–United States of America Trilateral Workshop Held in Honolulu, Hawaii, 18–22 July, 1988.* American Society of Agronomy, Crop Science Society of America and Soil Science Society of America, Madison, Wisconsin, pp. 277–291.

Stout, W.L., Byers, R.A., Leath, K.T., Bahler, C.C. and Hoffman, L.D. (1992) Effects of weed and invertebrate control on alfalfa establishment in oat stubble. *Journal of Production Agriculture* 5, 349–352.

Taylor, N.L. (1985) Clovers around the world. In: Taylor N.L. (ed.) *Clover Science and Technology*. Agronomy Series No. 25, American Society of Agronomy, Crop Science Society of America, Soil Science of America, Madison, Wisconsin, pp. 2–6.

Taylor, T.H., Smith, E.M. and Templeton, W.C. (1969) Use of minimum tillage and herbicide for establishing legumes in Kentucky bluegrass (*Poa pratensis* L.) swards. *Agronomy Journal* 61, 761–766.

Vance, C.P., Heichel, G.H. and Phillips, D.A. (1988) Nodulation and symbiotic dinitrogen fixation. In: Hanson, A.A., Barnes, D.K. and Hill, R.R. Jr (eds) *Alfalfa and Alfalfa Improvement*. Agronomy Series No. 29, American Society of Agronomy, Crop Science Society of America, Soil Science of America, Madison, Wisconsin, pp. 229–257.

Washko, J.R. (1974) Forages and grasslands in the Northeast. In: Sprague, M.A. (ed.) *Grasslands of the United States, Their Economic and Ecological Importance. A Symposium of the American Forage and Grassland Council.* Iowa State University Press, Ames, Iowa, pp. 98–112.

Wilson, M.J., Glen, D.M. and George, S.K. (1993) The rhabditid nematode *Phasmarhabditis hermaphrodita* as a potential biological control agent for slugs. *Biocontrol Science and Technology* 3, 503–511.

Welty, L.E., Anderson, R.L., Delany, R.H. and Hensleigh, P.F. (1981) Glyphosate timing effects on establishment of sod-seeded legumes and grasses. *Agronomy Journal* 73, 813–817.

16 Gastropods as Pests in Vegetable and Ornamental Crops in Western Europe

GORDON PORT[1] AND ALBERT ESTER[2]

[1]*Department of Agricultural and Environmental Science, University of Newcastle, Newcastle upon Tyne NE1 7RU, UK;* [2]*Applied Plant Research, PO Box 430, 8200 AK Lelystad, The Netherlands*

Diversity in Cropping Situations

Consistent with the cultural and geographical diversity of Western Europe there is a great diversity of vegetable and ornamental crops grown in the region. In the southern countries most of these crops are grown out of doors, but moving northwards an increasing proportion are grown as protected crops under glass and in polyethylene tunnels. To illustrate this diversity, 45 distinct commercial vegetable crop types were recorded as grown outdoors in England and Wales in 1991 (Thomas *et al.*, 1991a). Similarly there were 28 commercial bulb and flower crops in England and Wales in 1992 (Thomas *et al.*, 1993), and these figures include some minor crops that have been grouped together. There were more than 100 types of commercial edible and ornamental protected crops in England and Wales in 1991 (Thomas *et al.*, 1991b), although some of these would be the same types as the outdoor crops.

In addition to growers of commercial crops there are a large number of non-commercial growers, for whom gastropod pests may assume significance far beyond the economic damage they cause.

Gastropod Species Found in Vegetable and Ornamental Crops

The most serious gastropod pests of vegetable and ornamental crops are those widely distributed by human activities, mostly in association with movement of soil and plant material. Thus the pest problems are ubiquitous and differences in the pest fauna from site to site are usually the result of varying conditions affecting the chances of a species establishing and thriving at a particular location. In crops where there is much physical disturbance, such as soil cultivation, gastropods of the slug form tend to predominate, while in the more stable situations both slug and snail forms are generally present.

The most widespread pest species is the agriolimacid *Deroceras reticulatum* (Müller), although at particular times and locations other species may be more significant. Often found living in commercial and non-commercial plots are *Deroceras laeve* (Müller) and *Derocenas panormitonum* (Lessona & Pollonera), the arionids *Arion hortensis* de Férussac, *Arion rufus* (Linnaeus), *Arion silvaticus* Lohmander and *Arion circumscriptus* Johnston, the milacids *Tandonia budapestensis* (Hazay) and *Tandonia sowerbyi* (de Férussac), the helicids *Cantareus aspersus* (Müller), *Cepaea hortensis* (Müller) and *Helix pomatia* Linnaeus, the succineid *Oxyloma pfeifferi* (Rossmässler), and the zonitid *Zonitoides nitidus* (Müller), these may sometimes cause significant damage (Morzer Bruyns *et al.*, 1959; Foster, 1977; Atkinson *et al.*, 1979; Porcelli and Parenzan, 1988; Kurppa, 1989).

Economic Importance

Molluscicides are widely used in European agriculture and horticulture (Table 16.1). Detailed surveys of pesticide use in England and Wales are conducted by the Department for Environment, Food and Rural Affairs (DEFRA) (formerly Ministry of Agriculture, Fisheries and Food). Surveys of molluscicide use in 1991 (outdoor vegetables and protected crops) and 1993 (outdoor bulbs and flowers) give an indication of the extent of perceived gastropod problems (Table 16.2). As the data were collected in different years, when the pest status of the gastropods may have varied due to differing weather conditions, close comparison of the data is probably unwise. Growers may have applied molluscicides prophylactically, irrespective of the true pest status of gastropods, but usually the extent of molluscicide use is greatest in years or at sites where gastropods are frequently troublesome.

Clearly the crop type in which gastropods have greatest importance, judged by the use of molluscicides, is brassicas (Fig. 16.1). There are no data on the relative abundance of gastropods in the different crops, but the reason brassicas are perceived to be at risk is because of the low threshold for damage to these plants in general and Brussels sprouts (*Brassica*

Table 16.1. Estimated quantities of molluscicide used in European horticulture and agriculture in 1994 (courtesy of Lonza).

Location	Molluscicide use (kg × 1000)
France	30,000
Germany	450
The Netherlands	110
UK	10,000

Table 16.2. The area of vegetable and ornamental crops in England and Wales (surveyed between 1991 and 1993), together with the relative area of these crops treated with molluscicides.

Crop group	Area grown (ha)	Molluscicide-treated area (ha)	Area treated as % area grown
Outdoor vegetables*			
Leafy brassicas	41,067	2,881	7.02
Root brassicas	2,980	131	4.40
Peas and beans	50,469	15	0.03
Onions and leeks	11,500		0.00
Carrot, parsnip, celery	17,685		0.00
Lettuce, endive, etc.	8,305		0.00
Sweetcorn	1,533		0.00
Beetroot and root vegetables	2,426	31	1.28
Curcubits	970	4	0.41
Other vegetables	2,687	45	1.67
Subtotal	139,622	3,107	2.23
Outdoor bulbs and flowers†			
Bulbs	4,759		0.00
Chrysanthemums	137	15	10.95
Other flowers for cutting	896	319	35.60
Subtotal	5,792	334	5.77
Protected crops‡			
Tomato	422.4	18.71	4.43
Cucumber	238.58		0.00
Lettuce	1,039.82	192.1	18.47
Peppers	66.15	15.36	23.22
Celery	100.86	58.22	57.72
Other vegetables	374.36	43.11	11.52
Other fruit	23.87	2.58	10.81
Chrysanthemums	315.28	25.65	8.14
Pinks	20.72	2.43	11.73
Carnations	9.44		0.00
Alstroemeria	24.75	11.23	45.37
Roses	10.06		0.00
Flowers, foliage	120.66	9.4	7.79
Chrysanthemums, pot	76.11	18.54	24.36
Pot plants	176.13	67.77	38.48
Plants for propagation	960.61	143.86	14.98
Subtotal	3,979.8	608.96	15.30
Total	149,393.8	4,049.96	2.71

*Thomas *et al.*, 1991a.
†Thomas *et al.*, 1993.
‡Thomas *et al.*, 1991b.

oleracea Linnaeus var. *gemmifera* de Candolle) in particular. For example, if more than 5% of Brussels sprouts are damaged, whether by gastropods or other causes, the crop will attract a lower price or may be

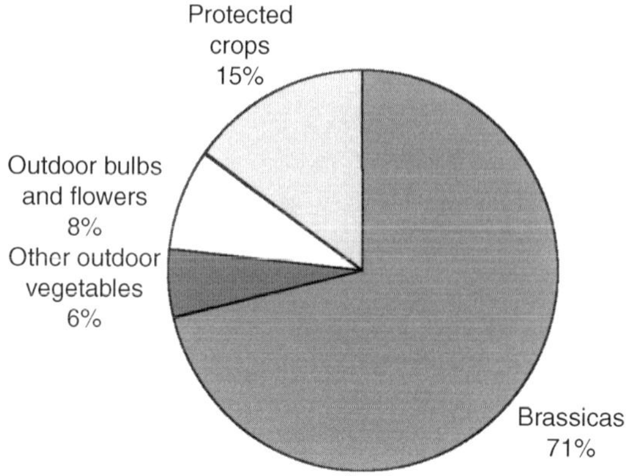

Fig. 16.1. The area of vegetable and ornamental crops treated with molluscicide in England and Wales (surveyed between 1991 and 1993), data from Table 9.1.

rejected altogether. The damage occurs close to harvest, in the late summer and autumn, when weather conditions usually favour gastropod activity. At that time, the canopy cover provided by the mature plants, together with fallen leaves on the ground, makes control with molluscicide baits very difficult and, in the absence of useful economic thresholds, many growers apply molluscicides prophylactically earlier in the season. We have estimated that a 20% decrease in gastropod damage, principally due to *D. reticulatum*, to the sprout crop would be worth £2 million each year in England alone.

In none of the crops surveyed by DEFRA were gastropods considered the most important pest group, but in some of the protected crops a large percentage of the crop area was treated with molluscicide. Examples of these include celery (*Apium graveolens* Linnaeus) (Apiaceae), *Alstroemeria* Linnaeus spp. (Liliaceae) and pot plants under protected cropping, and cut flowers grown outdoors. As with Brussels sprouts, the relatively high use of molluscicides on these crops reflects the difficulties of reducing damage to crops with very low thresholds before consumer acceptability is affected.

Estimates of the economic importance of gastropod pests of some horticultural crops in the Netherlands are shown in Table 16.3. These estimates are of damage over and above that already prevented by molluscicide use, but there are no detailed figures on molluscicide usage to complement the data in the table. However, 84% of the molluscicides used in commercial crops in 1995 were used in Brussels sprouts and the rest were mainly in protected crops, with 8% in chrysanthemums (*Chrysanthemum sensu lato* spp.) (Asteraceae).

In non-commercial situations, primarily amateur gardens, gastropods may often be a persistent problem of a wide range of crops, partly because of the diverse habitats in these areas, which provide ideal refuges for gastropods. There are few data with which to make an objective assessment, but, in 1995, 70% of molluscicide used in the Netherlands was for non-commercial crops.

Table 16.3. Estimates of the economic importance of gastropod pests of some horticultural crops in the Netherlands 1994. Estimated yield loss and value of loss shown in parentheses. (From Anon., 1995.)

Crop	Area grown (ha)	Estimated % damage	Yield (kg × 1000)	Value (million Dutch guilders)
Asparagus	80	20	32.418 (6.483)	0.1713 (0.0343)
Brussels sprouts	5,000	2–5	93,000 (1,900–4,600)	74 (1.5–3.7)
Brassicas and leaf vegetables	16,537	0.5–1.0	460,000 (2,300–4,600)	315 (1.6–3.2)
All field vegetables	43,294	0.5	1,928,000 (9,600)	1261 (6.3)

Sampling, Economic Thresholds and Crop Damage

Control strategies should ideally be based on information on numbers of pest animals present and the likelihood of crop damage. However, in practice, many growers apply molluscicide in response to observing plant damage or in anticipation of damage (prophylactic action). Abundance in most pest species may be monitored with simple shelter traps (pieces of board, ceramic tiles, plant-pot saucers) covering either food or molluscicide bait. The food may be pieces of vegetable, cereal bran or chicken food (Young *et al.*, 1996) and serves to attract the gastropods to the shelter. If the traps are checked during the early morning, many gastropods will still be taking refuge beneath the trap. Use of molluscicide bait usually causes gastropods to become paralysed under or near the trap, making positive results more likely. It must be remembered that gastropods will only be active in favourable weather conditions and the traps may not contain specimens after unfavourable conditions, even if gastropods are present at the site.

Given the dependence of trap captures on weather and other site effects, economic thresholds have not been defined. Furthermore, for many growers any damage inflicted on the crop by gastropods is of concern and they will generally consider applying molluscicides. For some crops there is information on the nature of damage and this is given below.

Pest Status in Specific Crops

Asparagus (Asparagus officinalis *Linnaeus)* (Liliaceae*)*

Asparagus remains commercially viable for 6–14 years on the same site. From the second year growers start to harvest the green asparagus stems. During autumn and winter all the stems die from frost or natural senescence and are removed, so the field is bare during these seasons. Rainfall during the winter causes a surface crust to appear on the soil, restricting gastropods to below the soil surface. In spring, as soil temperature increases, the gastropods become more active and move from the soil to the stems of the plants. Thus, around the roots and stem bases of the plants, there may be large numbers of gastropods present and they feed on the young stems. As temperatures continue to rise, the plants start to grow rapidly. When the stems emerge, the soil crust is broken and the gastropods may disperse over the field. Damage is thus most severe in the first few weeks of the harvesting season. Damaged stems are crooked and readily decay after harvest and are thus unmarketable.

*Brussels sprouts (*Brassica oleracea *Linnaeus var.* gemmifera *de Candolle)* *(*Brassicaceae*)*

Brussels sprouts seedlings are transplanted in the spring. The crop grows on a range of soil types, of which clays and loams provide a good habitat for agriolimacids, arionids and milacids. Favourability is particularly high in those soils that crack when dry, in that they provide crevices into which the gastropods can retreat and escape desiccation. During the summer, the crop provides sparse cover and the soil surface is not favourable for the gastropods. Hence, gastropods are not considered pests at that stage of crop growth. In late summer, the first buttons (sprouts) start to develop, the exact timing of which depends on the Brussels sprouts cultivar. In this period, the lowest leaves senesce and fall to the soil and the crop establishes a dense canopy, thus providing an environment at soil level ideal for *D. reticulatum* and other gastropods. During the evening and night the gastropods move on to the plants and attack the young buttons by eating the outer leaves. Gastropods are rarely found on the plants during daytime because the low humidity is unfavourable. For crops in Germany, Godan (1973) commonly found up to 20 *D. reticulatum* per plant in counts made at night. Often one *D. reticulatum* affects four to eight buttons per night. Fields with a moderate population density of gastropods may have 60–80% of the buttons attacked. The damage includes deformation, rot and contamination with slime and faeces. If more than 5% of the buttons are affected, the crop is downgraded. The affected sprouts have to be removed before sale, with a consequent increase in costs. Growers usually have to protect their crop until harvest, which means for a 6–8 week period.

*Lettuce (*Lactuca sativa *Linnaeus)* *(*Asteraceae*)*

Many genotypes of lettuce, such as iceberg and headed lettuce, are susceptible to gastropods, particular agriolimacids such as *D. reticulatum*.

Outdoors, the crop is grown mainly on sandy soils for a production season that runs from spring to autumn. The sandy soils, and minimal cover provided by the young crop, result in conditions that are too dry for *D. reticulatum* to be very active. Sometimes the outer leaves of iceberg lettuce are attacked by juvenile *D. reticulatum*, but this damage is not of economic consequence, because these outer leaves are always left in the field when the crop is harvested.

During crop growth, particularly the last 4 weeks before harvest, it is common practice to irrigate the lettuces several times per week. Gastropods assume importance after formation of the heads, when closure traps water between the leaves. *D. reticulatum* use the heads as a shelter (Symondson, 1993) and easily devalue the crop by their feeding, which encourages decay and is cosmetically unacceptable (Wilson *et al.*, 1995). For crops in Germany, Godan (1973) found up to five to ten *D. reticulatum* per plant. In the case of organic growers, gastropods are the most important pest problem in lettuce (Peacock and Norton, 1990).

*Curly kale (*Brassica oleracea *Linnaeus convar.* acephala *de Candolle) (*Brassicaceae*)*

Curly kale is sown in spring and transplanted in midsummer. The crop is harvested from autumn through to the following spring. From late summer, the plants have a mass of leaves, which provide a closed canopy, with the lower leaves of the plant usually touching the soil. These conditions are favourable for gastropods, which eat the leaves, particular the soft tissue between the veins, rendering the crop unmarketable. The gastropods may hide by day on the plants, between the curly leaves, and thus may be present at harvest.

*Broccoli and cauliflower (*Brassica oleracea *Linnaeus var.* botrytis *Linnaeus) (*Brassicaceae*)*

Broccoli and cauliflower are grown in a range of soils, but usually in loams and sandy loams rather than heavy clays. Gastropod damage to these crops varies from year to year, depending on weather and the type of soil. Severe damage usually occurs in years with high rainfall in summer and autumn. Gastropods feed on the crop from August and continue through the autumn. They feed on the leaves and the flowerheads. It is the feeding activity in the heads of the plants, manifest in holes and deposits of mucus and faeces, that leads to economic losses in these crops.

*Cabbage (*Brassica oleracea *Linnaeus var.* capitata *Linnaeus) (*Brassicaceae*)*

Cabbage crops are grown from spring until late autumn, with harvest in the autumn. A part of the harvest is for immediate sale, but most is stored for up to 7 months before sale the following year. Damage to seedling plants by gastropods generally occurs only in wet periods in spring or early summer. In late summer and autumn, the damage caused by gastropods is often more severe and consists of many holes in the outer leaves. At harvest, many of the damaged leaves must be removed to retain market appeal. However, this leaf removal leads to lower-weight cabbage and lower financial return to the grower. If too many damaged leaves have to be removed from the cabbage, the paler head of the cabbage is exposed and the crop can only be used for processing, at further cost to the grower. In storage the temperature is about 1°C, which optimizes shelf-life, but allows some gastropod activity. During the storage period the gastropods that do remain go deeper into the cabbage core. When the temperature increases, as the cabbage is taken out of storage, these animals leave their resting sites and resume feeding over the outside of the cabbage heads. Moreover, they contaminate the heads with mucus and faeces.

*Carrot (*Daucus carota *Linnaeus) (*Apiaceae*)*

Many kinds of carrot are susceptible to gastropods. The bush-carrot crop is grown mainly on sandy soil, with harvest usually within 3 months. The crops destined for harvest in winter are grown on ridges in sandy or sandy–clay soil. The period of gastropod damage is autumn, mainly from September until November. At that time, the carrot foliage is tall enough to provide a closed canopy. Gastropods are often found on the shaded side of the ridges, among the foliage of the plants and on that part of the root that is above soil level.

The damage may appear on the leaves, but is most important on the carrot roots. Gastropods such as *D. reticulatum* initially attack the carrot root from the top by excavating a hole, but later they feed move extensively. As a consequence of this damage, there is often a bacterial rotting of the inside part of the attacked roots (Dawkins *et al.*, 1985). Most carrot damage is found at the margins of the carrot field (about 5 m into the field). The sandy soil provides little shelter, so damage generally results from gastropods migrating in from neighbouring vegetation. The main problem for growers is to separate the damaged carrot roots from the undamaged ones at harvest. Some growers do not harvest the few metres of carrot rows around the perimeter of the field.

*Pea (*Pisum sativum *Linnaeus) (*Fabaceae*)*

The pea crop is usually established by drilling seed in early spring and the peas are harvested in summer. In rainy periods in late spring or early summer, gastropods can attack the foliage and pods. From the start of flowering the plants may lodge, which results in favourable soil-surface conditions for gastropods. Gastropod feeding on the leaves before flowering does not result in any loss of yield, because at that crop growth stage there are sufficient leaves at the top of the plant to compensate for leaf removal on the lower parts. Feeding on the plants at later crop growth stages is more critical. In particular, damage to the pods can lead to severe yield loss. Furthermore, infestation of the crop can lead to contamination of the machine-harvested peas.

*Bean (*Phaseolus vulgaris *Linnaeus) (*Fabaceae*)*

Bean crops are established by drilling seed in spring through to early summer. Because the crop is susceptible to frost, the pods are generally harvested before autumn. The seedlings, through to appearance of the first true leaves, are the crop stages most susceptible to gastropod damage. Below the soil surface, damage occurs to the emerging cotyledons, the hypocotyl and the first leaves. Damage to the foliage of older seedling and mature plants does occur, but is not of economic importance. Here the leaves are shredded.

*Chinese cabbage (*Brassica chinensis *Linnaeus) Fabaceae)*

Chinese cabbage is grown on sandy soils, with the cropping season extending from early spring to autumn. This crop is highly favoured by gastropods, which are most damaging during wet weather in spring and autumn. Leaf damage is most severe after formation of the heads, as *D. reticulatum* shelter within the developing cabbage.

Control

Control of gastropods in vegetable crops has always been difficult, due, on the one hand, to low economic thresholds and, on the other, to the variable life cycles of the main pest species and the interaction of many environmental factors, such as different crops, soil type, climate and timing of application of the molluscicidal baits (Port and Port, 1986). In this section the factors that predispose crops to damage are discussed and the specific control options considered.

History of the site

Gastropods tend to be of greater pest status on clay soils. The surface structure of the soil is very important in affecting gastropod activity, primarily because a coarse soil surface structure provides numerous refuges for the animals and makes it more difficult to cover and thus protect sown seeds. In soils where rainfall can result in a surface crust, this may provide similar protection to the gastropods. If crops are normally grown on sandy soils, the potential for losses due to gastropods is lower.

When fields are left fallow or have had a previous crop providing dense vegetation cover and/or crop residues, large numbers of gastropods are often observed in the subsequent crop. These conditions provide shelter and food for gastropods, increasing their ability to survive extreme weather conditions in either winter or summer. Similarly, when field margins have vegetation providing good cover, gastropods may move into the crop from the margins (Frank, 1998). Growers intending to produce a crop known to be at risk from gastropods should avoid using such sites. However, many factors must be considered when making decisions about cropping, and the potential of damage from gastropods may have a low priority, given the relatively low cost of chemical molluscicides.

Cultivations

When large numbers of gastropods occur, repeated cultivations, especially during dry periods, can be used as a strategy to reduce their numbers. Fine seed-beds restrict movement by gastropods and reduce the availability of shelter sites. Where vegetation is maintained in field margins and crop headlands, it should be mowed frequently. Stringer *et al.* (1974) found that, when weed cover was reduced to 5%, the number of *Cepaea nemoralis* (Linnaeus) decreased by 50%.

Molluscicides

Molluscicides, usually carbamates, metaldehyde or bensultap, are often used in baits in attempts to control gastropods. Glen and Orsman (1986) and Frain and Newell (1983), working at sites where *Deroceras* Rafinesque Schmaltz were predominant, found that these molluscicidal baits gave comparable results.

Given that the different molluscicides give similar protection to crops, the main variable factors in their use are the rate and timing of application. Moens (1980) recommended a lower (unspecified) application rate of molluscicide bait pellets, but with applications made at intervals of 1–3 weeks for vegetable seedlings in spring, instead of the single, higher-rate application that is used in arable crops. However, data to support this approach are not available from other studies.

As gastropods are usually only active in damp conditions, it may be best to delay application of molluscicides until such conditions are forecast (Young and Port, 1996). None the less, it is important for the baits to be applied before damage becomes serious and in a location where the pest population can encounter them. Growers of asparagus often use molluscicide baits, but they do not reach the gastropods at the base of the asparagus stem, where damage occurs. In Brussels sprouts, baits may be intercepted by the leaves of the crop or fall on to leaves on the ground, and thus most baits do not reach areas where gastropods will encounter them. Regular monitoring for gastropod activity, combined with careful timing and placement of molluscicide baits, may improve their efficiency in reducing damage, but again there are insufficient data to make judgements about the economic value of such approaches.

Whether more rational use of molluscicides can provide economic benefits remains to be proved, but there is now considerable pressure from customers for growers to use integrated crop management (ICM). ICM will focus attention on using all inputs, including molluscicides, in a manner that can be justified from economic, environmental and human-health perspectives. Thus growers will become more aware of the need to use a range of control options and to use molluscicides efficiently.

Biological controls

The parasitic rhabditid nematode, *Phasmarhabditis hermaphrodita* (Schneider) is considered by many growers and researchers to offer the best alternative to chemical control (Ester and Geleen, 1996; Glen *et al.*, 1996). Effective gastropod control with the nematode has been shown, however, to be highly dependent on optimal timing of application in relation to crop development. Further, placement of application of the nematodes is important. The comments relating to timing and placement of molluscicides apply similarly here.

In greenhouses, organic growers of asparagus and flowers, such as roses (*Rosa* spp.; Rosaceae), keep quail (*Coturnix coturnix* (Linnaeus); Phasianidae) and small chickens (*Gallus gallus* (Linnaeus); Phasianidae) to protect crops from gastropod damage, but there are no objective data on the efficacy of these approaches.

Repellents and barriers

A number of repellent compounds have been investigated over the years, but so far none have achieved commercial success. However, there remains much interest in the potential use of repellent compounds for protecting vegetable and ornamental crops. Ester and Nijënstein (1995) found that wheat (*Triticum aestivum* Linnaeus) (Gramineae) seeds treated with carvone, a product derived from caraway seeds (*Carum carvi* Linnaeus; Apiaceae) that prevents sprouting of products like potato (*Solanum tuberosum* Linnaeus) (Solanaceae) (Diepenhorst and Hartmans, 1993), achieved a similar level of protection against gastropods, to that provided by the molluscicide methiocarb. A possibility would be to use carvone (trade name Talent) as a gas, to provide postharvest protection of vegetables such as white cabbage. Other products with repellent and/or molluscicidal activity include Limonene-r, saponins from quinoa (*Chenopodium quinoa* von Willdenow) (Chenopodiaceae) seeds, extracts of spinach (*Spinacia oleracea* Linnaeus) (Chenopodiaceae) and azadirachtin from neem (*Azadirachta indica* de Jussieu) (Meliaceae) (Price *et al.*, 1987; Ruskin, 1992; West and Mordue, 1992; Achuthan *et al.*, 1994; Ester and Nijënstein, 1996).

A repellent that might be sprayed on to the stems of Brussels sprouts would prevent gastropods climbing the plant and damaging the buttons (Dawson *et al.*, 1996), but there are difficulties with application in the field. Repellent mulches that may be applied around susceptible plants and thus prevent gastropods reaching them may be more practical, but still remain to be exploited (G.L. Port, unpublished data).

Physiochemical barriers, made of zinc-plated steel, are used in Swiss organic horticulture and, despite the high initial investment, if used properly may be very effective.

Resistant genotypes

Some genotypes of vegetables and ornamentals are more resistant to attack than others. For example, resistant genotypes provide an important source of protection against gastropods in potato (Johnston *et al.*, 1989). However, this useful resistance may often be linked to other traits that reduce consumer acceptance of the product. For example, Moens (1989) and Glen *et al.* (1990) found that intensity of feeding activity of *D. reticulatum* was inversely related to the glucosinolate content of oilseed rape (*Brassica napus* Linnaeus var. *oleifera* Linnaeus) seedlings, and yet cultivars producing oil of low glucosinolate content command the greater market share. The bitterness associated with certain cultivars of Brussels sprout is linked to the presence of glucosinolates, sinigrin and progoitrin (Fenwick *et al.*, 1983). This variation in the glucosinolate content may provide resistance to gastropods, but at the same time may compromise the marketability of the product.

Integrated Management Programmes and Organic Systems

There are no published details of integrated management programmes against gastropods in vegetable or ornamental crops. Those growers who use molluscicides often rely largely on that single method of control. However, growers of organic crops and non-commercial growers who prefer to avoid pesticide use have no option other than to attempt to use one or more of the alternatives to molluscicides mentioned above. The recent availability of the nematode *P. hermaphrodita* as a biological control offers hope for better management of gastropod pests, but this method and many of the others mentioned need further research and development to ensure better control of gastropod pests in vegetable and ornamental crops.

References

Achuthan, N.G., Mohamed, A.I. and Nuruzzaman, M. (1994) Survival, body mass and feeding of the bulb-eating slug, *Milax rusticus*, treated with spinach homogenate filtrate. *Arid Soil Research and Rehabilitation* 8, 101–103.

Anon. (1995) *Tuinbouw Statistiek 1995.* Produktschap voor Siergewassen, Groenten en Fruit, The Hague.

Atkinson, H.J., Gibson, N. and Evans, H. (1979) A study of common crop pests in allotment gardens around Leeds. *Plant Pathology* 28, 169–177.

Dawkins, G., Hislop, G., Luxton, M. and Bishop, C. (1985) Transmission of liquorice rot of carrots by slugs. *Journal of Molluscan Studies* 51, 83–85.

Dawson, G.W., Henderson, I.F., Martin, A.P. and Pye, B.J. (1996) Physiochemical barriers as plant protectants against slugs (Gastropoda: Pulmonata). In: Henderson, I.F. (ed.) *Slug and Snail Pests in Agriculture.* Monograph No. 66, British Crop Protection Council, Farnham, pp. 439–444.

Diepenhorst, P.K. and Hartmans, K.J. (1993) Carvone, a natural sprout inhibitor. In: *12th Conference of European Association for Potato Research*, Paris, pp. 67–68.

Ester, A. and Geleen, P.M.T.M. (1996) Integrated control of slugs in sugar beet growing in a rye cover crop. In: Henderson, I.F. (ed.) *Slug and Snail Pests in Agriculture.* Monograph No. 66, British Crop Protection Council, Farnham, pp. 445–450.

Ester, A. and Nijënstein, J.H. (1995) Control of the field slug *Deroceras reticulatum* (Muller) (Pulmonata, Limacidae) by pesticides applied to winter-wheat seed. *Crop Protection* 14, 409–413.

Ester, A. and Nijënstein, J.H. (1996) Control of field slug (*Deroceras reticulatum* (Muller)) by seed- applied pesticides in perennial ryegrass assessed by laboratory tests. *Zeitschrift fur Pflanzenkrankheiten und Pflanzenschutz – Journal of Plant Diseases and Protection* 103, 42–49.

Fenwick, G.R., Griffiths, N.M. and Heaney, R.K. (1983) Bitterness in Brussels sprouts (*Brassica oleracea* L. var. *gemmifera*): the role of glucosinolates and their breakdown products. *Journal of the Science of Food and Agriculture* 34, 73–80.

Foster, G.N. (1977) Problems in cucumber crops caused by slugs, cuckoo-spit insect, mushroom cecid, hairy fungus beetle and the house mouse. *Plant Pathology* 26, 100–101.

Frain, J.M. and Newell, P.F (1983) Testing molluscicides against slugs – the importance of assessing the residual population. *Journal of Molluscan Studies* 49, 164–173.

Frank, T. (1998) Slug damage and numbers of the slug pests, *Arion lusitanicus* and *Deroceras reticulatum*, in oilseed rape grown beside sown wildflower strips. *Agriculture, Ecosystems and Environment* 67, 67–78.

Glen, D.M. and Orsman, I.A. (1986) Comparison of molluscicides available to gardeners, based on metaldehyde, methiocarb or aluminium sulphate. *Crop Protection* 5, 371–375.

Glen, D.M., Jones, H. and Fieldsend, J.K. (1990) Damage to oilseed rape seedlings by the field slug *Deroceras reticulatum* in relation to glucosinolate concentration. *Annals of Applied Biology* 117, 197–207.

Glen, D.M., Wilson, M.J., Hughes, L., Cargeeg, P. and Hajjjar, A. (1996) Exploring and exploiting the potential of the rhabditid nematode *Phasmarhabditis hermaphrodita* as a biocontrol agent for slugs. In: Henderson, I.F. (ed.) *Slug and Snail Pests in Agriculture.* Monograph No. 66, British Crop Protection Council, Farnham, pp. 271–280.

Godan, D. (1973) Injury and economic importance of molluscs in the German Federal Republic. *Nachrichtenblatt des Deutschen Pflanzenschutzdienstes* 25, 97–101.

Johnston, K.A., Kershaw, W.J.S. and Pearce, R.S. (1989) Biochemical mechanisms of resistance of potato cultivars to slug attack. In: Henderson, I.F. (ed.) *Slugs and Snails in World Agriculture.* Monograph No. 40, British Crop Protection Council, Thornton Heath, pp. 281–288.

Kurppa, S. (1989) Pests of cultivated plants in Finland during 1988. *Annales Agriculturae Fenniae* 28, 97–102.

Moens, R. (1980) Het slakkenprobleem in de plantenbescherming. *Landbouwtijdschrift* 33, 113–128.

Moens, R. (1989) Factors affecting slug damage and control measure decisions. In: Henderson, I.F. (ed.) *Slugs and Snails in World Agriculture.* Monograph No. 40, British Crop Protection Council, Thornton Heath, pp. 227–236.

Morzer Bruyns, M.F., van Regteren Altena, C.O. and Butot, L.J.M. (1959) The Netherlands as an environment for land mollusca. *Basteria* 23, 132–162.

Peacock, L. and Norton, G.A. (1990) A critical analysis of organic vegetable crop protection in the UK. *Agriculture, Ecosystems and Environment* 31, 187–197.

Porcelli, F. and Parenzan, P. (1988) *Zonitoides nitidus* O.F. Muller (Gastropoda – Stylommatophora – Zonitidae) harmful to orchids in a greenhouse. *Informatore Fitopatologico* 38, 47–50.

Port, C.M. and Port, G.R. (1986) The biology and behaviour of slugs in relation to crop damage and control. *Agricultural Zoology Reviews* 1, 253–299.

Price, K.R., Johnson, I.T. and Fenwick, G.R. (1987) The chemistry and biological significance of saponins in foods and feeding stuffs. In: *CRC Critical Reviews in Food Science and Nutrition*, Vol. 26. CRC Press, Boca Raton, Florida, pp. 27–135.

Ruskin, F.R. (1992) *Neem, a Tree for Solving Global Problems.* National Academy Press, Washington, DC.

Stringer, A., Lyons, C.H. and Morgan, N.C. (1974) *Report of Long Ashton Research Station for 1973.* Long Ashton Research Station, Bristol, pp. 122–123.

Symondson, W.O.C. (1993) The effects of crop development upon slug distribution and control by *Abax parallelepipedus* (Coleoptera, Carabidae). *Annals of Applied Biology* 123, 449–457.

Thomas, M.R., Davis, R.P. and Garthwaite, D.G. (1991a) *Pesticide Usage Survey Report 101 – Vegetables for Human Consumption.* Ministry of Agriculture, Fisheries and Food, London, 62 pp.

Thomas, M.R., Davis, R.P. and Garthwaite, D.G. (1991b) *Pesticide Usage Survey Report 102 – Protected Crops (Edible and Ornamental).* Ministry of Agriculture, Fisheries and Food, London, 104 pp.

Thomas, R.E., Garthwaite, D.G. and Thomas, M.R. (1993) *Pesticide Usage Survey Report 121 – Outdoor Bulbs and Flowers in Great Britain.* Ministry of Agriculture, Fisheries and Food, London, 36 pp.

West, A.J. and Mordue, A.J. (1992) The influence of azadirachtin on the feeding behaviour of cereal aphids and slugs. *Entomologia Experimentalis et Applicata* 63, 75–79.

Wilson, M.J., Glen, D.M., George, S.K. and Hughes, L.A. (1995) Biocontrol of slugs in protected lettuce using the rhabditid nematode *Phasmarhabditis hermaphrodita. Biocontrol Science and Technology* 5, 233–242.

Young, A.G. and Port, G.R. (1996) Forecasting slug activity in cereal fields. In: *Crop Protection in Northern Britain 1996.* Scottish Crop Research Institute, Dundee, pp. 151–156.

Young, A.G., Port, G.R., Craig, A.D., James, D.A. and Green, T. (1996) The use of refuge traps in assessing risk of slug damage: a comparison of trap material and bait. In: Henderson, I.F. (ed.) *Slug and Snail Pests in Agriculture.* Monograph No. 66, British Crop Protection Council, Farnham, pp. 133–140.

reflecting variation in orchard management. In high-rainfall years, *C. aspersus* is a potential pest in most orchards, but not all.

C. aspersus feed on the foliage of citrus. Mature trees usually do not incur significant damage, but 1- and 2-year-old trees can sustain heavy foliage losses and indeed may be completely defoliated. On rare occasions, heavy infestations can lead to damage of the woody stems of young trees. Pappas and Carman (1961) demonstrated that defoliation resulting from heavy infestations can reduce tree vigour and fruit yield.

Of more critical concern to growers, however, is the damage done to the fruit. *C. aspersus* will feed to a limited extent on green citrus fruits, but will feed extensively on ripe fruit. Damage to fruit initially occurs as small feeding holes, where the flavedo is consumed by the *C. aspersus* to expose the white albedo. Where *C. aspersus* feeding is intense or prolonged, large holes are produced in the fruit, with the gastropods consuming both the rind and the fruit flesh. This damaged fruit is either left in the orchard or, if picked, is eliminated in the packinghouse. This, of course, is a loss for the grower. Of equal concern is fruit that sustain minor *C. aspersus* damage to the rind before picking, the signs of which are not easily observed by pickers or packinghouse sorters. These small wounds are perfect entry points for *Penicillium* (Hyphomycetes) fungi, a major postharvest decay organism in citrus fruit. The fruit is packed and transported over a period up to 2 weeks to export markets. By the time the cartons of fruit arrive and are opened, the fruit are completely covered with the blue or green mould. Not only is the shipment lost, but also future shipments are placed in jeopardy because of customer dissatisfaction.

In the high-rainfall years, fruit losses are often in the order of 40–50% and sometimes reach 90–100% (Pappas and Carman, 1961; Fisher and Orth, 1985; Sakovich and Bailey, 1985).

Until the 1970s, most growers irrigated their orchards via furrows and used soil tillage for weed control. Running cultivation equipment through the orchard several times a year undoubtedly contributed to control of *C. aspersus* through destruction of egg nests and animals residing on the ground. Contemporary orchard management relies on sprinkler or drip irrigation and weed control by pre-emergence herbicides. As a consequence, the orchard floor remains largely undisturbed and *C. aspersus* has become of increasing concern to citrus growers (Morse and Sakovich, 1986; Sakovich, 1996).

Biology of *Cantareus aspersus* in the Citrus Orchards

Basinger (1931), Gammon (1943), Ingram (1946), Hanna (1966), Fisher and Orth (1985) and the University of California Statewide Integrated Pest Management Project (UCSIPMP, 1991) provide information on the biology of *C. aspersus* in California.

During winter *C. aspersus* generally hibernate in the soil to a depth of 10–20 mm (Basinger, 1931). With the onset of rains in spring, *C. aspersus* become active. Initial activity occurs on the ground, with mating and subsequent oviposition. Eggs are deposited at 30–40 mm depth in the soil (Basinger, 1931). A large proportion of the *C. aspersus* then move up into the trees, where they spend the bulk of the time in a resting position on the trunk, leaves and fruit but feed actively during moist periods. Feeding occurs principally at night but can extend into daylight hours under conditions of morning fog and light rain. During the hot, dry summer months, especially in orchards with long intervals between irrigation events, *C. aspersus* remain inactive and may aestivate, with the shell aperture sealed with an epiphragm. Under favourable conditions, where growth is not interrupted by an aestivatory period, *C. aspersus* may reach maturity in about 4 months.

Leaf and fruit damage is usually concentrated at the skirts of the trees. It is only during unusually wet springs that *C. aspersus* can be found high in the canopy of the trees.

Control

Chemical control

Since the early 1920s, control of *C. aspersus* in citrus has relied on various molluscicidal sprays and baits (Basinger, 1923; Tubbs, 1941; Pappas and Carman, 1955, 1961, 1980; Morse and Sakovich, 1986). Methiocarb was a molluscicide that was used extensively in California citrus, until the registration was revoked recently. At present, metaldehyde and chelated iron phosphate are the only chemicals registered in California for the control of gastropods in citrus. Application rate and frequency vary, depending upon the severity of the infestation. This in turn is generally dependent upon the amount of rainfall and morning fog.

Experience has shown that placement of molluscicide baits is very important for effective control. The amount of metaldehyde ingested may be sublethal if *C. aspersus* reach sheltered, high-humidity sites after feeding on the baits. Placement of the baits midway between the tree rows may prevent intoxicated *C. aspersus* from finding shelter, but control may be compromised by the distance the animals have to travel to locate the baits. Therefore, the current recommendation is that the baits be placed neither directly under the trees nor in the middle of the alleys between the tree rows, but near the drip line of the trees (Sakovich and Bailey, 1985). While it is recognized that *C. aspersus* activity and thus control are highest when baits are applied just after rainfall or irrigation, there is much dissatisfaction among growers about the loss of efficacy of metaldehyde bait formulations under very moist conditions. Furthermore, it is not always possible to get application equipment into the orchard when the soil is wet.

Deadline 40 (Pace International LP), a liquefied formulation of metaldehyde crystals, had gained much acceptance among growers as an alternative to bait formulations for *C. aspersus* control in citrus. This material had been shown to have good efficacy against *C. aspersus* (Morse and Sakovich, 1986). That Deadline 40 is very persistent under heavy rainfall was seen by growers as a major advantage. The highly viscous material, however, required special machinery for application to the orchard floor. Some growers had purchased custom-built applicators, which, when pulled by an all-terrain vehicle, enable strips of Deadline 40 to be applied to the ground close to the drip line of the trees. Unfortunately this product is no longer registered for use in commercial citrus. More recently, a dry pellet formulation of metaldehyde, Deadline MPs, has become available and has been shown to be very efficacious in *C. aspersus* control in the orchards (N. Sakovich, unpublished data).

Molluscicidal control of *C. aspersus* is expensive, especially in years of heavy rainfall, when heavy pest infestations demand multiple applications at high rates. Further, growers are always aware that, because of the stringent environmental protection requirements, continued availability of pesticides for use in citrus is not assured. These facts, coupled with an increasing awareness of labour safety issues, have led growers to seek alternatives to pesticides.

Skirt-pruning and mechanical barriers

Removal of the lower branches of the trees, by pruning, can minimize the contact of the foliage with the ground and thus reduce the numbers of *C. aspersus* on the trees. With mature trees that are skirt-pruned for the first time, fruit yields are obviously reduced in that first year. However, losses are disproportionately lower than would be expected from the amount of pruning, as the trees compensate somewhat by setting more fruit in the canopy. Records on fruit production in Valencia orange over a 7-year period indicate that yields are not compromised by skirt-pruning of the trees (Sakovich, 1984). Indeed, fruit located in the skirt area of unpruned trees are often badly bruised, due to wind-assisted impact on the ground, in addition to being most heavily damaged by *C. aspersus*.

If growers begin the skirt-pruning when the trees are young, as is now standard practice, there is no initial fruit loss. While first-time pruning of old trees must be done by hand, the majority of pruning operations in California citrus orchards are now done mechanically with specially designed equipment.

In addition to *C. aspersus* control, skirt-pruning offers advantages in reduced incidence of Argentine ant (*Iridomyrmex humilis* (Mayr); Formicidae) (when used in conjunction with chemical treatment of the trunk), Fuller's rose weevil (*Asynonychus godmani* Crotch; Curculionidae), phytophthora brown rot and gummosis (*Phytophthora* spp.; Pythiaceae), provides for easier inspection and maintenance of the irrigation system and reduces frost damage by improving air flow (Fisher *et al.*, 1983).

Once trees are skirt-pruned, usually at a height of 450–600 mm, the trunks provide the only access into the trees for *C. aspersus*, as well as for ants and Fuller's rose weevil. A barrier must be applied. Research conducted at the University of California, Riverside, found copper to be an excellent repellent to *C. aspersus* (Sakovich and Bailey, 1985). The mechanism of this repellency is not known. There are basically two ways copper can be applied as a barrier in citrus orchards. First, the trunks can be painted or sprayed with a Bordeaux slurry. Field trials show that the spray swath must be at least 150 mm, but can be applied anywhere between the ground level and the first branches. While Bordeaux is relatively soluble, repellency is maintained for about 1 year. A small percentage of white latex house paint or a good spreader/sticker may be added to increase the persistence of the Bordeaux.

Secondly, as a more permanent barrier, a copper sheeting (Snail-Barr™) is wrapped around the trunk: one end of the sheet is stapled to the tree and then the sheet is wrapped around and fastened with a nickel-plated paper-clip, leaving sufficient space for growth expansion of the trunk (Sakovich and Bailey, 1985). Applied in this way, copper sheet generally remains effective as a barrier in California citrus for 5 years. In a few cases, it has been reported that the sheet becomes ineffective, as the surface becomes coated with substances that allow the gastropods to pass over it. In some areas, birds and other vertebrates remove the bands from the trees, presumably while seeking food, such as insects, which may reside between the sheet and trunk. Regular inspection and maintenance is needed to ensure that the copper sheets provide an effective barrier to *C. aspersus*.

Biological control

Attempts during the late 1950s and early 1960s to establish predatory gastropods – *Gonaxis* Taylor spp. (Streptaxidae) from East Africa; *Euglandina rosea* (de Férussac) (Oleacinidae) from Florida – as biological agents of *C. aspersus* in California were not successful (Fisher and Orth, 1985).

Rumina decollata (Linnaeus) (Subulinidae), a facultative predatory gastropod introduced from Europe and widespread as a naturalized species in California, has been shown to attack, kill and consume *C. aspersus* (Fisher *et al.*, 1980; Fisher and Orth, 1985). Consequently, *R. decollata* has been reared commercially and sold for use in inoculative releases in citrus orchards (Sakovich *et al.*, 1984) and further disseminated by private individuals and grower cooperatives. By 1985, *R. decollata* had been employed as a control agent in approximately 20,000 ha of citrus in southern California (Fisher and Orth, 1985). Orchards where this predator has become well established harbour either low numbers of *C. aspersus* or none at all. Fisher and Orth (1985) summarize the available information on the biology of *R. decollata* and Barker and Efford (2002) summarize its predatory behaviour.

Depending upon how much effort growers are willing to expend, *R. decollata* may be distributed throughout the orchard or placed at strategic points from which natural and assisted dispersal may occur. Provision of supplemental food (e.g. pelletized plant-based animal foods; leafy vegetables), shelter (e.g. fertilizer bags) and irrigation has been found to increase the likelihood of successful establishment (Sakovich *et al.*, 1984).

R. decollata does not provide for rapid control of *C. aspersus* infestations. Because time is required to reach effective population size and because of their feeding preference for young *C. aspersus* prey, 4–6 years are required for *R. decollata* to effect control in California citrus orchards (Fisher and Orth, 1985; Sakovich, 1996). This process may be hastened by treating the orchard with molluscicides prior to the introduction of *R. decollata*: an interval of 60 days between molluscicide treatment and *R. decollata* introduction is required, as these gastropods are susceptible to metaldehyde (Sakovich, 1996) and chelated iron phosphate (N.J. Sakovich, pers. observ.).

While *R. decollata* will also feed on plant material and is known to be a pest of cultivated plants, it does not cause damage to citrus (Fisher and Orth, 1985).

The staphylinid beetle *Staphylinus* (*Ocypus*) *olens* Müller has been shown to be a potential control agent of *C. aspersus* in California citrus (Orth *et al.*, 1975; Fisher *et al.*, 1976). However, the use of this species as a biological control agent in orchards has not been actively pursued.

Ducks (*Anas* spp. and *Cairina moschata* (Linnaeus) Anatidae) have been used for many years in other countries for gastropod and weed control. Recently, growers in California have established flocks of ducks in citrus orchards, with excellent gastropod- and weed-control outcomes. There is, of course, extra labour involved in managing the ducks. Due to the risk of predation by dogs (*Canis domesticus* Linnaeus) and coyotes (*Canis latrans* Say) (Canidae), the ducks must be penned within sturdy enclosures whenever they are not being watched. Typically, a labourer will each morning take the ducks into the orchard for as little as half an hour to scavenge for food. They are then penned for the rest of the day. This routine will continue until the orchard is freed of *C. aspersus* infestation. Increasingly, flocks of ducks are shared among a group of growers, moving from one orchard to the next on a rotation.

Integrated pest management

The control of *C. aspersus* in California citrus orchards represents one of the few examples of a truly integrated pest-management approach, which, when properly applied, can bring about complete management of an invertebrate pest (Flint, 1984). Guidelines for integrated control of *C. aspersus* in citrus are available from the UCSIPMP (www.ipm.ucdavis.edu) (UCSIPMP, 1991). The first step in an integrated approach to control involves the application of molluscicide to reduce the gastropod population. This initial step is most critical in orchards that support high *C. aspersus* populations, and the molluscicide must be applied early in the season before the pest moves into the trees in significant numbers. Then, simultaneously, the trees are skirt-pruned, copper barriers are applied to the trunks and *R. decollata* is released into the orchard. The skirt-pruning and copper barriers keep *C. aspersus* from infesting the trees during the period in which *R. decollata* numbers increase. Once *R. decollata* populations have increased sufficiently to effect control of *C. aspersus*, maintenance of the copper barriers can cease. Once pest control is achieved, the *R. decollata* may be harvested and transferred to new areas.

References

Barker, G.M. and Efford, M.G. (2002) Predatory gastropods as natural enemies of terrestrial gastropods and other invertebrates. In: Barker, G.M. (ed.) *Natural Enemies of Terrestrial Molluscs*. CAB International, Wallingford.

Basinger, A.J. (1923) A valuable snail poison. *Journal of Economic Entomology* 16, 456–458.

Basinger, A.J. (1931) *The European Brown Garden Snail in California*. Agricultural Experimental Station Bulletin 515, University of California, Berkeley, 22 pp.

Fisher, T.W. and Orth, R.E (1985) *Biological Control of Snails. Observations of the snail* Rumina decollata *Linnaeus, 1758 (Stylommatophora: Subulinidae) with Particular Reference to its Effectiveness in the Biological Control of* Helix aspersa *Müller, 1775 (Stylommatophora: Helicidae) in California*. Occasional Papers No. 1, Department of Entomology, University of California, Riverside, 111 pp.

Fisher, T.W., Moore, I., Legner, E.F. and Orth, R.E. (1976) *Ocypus olens*, a predator of brown garden snail. *California Agriculture*, March, 20–21.

Fisher, T.W., Orth, R. and Swanson, S. (1980) Snail against snail. *California Agriculture* 34, 18–20.

Fisher, T.W., Bailey, J. and Sakovich, N.J. (1983) A new approach: skirt pruning, trunk treatment for snail control. *Citrograph* 68, 292–294, 296–297.

Flint, M.L. (1984) *Integrated Pest Management for Citrus*. Publication No. 3303, University of California Statewide IPM Project, Division of Agriculture and Natural Resources, 144 pp.

Forbes, E. (1850) On the species of Mollusca collected during the surveying voyages of the *Herold* and *Pandora*, by Capt. Kellett, R.N., C.B. and Lieut. Wood, R.N. *Proceedings of the Zoological Society of London* 18, 53.

Gammon, E.T. (1943) Helicid snails in California. *State of California Department of Agriculture Bulletin* 32, 173–187.

Hanna, G.D. (1966) Introduced molluscs of western North America. *California Academy of Sciences Occasional Papers* 48, 1–108.

Ingram, W.M. (1946) The European brown snail in Oakland, California. *Bulletin of the Southern California Academy of Science* 45, 152–158.

Morse, J.G. and Sakovich, N.J. (1986) Control of *Helix aspersa* on citrus and avocodo. *Journal of Agricultural Entomology* 3, 342–349.

Orth, R.E., Moore, I., Fisher, T.W. and Legner, E.F. (1975) A rove beetle, *Ocypus olens*, with potential for biological control of the brown garden snail, *Helix aspersa*, in California, including a key to the Nearctic species of *Ocyus*. *Canadian Entomologist* 107, 1111–1116.

Pappas, J.L. and Carman, G.E. (1955) Field screening trails with various materials against European brown snail on citrus in California. *Journal of Economic Entomology* 48, 698–700.

Pappas, J.L. and Carman, G.E. (1961) Control of European brown snail in citrus groves in southern California with guthion and metaldehyde sprays. *Journal of Economic Entomology* 54, 152–156.

Pappas, J.L. and Carman, G.E. (1980) New treatment programs for brown snail control. *Citograph* 65, 235–238.

Sakovich, N.J. (1984) New methods of snail control in southern California. In: *Proceedings of the 1984 International Citrus Congress, São Paulo.* International Society of Citriculture, São Paulo, pp. 486–487.

Sakovich, N.J. (1996) An integrated pest management (IPM) approach to the control of the brown garden snail (*Helix aspersa*) in California citrus orchards. In: Henderson, I.F. (ed.) *Slug and Snail Pests in Agriculture.* Symposium Proceedings No. 66, British Crop Protection Council, Farnham, pp. 283–287.

Sakovich, N.J. and Bailey, B. (1985) Skirt pruning and tree banding as snail controls. *Citrograph* 70, 18–21.

Sakovich, N.J., Bailey, B. and Fisher, T.W. (1984) *Decollate Snails for Control of Brown Garden Snails in Southern California Citrus Groves.* Cooperative Extension Bulletin 21384, University of California, Berkeley, 14 pp.

Stearns, R.E.C. (1900) Exotic Mollusca in California. *Science* 11, 655–659.

Tubbs, D.W. (1941) Tartar emetic – a new possibility in snail control. *Orange County Farm Bureau News* 23(12), 5.

UCSIPMP (1991) *Integrated Pest Management for Citrus*, 2nd edn. University of California Statewide Integrated Pest Management Project, Berkeley, 144 pp.

18 Gastropods as Pests in New Zealand Pastoral Agriculture, with Emphasis on Agriolimacidae, Arionidae and Milacidae

Gary M. Barker

Landcare Research, Private Bag 3127, Hamilton, New Zealand

New Zealand Pastures as an Environment for Gastropods

New Zealand's pastoral agriculture is based on effective use of 14.1 million ha of mixed grass and legume pastures by 46.2 million sheep (*Ovis aries* Linnaeus; Bovidae), 4.4 million beef and 4.4 million dairy cattle (*Bos taurus* Linnaeus; Bovidae) and 1.2 million deer (*Cervus elaphus* (Linnaeus), *Dama dama* (Linnaeus); Cervidae) (Statistics New Zealand, 2001). Yet agriculturally New Zealand is a very young country. At least 75% of New Zealand's land surface was naturally forested at the time of human colonization around AD 800 (Davidson, 1984); non-forested areas were largely restricted to alpine, coastal dune land and wetland areas. Since human colonization, the transformation of the New Zealand vegetation cover has been dramatic (Salmon, 1975; Cumberland, 1981; McGlone, 1983; O'Loughlin and Owens, 1987). Considerable modification and loss of forest cover occurred during the era of exclusive Polynesian occupation. After European contact and settlement, the rate of deforestation increased, reaching a peak in the decades around the beginning of the 20th century. Deforestation following European settlement occurred predominantly for establishment of pastoral agriculture, although large areas of forest were also heavily modified by timber extraction. Present land use is dominated by pastoralism: farmed land occupies over 60% of the total area of New Zealand and only 14% remains in indigenous forest.

This massive environmental shift has led to major changes in the distribution and abundance of the indigenous fauna. A few indigenous species have managed to bridge the gap between the old and new environments and succeed in the new agricultural conditions. Indeed, several herbivorous insects such as *Costelytra zealandica* White (Coleoptera: Scarabaeidae) and several species of *Wiseana* Viette (Lepidoptera: Hepialidae) have been particularly successful in colonizing the new grasslands and occur in such high abundance as to be recognized as major pasture pests. However, most indigenous species have declined markedly under intensive agricultural regimes. New Zealand has a remarkably rich indigenous terrestrial gastropod fauna (Barker and Mayhill, 1998), but less than 1% of species, in the families Athoracophoridae, Charopidae, Punctidae and Rhytididae, persist in low abundance in natural and unimproved, induced grasslands that are used for grazing in the high country. Even fewer indigenous gastropod species occur in improved pastures, comprising almost exclusively the athoracophorid *Athoracophorus bitentaculatus* (Quoy & Gaimard) and the punctid *Paralaoma caputspinulae* (Reeve) (Appendix 18.1).

The new environments created by human colonization of New Zealand have proved to be highly vulnerable to invasion by exotic invertebrate species. Indeed, over much of New Zealand, the invertebrate faunas of pastures are strongly dominated by exotic species, many of which are significant pests. There is no evidence that terrestrial gastropod species were introduced by the Polynesian settlers. However, the habitat disturbances caused by their cultural activities undoubtedly contributed to the ease with which foreign gastropods established during the early years of European colonization (Barker, 1992). Eighteen introduced gastropod species occur in New Zealand pastures (Appendix 18.1). *Deroceras laeve* (Müller) (Agriolimacidae), a recognized pest in pastures in North America, is established in New Zealand but has yet to colonize pastures. The gastropods of greatest concern in pastures due to their effects on plant productivity are comprised entirely of slug forms, namely *Arion intermedius* Normand, *Arion hortensis* de Férussac and *Arion distinctus* Mabille in Arionidae, *Milax gagates* (Draparnaud) and *Tandonia sowerbyi* (de Férussac) in Milacidae and *Deroceras reticulatum* (Müller) and *Deroceras panormitanum* (Lessona & Pollonera) in Agriolimacidae. It is these species that this chapter specifically addresses. Other species of potential pest status are mentioned below where relevant.

The gastropod pests most important in agricultural situations are those that are most flexible in their life-history traits. This is the case in New Zealand pastoral environments. The invasive species in the genus *Deroceras* Rafinesque Schmaltz are polyvoltine and semelparous: these opportunist, *r*-strategist species can reproduce whenever conditions are suitable. Although having wide tolerances, their life-history parameters are temperature-sensitive. In the warm to cool temperate climates that prevail in New Zealand pastoral regions, there is often considerable overlap in generations, with peak reproductive activity coincident with moist, warm conditions in spring and autumn. Elsewhere, in cold temperate to Arctic regions, these same species have an annual life cycle, with overwintering predominantly as eggs produced in autumn and maturation through spring and summer. *D. reticulatum* is a heterozygous outcrossing species (McCracken and Selander, 1980; Foltz *et al.*, 1984). It has successfully adapted to the drier, unstable conditions caused by disturbance from modern farming methods (Runham and Hunter, 1970; South, 1974; Godan, 1983). *D. panormitanum* is generally most prevalent in the more moist environments. It is, however, more restricted to strongly modified environments. From electrophoresis studies, Folt *et al.* (1984) concluded that *D. panormitanum* is primarily, if not exclusively, an outcrossing species. However, the substantial level of heterogeneity among material from different localities raises the possibility of a complex of species and breeding systems. The taxonomic status of *D. panormitanum* and its relationship to material assignable to *Deroceras pollonerae* (Simroth) and *Deroceras caruanae* (Pollonera) has long been disputed (the latter two nominal species are now generally regarded as junior synonyms of *D. panormitanum* (see Giusti, 1986; Barker, 1999)).

The *Arion* de Férussac species occurring in New Zealand pastures are generally univoltine and iteroparous. Their population phenology, like that of *Deroceras*, is dependent upon prevailing environmental conditions and hence varies from year to year and from locality to locality in any one year. There may be one or two periods of recruitment per annum (generally spring and autumn). Among *Arion* species, some are warm-season-active and others are cool-season-active: hence their population events are not synchronous. *A. distinctus* and *A. hortensis* are outcrossing species (McCracken and Selander, 1980; Foltz *et al.*, 1982; Backeljau and de Bruyn, 1990) that occur widely in New Zealand pastures, albeit most commonly in the lowlands of the North Island. Allozyme studies in various parts of the world show that *A. intermedius* consists of a number of homozygous multilocus genotypes or strains that have generally been regarded as constituting a single obligated self-fertilizing, monogenic species (McCracken and Selander, 1980; Foltz *et al.*, 1982, 1984; Dolan and Fleming, 1988; Backeljau and de Bruyn, 1991; Backeljau *et al.*, 1992). The lack of heterozygotes, combined with: (i) the apparent absence of mating behaviour, (ii) the production of viable offspring by specimens reared in isolation (Chichester and Getz, 1973; Davies, 1977) and (iii) the invariance in isoelectric focusing of non-specific digestive gland esterases (Backeljau, 1985), suggests that *A. intermedius* is a selfing species (or complex of species). However, the discovery of spermatophores in Spanish (Garrido *et al.*, 1995) and New Zealand (Barker, 1999) populations suggest that at least some *A. intermedius* may engage in mating behaviour and thus may have the capacity to outcross. This species occurs most frequently in relatively undisturbed habitats (Barker, 1979; McCracken and Selander, 1980; Foltz *et al.*, 1984). It is a characteristic element of grasslands (Lutman, 1978; Godan, 1983; Barker, 1989a; South, 1989a,b) but is not widely regarded as a pest in agricultural systems.

Among Milacidae, *M. gagates* and *T. sowerbyi* have lifespans of 12–14 months. Many populations exhibit both autumn and spring periods of recruitment but individuals contribute to one generation only. Therefore these species must be regarded as univoltine and iteroparous. Electrophoretic heterozygosity indicates that *M. gagates* and *T. sowerbyi* breed by outcrossing (Foltz *et al.*, 1984), although Karlin and Bacon (1960) reported self-fertilization in *M. gagates*. Both *M. gagates* and *T. sowerbyi* are evidently restricted to high-rainfall areas and are most prevalent in the lowland and hill pastures of the North Island.

The available information on the bionomics of these pestiferous gastropods in New Zealand is summarized by Barker (1999).

The physical features of the environment play a major role in determining the distribution of gastropods and orchestrating the underlying processes of population regulation. Moisture is paramount for survival and growth, and many behavioural traits can be interpreted in terms of water-balance requirements. The mitigating influence of shelter is important in producing local microclimates where conditions are more favourable than in the ambient environment. In this respect, the structure of the habitat is of prime importance in providing an insulating effect from, for example, exposure to the desiccating effects of wind or extreme fluctuations in temperature.

The location of New Zealand within a large ocean area exposes it to moisture-laden, predominantly westerly winds. Although these moderate air temperature, orography causes variation that is noticeable in the spatial patterning of a range of climate parameters. The steep and mountainous nature of the topography dominates the landscape of mainland New Zealand (North, South and Stewart Islands), some 50% of the terrain being classed as steep land (slope > 30°) and 20% as hilly (slope 18–30°). Of the remaining 30% of land with slope less than 18°, only 3.2 million ha, or about 12% of the total land area, is classified as ploughable. This landscape is largely the result of rapid uplifts in recent geological time, coupled with strong erosion and, particularly in the south, extensive glaciation. In the North Island, a hard sedimentary mountain chain is the dominating feature, flanked on either side by broad bands of hill country formed from dissected uplifts of more recent sedimentary rocks. The centre of the Island is dominated by an elevated region of active volcanoes, surrounded by a wide plateau formed from the ash of prehistoric eruptions. Low country is limited to alluvial valley and sandy coastal strips. The South Island is dominated by the Southern Alps, comprising hard metamorphoses and sedimentary rock mountains to 4000 m, flanked on either side by lower ranges of hills of similar material. Lowland is limited to intermontane basins, alluvial-valley plains and border plains, dominated by the Canterbury Plains built up on the eastern side of the Alps.

As illustrated in Plate 19, mainland New Zealand can be divided into seven broad agroclimatic zones. When considering the New Zealand region as a whole, two additional agroclimatic zones may be recognized, namely the sub-Antarctic islands and Chatham Islands. This diversity is a function of variable climate and land resource. Most important are variations in summer rainfall, winter temperature, soil drainage and nutrient status, and land contour. The composition of the gastropod fauna in pastures and the abundance and phenology in species vary according to these agroclimatic zones.

High-fertility, moderate–high-rainfall lowlands

In the North Island, dairy farms dominate about 1 million ha of flat to low undulating land, while sheep and beef finishing systems occupy a further 1 million ha of undulating land of < 15–20° slope. Within both systems, white clover (*Trifolium repens* Linnaeus; Fabaceae) is commonly used as the pasture legume to provide forage of high nutritive value for livestock and, more importantly, for development and maintenance of an improved nitrogen cycle. In the subtropical areas C_4 grasses, such as Kikuyu (*Pennisetum clandestinum* Chiov.; Gramineae) and paspalum (*Paspalum dilatatum* Poiret; Gramineae) predominate over C_3 grasses, principally ryegrass (*Lolium* Linnaeus spp.; Gramineae). In the central North Island areas, perennial ryegrass (*Lolium perenne* Linnaeus) is the dominant grass. The zone receives 1000–2500 mm rainfall $year^{-1}$, with precipitation least during summer and autumn. Pasture production ranges from 10 to 17 t dry matter (DM) ha^{-1} $year^{-1}$. Lucerne (*Medicago sativa* Linnaeus; Fabaceae) is used locally on free-draining soils either as a specialist crop for hay or as a dual-purpose, hay–grazing crop.

D. reticulatum is the most common gastropod species but *D. panormitanum* can be equally abundant locally, especially on wet soils. *A. hortensis*, *A. distinctus*, *A. intermedius*, *M. gagates* and *T. sowerbyi* occur throughout the zone and are most numerous on clay soils under sheep grazing. *Vallonia excentrica* Sterki (Valloniidae) occurs in high abundance in northern New Zealand pastures (Barker, 1985, 1999). Species of *Vallonia* Risso are not generally recognized as pests in pasture, although damage to seedlings has been observed (Muhle, 1941; G.M. Barker, personal observation). *Prietocella barbara* (Linnaeus) (Hygromiidae) and *Lehmannia nyctelia* (Bourguignat) (Limacidae) are becoming increasingly prevalent in northern areas (Barker, 1999), but their pest status in New Zealand pastures is currently not known. Working in Queensland, Australia, Yamashita *et al.* (1979) demonstrated that while *L. nyctelia* feeds readily on a range of pasture plants, including the forage legume Kenya white clover (*Trifolium semipilosus* Fernald), this gastropod was not particularly prevalent in grazed pastures. *P. barbara* is recognized as a pasture pest in South Australia (Baker, 1989).

The warm, humid conditions conducive to high gastropod populations are tempered by summer conditions of high temperatures and low soil moisture, coupled with high pasture utilization. At most locations populations of the agriolimacids are characterized by marked temporal instability (Fig. 18.1). Population density > 50 m^{-2}, common in spring following moist summers, is often followed by population collapse and stabilization at low density for several years. Epizootics of disease caused by *Microsporidium novacastriensis* (Jones & Selman) (Microspora) and especially *Tetrahymena rostrata* (Kahl) (Ciliophora) is responsible for the density-dependent regulation (Barker, 1993a; G.M. Barker, unpublished data) (Fig. 18.2). Both pathogens were shown in laboratory experiments to reduce the feeding, growth rate and fecundity of *D. reticulatum* and *D. panormitanum*. Studies on temperature relationships of pathogenicity supported the hypothesis that epizootic mortality in the *Deroceras* populations caused by *T. rostrata* was mediated by the onset of hot, drying weather in late spring–early summer (Barker, 1993a; G.M. Barker, unpublished data). At some sites, this pattern of mortality is apparently moderated by maintenance of humid, cool microhabitats over the summer, although the incidence of infection by *M. novacastriensis* and *T. rostrata* can remain high. These conditions are provided over large areas by dense vegetative cover, particularly that of Kikuyu, and by clay soils, which during summer remain moist or crack extensively. At a smaller scale, thistles (*Carduus nutans* Linnaeus, *Cirsium vulgare* (Savi) Tenore; Asteraceae), ragwort (*Senecio jacobaea* Linnaeus; Asteraceae) and

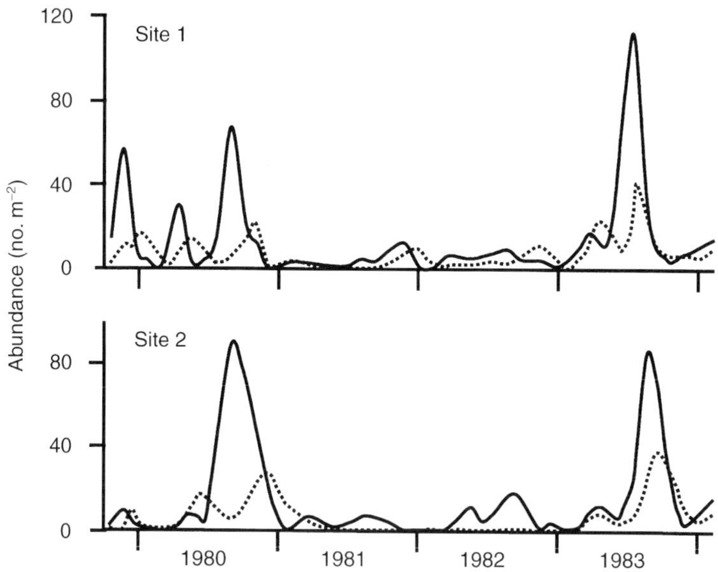

Fig. 18.1. Temporal variation in abundance of herbivorous gastropods in two lowland pastures in the Waikato, northern New Zealand. Data are presented only for the agriolimacids *Deroceras reticulatum* (Müller) (________) and *Deroceras panormitanum* (Lessona and Pollonera) (..........), as these two invasive species dominate gastropod communities in these intensively grazed pasture systems. (After Barker and Addison, 1992; Barker, 1993a; G.M. Barker, unpublished data.)

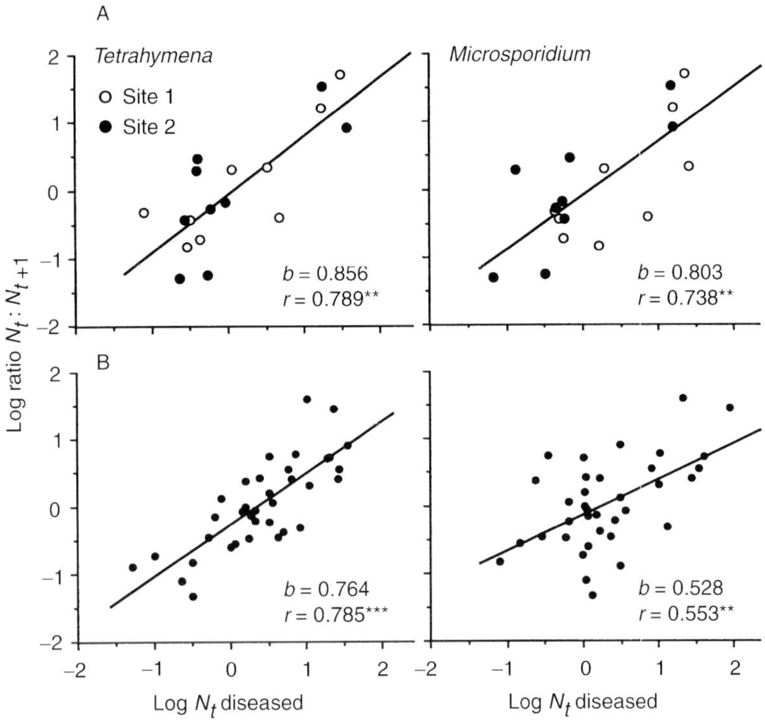

Fig. 18.2. Relationship between the incidence of disease caused by *Tetrahymena rostrata* (Kahl) (Ciliophora) or *Microsporidium novacastriensis* (Jones & Selman) (Microspora) and population growth in *Deroceras reticulatum* (Müller) (Agriolimacidae) in lowland pastures of the Waikato, northern New Zealand. The incidence of disease is expressed as the logarithm of numbers of *D. reticulatum* m^{-2} infected by *T. rostrata* or *M. novacastriensis*. Population trend is expressed as the logarithm of the ratio of the numbers of *D. reticulatum* m^{-2} at sampling times t and $t + 1$. Population increases are indicated by ratio values > 0 and population declines by ratio values < 0. **A.** Relationship between disease incidence and intergeneration population trend for two populations (see Fig. 18.1) sampled in each of ten consecutive generations. **B.** Relationship between disease incidence and annual population trend for 12 populations sampled in spring (September) in each of 4 consecutive years. (After Barker, 1993a; G.M. Barker, unpublished data.)

other robust weeds can provide shelter within otherwise closely grazed pasture. Furthermore, it is in this high-fertility, moderate- to high-rainfall zone that the introduced lumbricid earthworms are most prevalent. Their burrows provide important refugia for the non-borrowing gastropods, such as *D. reticulatum*.

In the South Island, sheep and beef finishing systems traditionally occupy this zone on flat and undulating land in Southland/southern Otago, and sheep, beef and dairying systems in Westland. In recent years there has been substantial conversion to dairying in Southland. The winters in these South Island areas are long and cold and the summers cool and moist, relative to that in the North Island. The pastures are comprised mainly of perennial ryegrass and white clover. Gastropod populations here are predominantly *D. reticulatum*, *D. panormitanum*, *A. intermedius* and *M. gagates*. Nothing is known of their bionomics or seasonal abundance in the region.

Throughout this lowland, high-rainfall zone, stock management has a direct influence on gastropod abundance. Gastropods are susceptible both to treading by farm livestock and to desiccation resulting from the rapid removal of vegetative cover effected by rotational grazing. Beginning in late autumn, dairy cattle are typically stocked at high densities (250–500 cows ha^{-1} day^{-1}) on a limited area of pasture in order to conserve and accumulate feed on the remainder of the farm for late winter–spring, when calving and milking commence. Pasture is not so rigidly rationed to herds in milk, which are rotated around the pastures at stocking rates of 60–80 cows ha^{-1} day^{-1}. Populations of gastropods in the areas subject to high stocking in autumn are greatly reduced (G.M. Barker, unpublished data). Populations in the winter-spelled pastures increase rapidly and can cause substantial damage to the clover. In pastures grazed by sheep, substantial mortality in gastropods can result during periods when large flocks of sheep are rotationally grazed at effective stocking rates of 1000–2000 sheep equivalents ha^{-1} day^{-1}. Under these circumstances, Ferguson *et al.* (1989a) demonstrated that the combined effect of treading and habitat alteration can cause 90% mortality in *D. reticulatum* (Fig. 18.3).

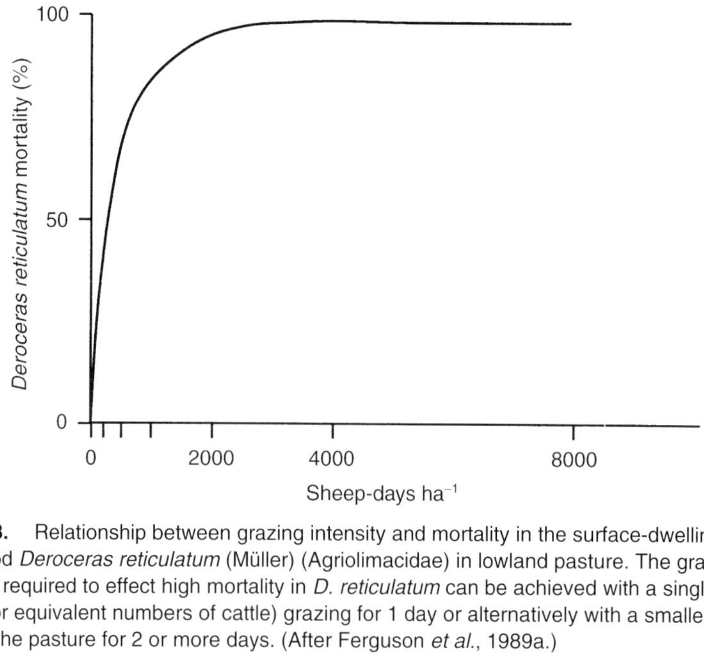

Fig. 18.3. Relationship between grazing intensity and mortality in the surface-dwelling gastropod *Deroceras reticulatum* (Müller) (Agriolimacidae) in lowland pasture. The grazing intensity required to effect high mortality in *D. reticulatum* can be achieved with a single flock of sheep (or equivalent numbers of cattle) grazing for 1 day or alternatively with a smaller flock grazing the pasture for 2 or more days. (After Ferguson *et al.*, 1989a.)

Low–moderate-fertility, hill land

This zone occupies 5 million ha and traditionally carries breeding sheep and beef cattle at stocking rates of 10–12 sheep equivalents ha^{-1}.

While summer drought is common in low-rainfall eastern regions (750–1000 mm rainfall per annum), localized drought can also occur within wetter, western regions (1000–2000 mm rainfall $year^{-1}$) as a result of slope and aspect factors. Within these contrasting hill environments, similar plant species are found, but their proportion varies. White clover is generally sought as the dominant legume, but, where conditions are unstable, opportunistic annual legumes (*Lotus* Linnaeus and *Trifolium* Linnaeus spp.) become abundant. The most common grasses are browntop (*Agrostis capillaris* Linnaeus; Gramineae), sweet vernal (*Anthoxanthum odoratum* Linnaeus; Gramineae) and perennial ryegrass. Except for some semiarid areas of eastern New Zealand, pasture is an unstable vegetation and the successional drift back to forest cover is stalled by fertilizer inputs and grazing pressure.

Common to all hill country are the dominating effects of slope and aspect on pasture production. Temperatures and potential pasture winter growth rates increase on steeper and more north-facing slopes, but the same physical factors accentuate summer moisture deficits through increased water runoff and evaporation. In addition, slope and aspect induce preference–rejection of land and vegetation type by grazing animals and result in areas of over- and undergrazing and nutrient transfer. Commonly, pasture on easier slopes, kerbs of contour tracks (see Fig. 18.4) and warmer aspects are grazed in preference to that on steep, intertrack and colder land. Division of paddocks into uniform aspects, slopes and vegetation types and rotational grazing with high animal densities of 300–400 sheep equivalents ha^{-1} are primary means of achieving pasture management.

These variations in climate, topography and grazing management have important effects on gastropod populations dominated by *D. reticulatum* and *A. intermedius*. While abundance is generally higher and fluctuates less from year to year in summer-moist areas, *D. reticulatum* and *A. intermedius* populations exhibit a marked spring peak in density (Barker, 1991a; B.E. Burridge and J.F.L. Charlton, personal communication). Furthermore, even within high-rainfall areas, their dispersion is aggregated. Working in summer-moist hill-country pastures of the Kaimai Range in the North Island, Barker (1991a) found populations of *D. reticulatum* and and *A. intermedius* are generally higher on the cooler southerly aspects than on the warmer north-facing aspects of the hill slopes. Furthermore, high abundances of *D. reticulatum* were found to be associated with the moist, undisturbed areas of track kerbs and upper intertrack slopes, while abundances of *A. intermedius* were highest for the intertrack slopes. Low abundance of both species occurred on the tracks and at sheep camp-sites (Fig. 18.4). Analysis of cohort life-tables indicated that these dispersion patterns were important in the dynamics

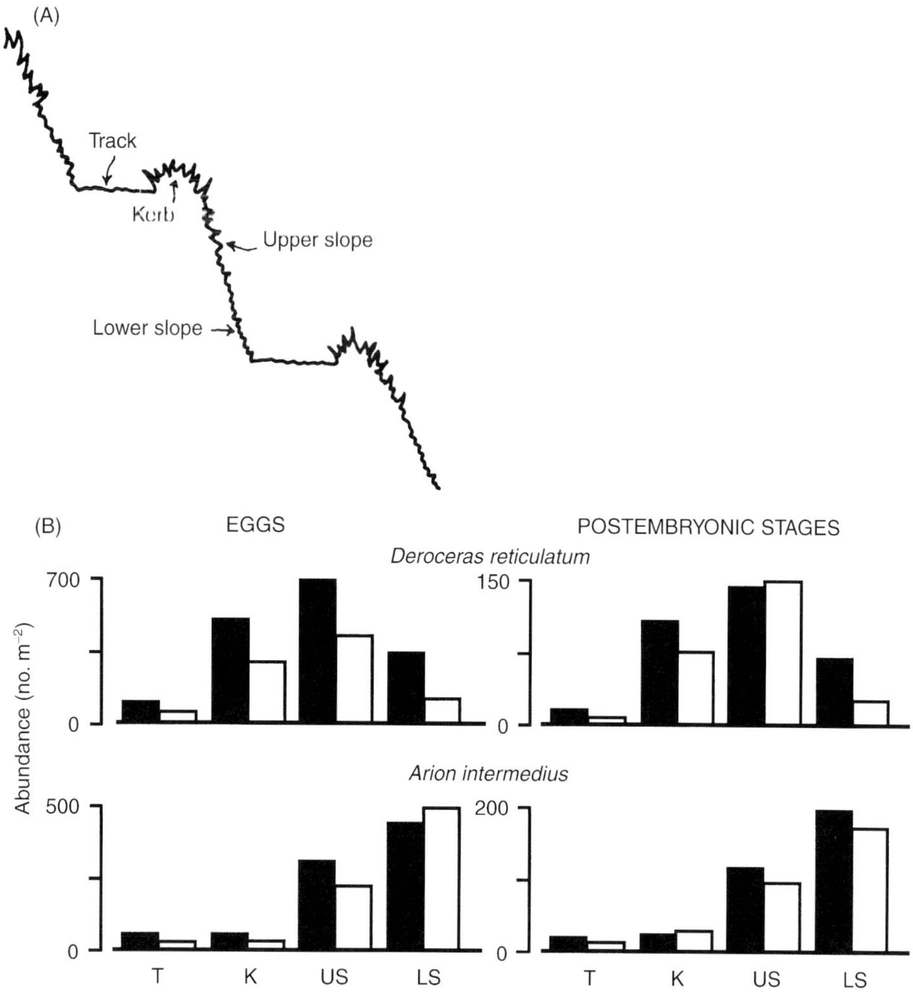

Fig. 18.4. Dispersion of *Deroceras reticulatum* (Müller) (Agriolimacidae) and *Arion intermedius* Normand (Arionidae) in moist hill-country pasture in the Kaimai range, northern New Zealand, in relation to microtopographic variation on the hill slopes. **A.** Illustration of the microtopographic features of the hill slopes, with nominal slope strata indicated. **B.** Mean abundance of egg and postembryonic stages of *D. reticulatum* and *A. intermedius* in track (T), kerb (K), upper-slope (US) and lower-slope (LS) strata. Shaded bars are for south-facing slopes and the open bars for north-facing slopes. (From Barker, 1991a.)

of the populations. The stage mortality contributing most to the variance in reproductive populations of both species was that between hatching and onset of reproduction; most of the mortality occurred shortly after hatching on south-facing slopes, but tended to occur later in the life cycle on north-facing slopes. For *D. reticulatum* this mortality was inversely related to initial density and associated with predation by carabids and birds and with treading by sheep. Density relationships indicated that variations in natality had a stabilizing influence on *D. reticulatum* populations. The prevalence of *M. novacastriensis* and *T. rostrata* was found to be very low in *D. reticulatum* in these cool, moist environments (G.M. Barker, unpublished data).

Charlton *et al.* (1985) concluded, from analyses of pitfall-trap catches in summer-moist hill country of southern North Island, that pastures set-stocked with sheep had lower gastropod numbers and activity than pastures rotationally grazed with sheep, while numbers and activity were highest in pastures rotationally grazed by cattle. They also concluded that abundance, as measured by trap catches, was lower on southerly than on northerly aspects. Their conclusions must be viewed with caution, however, as pitfall-trap catches could equally be interpreted as a measure of activity induced by periodic pasture defoliation. On moderate to steep hills, treading by animals is largely confined to contour tracks and therefore has less impact on gastropods than on flat land.

In summer-dry hill country, localized areas of dense vegetation and hillside seepage areas provide the moist refugia critical to gastropod population persistence.

Cold intermontane and high-altitude land

This zone occupies approximately 4.5 million ha. It is located at altitudes of 300–1000 m where winter temperatures are below the threshold for pasture growth for 3–5 months. Fine-wool sheep and store beef cattle are farmed on large holdings at stocking rates of 1–3 sheep equivalents ha^{-1}. 'Unimproved' native tussock grassland communities are principally dominated by indigenous grass species in the genera *Chionochloa* Zotov (Gramineae), *Festuca* Linnaeus (Gramineae), *Poa* Linnaeus (Gramineae) and *Rytidosperma* von Steudel (Gramineae) and include a diversity of forbs. These grasslands often now include adventive grasses, such as browntop and sweet vernal, and dicotyledonous weeds, such as *Hypochaeris radicata* Linnaeus, *Hieracium pilosella* Linnaeus and *Hieracium praealtum* Gochnat (all Asteraceae). These grasslands are characterized by high biomass and low productivity (2–3 t DM ha^{-1} $year^{-1}$). Improvements in herbage availability, particularly for winter, are sought through the use of fertilizers and the introduction of clovers (principally white clover and alsike clover (*Trifolium hybridum* Linnaeus)) and grasses (principally perennial ryegrass and cocksfoot (*Dactylis glomerata* Linnaeus; Gramineae)). Tussock grassland productivity and stock-carrying capacity can be increased considerably by the introduction of these exotic pasture species. It is normally recommended that a legume, commonly white clover, is established first in order to raise soil fertility levels before grasses can be successfully introduced. In their natural state, these soils generally have low pH and low fertility and are lacking in rhizobia. Therefore, distribution of the appropriate fertilizers, inoculation of seed with *Rhizobium* Frank (Eubacteriales) and coating with lime are required for successful oversowing. Those developments may increase annual production to 5–10 t DM ha^{-1}.

D. reticulatum and *A. intermedius* occur throughout this grassland zone in low abundance, often associated with tussock plants, rocks and moist areas adjacent to streams and seepage areas. Nothing is known of their bionomics in this zone.

Dry arable

Eastern livestock and cropping enterprises include a pasture component within rotating land-use patterns. These are generally short-term pastures, with longevity of < 10 years. Perennial ryegrass and white clover predominate in pastures, which may also include a wide range of adventive species, such as browntop, sweet vernal, crested dogstail (*Cynosurus cristatus* Linnaeus: Gramineae), *Poa* spp. and subterranean clover (*Trifolium subterraneum* Linnaeus). Within this subhydrous zone (500–800 mm rainfall year^{-1}), irrigation and lucerne can reduce many of the farming uncertainties. Recently, dairying has expanded on irrigated land on the Canterbury Plains.

Of the several gastropod species present, only *D. reticulatum* is common. In dryland pastures, populations are almost invariably low but can increase under irrigation and in pastures closed for seed crops.

Sub-Antarctic and Chatham islands

The main island groups within the territory of New Zealand, usually described as 'sub-Antarctic', are the Auckland (46,000 ha) and Campbell (11,330 ha) Islands, almost due south of Stewart Island, and the Antipodes and Bounty Islands farther out to the south-east of it. The climate is characterized by extreme wind, cloudiness, high humidity, uniform cool temperatures and moderate rainfall (Auckland – 1780 mm, Campbell – 1400 mm rainfall year^{-1}). The Auckland and Campbell Islands are in large part overlain by blanket peats. In the 1890s, cattle and sheep were introduced on the Auckland and Campbell Islands, but the isolation of the islands caused these farming ventures to be abandoned, as late as 1931 in the case of Campbell Island. Thus, while farming no longer occurs in the New Zealand sub-Antarctic, it is of interest that the adventive gastropods that characterize pastures in cool, moist environments on the mainland have established and persisted in these island systems: *D. reticulatum* occurs in the Auckland and Campbell Islands and *A. intermedius* in the Aucklands. On Campbell many of the distinctive vegetation communities – such as megaherb fields, tall *Chionchloa antarctica* (Hook) Zotov tussock and *C. antarctica–Dracophyllum scoparium* Hook (Epacridaceae) mosaics – were strongly modified by grazing, often leading to induced *Poa litorosa* Cheeseman

(Gramineae)–*Bulbinella rossii* (Hook) Cheeseman (Asphodelaceae) short-tussock meadows. Some of these induced meadows are now dominated by introduced European grasses. It is in these modified vegetation communities that *D. reticulatum* primarily occurs. The impacts of grazing on the vegetation of the Auckland Islands was less dramatic, although still significant locally, as on Enderby Island. The induced turf communities are dominated by indigenous plant species, including *P. litorosa*, *B. rossii*, *Dracophyllum longifolium* (Forster and Forster) Brown, *Cassinia leptophylla* (Forster) Brown (Asteraceae) and *Oreobolus pectinatus* Hook (Cyperaceae). *D. reticulatum* and *A. intermedius* persist in these modified systems.

The Chatham Islands are situated *c.* 800 km east of the New Zealand mainland and principally comprise two relatively large inhabited islands, Pitt (6200 ha) and Chatham (90,000 ha). The climate is temperate and oceanic, with rainfall generally less than 1250 mm year^{-1}. A large part of the land area of the Chatham Islands comprises blanket peats over basalt, derived from *Dracophyllum arboreum* Cockayne-dominated forests and *Sporadanthus traversii* (Müller) Kirk–*D. scoparium* wetland, or raised peat domes, derived from *S. traversii* (Restionaceae) and *D. scoparium* over consolidated sands. Much of the land area has been converted to pastures, which are similar in character to the summer-moist hill country of the North Island, being predominantly of browntop and sweet vernal. Perennial ryegrass and white clover occur where there have been improvements in soil fertility. *D. reticulatum* and *A. intermedius* occur throughout these pastures, with *D. panormitanum* and *M. gagates* primarly restricted to the more improved pastures.

Gastropods as Pests of Established Pasture

Nature of damage

Gastropod herbivory on established plants in New Zealand pastures generally comprises five forms: (i) damage to the leaf lamina of dicotyledonous plants; (ii) removal of meristem tissues of dicotyledonous plants; (iii) removal of the leaf and sheath lamina of monocotyledonous plants; (iv) damage to underground storage organs and fleshy roots of dicotyledonous plants; (v) damage to inflorescences of dicotyledonous plants. A very high proportion of white-clover leaves can exhibit damage to the leaf lamina in winter and spring (e.g. Barker and Addison, 1992) and it is this form of damage that is most widely recognized by farmers. Damage to inflorescences in dicotyledons has only been noted for the weed species creeping buttercup (*Ranunculus repens* Linnaeus; Ranunculaceae) and dandelion (*Taraxacum officinale* Weber; Asteraceae) and will not be considered further here.

Additionally, gastropods can be selective predators on seed and seedlings occurring in the established pasture, thus potentially interfering with natural regenerative processes. A wide range of dicotyledonous and monocotyledonous plants are damaged in the seed and seedling stages, but most farmers and agronomists have been concerned only with gastropod herbivory on forage legumes.

Risk areas and extent of losses

Gastropods have generally not been recognized as pests of white clover in established pasture in New Zealand, despite the common occurrence and the obvious symptoms of clover defoliation. However, some farmers in northern North Island have expressed concern about the loss of the nutritive quality of their dairy pastures as a result of the extensive removal of leaf lamina in winter and spring effected by gastropods and other defoliating pests, such as the collembolan *Sminthurus viridis* Linnaeus.

Working in lowland pastures of the Waikato region, grazed by sheep and beef cattle, Barker *et al.* (1985) and Barker and Addison (1992) demonstrated that agriolimacid infestations of 20–80 m^{-2} (biomass, 3–20 g m^{-2}, 30–200 kg ha^{-1}) were associated with extensive damage to leaf lamina from white clover (Fig. 18.5). Applications of molluscicides reduced gastropod numbers in these pastures and resulted in increases in pasture white-clover content, which was reflected in 12–40% annual increases in clover yield. The molluscicide treatments increased the number of white-clover stolon growing points, indicating that agriolimacids in these pastures influenced clover growth primarily by damage to stolon apical tissues and nodal buds. No other quantitative data are available on the importance of gastropod herbivory on white clover in New Zealand, despite the critical role of this legume in most pastoral environments. Barker and Addison (1992) suggested that, under conditions of sustained high gastropod abundance and damage, the productivity of white clover, and eventually that of the entire sward, would be adversely affected. In their particular study of infestations in two pastures of the Waikato, Barker and Addison (1992) found agriolimacid populations to vary markedly over a 26-month period and high levels of herbivory on white clover occurred principally in winter and spring. Other studies in the Waikato (Barker, 1993a; G.M. Barker, unpublished data) have demonstrated a marked instability in populations of agriolimacids at a number of sites, suggesting that instability is a feature of populations in this type of summer/autumn-dry environment, which occupies about 2 million ha in the North Island.

Greater stability and thus more sustained herbivory by gastropods are likely to occur in cool, summer-moist environments. This applies to extensive areas of hill country in the North Island and the lowlands of Westland and Southland in the South Island, such that about 2 million ha may be under sustained pressure from herbivory. The contribution of

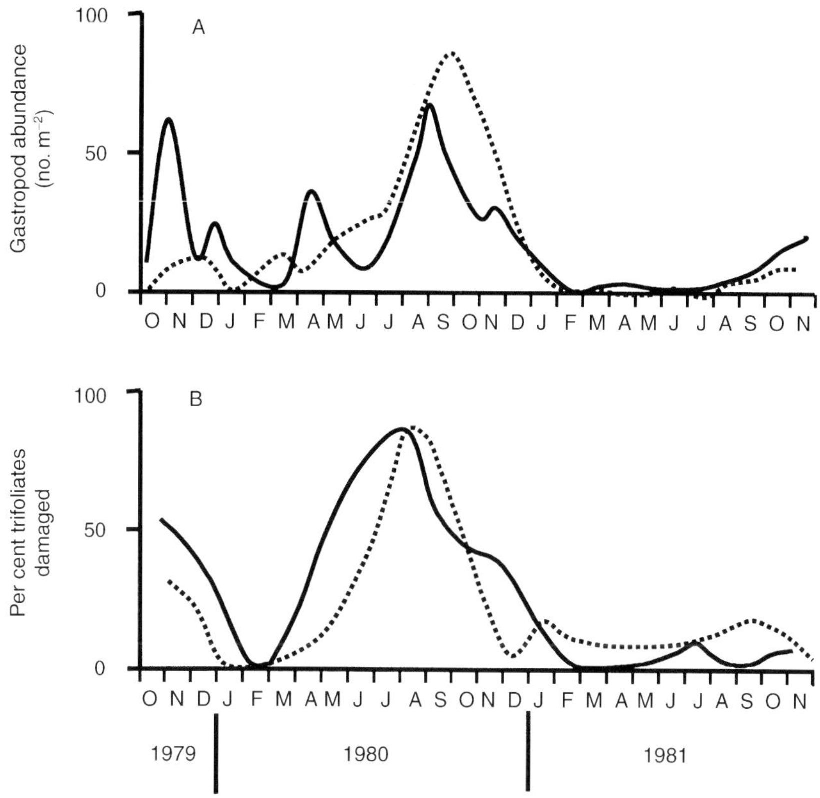

Fig. 18.5. **A**. Temporal variation in abundance of herbivorous gastropods in two lowland pastures in the Waikato, northern New Zealand. The herbivorous gastropod fauna at both sites were dominated by *Deroceras reticulatum* (Müller) (Agriolimacidae), but included also *Deroceras panormitanum* (Lessona & Pollonera) (Agriolimacidae), *Milax gagates* (Draparnaud) (Milacidae), *Arion hortensis* de Férussac and *Arion intermedius* Normand (Arionidae) at Site 1 and *Deroceras panormitanum* at Site 2. **B**. Temporal variation in the intensity of damage inflicted on white clover (*Trifolium repens* Linnaeus) (Fabaceae) foliage at these two sites. ____________, Site 1; ...., Site 2. (After Barker, 1993a.)

sustained, selective herbivory to the low content of white clover in summer-moist hill country has not been addressed.

The importance of the other forms of gastropod herbivory in established New Zealand pastures (itemized above) has yet to be adequately examined but is discussed below.

Ecology of established pastures and role of gastropod herbivory

Nitrogen (N) is the element that most limits the productivity of developed pasture. Where little artificial N is applied, as has been the case generally in New Zealand, pasture productive potential is dependent on N fixation by *Rhizobium* associated with the root systems of forage legumes. White clover is capable of fixing in excess of 500 kg N ha^{-1} annually under optimum conditions, although in reality much smaller quantities are usually fixed in New Zealand pastures (Hoglund *et al.*, 1979; Crush, 1987). Legumes also have the potential to contribute to animal performance directly through their superior digestibility and protein content compared with grasses and by improving voluntary intake in sheep and cattle. Furthermore, production in mixed-species pasture, in the absence of fertilizer N, is potentially greater than when either white clover or ryegrass is sown alone. This results from differences and complementarity in morphology, seasonal growth patterns and pest susceptibility, which allow one species to compensate for temporary setbacks in the growth of its companion. These attributes have provided for a lower cost of animal production in New Zealand compared with northern-hemisphere pastoral agriculture. None the less, a sensitive competitive balance exists between the legume and its companion grasses – white clover is usually the junior and more sensitive partner in this relationship. Furthermore, white clover is integrally associated with its rhizobial commensal, which is also directly sensitive to environmental changes (Chanway *et al.*, 1989). Many biotic and abiotic factors can affect the competitive relationship between white clover and grasses, and small changes in pest burdens may initiate large, temporal changes in pasture species composition by shifting the competitive balance in one direction. Similarly, small improvements in competitive vigour of white clover by reducing the burden of a major pest may improve the sustainable productive potential of pasture by improving both the level and the stability of N inputs.

A number of demographic studies have highlighted the dynamics of white clover during spring as critically important to the vegetative persistence of this legume and the annual levels of N fixation. Once past the seedling stage, white clover relies on stolons and their associated nodal roots to persist. The growth pattern of white clover in both lowland and hill pastures of New Zealand involves an annual minimum in the size of the stolon populations in spring, when large amounts of stolon die and decay, followed by a rebuilding of the stolon pool over the following summer. Changes in plant morphology parallel this pattern, with the death of older stolons causing larger plants to break up during spring, with resulting daughter plants rapidly increasing in size over the subsequent summer (Chapman, 1983; Hay, 1983; Hay *et al.*, 1983, 1989a; Brock *et al.*, 1988; Harris, 1993). Thus, individual white-clover plants are smallest and most vulnerable to stress in spring (Brock *et al.*, 1988). White-clover stolons in spring have a low starch content, which means their energy reserves are only 10–15% of those in autumn (Hay *et al.*, 1989b) and plants are slow to recover from damage. The results of Barker *et al.* (1985) and Barker and Addison (1992) point to gastropods, through selective herbivory on plants at their most vulnerable stage, making a substantial contribution to the pest burden and associated losses of white clover in New Zealand pastures. The work of Barker *et al.* (1985) and Barker and Addison (1992) needs to be extended over a longer time frame to demonstrate that selective herbivory by gastropods is manifest in longer-term reductions in pasture productivity and quality. Under conditions of sustained high gastropod populations, as in wet years or districts, the competitive ability and persistence of white clover may be seriously reduced, as noted by Howitt (1961) for the cultivar Ladino in Washington.

Over winter, the majority of white-clover stolons are buried below the soil surface (Hay *et al.*, 1983, 1987; Hay, 1985) and the plant populations remain relatively stable. At that time, most stolon apical tissues and nodal buds are available only to burrowing gastropod species, such as *A. hortensis* and *M. gagates*, which are generally less abundant that the surface-dwelling agriolimacids.

The percentage of legume required to have any significant effect on animal performance is generally considered to be much higher than the current low average of 10–20% found in many New Zealand white clover/ryegrass pastures (Chapman *et al.*, 1995; Caradus *et al.*, 1996). Harris *et al.* (1997) showed that summer-pasture legume contents of 50–65% are required to achieve near-maximum, per cow, milk production. Dairy farmers are increasingly less willing to rely on the white-clover–*Rhizobium* association to fix sufficient N to meet the demands of high-producing pastures or the nutritional requirements of high-producing dairy herds. The use of N fertilizer has increased markedly in recent years. This reduced reliance on white clover has been heightened by both the increased returns for milk products and thus higher yield expectations from pastures and the realization that chronic pest burdens substantially reduce the legume content and N-fixation capacity of pastures (Watson *et al.*, 1985, 1993; Watson and Mercer, 2000). While some progress has been made in development of clovers resistant to root-knot nematodes (*Meloidogyne hapla* Chitwood, *Meloidogyne trifoliophila* Bernard & Eisenback; Heteroderidae) and clover cyst nematodes (*Heterodera trifolii* Goffart; Heteroderidae), the principal pests involved (van den Bosch and Mercer, 1996; Mercer and van den Bosch, 1997; Mercer *et al.*, 1999; Watson and Mercer, 2000), the willingness to rely on N fixation has been further undermined by the establishment of new pests attacking clovers, such as clover root weevil (*Sitona lepidus* Gyllenhal; Coleoptera, Curculionidae) (Barker *et al.*, 1996; Barratt *et al.*, 1996; Willoughby *et al.*, 1999). The limitations of white clover are also of increasing concern to pastoral farmers in other environments. Unfortunately, the use of N fertilizers by farmers as a strategy to alleviate N stress in pastures further compromises the persistence of white clover in mixed swards.

White clover is polymorphic in respect to cyanogenesis. Its ability to release hydrogen cyanide (HCN) from damaged leaves demands the presence of both the cyanogenetic glucosides linamarin and lotaustralin and the β-glucosidase enzyme linamarase, which is capable of hydrolysing the glucosides. In white clover, this system is genetically controlled by genes at two unlinked loci, such that one allelic pair (generally designated *Ac*/*ac*) controls the production of the glucosides and the other pair (designated *Li*/*li*) controls production of the enyzme. The alleles *Ac* and *Li* are completely dominant to *ac* and *li*, respectively. Thus populations polymorphic at both loci include four phenotypes: (i) *Ac/Li* plants, with glucosides and enzyme, are cyanogenetic; (ii) *Ac/li* plants, with glucosides but no enzyme, are acyanogenetic; (iii) *ac/Li* plants, with enzyme but no glucosides, are acyanogenetic; (iv) *ac/li* plants, with neither glucosides nor enzyme, are acyanogenetic (Corkill, 1942; Attwood and Sullivan, 1943; Nass, 1972; Hughes, 1991). Corkill (1952) found that gastropod damage was often more severe in New Zealand on glucoside-free white-clover plants, which is generally consistent with subsequent studies indicating that herbivorous gastropods prefer acyanogenic plants and thus selectively graze more on acyanogenetic plants than on cyanogenetic ones (e.g. Angseesing, 1972; Dirzo and Harper, 1982a,b). However, it is clear from a number of laboratory and field experiments that the defence against gastropods conferred by cyanogensis is not absolute. Where the gastropods have a choice, the defensive role of cyanogenesis is expressed both as an under-representation of heavily damaged cyanogenetic plant tissues and as an excess of undamaged and no-more-than-nibbled cyanogenetic tissues. The excess of nibbled tissues of the cyanogenetic phenotypes is indicative of 'sampling' made by the gastropods, as cyanogenesis does not appear to be associated with any aposematic feature and defence against herbivory is dependent on immediate release of HCN upon damage of the tissue (Dirzo and Harper, 1982a). Where gastropods do not have a choice, as in pasture situations where the white-clover component of the sward is comprised of cyanogenetic plants, substantial feeding may occur on the clover.

Working in Scotland, Burgess and Ennos (1987) determined that *D. reticulatum* obtained from pastures with a low frequency of cyanogenetic white clover showed a significantly greater degree of selective herbivory of acyanogenetic morphs than animals taken from a site containing a high frequency of cyanogenetic clover. Differences in selectivity between *D. reticulatum* populations were caused both by differences in the rate of initiation of feeding on cyanogenetic morphs and by differences in the extent of damage once feeding had been initiated. Differences in feeding behaviour among *D. reticulatum* were apparently genetically based rather than being dependent on prior feeding experiences. Dirzo and Harper (1982a) found some evidence for polymorphism in detoxification of HCN among *Deroceras*, although this has yet to be examined thoroughly. The implication from the study of Burgess and Ennos (1987) is that the selective advantage enjoyed by the cyanogenetic morph under herbivory by *D. reticulatum* is likely to be frequency-dependent. At sites where cyanogenetic frequency is low, cyanogenetic morphs enjoy substantial protection from herbivory, whereas, where cyanogenetic morph frequency is high, this selective advantage is much reduced. It is likely that, when cyanogenetic morphs occur at high frequency, they may enjoy very little protection from gastropod herbivory. As white clover in New Zealand is predominantly cyanogenetic (Caradus *et al.*, 1989), we would predict from the results of Burgess and Ennos (1987) that no advantage in pest control would accrue from sowing cyanogenetic cultivars into gastropod-infested pastures. Observations of damage to white clover in New Zealand pastures (G.M. Barker, unpublished data) support this contention. Indeed, assays of white clover selections from New Zealand hill-country pastures (see Macfarlane and Sheath, 1984; Macfarlane *et al.*, 1990) indicated high rates of acceptance of cyanogenetic ecotypes by *D. reticulatum* sourced from pastures containing a predominance of cyanogenetic white clover (G.M. Barker, unpublished data), suggesting that the situation observed by Burgess and Ennos (1987) is applicable to the New Zealand pastoral situation.

Red clover (*Trifolium pratense* Linnaeus) is the second most commonly sown legume in New Zealand pastures. It is valued for its ability to produce large quantities of high-quality forage for grazing and hay. Further, red clover is an important pioneer legume in oversowing development of moist hill and high country. All red-clover cultivars currently used in New Zealand grow from crowns, supported by a deeply penetrating tap root, which helps maintain an adequate water-supply in dry conditions and is the major store of non-structural carbohydrate for rapid regrowth after defoliation. Unlike white clover, red clover does not persist indefinitely in permanent pasture and the economic benefits of sowing red clover depend largely on the number of years that production can be maintained at a satisfactory level. This is determined by the longevity of individual plants in the sward and their sustained competitiveness and vigour. Red clover can remain productive for 5 or more years, but it is not uncommon for stands to persist for only 1–2 years. Red-clover plants usually die as a result of deterioration of their primary axis, i.e. the tap root and crown, due to fungal diseases, infection by the stem nematode *Ditylenchus dipsaci* (Kuhn) Filipjev (Tylenchidae) (Skipp and Christensen, 1990) and treading damage. Tap-root injury by invertebrates is likely to contribute to the establishment of fungal pathogens. Among the gastropods residing in New Zealand pastures, *M. gagates* and *A. hortensis* in particular occur below the soil surface for considerable periods of time and are known to feed on the tap root of red clover (G.M. Barker, personal observation). As discussed below, gastropods also have the potential to vector stem nematodes, thus further contributing to the decline epidemiology of this legume. There has been recent renewed interest in red clover with the advent of nodal rooting, prostrate selections (Smith and Bishop, 1993; Hyslop *et al.*, 1996) that offer the possibility of long-term persistence in pastures. Whether gastropods will feed on the stolon apical tissues and nodal buds of these red-clover selections in a manner similar to that seen with white clover (Barker and Addison, 1992) is yet to be determined.

A number of other legume species are used in or have been evaluated for environments to which red and white clovers are ill suited. Subterranean clover is most evident in dry environments, where persistence of white clover is marginal to non-existent. Most of these dry environments are also marginal for gastropods and their herbivory is not prevalent on subterranean clover. However, in hill country, subterranean clover can often occur at dry microsites within a generally moist environment and these clovers can sustain heavy grazing by gastropods that reside primarily in adjacent moist microsites (G.M. Barker, personal observation). In laboratory feeding assays, subterranean clover was found to be moderately acceptable to *D. reticulatum* (Barker *et al.*, 1983) and to several species of helicids and hygromiids (Baker, 1989).

There is an increasing interest in Caucasian clover (*Trifolium ambiguum* von Bierberstein) as an alternative to white clover in lowland, hill and high-country environments, primarily because of its superior persistence and productivity in summer-dry conditions (e.g. Woodman *et al.*, 1992; Allan and Keoghan, 1994; Moss *et al.*, 1996; Watson *et al.*, 1996a,b, 1998; Scott, 1998; Black and Lucas, 2000; Black *et al.*, 2000). None the less, Caucasian clover is subject to heavy pest burdens, which threaten its persistence and productivity in northern areas of New Zealand (e.g. Watson *et al.*, 1996a, 2000). Nothing is currently known of the pest status of gastropods in Caucasian clover in pastures. This species is cyanogenetic.

Many legumes of warm temperate and subtropical origin have been evaluated in Northland, the northernmost part of North Island. Only two, Kenya white clover and lotononis (*Lotononis bainesii* Baker; Fabaceae), appear to have potential under grazing (Rumball and Lambert, 1978). No data are currently available on the potential importance of gastropod herbivory on these legumes under New Zealand conditions, but the studies of Yamashita *et al.* (1979) in Queensland, Australia, point to their high susceptibility. Kenya white clover is cyanogenetic but damage to its tissues evidently releases less HCN than cyanogenetic morphs of white clover (Yamashita *et al.*, 1979).

The role of lotus major (*Lotus pedunculatus* Cavanilles) as a pioneer legume is well established in New Zealand (Montgomery, 1938; Suckling, 1966; Levy, 1970), and with the increased cost of fertilizers there has recently been renewed interest in this species for the development of moist, acid and infertile soils. Under these conditions, lotus major is persistent, even when intensely grazed. Where soil conditions are modified and the competitive ability of companion species improved, then the persistence of lotus major becomes very sensitive to grazing management. Barker *et al.* (1983) found lotus major to be largely rejected by *D. reticulatum* in laboratory non-choice experiments, which is consistent with the low prevalence of gastropod feeding on this plant species in New Zealand hill-country pastures (G.M. Barker, personal observation). Cultivars of bird's-foot trefoil (*Lotus corniculatus* Linnaeus) are primarily used in dry environments and therefore subject to low levels of gastropod herbivory in New Zealand pastures. Bird's-foot trefoil is polymorphic with respect to cyanogenesis, with essentially the same genetic control as white clover. Evidently the degree to which cyanogenesis confers defence against herbivorous gastropods is more pronounced in bird's-foot trefoil than in white clover (Crawford-Sidebotham, 1972).

Lucerne is primarily used in dry environments and therefore gastropods are not generally recognized as pests of this species (Pottinger and Macfarlane, 1967). However, lucerne is also used as specialist forage on free-draining soils within regions that are less arid, such as the Waikato and Bay of Plenty in northern New Zealand, either in monoculture or as mixed swards with ryegrasses. In wetter-than-average years, these crops can sustain severe defoliation by gastropods in early spring. No quantitative data are currently available on the impact of this early-season damage on yield and nutritive quality. Despite the great ecological adaptations available within lucerne, only a very narrow range of genetic material is cultivated in New Zealand thus far. The susceptibility of lucerne to gastropod herbivory may prevent greater utilization of this species in moist environments.

Several forage legumes contain oestrogenic compounds, which can cause infertility and other reproductive disorders in grazing livestock (Collins and Cox, 1984; Davies, 1987). Two groups of compounds are involved, namely isoflavones and coumestans, which are phenolics of the isoflavonoid group. White clover is generally considered to be non-oestrogenic, although there have been occasional reports of oestrogenic effects in sheep and cattle. Coumestans are present in only small amounts in healthy white clover and lucerne but can increase very markedly in response to attack by fungi and insects (Wong and Latch, 1971; Kain and Biggs, 1980). Of the forage legumes used in New Zealand pastures, oestrogenic activity has primarily been associated with subterranean clover and red clover. The principal phyto-oestrogens in these latter legumes are isoflavones, of which formononetin occurs in greatest concentrations and is the main oestrogenic compound. Red-clover cultivars in use in New Zealand are oestrogenically active at all times of the year and the impact on grazing livestock is dependent on the amounts of legume material ingested. Barker *et al.* (1983) found red clover to be highly acceptable to *D. reticulatum*. However, reproductive activity in this gastropod was arrested when animals were maintained on a pure red-clover diet. Subsequent laboratory experiments demonstrated that the reproductive development and fecundity of *D. reticulatum* were impaired when these gastropods were maintained in microswards comprising 10–30% red clover by dry weight and when maintained on agar-based artificial diets containing isoflavones (G.M. Barker, unpublished data). The implications are that pastures with red clover as the sole legume are unlikely to support high gastropod infestations due to the oestrogenic effects. It is likely that the isoflavones effecting infertility in gastropods are different from these operating in livestock, due to differences in the enzymatic properties of the gastropod and ruminant digestive systems.

The role of gastropods as herbivores of grasses has not been seriously addressed. Speiser and Rowell-Rahier (1991) concluded that 'grasses are usually avoided by snails and slugs', and yet published data on the acceptability of grasses to gastropods and their importance in the diet of gastropods in the field are unequivocal (see Ingram and Peterson, 1947; Grime *et al.*, 1968; Grime and Blythe, 1969; Pallant, 1969, 1972; Duval, 1973; Williamson and Cameron, 1976; Cater *et al.*, 1979; Dirzo, 1980; Barker *et al.*, 1983; Cottam, 1985; Chang, 1991; Iglesias and Castillejo, 1999). Several studies point to grasses as the main diet components of gastropods when these plants are abundant in the habitat (Grime and Blythe, 1969; Pallant, 1972; Dan, 1978; Iglesias and Castillejo, 1999). Furthermore, in agricultural habitats, several gastropods are well recognized as pests of gramineous grain and forage crops.

Forage grasses are highly tolerant of, and indeed stimulated by, periodic defoliation. Invertebrates that defoliate grasses are only pestiferous to the extent that they either compete directly with livestock for the foliage or reduce the fitness of the plants by preventing recovery following defoliation by livestock. The laboratory studies of Barker *et al.* (1983) point to poor performance of *D. reticulatum* maintained on grass diets, relative to a white-clover diet, despite several grasses being of high relative acceptability in both choice and non-choice preference tests. The strong selectivity of gastropod herbivory on white clover and several other forbs and the relatively low incidence of evidence of feeding damage on leaves of forage grasses evident in the field suggest that gastropods are not particularly important as defoliators of grasses in New Zealand pastures. That Barker and Addison (1992) found no evidence for increase in ryegrass yield associated with gastropod control in established Waikato pastures supports these contentions.

These gross interpretations, however, may belittle the subtle interactions between grasses and herbivorous gastropods. Gastropods prefer a mixed diet – unpalatable species may be eaten even if palatable species are abundant. There is evidence that gastropods cannot be sustained successfully on monotonous diets of palatable foods and that they require a varied food intake (Frain and Newell, 1982; Speiser and Rowell-Rahier, 1993; Peters *et al.*, 2000). Rollo (1987) and Rollo and Shibata (1991) demonstrated that food quality has an enormous impact on growth and maturation rates, maturation sizes, reproduction and longevity. We can speculate that acceptable grass species may play a role in providing the required diet variation. Indeed, close inspection of lowland pastures in the high-rainfall areas of New Zealand reveals significant levels of gastropod herbivory on the weed grass *Poa annua* Linnaeus. This herbivory is effected primarily by *D. reticulatum*. While contributing little to pasture yield, *P. annua* has a key ecological role in these pastures as the cool-season colonizer of gaps created by microsite disturbances. Oscillations between *P. annua* and annual warm-season

(C_4) grasses monopolize gaps in the pasture to the exclusion of perennial forage species, such as ryegrasses and white clover (Wardle *et al.*, 1994). Furthermore, the *P. annua* populations serve as a key secondary host of the Argentine stem weevil (*Listronotus bonariensis* (Kuschel); Coleoptera, Curculionidae), a ryegrass pest (Barker, 1993b). In northern New Zealand *D. reticulatum* and *D. panormitanum* have also been observed feeding on the succulent stem bases and leaves of annual ryegrasses (*Lolium multiflorum* de Lamarck), sown as a specialist, winter forage for dairy cows (G.M. Barker, personal observation).

Gastropod herbivory on perennial ryegrasses is mush less pronounced and generally of no practical consequence in the productivity of New Zealand pastures. However, perennial ryegrass in New Zealand pastures is predominantly infected by the clavicipitaceous fungus *Neotyphodium lolii* (Latch, Christensen & Samuels) Glen, Bacon & Hanlin and the low level of herbivory may in part reflect the reduced acceptability of these infected plants. As with closely related fungi in other Gramineae, *N. lolii* occurs as mutualistic, systemic, seed-borne infections in perennial ryegrass and is associated with the presence of a range of alkaloids with demonstrated roles in conferring plant resistance against a range of herbivorous vertebrates and invertebrates (see Cheplick and Clay, 1988; Clay, 1988, 1990, 1994; Clement *et al.*, 1994; Rowan and Latch, 1994). At present, considerable investment is occurring in New Zealand in the development and utilization of novel combinations of *Neotyphodium* Glen, Bacon & Hanlin strains and various forage grasses to provide for enhanced insect-pest resistance and reduced adverse effects on livestock. While attempts are being made to optimize the alkaloid spectra in infected grasses in relation to resistance against major insect pests and health in livestock, little consideration has been given to the consequences for the ecology of lesser pests.

In a series of experiments Barker and Raymond (2002) demonstrated that infection by *Neotyphodium* endophytes can indeed influence gastropod herbivory on grasses, affecting feeding preferences and extent of damage to foliar tissues. However, the interaction of *Neotyphodium*-infected grasses and their gastropod herbivores was shown to be complex, being dependent on gastropod species and the *Neotyphodium* strain and associated alkaloid spectra present in infected plants. Among the indole dipterpenoid alkaloids evaluated in artificial diets, lolitrem B was demonstrated to reduce feeding in *D. reticulatum* and *D. panormitanum*, while diets containing paxilline, lolitroil, α-paxitriol or β-paxitriol tended to be preferred over the untreated diet. The pyrrolopyrazine alkaloid peramine was neutral in its effect. Among the ergopeptine alkaloids tested, ergotamine and ergovaline were demonstrated to be phagostimulatory. Furthermore, Barker and Raymond (2002) were able to demonstrate that *D. reticulatum* and *D. panormitanum*, obtained from sites with contrasting frequencies of infected grasses, exhibited differential responses to lolitrem B and ergovaline alkaloids incorporated into artificial diet. *Neotyphodium* alkaloids encountered by these gastropods in their natural food plants were less deterrent or antifeedant than were novel alkaloids. This is contrary to the situation for *Arianta arbustorum* (Linnaeus) (Helicidae) feeding on pyrrolizidine-rich plants, where Speiser *et al.* (1992) suggested that the animals have mechanisms for rejection of allochemicals that they encounter in their natural food plants, but not for novel allochemicals. The implication of these differential population responses to *Neotyphodium* alkaloids is either phenotypic plasticity or local genetic adaptation among populations. Learned food aversion has been amply demonstrated in gastropods, but Barker and Raymond (2002) concluded that such learning did not apply to the *Deroceras* populations in their study. The possible involvement of phenotypic plasticity in induction of detoxification enzymes (Brattsten, 1988) has yet to be experimentally examined.

In recent years, complex pasture swards have been promoted, primarily in terms of the stability of composition and production in marginal and variable environments and the contribution to a 'balanced' animal diet. Complex swards are perceived to contribute towards increasing biodiversity in grasslands, the value of which is currently hotly debated. As a result of this conceptual shift, a number of species that pastoralists formerly thought of as weeds have become recognized as having agronomic value and are increasingly being developed as forage cultivars (Charlton and Stewart, 1999). For example, two cultivars have been developed for narrow-leaved plantain (*Plantago lanceolata* Linnaeus; Plantaginaceae), a species occurring as a ubiquitous adventive in pastures throughout New Zealand. The acceptability of narrow-leaved plantain to gastropods in its native European range varies (Grime *et al.*, 1968; de Nooij and Mook, 1992; Molgaard, 1992; Hulme, 1996), evidently as a function of variation in leaf allelochemistry. Gastropod herbivory on adventive populations of narrow-leaved plantain is not uncommon but none the less of minor consequence in New Zealand pastures. The narrow-leaved plantain cultivars can suffer more severe defoliation damage by *D. reticulatum* (G.M. Barker, personal observation), which may be related to the microenvironments under the dense canopies of the sown cultivars being more conducive to gastropod populations than the sparse, open canopies of adventive plantain populations. No data are currently available on the losses attributable to gastropods in pastures sown to narrow-leaved plantain cultivars. Chicory (*Cichorium intybus* Linnaeus; Asteraceae) occurs widely as an adventive weed in New Zealand and has also been developed as a forage. The single available forage cultivar of chicory is known to be susceptible to gastropod herbivory (Moloney and Milne, 1993), but the extent and agronomic importance of the losses have not been quantified.

While some seedling regeneration occurs in grazed, moist lowland and hill-country pastures, it is not a significant mechanism of persistence of white clover in these environments. Mean establishment of one plant 5.5 m^{-2} $year^{-1}$ in moist hill country is negligible when compared with background stolon populations of 3000–4000 m^{-2} (Chapman, 1987; Chapman and Anderson, 1987). The main factor in the low rates of clover establishment is the competition from resident species (Chapman, 1987). None the less, regeneration by natural seeding is important to pasture resilience following disturbances such as that caused by pest attack, over-grazing and drought. The preference of gastropods for seedling clovers over mature plants (e.g. Charlton, 1978a) may limit regeneration from naturally shed seed, thus further limiting persistence when vegetative continuance is reduced by pests or other factors. The research of Blank and Bell (1982) clearly implicates gastropods as a factor in seed loss from the soil surface over the wet, cool winter and spring months in northern New Zealand. None the less, the role of seed and seedling herbivory by gastropods in shaping plant-community composition in established New Zealand pastures has not been studied. Even in summer-dry hill-country areas, where clovers and other species re-establish annually from the soil seed bank, gastropods may be important in regulating community composition if seedling recruitment occurs near the refugia in which the gastropods persisted over summer. Working in England, Hanley *et al.* (1995a, 1996a,b) showed that selective herbivory by gastropods influences the outcome of recruitment of seedlings in artificially created gaps in grassland.

Hume and Barker (1991) have discounted natural reseeding as making any significant contribution to ryegrass perenniality in both summer-moist and summer-dry hill country on account of low survival of seedlings, although these authors considered there could be a cumulative benefit over a number of years. Evidence for the involvement of herbivory by gastropods in the apparent low rates of natural recruitment in grass seedlings is at present lacking. Reseeding in perennial ryegrass is common in lowland pastures in New Zealand, especially in those pastures harvested for hay or otherwise fallowed over summer, as has been recorded for pastures in the Waikato by L'Huillier and Aislabie (1988). Any effects of pests, should they occur, would generally be masked in these lowland situations by the often extremely high initial seedling numbers and the natural processes of self-thinning in the seedling stands.

Management of gastropods

The current widespread intensive management practices, including high stocking rates and rotational grazing, effect a high level of gastropod control on many farms. Further, in lowland pastures of northern New Zealand at least, naturally occurring pathogens play a significant role in regulating agriolimacid numbers. Both factors operate unbeknown to farmers. It remains to be demonstrated that the high levels of cyanogenesis expressed in New Zealand white-clover cultivars are also a factor in suppressing gastropod numbers in pastures.

Few farmers actively manage gastropods in established pastures in New Zealand. This situation is unlikely to change unless there is further, clear demonstration that gastropods are important impediments to farm profitability. Farmers will only implement controls where there is a high benefit : cost ratio. As indicated in the above paragraphs, there is already some evidence that gastropods are significant contributors to the pest burden that undermines the productive potential of white clover in New Zealand lowland pastures. It has yet to be shown that the level of losses justifies adoption of pest-management strategies, although such losses are likely to come under increased scrutiny as management of the lowlands is further intensified. Other components of the pest complex, such as root-knot and cyst nematodes and clover-root weevil, are of higher priority. In the longer term, control options for the gastropod pests include: (i) development of farming systems that are less reliant on biological N fixation; (ii) development of white-clover cultivars resistant to gastropods, including genetic modification for strategic expression of allelochemicals; (iii) use of alternative legume species with resistance to gastropods; (iv) exploitation of *M. novacastriensis* and *T. rostrata* as inundative biological control agents of gastropods; and (v) use of indigenous parasites, such as malacophagous nematodes, as biological control agents.

There is also circumstantial evidence that gastropods may contribute to the poor persistence of white clover in moist hill-country pastures. There, however, the low marginal returns make it unlikely that the investment in research and development necessary to implement effective controls will ever be made available. Thus, the development of gastropod controls for hill-country use must get a 'piggy-back' on investment in the lowlands.

Gastropods as Pests During Pasture Establishment

Nature of damage

During pasture establishment, gastropods can be regarded as 'plant-population reducers'. Damage is caused by: (i) ingestion of entire seeds or hollowing out of seeds; (ii) severing the young shoots at or below ground level; (iii) removal of cotyledons; and (iv) shredding or removal of leaves of young seedlings. Seed removal/hollowing and severance of young shoots are the most widely recognized forms of damage as they result in immediate reduction of seedling densities. Of equal importance, though of more insidious nature, are the effects of cotyledon and leaf removal during early seedling growth. While sublethal, these latter forms of damage can reduce seedling vigour, which in turn reduces seedling competitive ability and long-term survival in mixed-species swards. Most affected by sublethal damage are species with inherently poor seedling vigour, and where growth is also restricted by low soil fertility, suboptimal temperatures, attack from other pests or competition from resident plants.

Risk areas and extent of losses

Large areas of New Zealand pastoral land are renewed or renovated annually. A survey by Sangakkara *et al.* (1982) indicates that 60–80% of farmers resow a proportion of their farms each year. This annual pasture renewal or renovation represents 5–6% of the grassland area, or 0.75–0.90 million ha.

In hill and high country aerial oversowing is the only practical means of introducing new species or cultivars, while on undulating to flat land cultivation and conservation-tillage practices, such as direct drilling, are viable options. Despite the proved advantages of placing the seed in the soil rather than on the surface, broadcasting still remains a common method of pasture establishment in all land classes (Sangakkara *et al.*, 1982). Only in areas of intensive lowland farming has there been a substantial rate of adoption of direct drilling techniques.

Gastropods are rarely of pest status in seed-beds prepared by cultivation, because populations are substantially reduced by the mechanical action of cultivation equipment. Only in situations where cultivation passes are reduced, producing an unconsolidated, cloddy seed-bed, are gastropods potential pests. The lesser habitat destruction with broadcast seeding and direct-drilling methods favours greater carry-over of gastropod populations from the previous pasture, with consequent high risk of damage to seedlings.

There have been considerable increases in the productivity of New Zealand hill country over the past 50 years, through fertilizer, subdivision and oversowing. However, the virtual absence of legumes from large areas of hill-country pasture remains a primary limitation to achieving yield potentials. Despite many studies on pasture management for optimizing the seedling environment, establishment of oversown legumes remains very poor. For example, the white-clover establishment rate (% sown seed present as plants at 2 months) is commonly 0–25%. Agronomists working in summer-moist hill country have long implicated the presence of agriolimacids and arionids as a factor in the poor legume establishment following oversowing (Suckling, 1949, 1951; Madden, 1952; Blackmore, 1964). In experiments with pasture turfs removed to the glasshouse, Charlton (1978a,b, 1979) confirmed the potential for arionids and agriolimacids to severely restrict survival of sown seed and young seedlings (Fig. 18.6). Charlton (1978b, 1981) was unable to demonstrate benefits from protecting oversown legumes from gastropods in field trials, however, as the levels of legume establishment remained low. This suggests that the molluscicide treatments used by Charlton were ineffective in providing seedling protection over the 6–8-week period needed to attain plant size tolerant of gastropod grazing and that, more importantly,

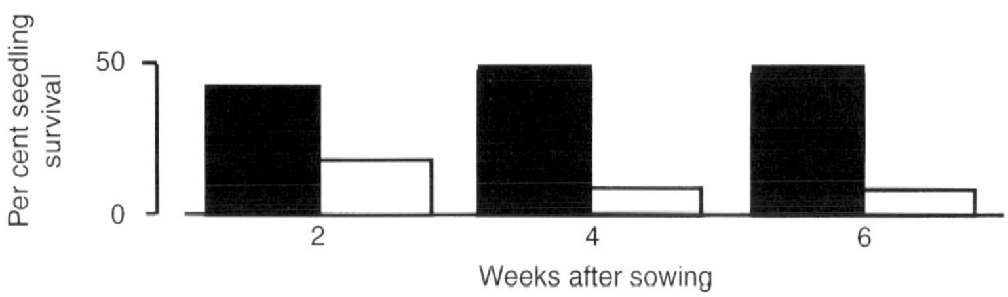

Fig. 18.6. Survival of white clover (*Trifolium repens* Linnaeus) (Fabaceae) oversown into turfs removed from summer-moist hill pastures in Manawatu, North Island, New Zealand, and maintained in the glasshouse with (equivalent to 100 m^{-2}) (❐) and without (■) the introduction of the herbivorous gastropods *Deroceras reticulatum* (Müller) (Agriolimacidae) and *Arion hortensis* de Férussac (Arionidae). The reduction in seedling survival was significant at $P < 0.05$ for all assessment dates after sowing. (After Charlton, 1979.)

other factors had adverse effects on seedling survival. Blank and Bell (1982) found that *D. reticulatum* and *D. panormitanum* destroyed significant quantities of legume seed placed on the soil surface of pastures in northern North Island during the winter and spring periods.

Despite the potential for gastropods to impair the establishment of oversown legumes on some 2 million ha of hill country (and 2 million ha of lowland), their pest status has not been adequately quantified. There remains a need to construct quantitative relationships between the intensity of infestations and seedling survival and yield following broadcast seeding.

The one aspect of current direct-drilling technology that places new sowings at risk from gastropods is the rapid old-pasture-to-new-pasture rotation. This effect is accentuated when: (i) seed is overdrilled into an existing sward; (ii) the drill coulters form open furrows in the soil; and (iii) adequate seed cover is not achieved because the seed drill is not equipped with press-wheels or harrows. Experiments that seek to quantify the damage potential of gastropods tend to adopt best practice and minimize the risk factors. None the less, such assessments of damage potential have confirmed that gastropods do constitute a significant threat to the successful establishment of pastures by direct drilling. Ferguson and Barratt (1983), for instance, recorded significant losses following direct drilling of a ryegrass/white-clover seed mix in pasture plots differentially infested with *D. reticulatum*. Losses in ryegrass production at 6 weeks after sowing occurred at all *D. reticulatum* infestations ranging from 5 to 80 m^{-2}. By the third harvest at 14 weeks, these effects on ryegrass were absent. A similar pattern was recorded for clover production, though an effect from 80 *D. reticulatum* m^{-2} was still evident at 14 weeks. Ferguson and Barratt (1983) calculated greatest proportional reductions in seedling numbers at five to ten *D. reticulatum* m^{-2} (Fig. 18.7) and suggested that the economic threshold for this species in direct-drilled pastures was within this range. Working in pastures

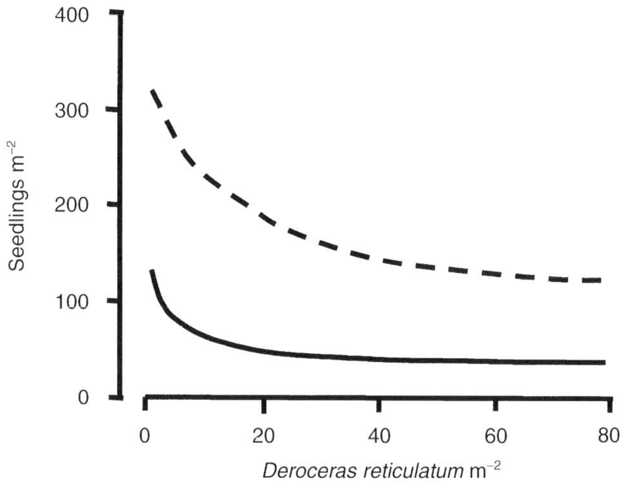

Fig. 18.7. Numbers of perennial ryegrass (*Lolium perenne* Linnaeus) (Gramineae) and white clover (*Trifolium repens* Linnaeus) (Fabaceae) seedlings established in lowland pasture in southern New Zealand following direct drilling, as a function of *Deroceras reticulatum* (Müller) (Agriolimacidae) infestation level. (After Ferguson and Barratt, 1983.)

infested with *D. reticulatum*, *D. panormitanum*, *A. hortensis*, *A. intermedius* and *M. gagates*, Barker (1991b) demonstrated that, following direct drilling, numbers of seedlings lost to herbivory increased asymptotically with numbers of these gastropods. The effects observed on seedling numbers and yield were negligible in plots with the lowest gastropod numbers (mean 0.4 per m row; 2.9 m^{-2}) but were marked at the higher gastropod numbers (3.8–11.2 per m row; 25.7–75.1 m^{-2}) (Fig. 18.8). Greatest proportional effect on seedlings occurred at 3.8 gastropods per m row, indicating a threshold for damage at or below this level of infestation. Thus, in situations where gastropods such as *D. reticulatum* are present at densities > 10 m^{-2}, substantial seedling losses can occur to both grasses and legumes (Ferguson and Barratt, 1983; Barker, 1991b) unless appropriate control measures are taken. For lowland pastures, an estimated 0.25 million and 0.5 million ha are infested to this level in autumn and spring, respectively. Losses are expected to be higher and more extensive where best practice is not adopted when direct-drilling ryegrass and white clover.

Establishment failures with both oversowing and direct drilling sometimes occur at the paddock scale. More often, the damage is uneven, with localized areas within the paddock that support high gastropod numbers and where the growth of the seedlings is checked by environmental conditions sustaining the highest losses. Furthermore, one sown species may be adversely affected, while others, by virtue of their growth form or chemistry, escape predation by the gastropods.

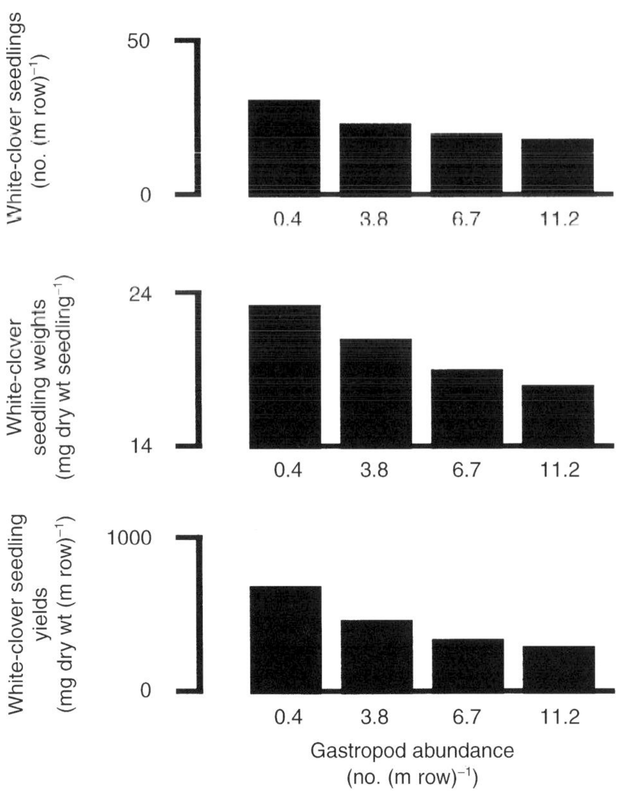

Fig. 18.8. Numbers, weights and yield of white clover (*Trifolium repens* Linnaeus) (Fabaceae) seedlings 8 weeks after direct drilling in lowland pasture in northern New Zealand as a function of herbivorous gastropod infestation level. The herbivorous gastropod fauna at the site comprised predominantly *Deroceras reticulatum* (Müller) (Agriolimacidae), but also included *Deroceras panormitanum* (Lessona & Pollonera) (Agriolimacidae), *Milax gagates* (Draparnaud) (Milacidae), *Arion hortensis* de Férussac and *Arion intermedius* Normand (Arionidae). (After Barker, 1991b.)

The importance of *P. barbara* in seedling losses during renovation of pastures in northern New Zealand pastures has not been investigated. While primarily a detritivore, *P. barbara* is known to feed on seedlings (e.g. Baker, 1989) and, given that its abundance in many pastures in Northland often exceeds 500 m^{-2}, damage potential is evident.

Ecology of establishing pastures and role of gastropod herbivory

Although gastropods alone can cause establishment failures, they are generally only one facet of a larger problem – the use of inappropriate technology by farmers. Failure to apply good plant-husbandry skills is the single most important reason for poor seedling establishment. Damage by gastropods is accentuated by poor management, whether it be in the preparation of the pasture for sowing, the actual drilling or oversowing operation or postsowing grazing management. It is expensive and ill-advised to apply molluscicides when other factors impair seedling establishment or when the effects of gastropods could be reduced by simple cultural practices. The solution to gastropod problems in pasture renovation, as in any cropping system, demands integration of gastropod control options into an overall biologically and economically sound management system.

A common farming practice is to direct-drill (overdrilling) or broadcast seed without actively managing the old pasture to suppress competition from resident plants. In early work on oversowing, before the advent of herbicides, it was quickly realized that intensive grazing of pastures pre- and postsowing was needed if any measure of success in establishment of the sown species was to be achieved. Later it was demonstrated that the most efficacious approach to pasture renovation by oversowing or direct drilling is to eliminate all potentially competitive vegetation by blanket application of broad-spectrum herbicides, such as glyphosate (e.g. Kunelius *et al.*, 1982; Sithamaranathan *et al.*, 1986; Thom, 1988; Barker and Dymock, 1993). None the less, farmers have been reluctant to use herbicides, because of the extra cost and management inputs required. This herbicide-free approach to pasture renovation represents a substantial waste of investment by farmers, as seedling establishment is usually very poor (Suckling, 1949; Madden, 1952; Barker *et al.*, 1988, 1990b). Although gastropods may contribute to seedling losses, the failure to eliminate competition from resident plants is the primary limitation to successful pasture establishment in New Zealand. Band spraying of herbicides over the drilled rows of seed represents a compromise approach to pasture renovation by direct drilling (Blackmore, 1962). The cost of herbicide is approximately halved, while providing a temporary suppression of competition. Seedling establishment under a banded-herbicide treatment is, however, intermediate to no-herbicide and blanket-herbicide treatments (Barker *et al.*, 1988, 1990a). The technique is not at present widely used, because the necessary equipment is not readily available.

The use of herbicides to prepare pasture for renovation has important implications for pest management (Robertson *et al.*, 1981; Barker *et al.*, 1988, 1990b; Fig. 18.9). Killing all vegetation, as in blanket treatments with broad-spectrum herbicides, greatly increases insect, nematode and gastropod burden on establishing seedlings, through their aggregation in the drill furrows and their feeding on the rows of seedlings. When no herbicide or banded herbicide is used, the attractiveness of germinating seed and seedlings to herbivorous invertebrates is largely mitigated by the presence of established plants. In the case of banded herbicide, aggregation in the bands of residual old pasture reduces pest pressure on the developing seedlings. Pesticides applied as granular or bait treatments in the furrows with the seed are most effective when blanket-herbicide treatments are used, because the aggregation of pests maximizes their contact with the pesticide (Barker *et al.*, 1988; Barker, 1990). Barker *et al.* (1988) developed the concept of a pest fauna, in recognition of the multispecies nature of most pasture infestations and the complex pest burden affecting seedlings in pasture-renovation situations. The contribution of gastropods to the pest fauna varies greatly. Gastropod survival and the degree of aggregation in the drill furrows are governed by the degree of physical disturbance of their habitat. The more severe the depletion of shelter and food on the soil surface, the greater the need for gastropods to seek out alternative niches. The furrows made by drill coulters provide ready

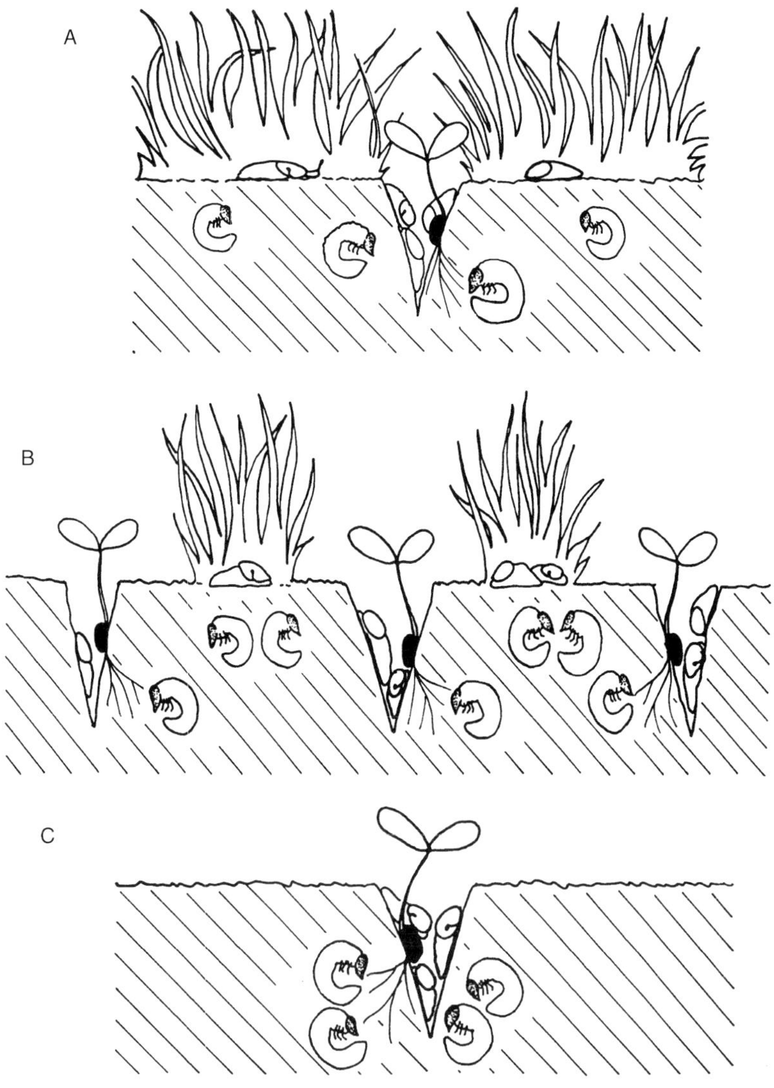

alternatives to surface shelters and sown seed provides an alternative food source. During the initial stages of cover depletion, whether through grazing or herbicide action, most of the surface activity by gastropods reflects a need for such shelter. Severe population depletions can occur at this stage if alternative shelters cannot be found (Barker, 1986). Retention of inter-row vegetation (e.g. by band spraying) is less disruptive than total shelter/food removal. In these situations, gastropod sheltering and foraging activity may be proportionally higher in the inter-row region than in the vicinity of drilled seed rows.

The interaction between herbicide treatments and the dispersion and pest status of gastropods is illustrated by the results of Barker (1990) (Fig. 18.10). With direct drilling of seed into the pasture without manipulation of the vegetative cover or with banding of glyphosate herbicide, agriolimacid and milacid numbers were similar on and between the drill rows. With band-spraying of paraquat herbicide and the rapid depletion of vegetative cover over the drill furrows, numbers of these gastropods were higher in the rows of residual vegetation (old pasture) than on the seedling rows. Where the old pasture had been removed by broadcast-spraying of either paraquat or glyphosate, agriolimacids and milacids were largely confined to the seedling rows by 2 weeks after drilling. Seedling vigour, as indicated by mean seedling weight at 8 weeks, was influenced by herbicide and molluscicide applications. For both ryegrass and white-clover, seedling weight was increased by band-spraying and especially by broadcast herbicide treatments. For ryegrass, there was a significant interaction between herbicide and molluscicide treatments: the molluscicide did not increase ryegrass seedling weight in plots direct-drilled without herbicide but enhanced seedling growth by 14–20% where herbicide was used. White-clover seedling weights were increased by an average of 16% with molluscicide application, irrespective of the herbicide treatment.

Fig. 18.9. (Opposite) Diagrammatic illustration of the influence of vegetative cover, effected by different herbicide treatments, on the dispersion and impacts of pestiferous invertebrates in relation to rows of seedlings developing in direct-drilled pasture. **A.** No herbicide. The resident pasture is retained. Surface-dwelling pest invertebrates, illustrated by agriolimacid gastropods, are dispersed throughout the pasture and in the furrows created by the drill coulters. Soil-dwelling pests, illustrated by scarabaeid larvae, are dispersed throughout the topsoil. Some herbivory occurs on seedlings developing in the drill furrows, but seedling survival is most strongly influenced by competition from resident pasture plants. **B.** Herbicide banded on the seed rows. Resident pasture is only retained in the zones between the seed rows. Surface-dwelling pests aggregate in the residual bands of resident pasture and in the furrows created by the drill coulters. Soil-dwelling pests are primarily aggregated in the root zone of the bands of resident pasture and around the roots of developing seedlings. Herbivory on the seedlings developing in the drill furrows is reduced compared with the no-herbicide situation, but seedling survival is enhanced by the reduction in competition from resident pasture plants. **C.** Blanket herbicide. No resident pasture retained. Surface-dwelling pests are largely confined to the furrows created by the drill coulters and soil-dwelling pests aggregate around the roots of developing seedlings in the drill furrows. Herbivory on seedlings developing in the drill furrows is substantial as the pests have few alternative food sources. If pests are controlled, seedling survival is high, due to the absence of competition from resident pasture plants.

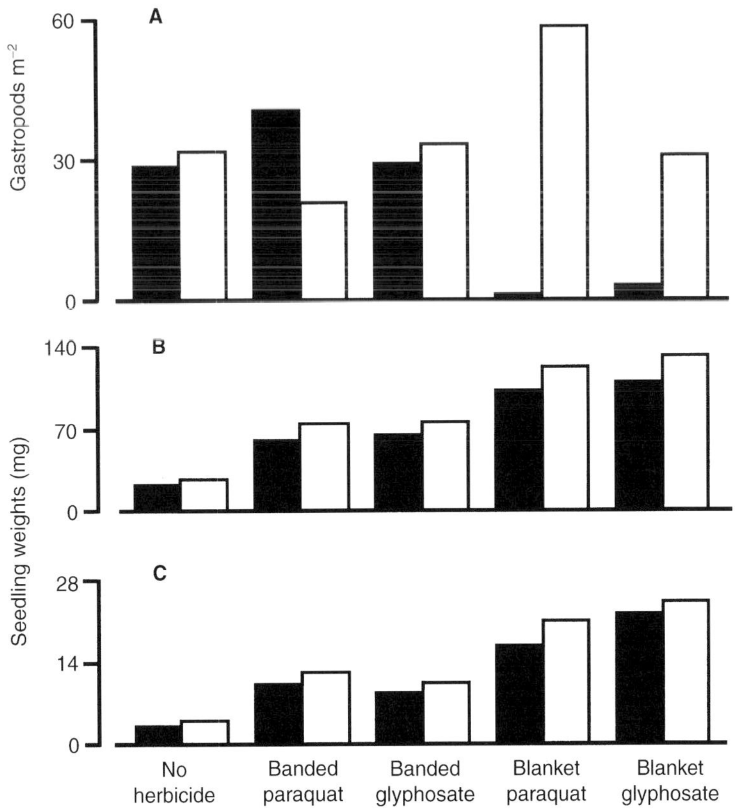

Fig. 18.10. Influence of herbicide treatments on dispersion of herbivorous gastropods and vigour of seedlings in a direct-drilled pasture in lowland northern New Zealand. **A.** Numbers of herbivorous gastropods, comprising *Deroceras reticulatum* (Müller), *Deroceras panormitanum* (Lessona & Pollonera) (Agriolimacidae) and *Milax gagates* (Draparnaud) (Milacidae), on (■) and off (❒) seedling rows. **B.** Mean size of perennial ryegrass (*Lolium perenne* Linnaeus) (Gramineae) seedlings in plots with (❒) and without (■) molluscicide treatment. C. Mean size of white-clover (*Trifolium repens* Linnaeus) (Fabaceae) seedlings in plots with (❒) and without (■) molluscicide treatment. (After Barker, 1990.)

Contrary to popular farmer belief, increasing seeding rate to compensate for anticipated seed and seedling losses as a result of pests is not a sound approach to pasture renovation. First, high seeding rates, as commonly used by New Zealand farmers (Sangakkara *et al.*, 1982), do not result in long-term increases in yields (Cullen, 1964; Barker *et al.*, 1990b) because of high interplant competition and self-thinning of seedling populations (e.g. Kays and Harper, 1974). Secondly, the anticipated benefits of more sown seed can be negated by increased pest aggregation and feeding on seedling rows. Indeed, excessive seeding rates promote pest problems, as herbivorous insects and gastropods aggregate in the seedling rows and their feeding is stimulated by high numbers of seeds and seedlings (Robertson *et al.*, 1981; Barker *et al.*, 1990b; Barker, 1991b; Fig. 18.11).

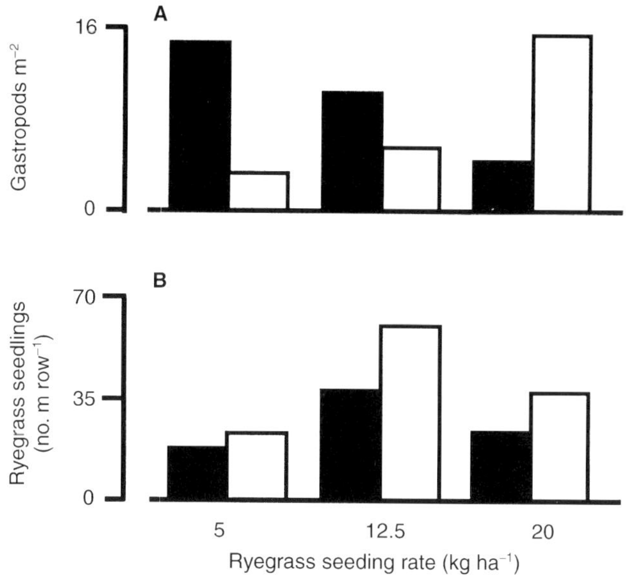

Fig. 18.11. Influence of perennial ryegrass (*Lolium perenne* Linnaeus) (Gramineae) seeding rates on dispersal of herbivorous gastropods and numbers of ryegrass seedlings established in a direct-drilled pasture in lowland northern New Zealand. **A.** Numbers of herbivorous gastropods, comprising *Deroceras reticulatum* (Müller), *Deroceras panormitanum* (Lessona & Pollonera) (Agriolimacidae) and *Arion intermedius* Normand (Arionidae), on (■) and off (❐) seedling rows. **B.** Mean number of perennial ryegrass seedlings established in plots with (❐) and without (■) molluscicide treatment. (After Barker *et al.*, 1990b.)

Intensive grazing of pasture in advance of or immediately after sowing seed not only reduces competition, but can also reduce gastropod numbers by a combined effect of treading and alteration of the habitat. Ferguson *et al.* (1989a) showed, for example, that stocking at 1500 sheep-days ha^{-1} reduced *D. reticulatum* populations by 90%, while under the same conditions methiocarb baits (2% a.i., 10 kg ha^{-1}) effected only 50% control. On steep hill country such intensive stocking rates are not readily achieved, due to inadequate subdivision, and will be less effective in controlling gastropods, because treading is not uniform. However, stocking at 300–400 sheep equivalents ha^{-1} can greatly improve seed-to-soil contact and thus establishment of broadcast seed, and will undoubtedly effect some measure of gastropod control. The maximum benefit of these treading treatments for longer-term seedling survival is, however, dependent on competition control through herbicide use and/or controlled grazing.

Drill coulters producing furrows that retain moisture vapour for the few critical days between drilling and seedling emergence have been shown to improve seedling establishment in dry soils (Baker, 1985). Experience in New Zealand indicates that the most complete devastation of direct-drilled pastures by gastropods occurs with the moisture-retentive inverted-T-shaped furrow produced by winged chisel coulters (Baker, 1985; G.M. Barker, unpublished data). Furrow cover by harrowing and/or furrow closure by press-wheels or stock treading improves seedling emergence and survival, due to improved moisture retention (Baker, 1985) and reduced exposure to gastropods (Hughes and Gaynor, 1984). Excessive sowing depth is particularly detrimental to small-seeded species (e.g. white clover) and increases the favourability of furrows for gastropods, such as *D. reticulatum*.

Gastropods can be reduced to a subordinate role in pest faunas affecting clover establishment, because other pests are more numerous, have greater per capita effect and/or act more quickly than the gastropods. While gastropods, such as *D. reticulatum* and *A. intermedius*, occur in unimproved tussock grassland, they are minor contributors to the substantial seed and seedling losses in oversown white clover. Barratt (1982, 1985), Barratt and Johnstone (1984), Barratt and Ferguson (1992) and Barratt *et al.* (1992) have shown that native broad-nosed weevils (e.g. *Irenimus* Pascoe spp.; Coleoptera, Curculionidae) damage and destroy white-clover seedlings. On average, 5% of sown seed survives to establish successfully in the presence of these insects and the greater proportion of these losses occur before and during germination rather than once seedlings reach the cotyledon stage (Barratt *et al.*, 1992). Evans *et al.* (1994) and Ferguson and Evans (1994) have reported similar results for Caucasian clover and bird's-foot trefoil.

In hill country, Charlton (1978b) obtained better establishment of oversown seed in autumn, compared with spring oversowings, which were subject to more severe gastropod pressure. The higher pest burden in spring is prevalent throughout much of New Zealand, because the preceding autumn and winter conditions provide for recruitment into gastropod populations. In all but the coldest districts, autumn is the preferred time of year for pasture renovation, because the special management requirements of new pastures are most easily accommodated within the context of whole-farm management at that time. Additionally, with autumn sowings, seedlings have available a longer period suitable for growth prior to the moisture- and heat-stress conditions of summer, and are at less risk of competition from dicotyledonous weeds, compared with those sown in spring. In many areas of New Zealand, there is often a window of opportunity in autumn, with a period suitable for seedling growth and yet largely free from gastropod infestation. This results because gastropods are generally depleted during summer and a lag occurs between the onset of autumn rains and recruitment into populations of species such as *D. reticulatum*. On some soils, however, gastropod populations can recover quickly in autumn through emergence of animals from refugia, such as earthworm burrows.

Experiments have shown that the threat represented by gastropods to pasture seeds in or on the soil varies according to the species of plants and the species of gastropod. In general, monocotyledon seeds are less susceptible to damage from gastropods than seeds of dicotyledons (e.g. Ferguson, 1984). Legumes are often destroyed as soon as the seeds imbibe water, while with grasses damage occurs after emergence of the radicle as the seed-coat provides initial protection. Plant species contain varying amounts of starch in their seeds. Starch is a potent feeding stimulant, as shown in *Ariolimax californicus* Cooper (Arionidae) by Senseman (1977). Ferguson (1984) demonstrated that water-soluble substances in seeds of both perennial ryegrass and white clover stimulated *D. reticulatum* to feed. It is therefore important to limit gastropod access to seeds by a cover of fine-soil tilth and/or repellent seed treatment.

The effect of seedling age (size) on the likelihood and impact of herbivory has been highlighted in a number of studies involving gastropods and grassland plants (e.g. Byers and Bierlein, 1982; Hulme, 1994; Hanley *et al.*, 1995b). In respect of pasture renovation, Charlton (1978a, 1979) recorded differential survival of legume seeds and seedlings in the presence of *D. reticulatum* and *A. hortensis* (Fig. 18.12). However, the relative susceptibility of legumes varied with seedling age and gastropod species. Most damage occurred on seedlings at the cotyledon stage. The seedlings were more tolerant of gastropod feeding once the true leaves developed. Lotus major was the least susceptible of the legumes tested and was totally rejected by gastropods once the seedlings reached the first-true-leaf stage. *D. reticulatum* caused more damage than *A. hortensis*, because of its more extensive feeding and its willingness to

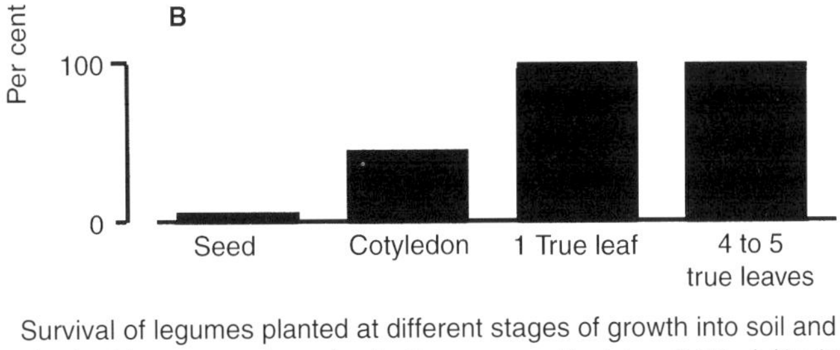

Fig. 18.12. Survival of legumes planted at different stages of growth into soil and exposed for 14 days to the herbivorous gastropods: **A.** *Deroceras reticulatum* (Müller) (Agriolimacidae) and **B** *Arion hortensis* de Férussac (Arionidae). Data are presented for the mean of four legume species, namely white clover (*Trifolium repens* Linnaeus), red clover (*Trifolium pratense* Linnaeus), subterranean clover (*Trifolium subterraneum* Linnaeus) and lotus major (*Lotus pedunculatus* Cavanilles) (all Fabaceae). The reduction in survival for seedlings exposed at the seed and cotyledon stages was significant greater ($P < 0.05$) than for seedlings exposed to the gastropods at the 1 true-leaf and 4–5 true-leaf stages. (After Charlton, 1979.)

attack all test legumes. *A. hortensis* fed most extensively on white clover once the plants reached the true-leaf stage. Working in North America, Byers and Bierlein (1982) found that, when offered a choice between seedlings of three legume species, *D. reticulatum* and *Arion fasciatus* (Nilsson) preferred red clover and lucerne over bird's-foot trefoil. *D. laeve* preferred red clover over both lucerne and bird's-foot trefoil. Byers and Bierlein (1982) noted that these gastropods preferred seedlings in the cotyledon stage. As plants matured, the gastropods stopped cutting the stems and fed exclusively on leaflets.

Seed from cultivars of white clover are evidently equally susceptible to gastropod feeding (Horrill and Richards, 1986; Mowat and Shakeel, 1988). The differences in cultivar susceptibility become apparent at the seedling stage, with gastropods feeding significantly more on the seedlings of acyanogenetic cultivars (e.g. Horrill and Richards, 1986; Burgess and Ennos, 1987; Mowat and Shakeel, 1988, 1989a,b; Raffaelli and Mordue, 1990; Glen *et al.*, 1991). Horrill and Richards (1986) found that in the cultivar S100, cyanogenetic potential in 5-day-old seedlings was higher in cotyledons (*c.* 240 μg CN^- g^{-1} fresh tissue) than in the stems (100 μg CN^- g^{-1} fresh tissue). By day 35, stem and leaf tissues had cyanogenetic potential similar to that initially present in the cotyledons (i.e. *c.* 240 μg CN^- g^{-1} fresh tissue). These authors found that *A. hortensis* did not discriminate between phenotypes when seedlings were 5 or 11 days old. However, for seedlings 16 to 35 days old, few grazed cyanogenetic seedlings suffered lethal damage, while most grazed acyanogenic seedlings were lethally damaged. These results suggest that only older seedlings and mature plants are likely to be protected by cyanogenesis, leaving a crucial 'window' during which time seedlings of all white-clover phenotypes are equally susceptible to gastropod herbivory. As noted above, gastropods must 'sample' a plant before they can recognize that it is distasteful. In the case of very young white-clover seedlings, this 'sampling' may often be fatal for the plant, and therefore cyanogenesis may have limited practical use as a pest-management strategy in pasture renovation. Indeed, the failure of cyanogenesis to inhibit gastropod feeding and provide for white-clover seedling survival has been noted by Miller *et al.* (1975) and Charlton (1979).

Barker (1989b) demonstrated that variation exists among lucerne cultivars in susceptibility to seedling damage by *D. reticulatum*.

Barratt (1980, 1985) and Ferguson *et al.* (1995) showed that sublethal damage to seedlings through partial or complete removal of the cotyledons can adversely affect survival and productivity of white clover and other leguminous plants. This sublethal effect is accentuated in low-stress environments but relegated to subordinate importance when seedling growth is reduced by stress, such as nutrient deficiency.

Blank and Bell (1982) found that there was no predation by *D. reticulatum* and *D. panormitanum* of grass seed distributed on the soil surface. In contrast, seedling grasses are readily eaten by gastropods.

Harper and Hatto (1968) demonstrated a considerable impact of *D. reticulatum* feeding at the base of grass seedlings, cutting through them at the soil surface and below the meristematic zone. This is consistent with the observed impacts of gastropods on ryegrass establishment in direct-drilled lowland pastures in New Zealand (Ferguson and Barratt, 1983; Barker, 1990, 1991b; Barker *et al.*, 1990b; McCallum and Thomson, 1990). For the most part, grass seedlings are very tolerant of damage, as they rapidly attain a size in which defoliation generally occurs above the meristem tissues.

Legumes such as white clover are at a competitive disadvantage in mixtures with ryegrass (Haynes, 1980). Damage by pests can have a marked effect on legume-seedling survival by accentuating their subordinate position within the hierarchy of a dense seedling population. Smaller plants, whether weakened by interplant competition or by pest damage, are at an accumulating disadvantage in pastures and eventually die (White and Harper, 1970). The competitive disadvantage of legumes should be minimized by practices that ensure maximum seedling vigour.

Reflecting the preoccupation with the effects on sown forage species, there has to date been little consideration given to the importance of gastropod herbivory on seedlings in shaping the weed flora developing in New Zealand pastures. Given the polyphagous and yet selective nature of gastropod herbivory, we can predict important effects on the success rate of weed species establishing from the soil seed bank. Wardle and Barker (1997) demonstrated that herbivory by *D. reticulatum* during winter, at population levels equivalent to 80 m^{-2}, had marked effects on the composition of plant communities developing from the seed bank in each of four soils. Despite differences in the composition of the seed bank, the total biomass of dicotyledonous plants was consistently reduced by herbivory, while monocotyledonous biomass was unaffected. In a summer experiment, herbivory by *Cantareus aspersus* (Müller) (Helicidae) did not influence total biomass of either dicotyledonous or monocotyledonous plants, although productivity in individual dicotyledonous plant species was reduced. These results are consistent with European studies (e.g. Brown and Gange, 1989; Silva and Teresa, 1992; Ramsell *et al.*, 1993; Hulme, 1994), which indicate that invertebrate herbivores are often selective against dicotyledonous plants. Wardle *et al.* (1998) recorded high susceptibility to herbivory by *D. reticulatum* among seedlings of the herbaceous dicotyledonous plants that commonly occur as weeds in New Zealand pastures. Of the 19 species tested, only *Rumex pulcher* Linnaeus (Polygonaceae) exhibited a significant level of resistance to herbivory.

Management of gastropods

Deliberate gastropod-control programmes during pasture establishment are not widely practised. In some districts 'best-practice' pasture

stem nematodes are readily dispersed on seed, machinery and surface water, Cook *et al.* (1989) concluded that gastropods such as *D. reticulatum* may well contribute to local spread within pastures. The moist conditions in which gastropods are most active would also be most suitable for nematode transmission.

In New Zealand stem nematode is often noted in spaced plantings or plots of pure clover, and West and Steele (1986) and Watson (1988) have demonstrated variation among white-clover cultivars in susceptibility under these conditions. However, stem nematode rarely occurs at levels producing symptoms in New Zealand mixed-species pasture (Watson and Barker, 1993). That resistant plants are relatively easily selected for suggest, however, that stem nematode may have significance in the ecology of white clover. Williams and Barclay (1972) showed that stem nematode can severely affect the establishment of white clover and, indeed, Yeates (1977) considered that stem nematode was primarily a problem in the establishment of New Zealand pastures. Stem nematode is evidently of much significance in the ecology of red clover in New Zealand, contributing substantially to the poor presence of the tap-rooted cultivars (Skipp and Christensen, 1990).

Conclusions

Invertebrate pests can have a significant impact on pasture productivity and, if uncontrolled, on the profitability of pastoral farming. However, pastoral farmers tend to be more concerned with the product they sell (animals and their products) than with indirect processes, such as pest damage, that affect the productivity of the pastures on which the animals feed. Because of the virtual day-to-day handling of animals and legislative requirements, husbandry practices that maintain good livestock health are well understood and effectively applied. Farmers are less familiar with and less interested in pasture pests and the correct application of controls against them. For most pastoral farmers, pests are only important when they interfere with ability to feed livestock. Also, pest problems are often sporadic and do not justify farmers' constant attention or are chronic and go unnoticed. Thus farmers generally take a curative approach to pest problems rather than a preventive one. Pasture pests are therefore a psychological problem as much as an economic one.

Gastropods have either not featured at all in rankings of importance of pasture pests (e.g. NRAC, 1974) or are ranked as minor pests (e.g. Watson *et al.*, 1989). This reflects the recognition that other pests interfere more with feed supply on farms, but is also to some degree the result of a general ignorance of the contribution of herbivorous gastropods to losses in primary productivity. None the less, the situation in New Zealand pastures, as summarized above, points to the significant pest status of gastropods in both established and establishing pastures in environments favourable for their population increase and activity.

For most New Zealand environments, pasture is an unstable vegetation requiring maintenance inputs if reversion to the natural forest climax vegetation is to be prevented. Farming, however, is vulnerable to fiscal and social changes, and new input/output equilibria are constantly being established. The maintenance inputs (e.g. labour, fertilizer) required to prevent reversion to forest are necessarily targeted at areas that will give the greatest response in economic production. The continued decline in returns for commodities and recent emphases on sustainability have led farmers to respond by: (i) further intensification of farming in the high-fertility lowland areas, particularly in dairying; (ii) further lowering of the input equilibrium, through reductions in development and maintenance expenditure and in production targets; and (iii) searching for new avenues and levels of income, e.g. diversification, product specification and off-farm investment. This has further accentuated the trichotomy in the underlying strategy and level of inputs in the different land classes – namely, nil to very low input in high-country sheep-grazing systems, moderate input in sheep and beef enterprises in hill country, and high-input enterprises in lowland dairying systems. For parts of the South Island high country, recreation and tourism have become key land-use activities. In hill country, plantation forestry has become a major land use. The net result has been a substantial reduction in pasture-development investment in hill and high country over the past decade or two, reducing the opportunities or situations in which gastropods can be damaging to seed and seedlings of sown pasture-plant species. Thus, gastropods remain potential pests but the changing circumstances in hill and high country mean that their pest status is rarely realized.

In the lowlands, the situation is markedly different. Increased grazing intensity increases pressure on pasture-plant species, leading to greater vulnerability to pest burdens and the need for frequent pasture renewal. While gastropods are generally less abundant under highly intensive grazing systems, their status as pests in established pasture is accentuated in farming systems that include paddocks that are afforded lengthy spelling from grazing in winter and spring, when gastropod numbers can increase. During pasture establishment, the pest status of the herbivorous gastropods is accentuated by practices that maximize the persistence of gastropod numbers from the old pasture into the early phases of the new pasture. These practices include overdrilling, where seed is simply sown into the old pasture, and where the furrows in the soil made by the drill coulters provide adequate shelter for the gastropods irrespective of the level of removal of the old pasture by herbicides.

While there is ample evidence that gastropods can adversely affect the establishment of grasses, it is the losses sustained by legumes in both established and establishing pastures that are of primary concern. Legumes such as white clover are critical to the success of New Zealand pastoral farming. However, their persistence and productivity is sensitive to the competitive dynamics that operate in mixed swards with grasses (Fig. 18.13). Along the stress–competition continuum, pastoralists seek to

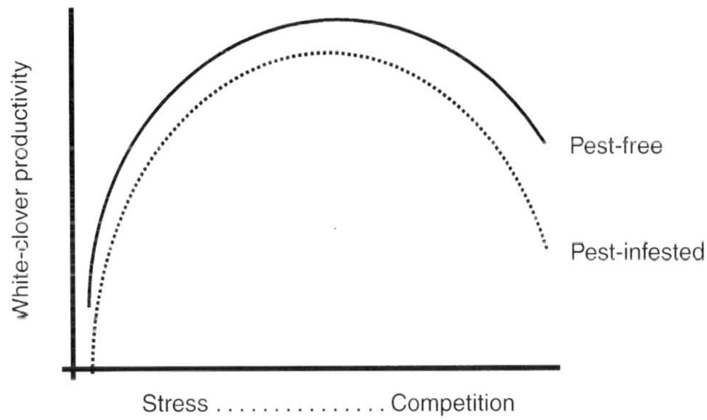

Fig. 18.13. Conceptual model of the productivity of white clover (*Trifolium repens* Linnaeus) (Fabaceae) in mixed swards with grasses along a gradient from highly stressful environments through to highly competitive environments and in situations with and without selective herbivory by invertebrate pests. The fitness of white clover is severely constrained in high-stress environments and, while subordinate in effect, herbivorous pests can drive the legume to extinction. In highly competitive environments, the fitness of white clover is reduced by the space-dominating effects of the companion grasses. In these environments, the selective herbivory by pests can substantially further reduce the fitness of the legume. White clover is most productive and most tolerant of herbivory in intermediate environments. The benefits of pest control thus vary along the stress–competition gradient.

define and operate within a zone of environmental conditions optimal for productivity from mixed legume–grass swards. Under optimal conditions, selective herbivory can diminish the contribution of white clover, but the plant exhibits high levels of tolerance and resilience. With increasing stress conditions, such as that imposed by low soil phosphorus or low soil moisture, the persistence and productivity of white clover are strongly compromised and, indeed, the species may be eliminated from the sward. Pests can accentuate these stress conditions and hasten the demise of white clover. Selective herbivory by gastropods can operate in this way. However, the alleviation of pest pressure will not result in marked improvements in white-clover performance, because growth is primarily constrained by environmental stress. At the other extreme, with an increasingly competitive environment, such as that imposed by high soil N, white clover is less able to compete with the companion grasses for space and light. The contribution of white clover will decline under these conditions, and this effect can be accentuated by selective herbivory by pests such as gastropods. The quantitative relationship between gastropod numbers and yield loss from mixed-species pasture will thus vary along the stress–competition gradient and is less easily defined than for single-species crop situations. None the less, an understanding of the dynamics in legume–grass swards is critical to effective management of white clover and its pest complex.

References

Allan, B.E. and Keoghan, J.M. (1994) More persistent legumes and grasses for oversown tussock country. *Proceedings of the New Zealand Grassland Association* 56, 143–147.

Angseesing, J.P.A. (1972) Selective eating of the acyanogenic form of *Trifolium repens*. *Heredity* 32, 73–83.

Ashby, J.W., The, P.B. and Close, R.C. (1979) Symptomatology of subterranean clover red leaf virus and its incidence in some legume crops, weed hosts, and certain alate aphids in Canterbury, New Zealand. *New Zealand Journal of Agricultural Research* 22, 361–365.

Attwood, S.S. and Sullivan, J.T. (1943) Inheritance of a cyanogenetic glucoside and its hydrolysing enzyme in *Trifolium repens*. *Journal of Heredity* 34, 311–320.

Backeljau, T. (1985) Estimation of genic similarity within and between *Arion hortensis* s.l. and *A. intermedius* by means of isoelectric focused esterase patterns in hepatopancreas homogenates (Mollusca, Pulmonata: Arionidae). *Zeitschrift für Zoologische Systematik und Evolutionsforschung* 23, 38–49.

Backeljau, T. and de Bruyn, L. (1990) On the infrageneric systematics of the genus *Arion* Férussac, 1819 (Mollusca, Pulmonata). *Bulletin de l'Institut Royal des Sciences Naturelles de Belgique, Biologie* 60, 35–68.

Backeljau, T. and de Bruyn, L. (1991) Preliminary report on the genetic variability of *Arion intermedius* in Europe (Pulmonata). *Journal of Medical and Applied Malacology* 3, 19–29.

Backeljau, T., de Brito, C.P., Tristão da Cunha, R.M., Frias Martins, A.M. and de Bruyn, L. (1992) Colour polymorphism and genetic strains in *Arion intermedius* from Flores, Azores (Mollusca: Pulmonata). *Biological Journal of the Linnean Society* 46, 131–143.

Baker, C.J. (1985) Technical potentialities of overdrilling for hill pasture improvement and renovation. In: *Proceedings of the 3rd AAAP Animal Science Congress*, Vol. 1, pp. 211–218.

Baker, G.H. (1989) Damage, population dynamics, movement and control of pest helicid snails in southern Australia. In: Henderson, I.F. (ed.) *Slugs and Snails in World Agriculture*. Monograph 41, British Crop Protection Council, Thornton Heath, pp. 175–185.

Barker, D.J. and Dymock, N. (1993) Effects of pre-sowing herbicide and subsequent sward mass on survival, development, and production of autumn oversown Wana cocksfoot and Tahora white clover seedlings. *New Zealand Journal of Agricultural Research* 36, 67–77.

Barker, G.M. (1979) The introduced slugs of New Zealand (Gastropoda: Pulmonata). *New Zealand Journal of Zoology* 6, 411–437.

Barker, G.M. (1985) Aspects of the biology of *Vallonia excentrica* (Mollusca-Vallonidae) in Waikato pastures. In: Chapman, R.B. (ed.) *Proceedings of the 4th Australasian Conference on Grassland Invertebrate Ecology*. Caxton Press, Christchurch, pp. 64–70.

Barker, G.M. (1986) Biology of pest slugs and their significance in conservation tillage systems. In: Hill, R.R., Clements, R.O., Hower, A.A., Jordan, T.A. and Zeiders, K.E. (eds) *Proceedings of an International Symposium on Establishment of Forage Crops by Conservation Tillage: Pest Management*. United States Regional Pasture Research Laboratory, University Park, Pennsylvania, pp. 83–106.

Barker, G.M. (1989a) Slug problems in New Zealand pastoral agriculture. In: Henderson, I.F. (ed.) *Slugs and Snails in World Agriculture*. Monograph 41, British Crop Protection Council, Thornton Heath, pp. 59–68.
Barker, G.M. (1989b) Susceptibility of lucerne cultivars and germplasm as seedlings to slugs. *Tests of Agrochemicals and Cultivars* 10, 146–147.
Barker, G.M. (1990) Pasture renovation: interactions of vegetation control with slug and insect infestations. *Journal of Agricultural Science* 115, 195–202.
Barker, G.M. (1991a) Biology of slugs (Agriolimacidae and Arionidae: Mollusca) in New Zealand hill country pastures. *Oecologia* 85, 581–595.
Barker, G.M. (1991b) Slug density–seedling establishment relationships in a pasture renovated by direct drilling. *Grass and Forage Science* 46, 113–120.
Barker, G.M. (1992) Naturalised terrestrial molluscs in New Zealand: origins and establishment. In: Heath, A.C.G. (compiler) *Proceedings of the 41st Annual Conference of the Entomological Society of New Zealand*. Entomological Society of New Zealand, Heretaunga.
Barker, G.M. (1993a) Population regulation of *Deroceras* slugs (Agriolimacidae) in northern New Zealand pastures with particular reference to the role of *Tetrahymena rostrata* (Kahl) (Ciliata) and *Microsporidium novacastriensis* (Jones & Selman) (Microspora). In: *Proceedings of the 3rd International Congress of Medical and Applied Malacology, Camden, 18–22 October, 1993*. Elizabeth McArthur Agricultural Institute, Camden.
Barker, G.M. (1993b) Contribution of the weed grass *Poa annua* to population phenology and recruitment of Argentine stem weevil in northern New Zealand pastures. In: Prestidge, R.A. (ed.) *Proceedings of the 6th Australasian Conference on Grassland Invertebrate Ecology*. AgResearch, Hamilton, pp. 155–187.
Barker, G.M. (1999) *Naturalised Terrestrial Stylommatophora (Mollusca: Gastropoda)*. Fauna of New Zealand 38, Manaaki Whenua Press, Lincoln, 253 pp.
Barker, G.M. and Addison, P.J. (1992) Pest status of slugs (Stylommatophora : Mollusca) in two New Zealand pastures. *Crop Protection* 11, 439–442.
Barker, G.M. and Mayhill, P.C. (1998) Patterns of diversity and habitat relationships in terrestrial mollusc communities of the Pukeamaru Ecological District, northeastern New Zealand. *Journal of Biogeography* 25, 215–238.
Barker, G.M. and Raymond, B.de H. (2002) Mollusc herbivory influenced by endophytic clavicipitaceous fungal infection in grasses. *Journal of Medical and Applied Malacology* (in press).
Barker, G.M., Willoughby, B.E. and Pottinger, R.P. (1983) Feeding by slugs on some pasture plant species: laboratory experiments. *Proceedings of the New Zealand Weed and Pest Control Conference* 36, 207–211.
Barker, G.M., Addison, P.J. and Pottinger, R.P. (1985) Biology and pest status of slugs (Mollusca) in two Waikato pastures. In: Chapman, R.B. (ed.) *Proceedings of the 4th Australasian Conference on Grassland Invertebrate Ecology*. Caxton, Press, Christchurch, pp. 18–25.
Barker, G.M., Robertson, L.N., Watson, R.N. and Willoughby, B.E. (1988) Pasture renovation – interactions of vegetation control on pest infestations. In: Stahle, P. (ed.) *Proceedings of the 5th Australasian Conference on Grassland Invertebrate Ecology*. D & D Printing, Victoria, pp. 153–162.
Barker, G.M., Addison, P.J., Firth, A.C. and Robertson, L.N. (1990a) Effect of vegetation suppression on ryegrass establishment by direct drilling. *Proceedings of the New Zealand Weed and Pest Control Conference* 43, 289–291.

Barker, G.M., Addison, P.J., Robertson, L.N. and Willoughby, B.E. (1990b) Interactions of seeding rate with pesticide treatments in pasture renovation by direct drilling. *New Zealand Journal of Agricultural Research* 33, 595–605.

Barker, G.M., Addison, P.J., Firth, A.C. and Barratt, B.I.P. (1996) *Sitona lepidus* Gyllenhal newly established in New Zealand: assessment of distribution in the North Island. *Proceedings of the New Zealand Plant Protection Conference* 49, 266–269.

Barratt, B.I.P. (1980) Effect of simulated insect damage to white clover seedlings. *Proceedings of the New Zealand Weed and Pest Control Conference* 33, 49–51.

Barratt, B.I.P. (1982) Insect damage to white clover oversown into New Zealand native tussock grassland. In: Lee, K.E. (ed.) *Proceedings of the 3rd Australasian Conference on Grassland Invertebrate Ecology*. Commonwealth Scientific and Industrial Research Organization, Adelaide, pp. 25–31.

Barratt, B.I.P. (1985) Effect of cotyledon damage on nodulation and growth of white clover oversown into native grassland in Central Otago, New Zealand. In: Chapman, R.B. (ed.) *Proceedings of the 4th Australasian Conference on Grassland Invertebrate Ecology*. Caxton Press, Christchurch, pp. 127–132.

Barratt, B.I.P. and Ferguson, C.M. (1992) Oversown white clover seed and seedling loss from insects and fungal pathogens. *Proceedings of the New Zealand Weed and Pest Control Conference* 45, 215–219.

Barratt, B.I.P. and Johnstone, P.D. (1984) Effects of insects, seeding rate, and insecticide seed dressing on white clover and Maku lotus in tussock grassland. *New Zealand Journal of Agricultural Research* 27, 13–18.

Barratt, B.I.P., Ferguson, C.M., Jones, P.A. and Johnstone, P.D. (1992) Effect of native weevils (Coleoptera: Curculionidae) on white clover establishment and yield in tussock grassland. *New Zealand Journal of Agricultural Research* 35, 63–73.

Barratt, B.I.P., Barker, G.M. and Addison, P.J. (1996) *Sitona lepidus* Gyllenhal (Coleoptera: Curculionidae), a potential clover pest new to New Zealand. *New Zealand Entomologist* 19, 23–30.

Bejakovich, D., Pearson, W.D. and O'Donnell, M.R. (1998) Nationwide survey of pests and diseases of cereal and grass seed crops in New Zealand. 1. Arthropods and molluscs. *Proceedings of the New Zealand Plant Protection Conference* 51, 38–50.

Black, A.D. and Lucas, R.J. (2000) Caucasian clover was more productive than white clover in grass mixtures under drought conditions. *Proceedings of the New Zealand Grassland Association* 62, 183–188.

Black, A.D., Pollock, K.M., Lucas, R.J., Amyes, J.M., Pownall, D.B. and Sedcole, J.R. (2000) Caucasian clover/ryegrass produced more legume than white clover/ryegrass pasture in a grazed comparison. *Proceedings of the New Zealand Grassland Association* 62, 69–74.

Blackmore, L.W. (1962) Band spraying – a new overdrilling technique. *New Zealand Journal of Agriculture* 104, 13–19.

Blackmore, L.W. (1964) Chemical renovation of pastures on hill country in Southern Hawkes Bay and Northern Wairarapa. *New Zealand Journal of Agriculture* 108, 122–135, 223–229, 233–234.

Blank, R.H. and Bell, D.S. (1982) Seasonal patterns of pasture seed loss to black field crickets (*Teleogryllus commodus*) and other invertebrates. *New Zealand Journal of Agricultural Research* 25, 123–129.

Brattsten, L.B. (1988) Enzymic adaptations in leaf-feeding insects to host-plant allelochemicals. *Journal of Chemical Ecology* 14, 1919–1939.

Brock, J.L., Hay, M.J.M., Thomas, V.J. and Sedcole, J.R. (1988) Morphology of white clover (*Trifolium repens* L.) plants in pastures under intensive sheep grazing. *Journal of Agricultural Science* 111, 273–283.

Brown, V.K. and Gange, A.L. (1989) Differential effects of above-ground and below-ground insect herbivory during early plant succession. *Oikos* 54, 67–76.

Burgess, R.S.L. and Ennos, R.A. (1987) Selective grazing of acyanogenic white clover: variation in behaviour among populations of the slug *Deroceras reticulatum*. *Oecologia* 73, 432–435.

Byers, R.A. and Bierlein, D.L. (1982) Feeding preferences of three slug species in the laboratory. *Melscheimer Entomological Series* 32, 5–11.

Byers, R.A., Barratt, B.I.P. and Calvin, D. (1989) Comparison between defined-area traps and refuge traps for sampling slugs in conservation-tillage crop environments. In: Henderson, I.F. (ed.) *Slugs and Snails in World Agriculture*. Monograph 41, British Crop Protection Council, Thornton Heath, pp. 187–192.

Caradus, J.R., MacKay, A.C., Woodfield, D.R., van der Bisch, J. and Wewala, S. (1989) Classification of a world collection of white clover cultivars. *Euphytica* 42, 183–196.

Caradus, J.R., Harris, S.L. and Johnson, R.J. (1996) Increased clover content for increased milk production. *Proceedings of the Ruakura Farmers' Conference* 48, 42–49.

Cater, M.A., Jeffery, R.C.V. and Williamson, P. (1979) Food overlap in co-existing populations of the land snails *Cepaea nemoralis* (L.) and *Cepaea hortensis* (Müll.). *Biological Journal of the Linnaean Society* 11, 169–176.

Catherall, P.L. (1970) *Cocksfoot Mottle Virus*. Descriptions of Plant Viruses 23, Commonwealth Mycological Institute/Association of Applied Biologists, Wellesbourne.

Catherall, P.L. (1987) Reaction of some white clover (*Trifolium repens*) varieties to inoculation with white clover mosaic virus. *Tests of Agrochemicals and Cultivars* 8, 142–143.

Chang, H.-W. (1991) Food preference of the land snail *Cepaea nemoralis* in a North American population. *Malacological Review* 24, 107–114.

Chanway, C.P., Holl, F.B. and Turkington, R. (1989) Effect of *Rhizobium leguminosarium* biovar *trifolii* genotype on specificity between *Trifolium repens* and *Lolium perenne*. *Journal of Ecology* 77, 1150–1160.

Chapman, D.F. (1983) Growth and demography of *Trifolium repens* stolons in grazed hill swards. *Journal of Applied Ecology* 20, 590–608.

Chapman, D.F. (1987) Natural re-seeding and *Trifolium repens* demography in grazed hill pastures. II. Seedling appearance and survival. *Journal of Applied Ecology* 24, 1037–1043.

Chapman, D.F. and Anderson, C.B. (1987) Natural re-seeding and *Trifolium repens* demography in grazed hill pastures. I. Flowerhead appearance and fate, and seed dynamics. *Journal of Applied Ecology* 24, 1025–1035.

Chapman, D.F., Parsons, A.J. and Schwinning, S. (1995) Management of clover in grazed pastures: expectations, limitations and opportunities. In: *White Clover: New Zealand's Competitive Edge*. Special Publication 11, Grassland Research and Practice Series 6, Agronomy Society of New Zealand, Christchurch, pp. 55–64.

Charlton, J.F.L. (1978a) Slugs as a possible cause of establishment failure in pasture legumes oversown in boxes. *New Zealand Journal of Experimental Agriculture* 6, 313–317.

Charlton, J.F.L. (1978b) Control of slug damage in oversown pasture legumes. *Proceedings of the New Zealand Weed and Pest Control Conference* 31, 127–130.

Charlton, J.F.L. (1979) Effect of slugs during establishment of oversown legumes in box experiments. In: Crosby, T.K. and Pottinger, R.P. (eds) *Proceedings of the 2nd Australasian Conference on Grassland Invertebrate Ecology*. Government Printer, Wellington, pp. 253–255.

Charlton, J.F.L. (1981) Initial trials for improvement of North Island hill country by conservation tillage. In: *Proceedings of the Conservation Tillage Seminar*, Vol. 2. Monsanto, Christchurch, pp. 27–34.

Charlton, J.F.L. and Stewart, A.V. (1999) Pasture species and cultivars used in New Zealand – a list. *Proceedings of the New Zealand Grassland Association* 61, 147–166.

Charlton, J.F.L., Watts, P.J. and Sedcole, J.R. (1985) Seasonal activity of slugs in New Zealand hill pasture. In: Chapman, R.B. (ed.) *Proceedings of the 4th Australasian Conference on Grassland Invertebrate Ecology*. Caxton Press, Christchurch, pp. 58–63.

Cheplick, G.P. and Clay, K. (1988) Acquired chemical defences in grasses: the role of fungal endophytes. *Oikos* 52, 309–318.

Chichester, L.F. and Getz, L.L. (1973) The terrestrial slugs of Northeastern North America. *Sterkiana* 51, 11–42.

Clay, K. (1988) Clavicipitaceous fungal endophytes of grasses: coevolution and the change from parasitisam to mutualism. In: Hawksworth, D.L. and Pirozynski, K. (eds) *Coevolution of Fungi with Plants and Animals*. Academic Press, London, pp. 79–105.

Clay, K. (1990) Fungal endophytes of grasses. *Annual Review of Ecology and Systematics* 21, 275–297.

Clay, K. (1994) The potential role of endophytes in ecosystems. In: Bacon, C.W. and White, J.F. (eds) *Biotechnology of Endophytic Fungi of Grasses*. CRC Press, Boca Raton, Florida, pp. 73–86.

Clement, S.L., Kaiser, W.J. and Eichenseer, H. (1994) *Acremonium* endophytes in germplasms of major grasses and their utilization for insect resistance. In: Bacon, C.W. and White, J.F. (eds) *Biotechnology of Endophytic Fungi of Grasses*. CRC Press, Boca Raton, Florida, pp. 185–199.

Collins, W.J. and Cox, R.I. (1984) Oestrogenic activity in forage legumes. In: Barnes, R.F., Ball, P.R., Brougham, R.W., Marten, G.C. and Minson, D.J. (eds) *Forage Legumes for Energy-efficient Animal Production*. United States Department of Agriculture, Washington, DC, pp. 268–276.

Cook, R., Thomas, B.T. and Mizen, K.A. (1989) Dissemination of white clover mosaic virus and stem nematode, *Ditylenchus dipsaci*, by the slug, *Deroceras reticulatum*. In: Henderson, I.F. (ed.) *Slugs and Snails in World Agriculture*. Monograph 41, British Crop Protection Council, Thornton Heath, pp. 107–112.

Corkill, L. (1942) Cyanogenesis in white clover (*Trifolium repens* L.). V. The inheritance of cyanogenesis. *New Zealand Journal of Science and Technology* 23B, 178–193.

Corkill, L. (1952) Cyanogenesis in white clover (*Trifolium repens* L.). VI. Experiments with high-glucoside and glucoside free strains. *New Zealand Journal of Science and Technology* 34A, 1–16.

Cottam, D.A. (1985) Frequency-dependent grazing by slugs and grasshoppers. *Journal of Ecology* 73, 925–933.

Crawford-Sidebotham, T.J. (1972) The role of slugs and snails in the maintenance of the cyanogenesis polymorphisms of *Lotus corniculatus* and *Trifolium repens*. *Heredity* 28, 405–411.

Crush, J.R. (1987) Nitrogen fixation. In: Baker, M.J. and Williams, W.M. (eds) *White Clover*. CAB International, Wallingford, UK, pp. 185–201.

Cullen, N.A. (1964) Species competition in establishing swards: suppression effects of ryegrass on establishment and production of associated grasses and clovers. *New Zealand Journal of Agricultural Research* 7, 678–693.

Cumberland, K.B. (1981) *Landmarks*. Reader's Digest Services, Sydney, 304 pp.

Dan, N.A. (1978) Studies on the growth and ecology of *Helix aspersa* Müller. PhD thesis, University of Manchester.

Davidson, J. (1984) *The Prehistory of New Zealand*. Longman Paul, Auckland, 207 pp.

Davies, H.L. (1987) Limitations to livestock production associated with phytoestrogens and bloat. In: Wheeler, J.L., Pearson, C.J. and Robards, G.E. (eds) *Temperate Pastures. Their Production, Use and Management*. Australian Wool Corporation/Commonwealth Scientific and Industrial Research Organization, Melbourne, pp. 446–456.

Davies, S.M. (1977) The *Arion hortensis* complex, with notes on *A. intermedius* Normand (Pulmonata: Arionidae). *Journal of Conchology* 29, 173–187.

de Nooij, M.P. and Mook, J.H. (1992) Interactions with other organisms. In: Kuiper, P.J.C. and Bos, M. (eds) *Plantago: a Multidisciplinary Study*. Ecological Studies Analysis and Synthesis, Vol. 89, Springer-Verlag, Berlin, pp. 52–68.

Dirzo, R. (1980) Experimental studies on slug–plant interactions. I. The acceptability of thirty plant species to the slug *Agriolimax caruanae*. *Journal of Ecology* 68, 981–998.

Dirzo, R. and Harper, J.L. (1982a) Experimental studies on slug–plant interactions. III. Differences in the acceptability of individual plants of *Trifolium repens* to slugs and snails. *Journal of Ecology* 70, 101–117.

Dirzo, R. and Harper, J.L. (1982b) Experimental studies on slug–plant interactions. IV. The performance of cyanogenic and acyanogenic morphs of *Trifolium repens* in the field. *Journal of Ecology* 70, 119–138.

Dolan, S. and Fleming, C.C. (1988) Isoenzymes in the identification and systematics of terrestrial slugs of the *Arion hortensis* complex. *Biochemical Systematics and Ecology* 16, 195–198.

Duval, D.M. (1973) A note on the acceptability of various weeds as food for *Arion hortensis* Férussac. *Journal of Conchology* 28, 37–39.

Edwardson, J.R. and Christie, R.G.T. (1986) *Viruses Infecting Forage Legumes*, Vol. 2. Monograph 14. Institute of Food and Agricultural Sciences, University of Florida, Gainesville.

Ester, A. and Nijënstein, J.H. (1995) Control of the field slug *Deroceras reticulatum* (Müller) (Pulmonata: Limacidae) by pesticides applied to winter wheat seed. *Crop Protection* 5, 409–413.

Ester, A., Darwinkel, A., Floot, H.W.G. and Nijënstein, J.H. (1996) Control of field slugs (*Deroceras reticulatum*) in winter wheat using seeds treated with pesticides. In: Henderson, I.F. (ed.) *Slug and Snail Pests in Agriculture*. Symposium Proceedings 66, British Crop Protection Council, Farnham, pp. 165–172.

Evans, A.A., Barratt, B.I.P. and Ferguson, C.M. (1994) Susceptibility of legume and *Hieracium* spp. seedlings to feeding by native broad-nosed weevils (Coleoptera: Curculionidae). *Proceedings of the New Zealand Plant Protection Conference* 47, 206–209.

Ferguson, C. (1984) Slug feeding on seeds and seedlings of ryegrass and white clover. *Proceedings of the New Zealand Weed and Pest Control Conference* 37, 64–67.

Ferguson, C. and Barratt, B.I.P. (1983) Slug damage to pasture renovated by direct drilling. *Proceedings of the New Zealand Weed and Pest Control Conference* 36, 212–215.

Ferguson, C. and Evans, A.A. (1994) Field screening of seedling dryling legume species to insect feeding. In: *Proceedings of the New Zealand Sustainable Management Conference.* Lincoln University, Canterbury, pp. 227–231.

Ferguson, C., Barratt, B.I.P. and Jones, P.A. (1989a) Control of the grey field slug (*Deroceras reticulatum* (Muller)) by stock management prior to direct-drilled pasture establishment. *Journal of Agricultural Science* 111, 443–449.

Ferguson, C., Barratt, B.I.P. and Jones, P.A. (1989b) A new technique for estimating density of the field slug (*Deroceras reticulatum* (Müller)). In: Henderson, I.F. (ed.) *Slugs and Snails in World Agriculture.* Monograph 41, British Crop Protection Council, Thornton Heath, pp. 331–336.

Ferguson, C., Shand, J.A. and Evans, A.A. (1995) Yield response of legume seedlings to simulated insect cotyledon damage and nutrient availability. *Proceedings of the New Zealand Plant Protection Conference* 48, 213–218.

Foltz, D.W., Ochman, H., Jones, J.S., Evangelist, M. and Selander, R.K. (1982) Genetic population structure and breeding systems in arionid slugs (Mollusca: Pulmonata). *Biological Journal of the Linnaean Society* 17, 225–241.

Foltz, D.W., Ochman, H. and Selander, R.K. (1984) Genetic diversity and breeding systems in terrestrial slugs of the families Limacidae and Arionidae. *Malacologia* 25, 593–605.

Forster, R.L.S., Morris-Krsinich, B.A.M. and Musgrave, D.R. (1985) Incidence of alfalfa mosaic virus, lucerne Australian latent virus, and lucerne transient streak virus in lucerne crops in the North Island of New Zealand. *New Zealand Journal of Agricultural Research* 28, 279–282.

Frain, J.M. and Newell, P.F. (1982) Meal size and feeding assay for *Deroceras reticulatum* (Müll.). *Journal of Molluscan Studies* 48, 98–99.

Fry, P.R. (1959) A clover mosaic virus in New Zealand pastures. *New Zealand Journal of Agricultural Research* 2, 971–981.

Garrido, C., Castillejo, J. and Iglesias, J. (1995) The spermatophore of *Arion intermedius* (Pulmonata: Arionidae). *Journal of Molluscan Studies* 61, 127–133.

Giusti, F. (1986) Notulae Malacologicae XXXIV. Again on the taxonomic status of *Deroceras panormitanum* (Lessona & Pollonera, 1882), *Deroceras pollonerai* (Simroth, 1889) and *Deroceras caruanai* (Pollonera, 1891) (Gastropoda: Pulmonata). *Bollettino Malacologico* 22, 57–64.

Glen, D.M., Cuerden, R. and Butler, R.C. (1991) Impact of the field slug *Deroceras reticulatum* on establishment of ryegrass and white clover in mixed swards. *Annals of Applied Biology* 119, 157–162.

Godan, D. (1983) *Pest Slugs and Snails. Biology and Control.* Springer, Berlin, 445 pp.

Grime, G.P. and Blythe, G.M. (1969) Relationships between snails and vegetation at the Winnats Pass. *Journal of Ecology* 57, 45–66.

Grime, G.P., MacPherson-Stewart, S.F. and Dearman, R.S. (1968) An investigation of leaf palatability using the snail *Cepaea nemoralis* L. *Journal of Ecology* 56, 405–420.

Guy, P.L. (1993) First record of ryegrass mosaic virus and its mite vector *Abacarus hystrix* (Nal.) in New Zealand. *New Zealand Journal of Agricultural Research* 36, 377–379.

Guy, P.L., Gibbs, A. and Harrower, K. (1980) The effect of white clover mosaic virus on nodulation of white clover (*Trifolium repens* L. cv. Ladino). *Australian Journal of Agricultural Research* 31, 307–311.

Hanley, M.E., Fenner, M. and Edwards, P.J. (1995a) An experimental field study of the effects of mollusc grazing on seedling recruitment and survival in grassland. *Journal of Ecology* 83, 621–627.

Hanley, M.E., Fenner, M. and Edwards, P.J. (1995b) The effect of seedling age on the likelihood of herbivory by the slug *Deroceras reticulatum. Functional Ecology* 9, 754–759.

Hanley, M.E., Fenner, M. and Edwards, P.J. (1996a) The effect of seedling size on the likelihood of herbivory by the slug *Deroceras reticulatum. Functional Ecology* 9, 745–759.

Hanley, M.E., Fenner, M. and Edwards, P.J. (1996b) The effect of mollusc grazing on seedling recruitment in artificially created grassland gaps. *Oecologia* 106, 240–246.

Harper, J.L. and Hatto, J. (1968) *First Annual Report on the Research Project 'The Control of Slugs and Snails in British Cropping Systems, Especially Grassland'.* School of Plant Biology, University College of North Wales, Bangor.

Harris, S.L. (1993) White clover plant morphology in Waikato dairy pastures. *Proceedings of the New Zealand Grassland Association* 55, 147–150.

Harris, S.L., Clark, D.A., Auldist, M.J., Waugh, C.D. and Laboyrie, P.G. (1997) Optimum white clover content for dairy pastures. *Proceedings of the New Zealand Grassland Association* 59, 29–33.

Hay, M.J.M. (1983) Seasonal variation in the distribution of white clover *Trifolium repens* L. stolons among 3 horizontal strata in 2 grazed swards. *New Zealand Journal of Agricultural Research* 26, 29–34.

Hay, M.J.M. (1985) Seasonal variation in the vertical distribution of white clover (*Trifolium repens* L.) stolons in two contrasting pastures. *Proceedings of the New Zealand Grassland Association* 46, 195–198.

Hay, M.J.M., Brock, J. and Fletcher, R.H. (1983) Effect of sheep grazing management on distribution of white clover stolons among 3 horizontal strata in ryegrass/white clover swards. *New Zealand Journal of Experimental Agriculture* 11, 215–218.

Hay, M.J.M., Chapman, D.F., Hay, R.J.M., Pennell, C.G.L., Woods, P.W. and Fletcher, R.H. (1987) Seasonal variation in the vertical distribution of white clover stolons in grazed swards. *New Zealand Journal of Agricultural Research* 30, 1–8.

Hay, M.J.M., Brock, J. and Thomas, V.J. (1989a) Characteristics of individual white clover plants in grazed swards. In: *Proceedings of the XVI International Grassland Congress.* Association Française pour la Production Fourragère, Versailles, pp. 1051–1052.

Hay, M.J.M., Chu, A.C.P., Knight, M.V. and Wewala, S. (1989b) Variation with season and node position in carbohydrate content of white clover stolons. In: *Proceedings of the XVI International Grassland Congress*. Association Française pour la Production Fourragère, Versailles, pp. 1059–1060.

Haynes, R.J. (1980) Competitive aspects of the grass–legume association. *Advances in Agronomy* 33, 227–261.

Hoglund, J.H., Crush, J.R., Brock, J.L., Ball, R. and Carran, R.A. (1979) Nitrogen fixation in pasture. XII. General discussion. *New Zealand Journal of Experimental Agriculture* 7, 45–51.

Horrill, J.C. and Richards, A.J. (1986) Differential grazing by the mollusc *Arion hortensis* on cyanogenic and acyanogenic seedlings of the white clover, *Trifolium repens*. *Heredity* 56, 277–281.

Howitt, A.J. (1961) Chemical control of slugs in orchard grass–ladino white clover pastures in the Pacific north-west. *Journal of Economic Entomology* 54, 778–781.

Hughes, K.A. and Gaynor, D.L. (1984) Comparison of Argentine stem weevil and slug damage in maize direct-drilled into pasture or following winter oats. *New Zealand Journal of Experimental Agriculture* 12, 47–53.

Hughes, M.A. (1991) The cyanogenic polymorphism in *Trifolium repens* L. (white clover). *Heredity* 66, 105–115.

Hulme, P.E. (1994) Seedling herbivory in grassland: relative impact of vertebrate and invertebrate herbivores. *Journal of Ecology* 82, 873–880.

Hulme, P.E. (1996) Herbivores and the performance of grassland plants: a comparison of arthropod, mollusc and rodent herbivory. *Journal of Ecology* 84, 43–51.

Hume, D.E. and Barker, D.J. (1991) Natural reseeding of five grass species in summer dry hill country. *Proceedings of the New Zealand Grassland Association* 53, 97–104.

Hyslop, M.G., Kemp, P.D., Barker, D.J. and Hodgson, J. (1996) Incidence of nodal rooting and plantlet survival in spreading red clovers. *Proceedings of the New Zealand Grassland Association* 58, 243–246.

Iglesias, J. and Castillejo, J. (1999) Field observations on feeding of the land snail *Helix aspersa* Müller. *Journal of Molluscan Studies* 65, 411–423.

Ingram, W.M. and Peterson, A. (1947) Food of the giant western slug, *Ariolimax columbianus* (Gould). *The Nautilus* 61, 49–51.

Kain, W.M. and Biggs, D.R. (1980) Effect of pea aphid and bluegreen lucerne aphid (*Acyrthosiphon* spp.) on coumestrol levels in herbage of lucerne (*Medicago sativa*). *New Zealand Journal of Agricultural Research* 23, 563–568.

Karlin, E.J. and Bacon, C. (1960) Courtship, mating, and egg-laying behaviour in the Limacidae (Mollusca). *Transactions of the American Microscopical Society* 80, 399–406.

Kays, S. and Harper, J.L. (1974) The regulation of plant and tiller density in a grass sward. *Journal of Ecology* 62, 97–105.

Khadhair, A.H., Sinha, R.C. and Peterson, J.F. (1984) Effects of white clover mosaic virus on various processes to symbiotic N_2 fixation in red clover. *Canadian Journal of Botany* 62, 38–43.

Koenig, R. (1978) *Potexvirus Group*. Descriptions of Plant Viruses 200, Commonwealth Mycological Institute/Association of Applied Biologists, Wellesbourne, 5 pp.

Kunelius, H.T., Harris, W., Henderson, J.D. and Baker, C.J. (1982) Comparison of tillage methods on red clover and ryegrass establishment and production under grazing in the establishment year. *New Zealand Journal of Experimental Agriculture* 10, 252–263.

Latch, G.C.M. and Skipp, R.A. (1987) Diseases. In: Baker, M.J. and Williams, W.M. (eds) *White Clover*. CAB International, Wallingford, pp. 421–460.

Ledgard, S.F., Briar, G.J. and Watson, R.N. (1988) New cultivars for Waikato dairy pastures: establishment, production and nitrogen fixation during the first year. *Proceedings of the New Zealand Grassland Association* 49, 207–211.

Levy, E.B. (1970) *Grasslands of New Zealand*. Government Printer, Wellington, 374 pp.

L'Huillier, P.J. and Aislabie, D.W. (1988) Natural reseeding in perennial ryegrass/white clover dairy pastures. *Proceedings of the New Zealand Grassland Association* 49, 111–115.

Lutman, J. (1978) The role of slugs in an *Agrostis–Festuca* grassland. In: Heal, O.W. and Perkins, D.F. (eds) *Production Ecology of British Moors and Montane Grasslands*. Springer, Berlin, pp. 332–416.

McCallum, D.A. and Thomson, N.A. (1990) Effect of a molluscicide and an insecticide on the establishment of direct-drilled ryegrass, tall fescue and phalaris. *Proceedings of the New Zealand Grassland Association* 52, 237–240.

McCracken, G.F. and Selander, R.K. (1980) Self-fertilisation and monogenic strains in natural populations of terrestrial slugs. *Proceedings of the National Academy of Sciences of the USA* 77, 684–688.

Macfarlane, M.J. and Sheath, G.W. (1984) Clover – what types for dry hill country? *Proceedings of the New Zealand Grassland Association* 45, 140–150.

Macfarlane, M.J., Sheath, G.W. and McGowan, A.W. (1990) Evaluation of clovers in dry hill country 5. White clover at Whatawhata. *New Zealand Journal of Agricultural Research* 33, 549–556.

McGlone, M.S. (1983) Polynesian deforestation of New Zealand: a preliminary synthesis. *Archaeology in Oceania* 18, 11–25.

McLaughlin, M.R. and Boykin, D.L. (1988) Virus diseases of seven species of forage legumes in the south-eastern United States. *Plant Diseases* 72, 539–542.

McLaughlin, M.R., Pederson, G.A., Evans, R.R. and Ivy, R.L. (1992) Virus diseases and stand decline in a white clover pasture. *Plant Disease* 76, 158–162.

Madden, E.A. (1952) Establishment of clovers on unploughable hill country. *Sheep Farming Annual, Massey University* 15, 37–48.

Mercer, C.F. and van den Bosch, J. (1997) Fourth and fifth generation progress in breeding white clover for resistance to root-knot nematode. *Proceedings of the New Zealand Plant Protection Conference* 50, 61–64.

Mercer, C.F., Watson, R.N., van den Bosch, J., Widdup, K.H., Miller, K.J. and Pennell, C.J. (1999) Resistence in white clover to the nematodes *Heterodera trifolii* and *Meloidogyne trifoliophila* – progress in breeding and early results from field tests. In: Matthiessen, J.N. (ed.) *Proceedings of the 7th Australasian Conference on Grassland Invertebrate Ecology*. CSIRO, Wembley, pp. 272–278.

Miller, J.D., Gibson, P.B., Cope, W.A. and Knight, W.E. (1975) Herbivore feeding on cyanogenic and acyanogenic white clover seedlings. *Crop Science* 15, 90–91.

Molgaard, P. (1992) Polymorphism for caffeic acid esters in populations of *Plantago major* ssp *pleisperma*. In: Kuiper, P.J.C. and Bos, M. (eds) *Plantago:*

a Multidisciplinary Study. Ecological Studies Analysis and Synthesis, Vol. 89, Springer-Verlag, Berlin, pp. 192–203.

Moloney, S.C. and Milne, G.D. (1993) Establishment and management of Grasslands Puna chicory used as a specialist, high quality forage herb. *Proceedings of the New Zealand Grassland Association* 55, 113–118.

Montgomery, K.M. (1938) *Lotus major* in surface-sown pastures. *New Zealand Journal of Agriculture* 57, 33–34.

Moss, R.A., Burton, R.N. and Allan, B.E. (1996) Productivity of Caucasian clover based pastures under irrigation. *Proceedings of the New Zealand Grassland Association* 58, 177–181.

Mowat, D.J. and Shakeel, M.A. (1988) The effect of pesticide application on the establishment of white clover in a newly-sown ryegrass/white clover sward. *Grass and Forage Science* 43, 371–375.

Mowat, D.J. and Shakeel, M.A. (1989a) The effect of some invertebrate species on persistence of white clover in ryegrass swards. *Grass and Forage Science* 44, 117–124.

Mowat, D.J. and Shakeel, M.A. (1989b) The effect of some invertebrate species on the growth of white clover (*Trifolium repens* L.) in the laboratory. *Grass and Forage Science* 44, 405–409.

Muhle, E. (1941) Die Gehauseschnecke *Vallonia pulchella* O.F. Muller als Schadling des Grassamenbaues. *Nachrichtenbl. Dt. Pflanzenschutzd* 21, 13–14.

Nass, H.G. (1972) Cyanogenesis: its inheritance in *Sorghum bicolor, Sorghum sudanense, Lotus*, and *Trifolium repens* – a review. *Crop Science* 12, 503–506.

NRAC (1974) *Report of the National Research Advisory Council for the Year Ended 31 March 1974*. WRAC, Wellington.

O'Loughlin, C.L. and Owens, I.F. (1987) Our dynamic environment. In: Holland, P.G. (ed.) *Southern Approaches*. New Zealand Geographical Society, Christchurch, pp. 59–90.

Pallant, D. (1969) The food of the grey field slug (*Agriolimax reticulatus* (Müller)) in grassland. *Journal of Animal Ecology* 41, 761–769.

Pallant, D. (1972) The food of the grey field slug (*Agriolimax reticulatus* (Müller)) in woodland. *Journal of Animal Ecology* 38, 391–398.

Peters, H.A., Baur, B., Bazzaz, F. and Körner, C. (2000) Consumption rates and food preferences of slugs in a calcareous grassland under current and future CO_2 conditions. *Oecologia* 125, 72–81.

Pottinger, R.P. (1973) Pests of grassland and small seeds crops. In: Langer, R.H.M. (ed.) *Pastures and Pasture Plants*. A.H. & A.W. Reed, Wellington, pp. 365–406.

Pottinger, R.P. and Macfarlane, R.P. (1967) Insect pests and nematodes of lucerne. In: Langer, R.H.M. (ed.) *The Lucerne Crop*. A.H. & A.W. Reed, Wellington, pp. 229–247.

Powell, A.L. and Bowen, I.D. (1996) The screening of naturally occurring compounds for use as seed treatments for the protection of winter wheat against slug damage. In: Henderson, I.F. (ed.) *Slug and Snail Pests in Agriculture*. Symposium Proceedings 66, British Crop Protection Council, Farnham, pp. 231–236.

Raffaelli, D. and Mordue, A.J. (1990) The relative importance of molluscs and insects as selective grazers of acyanogenic white clover (*Trifolium repens*). *Journal of Molluscan Studies* 56, 37–45.

Ramsell, J., Malloch, A.J.C. and Whittaker, J.B. (1993) When grazed by *Tipula paludosa*, *Lolium perenne* is a stronger competitor than *Rumex obtusifolius*. *Journal of Ecology* 81, 777–786.
Robertson, L.N., Davison, R.H. and Firth, A.C. (1981) Insecticide rates, seeding rates and Australian soldier fly control in pasture. *Proceedings of the New Zealand Weed and Pest Control Conference* 34, 161–163.
Rollo, C.D. (1987) The feeding of terrestrial slugs in the relation to food characteristics, starvation, maturation and life history. *Malacologia* 28, 29–39.
Rollo, C.D. and Shibata, D.M. (1991) Resilience, robustness, and plasticity in a terrestrial slug, with particular reference to food quality. *Canadian Journal of Zoology* 69, 978–987.
Rolston, M.P., Clifford, P.T.P., Charlton, J.F.L., Hampton, J.G., White, J.G.H., Wright, A.G. and Knox, D. (1990) New Zealand's herbage seed industry: an overview. *Proceedings of the New Zealand Grassland Association* 52, 55–58.
Rowan, D.D. and Latch, G.C.M. (1994) Utilization of endophyte-infected perennial ryegrasses for increased insect resistance. In: Bacon, C.W. and White, J.F. (eds) *Biotechnology of Endophytic Fungi of Grasses*. CRC Press, Boca Raton, pp. 169–183.
Rumball, P.J. and Lambert, J.P. (1978) Plant introduction trials: performance of some forage legumes in Northland. *New Zealand Journal of Experimental Agriculture* 8, 179–183.
Runham, N.W. and Hunter, P.J. (1970) *Terrestrial Slugs*. Hutchinson, London, 184 pp.
Salmon, J.T. (1975) The influence of man on biota. In: Kuschel, G. (ed.) *Biogeography and Ecology in New Zealand*. Dr W. Junk, The Hague, pp. 643–661.
Sangakkara, R., Roberts, E. and Watkin, B.R. (1982) Grass species used and pasture establishment practices in central New Zealand. *New Zealand Journal of Experimental Agriculture* 10, 359–364.
Scott, D. (1998) Sixteen years of Caucasian clover under contrasting managements. *Proceedings of the New Zealand Grassland Association* 60, 115–118.
Scott, S.W. (1982) The effects of white clover mosaic virus infection on the yield of red clover (*Trifolium pratense* L.) in mixtures and in pure stands. *Journal of Agricultural Science* 98, 455–460.
Scott, S.W. (1984) Clover rot. *Botanical Review* 50, 491–504.
Senseman, D.M. (1977) Starch: a potent feeding stimulant for the terrestrial slug *Ariolimax californicus*. *Journal of Chemical Ecology* 3, 707–715.
Shakeel, M.A. and Mowat, D.J. (1992) The potential of clover rot, *Sclerotinia trifolium* Erikss., by slugs. *Grass and Forage Science* 47, 199–202.
Silva, O. and Teresa, M. (1992) Effects of mollusc grazing on the development of grassland species. *Journal of Vegetation Science* 3, 267–270.
Sithamparanathan, J., Macfarlane, M.J. and Richardson, S. (1986) Effect of treading, herbicides, season, and seed coating on oversown grass and legume establishment in easy North Island hill country. *New Zealand Journal of Agricultural Research* 14, 173–182.
Skipp, R.A. and Christensen, M.J. (1990) Selection for persistence in red clover: influence of root disease and stem nematode. *New Zealand Journal of Agricultural Research* 33, 319–333.

Slykhuis, J.T. and Paliwal, Y.C. (1972) *Ryegrass Mosaic Virus*. Descriptions of Plant Viruses 86, Commonwealth Mycological Institute/Association of Applied Biologists.

Smales, T.E., Ferguson, C.M. and Guy, P.L. (1995) Invertebrate pests of pasture as potential plant virus vectors. *Proceedings of the New Zealand Plant Protection Conference* 48, 194–198.

Smales, T.E., Ferguson, C.M. and Guy, P.L. (1996) Pasture and lucerne pests as virus vectors. *Proceedings of the New Zealand Plant Protection Conference* 49, 275–279.

Smith, R.S. and Bishop, D.J. (1993) Astred – a stoloniferous red clover. In: *Proceedings of the XVII International Grassland Congress 1993*. Association Française pour la Production Fourragère, Versailles, pp. 421–423.

South, A. (1974) Changes in composition of the terrestrial mollusc fauna. In: Hawksworth, D.L. (ed.) *The Changing Flora and Fauna*. Academic Press, London, pp. 255–274.

South, A. (1989a) The effect of weather and other factors on numbers of slugs on permanent pasture. In: Henderson, I.F. (ed.) *Slugs and Snails in World Agriculture*. Monograph 41, British Crop Protection Council, Thornton Heath, pp. 355–360.

South, A. (1989b) A comparison of the life cycles of the slugs *Deroceras reticulatum* (Muller) and *Arion intermedius* Normand on permanent pasture. *Journal of Molluscan Studies* 55, 9–22.

Speiser, B. and Rowell-Rahier, M. (1991) Effects of food availability, nutritional value, and alkaloids on food choice in the generalist herbivore *Arianta arbustorum* (Gastropoda: Helicidae). *Oikos* 62, 306–318.

Speiser, B. and Rowell-Rahier, M. (1993) Does the land snail *Arianta arbustorum* prefer sequentially mixed over pure diets? *Functional Ecology* 7, 403–410.

Speiser, B., Harmatha, J. and Rowell-Rahier, M. (1992) Effects of pyrrolizidine alkaloids and sequiterpens on snail feeding. *Oecologia* 92, 257–265.

Statistics New Zealand (2001) Pastoral agriculture. http://www.stats.govt.nz/

Suckling, F.E.T. (1949) Improvement of hill country pastures in the Wellington Province. *Proceedings of the New Zealand Grassland Association* 11, 89–117.

Suckling, F.E.T. (1951) Results of recent experiments on surface sowing. *Proceedings of the New Zealand Grassland Association* 13, 119–127.

Suckling, F.E.T. (1966) *Hill Pasture Improvement*. Newton King Group/Department of Scientific and Industrial Research, Wanganui, 58 pp.

Thom, E.R. (1988) Changing seasonal growth of paspalum pastures by overdrilling ryegrass and white clover. *Proceedings of the New Zealand Grassland Association* 49, 135–140.

Upstone, M.E. (1969) Epidemiology of cocksfoot mottle virus. *Annals of Applied Biology* 64, 49–55.

van den Bosch, J. and Mercer, C.F. (1996) Progress towards clover cyst nematode resistant white clover. *Proceedings of the New Zealand Grassland Association* 58, 237–241.

Wardle, D.A. and Barker, G.M. (1997) Competition and herbivory in establishing grassland communities: implications for plant biomass, species diversity and soil microbial activity. *Oikos* 80, 470–480.

Wardle, D.A., Barker, G.M., Nicholson, K.S. and Addison, P.J. (1994) Cyclic oscillations between a summer annual (C_4) and a winter annual weed grass in

Waikato dairy pastures. *Proceedings of the New Zealand Plant Protection Conference* 47, 34–37.

Wardle, D.A., Barker, G.M., Bonner, K.I. and Nicholson, K.S. (1998) Can comparative approaches based on plant ecophysiological traits predict the nature of biotic interactions and individual plant species effects in ecosystems? *Journal of Ecology* 86, 405–420.

Watson, R.N. (1988) Infection of nine white clover cultivars by *Sclerotinia trifoliorum* and *Ditylenchus dipsaci*. *Proceedings of the New Zealand Weed and Pest Control Conference* 41, 121–125.

Watson, R.N. and Barker, G.M. (1993) Towards improving the role of legumes for grassland sustainability. In: Prestidge, R.A. (ed.) *Proceedings of the 6th Australasian Conference on Grassland Invertebrate Ecology*. AgResearch, Hamilton, pp. 213–226.

Watson, R.N. and Mercer, C.F. (2000) Pasture nematodes: the major scourge of white clover. *Proceedings of the New Zealand Grassland Association* 62, 195–199.

Watson, R.N., Yeates, G.W., Littler, R.A. and Steele, K.W. (1985) Responses in nitrogen fixation and herbage production following pesticide applications on temperate pastures. In: Chapman, R.B. (ed.) *Proceedings of the 4th Australasian Conference on Grassland Invertebrate Ecology*. Caxton Press, Christchurch, pp. 103–113.

Watson, R.N., Skipp, R.A. and Barratt, B.I.P. (1989) Initiatives in pest and disease control in New Zealand towards improving legume production and persistence. In: Marten, G.C., Matches, A.G., Barnes, R.F., Brougham, R.W., Clements, R.J. and Sheath, G.W. (eds) *Persistence of Forage Legumes*. American Society of Agronomy/Crop Science Society of America/Soil Science Society of America, Madison, Wisconsin, pp. 441–464.

Watson, R.N., Harris, S.L., Bell, N.L. and Neville, F.J. (1993) Pasture pests reduce white clover performance. *Proceedings of the Ruakura Farmers' Conference* 45, 57–61.

Watson, R.N., Neville, F.J. and Bell, N.L. (1996a) Insect pests associated with white and Caucasian clover in Bay of Plenty dairy pasture. *Proceedings of the New Zealand Plant Protection Conference* 49, 234–238.

Watson, R.N., Neville, F.J., Bell, N.L. and Harris, S.L. (1996b) Caucasian clover as a pasture legume for dryland dairying in the coastal Bay of Plenty. *Proceedings of the New Zealand Grassland Association* 58, 183–188.

Watson, R.N., Neville, F.J. and Bell, N.L. (1998) Caucasian clover performance in a year of severe drought. *Proceedings of the New Zealand Grassland Association* 60, 119–125.

Watson, R.N., Bell, N.L., Neville, F.J. and Davis, L.T. (2000) Pest populations during the first six years in ryegrass pastures containing white or Caucasian clover. *New Zealand Plant Protection* 53, 410–414.

West, C.P. and Steele, K.W. (1986) Tolerance of white clover cultivars to stem nematode (*Ditylenchus dipsaci*). *New Zealand Journal of Experimental Agriculture* 14, 227–229.

White, J. and Harper, J.L. (1970) Correlated changes in plant size and number of plant populations. *Journal of Ecology* 58, 467–485.

Williams, W.M. and Barclay, P.C. (1972) The effect of clover stem eelworm on the establishment of pure swards of white clover. *New Zealand Journal of Agricultural Research* 15, 356–362.

Williamson, P. and Cameron, R.A.D. (1976) Natural diet of the landsnail *Cepaea nemoralis*. *Oikos* 27, 493–500.

Willoughby, B.E., Goldson, S.L., Addison, P.J., Hardwick, S., Gerard, P.J. and Eerens, J.P.J. (1999) Clover root weevil (*Sitona lepidus*): the New Zealand response to a major biosecurity breach. In: Matthiessen, J.N. (ed.) *Proceedings of the 7th Australasian Conference on Grassland Invertebrate Ecology*. CSIRO, Wembley, pp. 35–42.

Wong, E. and Latch, G.C.M. (1971) Effect of fungal diseases on phenolic contents of white clover. *New Zealand Journal of Agricultural Research* 14, 633–638.

Woodman, R.F., Keoghan, J.M. and Allan, B.E. (1992) Pasture species for drought-prone lower slopes in the South Island high country. *Proceedings of the New Zealand Grassland Association* 54, 115–120.

Yamashita, Y., Jones, R.M. and Nicholson, C.H.L. (1979) Feeding of slugs (*Deroceras* sp. and *Lehmannia nyctelia*) on subtropical pasture species, particularly Kenya white clover (*Trifolium semipilosum*) cv. Safari. *Journal of Applied Ecology* 16, 307–318.

Yeates, G.W. (1977) Soil nematodes in New Zealand pastures. *Soil Science* 123, 415–422.

Appendix 18.1. Gastropod species occurring in New Zealand pastures.

Indigenous gastropods	
Athoracophoridae	
Athoracophorus bitentaculatus (Quoy & Gaimard)	Endemic to New Zealand. Widespread and common in indigenous and many modified environments in the North, South and Stewart Islands. While most abundant in indigenous forest and scrubland, populations often persist in unimproved and improved pasture
Punctidae	
Paralaoma caputspinulae (Reeve)	Widespread globally. In New Zealand this species occurs widely in the North Island, in the South Island south to mid-Canterbury, and in the Chatham Islands. Common in a variety of environments, especially dry coastal dunes and scrublands and dry lowland pastures
Introduced gastropods	
Agriolimacidae	
Deroceras reticulatum (Müller)	Present in disturbed environments throughout New Zealand, including Chatham, Auckland and Campbell Islands. Abundant in improved pastures and common in unimproved pastures, especially in moist districts
Deroceras panormitanum (Lessona & Pollonera)	Present in strongly disturbed environments throughout North, South and Stewart Islands, and the Kermadec and Chatham Islands. Abundant in improved pastures in moist districts, especially in northern North Island
Arionidae	
Arion distinctus Mabille	Locally common in disturbed habitats in North and South Islands. Occasionally of moderate abundance in improved pastures in the northern North Island
Arion hortensis de Férussac	Common in disturbed habitats, especially urban areas, but also penetrating pastures and modified indigenous forests in the North Island. Locally abundant in improved pastures in the northern North Island
Arion intermedius Normand	Present in the majority of indigenous and modified environments in the North, South, Stewart, Chatham and Auckland Islands. Common in both improved and unimproved pastures
Cochlicopidae	
Cochlicopa lubrica (Müller)	Common and locally abundant in modified environments in Kermedec and North Islands, and in the Nelson region of the South Island. Commonly present in pastures in the northern North Island but rarely abundant
Helicidae	
Cantareus aspersus (Müller)	Common and locally abundant in modified environments throughout North, South and Chatham Islands. Commonly present in pastures, especially in coastal North Island, but rarely abundant
Hygromiidae	
Candidula intersecta (Poiret)	Locally abundant in open, dry environments, including pastures, in North, South and Stewart Islands
Prietocella barbara (Linnaeus)	Widespread and locally abundant in open, dry environments, including pastures, in Northland, North Island. Recently established in improved pasture in the Waikato in central North Island

Appendix 18.1. Continued

Limacidae	
Lehmannia nyctelia (Bourguignat)	Widespread in modified environments in the North Island and the Nelson area of the South Island. Becoming increasingly prevalent in improved pastures of northern North Island
Limax maximus Linnaeus	Common in modified environments of North and South Islands. Not uncommon in improved pastures, especially in northern North Island, but rarely abundant
Milacidae	
Milax gagates (Draparnaud)	Common and locally abundant in improved pasture and arable land in North, South and Chatham Islands
Tandonia sowerbyi (de Férussac)	Widespread but uncommon in improved pasture in North and South Islands
Valloniidae	
Vallonia excentrica Sterki	Widespread and locally abundant in open, modified environments in the North Island and northern South Island. Especially abundant in improved pasture in northern North Island
Vertiginidae	
Vertigo ovata (Say)	Present in the North Island. Locally abundant in open, modified environments, including improved pasture
Zonitidae	
Oxychilus alliarius (Miller)	Common and locally abundant in modified environments, including improved pastures, in the North Island. Apparently restricted to urban areas in the South Island
Oxychilus cellarius (Müller)	Common and locally abundant in modified environments, including improved pastures, in the North and South Islands. Less common in unimproved grassland
Vitrea crystallina (Müller)	Widespread in North and South Islands. Rarely abundant, evidently restricted to calcareous and open environments, including pastures

Gastropod damage to oilseed rape, as with most other agricultural crops in western Europe, occurs mainly at establishment. Since gastropods were first identified as important pests of oilseed rape (Mallet and Bougaran, 1970), the extent of gastropod problems and their economic importance have increased considerably in this crop. The principal reasons for this are thought to be largely agronomic, including the growing of more susceptible cultivars, incorporation of crop residues into soil rather than removal by baling or burning, reduced cultivations and the tendency towards autumn sowing. In Great Britain, surveys by the Ministry of Agriculture show that the percentage of the area of oilseed rape treated with molluscicide ranged from 6% to 58% between 1977 and 1996 (Port and Port, 1986; Davis *et al.*, 1993; Garthwaite *et al.*, 1995; Thomas *et al.*, 1997). In western France, rape is routinely treated with molluscicide by a majority of farmers (Mouchart, 1984). The oilseed rape crop is second only to cereals in terms of the area treated with molluscicides in Britain (see Glen and Moens, Chapter 12, Table 12.1, this volume).

The consequences of failure of establishment in winter oilseed rape are more severe than in winter cereals because redrilling of rape is not a viable option, due to the fact that the seed must be sown no later than in the first 2 weeks of September. Thus it is particularly important in rape to maximize establishment success. It is critical that the understanding of gastropod biology and all the factors that could influence damage, are integrated and applied to minimize the seedling losses. Another important aspect of the gastropod problem in oilseed rape is that, as the crop develops, the dense canopy provides gastropods with a moist environment and an abundant food supply. Thus, gastropod populations can increase greatly during a single season when oilseed rape is grown. Winter wheat (*Triticum aestivum* Linnaeus) (Gramineae) normally follows oilseed rape in the crop rotation in western Europe and the first wheat crop after rape is at high risk from gastropod damage, as described by Glen and Moens (Chapter 12, this volume). Spring barley (*Hordeum vulgare* Linnaeus) (Gramineae) after oilseed rape in Norway can be totally defoliated by gastropods (Andersen, 1996).

Gastropod Species Responsible for Damage

As oilseed rape is grown mainly in rotation with cereals in western Europe, the main gastropod species responsible for damage are similar to those in cereals (see Glen and Moens, Chapter 12, this volume). Gastropods of the slug body form predominate and shelled gastropods (snails) are not considered to be pests, presumably because the latter have lower fitness in the disturbed agricultural ecosystems. The most widespread and abundant species in oilseed rape is *Deroceras reticulatum* (Müller) (Agriolimacideae). In some parts of Western Europe, such as Belgium, this species usually occurs alone in arable systems. Less frequently it is associated with small arionids, such as *Arion distinctus* Mabille and *Arion circumscriptus* Johnston. However, in England (Glen and Wiltshire, 1988) and France (Hommay, 1986, 1994; Chabert *et al.*, 1997) *D. reticulatum* is usually sympatric with one or more species of Arionidae in arable fields. Milacidae were found to be more localized in their distribution (Glen and Wiltshire, 1988; Chabert *et al.*, 1997).

Factors Affecting Gastropod Damage to Oilseed Rape

Gastropod damage to oilseed rape depends on three main factors; gastropod population density at establishment, gastropod feeding intensity activity in the seed-bed, and crop vulnerability during germination, emergence and the first-leaf stages.

Gastropod population density

Glen and Moens (Chapter 12, this volume) provide a detailed account of the factors influencing gastropod population density in West European arable fields. *D. reticulatum* and other gastropod species in these arable fields generally live in the upper 8–10 cm of soil, but retreat to greater deapth in the soil in order to survive when the upper layers of soil dry out (see Glen and Moens, Chapter 12, this volume). This can explain gastropod damage to rape seedlings even after dry seasons. The important influence of cultivation on gastropod population density and the percentage of rape seedlings subsequently killed by gastropod feeding has been demonstrated by Glen *et al.* (1996b) and Voss *et al.* (1998). Higher populations were recorded where there was no tillage or where non-inversion tillage was used than where soil was ploughed followed by subsequent cultvations to prepare a seed-bed. In experiments where each plot was subdivided and one subplot on each was treated with molluscicide while the other subplot was left untreated, molluscicide applications generally reduced gastropod activity and increased plant survival. On subplots treated with molluscicide in the experiment described by Glen *et al.* (1996b), there were no significant differences between different cultivation treatments in the numbers of rape seedlings that established. However, on untreated subplots, highly significant differences were observed, with the greatest numbers of plants on ploughed subplots (not significantly lower than on treated subplots), intermediate numbers on untreated subplots cultivated by non-inversion tillage and significantly fewest plants on untreated direct-drilled subplots. In accordance with these results, rape sown preharvest into cereal crops is at high risk of damage, because of high densities of gastropods in this situation (Tebrugge and Wagner, 1996).

In the experiment by Glen *et al.* (1996b), the reduction in numbers and biomass of rape seedlings on the untreated subplots, as a percentage of the value on treated subplots, was positively correlated with the biomass of gastropods m^{-2} recorded in samples from the upper 10 cm of soil in untreated subplots. Glen *et al.* (1996b) compared this relationship with that found for winter wheat in an earlier experiment (Glen *et al.*, 1989). This comparison is of interest because winter wheat is the most important crop affected by gastropod damage in western Europe and is thus the crop in which farmers have most experience in managing damage. In wheat, the relationship between % seed kill and gastropod biomass was strongly influenced by the percentage of fine-soil aggregates in the seed-bed and the depth at which the wheat seeds were sown. However, even in the conditions most conducive to damage to wheat (low % fine-soil aggregates and shallow sowing), the mortality caused by a given biomass of gastropods was substantially less than that caused to oilseed rape. Of course, the level of damage in each study was influenced by a variety of factors and the two studies are not, strictly speaking, directly comparable, However, the results point to a greater susceptibility of rape to gastropod damage in comparison with wheat.

Arion lusitanicus Mabille has been found in high abundance in wild-flower strips and grass headlands, with the large adults being present at the time of rape establishment in late August–early September (Frank, 1998b,c,d). Adults of this species can travel net mean distances of 3–8 m per night (Grimm *et al.*, 2000) while abundance declines with increasing distance from headlands (Frank, 1998a,b,c), they can be prevalent in rape during crop establishment. *A. lusitanicus* forage into rape fields at night, causing severe damage within 1–2 m of the edge of wild-flower strips (Frank, 1998c,d). By the time that wheat is sown, later in the autumn, the *A. lusitanicus* populations comprise, only juveniles (which are thought to be less mobile) and the trend of greater damage close to wild-flower strips is no longer apparent (Frank, 1998a). In most cases, *D. reticulatum* did not show a decline in densities at increasing distances from strips of semi-natural vegetation retained within the field (Frank, 1998a,c,d) except in one study where rape bordered two grass strips (Frank, 1998b).

Davies *et al.* (1997) found a positive relationship between the severity of gastropod damage to oilseed rape in autumn and overall weed ground cover. They also reported positive correlations of such damage with 5 years' set-aside and oilseed rape as a previous crop. It is likely that all three correlations with previous vegetative cover are indicative of conditions that favour high gastropod population densities.

Gastropod feeding activity

As explained earlier, modern cultivars of oilseed rape have been intentionally bred with low concentrations of glucosinolates in the seeds, in order to increase the palatability of the residue left after oil extraction as feed for farm animals (Ramans, 1989). For oilseed rape, the number of seedlings attacked and the leaf area destroyed by *D. reticulatum* has been shown to be inversely related to the concentration of glucosinolates in the seeds and their concentration in seedlings at the cotyledon stage (Glen *et al.*, 1990a; Moens *et al.*, 1992a). Glen *et al.* (1990a) found a strong correlation between the total concentration of glucosinolates in seeds and that in seedlings (hypocotyl and cotyledons). This suggests that glucosinolates in the seeds were transferred to the seedlings with little change in overall concentration, although there were increases and decreases in concentrations of certain individual glucosinolates (Glen *et al.*, 1990a). It is possible that certain individual glucosinolates may have more or less influence than others on palatability to *D. reticulatum*. However, cultivars with low overall concentrations of glucosinolates in the seed (6–12 mg g^{-1} of seed) were highly susceptible to damage, whereas cultivars with high glucosinolate concentrations (70–89 mg g^{-1} seed) were highly resistant. These observations strongly suggest that an important function of glucosinolates in rape seeds is to protect the young seedlings from being eaten by generalist herbivores, such as *D. reticulatum* and rabbits (*Oryctolagus cunniculus* Linnaeus; Leporidae) (Giamoustaris and Mithen, 1995).

Interestingly, Glen *et al.* (1990a) noted that counts of seedlings emerging in the presence or absence of gastropods indicated that the number of seedlings consumed in the initial stages of germination was not influenced by glucosinolate concentration. They proposed that gastropods did not initially avoid feeding on seedlings with a high concentration of glucosinolates, although they did so after a few days and with a high degree of discrimination. An alternative explanation for this initial feeding, that concentrations of glucosinolates were initially low in seedlings, is unlikely, because both seeds and young seedlings consist mainly of the cotyledons, which contain glucosinolates. The strength of the feeding deterrence to gastropods provided by glucosinolates in oilseed rape is indicated by the fact that, in the experiments of Glen *et al.* (1990a) and Moens *et al.* (1992a), *D. reticulatum* largely avoided feeding on seedlings with high glucosinolate concentrations, even when no alternative food was available. Indeed, damage to rape was not influenced by the presence of alternative food in the form of barley seedlings, which would often be available to gastropods feeding on winter oilseed rape in the field, as barley often precedes winter rape in the crop rotation (Glen *et al.*, 1990a). Byrne and Jones (1996) noted that high concentrations of glucosinolates in rape reduced feeding by *D. reticulatum* more rapidly than feeding by the non-pest species *Limacus pseudoflavus* (Evans) (Limacidae). *D. reticulatum* gained weight when fed on seedlings of rape with low concentrations of glucosinolates, but lost weight when fed on seedlings with high concentrations. *L. pseudoflavus* lost weight when fed on either type of rape.

In later stages (Zadok's growth stages > 25), older leaves of oilseed rape become less attractive and are less palatable to gastropods. Seedling vulnerability is very high during early stages of germination and emergence (GS 02–14). Indeed, at this stage, cutting of the hypocotyl or consumption of small quantities of plant tissue from the growing point or cotyledons causes the death of seedlings and gastropods can destroy many plants. Vulnerability decreases progressively during the first-true-leaf stages, because gastropods no longer attack vital organs, but they make holes in the leaves. Damage at this stage is serious when grazing is not compensated for by growth. In later stages GS (> 25 = formation of large leaves), seedlings are less palatable and attacks are generally limited to older leaves, so that plants are less susceptible to damage.

The duration of vulnerable seedling stages (germination, emergence and first-leaf stages) is prolonged by adverse weather and poor soil conditions (coarse macrostructure or 'capped' soils), but can be influenced by cultivar; slow-growing cultivars are at greater risk than cultivars that develop more quickly through susceptible stages (Moens *et al.*, 1992a). Crops with a low plant density (crops sown at a low drilling rate, or thinned by adverse weather or disease) are more vulnerable than densely sown healthy crops, which have the capacity to compensate for seedling losses. Thus, factors compromising plant fitness, such as late drilling or adverse growing conditions, will have an important influence on the severity of gastropod damage.

For cereal crops, especially wheat, accessibility of the seeds and the growing points of young shoots can be diminished if the seed is sown in a fine seed-bed favouring closure of the drill furrow. Under these conditions the gastropods are unable to readily locate the most vulnerable stages of the wheat crop (Glen and Moens, Chapter 12, this volume). Rape seeds, however, are protected in the first stages of germination (GS 00–02) by a strong outer coat; but the seedlings become progressively more accessible when the outer seed-coat is broken and the cotyledons and the growing point are pushed up to and above the soil surface (GS 03–14) (Moens, 1989). This pattern of vulnerability is consistent with the observation by Frank (1998d) that above-ground feeding by gastropods was of greater consequence than initial feeding below ground at the time of rape establishment. In contrast to monocotyledons, such as cereals, little can be done in terms of cultural measures to avoid gastropods gaining access to the vital organs, once seeds have germinated.

In laboratory experiments with 77 plant species, Briner and Frank (1998) found *A. lusitanicus* to prefer rape to all other species. Subsequent laboratory tests by Frank and Friedli (1999), with both *A. lusitanicus* and *D. reticulatum*, showed that rape was equalled only by *Capsella bursa-pastoris* (Linnaeus) Medikus (Brassicaceae) in vulnerability to defoliation. For *A. lusitanicus* only, *Taraxacum officionale* Weber (Asteraceae) was similar to rape in seedling vulnerability. In addition, the presence of *Veronica persica* Poiret (Scrophulariaceae) and *G. bursa-pastoris* significantly reduced the number of rape plants killed by both *A. lusitanicus* and *D. reticulatum* (Frank and Friedli, 1999). When exposed to a low density of *D. reticulatum* (10 m^{-2}) in the field, significantly more rape seedlings survived the first 4 weeks after emergence when seedlings of either

C. bursa-pastoris or *Stellaria media* (Linnaeus) Cirillo (Caryophyllaceae) were present than when rape seedlings were present alone (Frank and Barone, 1999). However, with *D. reticulatum* at a density of 20 m^{-2}, almost no rape plants survived, even in the presence of weeds.

Management of Gastropod Damage to Oilseed Rape

Integrated control of gastropod damage to rape seedlings can be achieved by means of cultural practices supported by chemical treatments. Biological control, especially predation by carabid beetles, is also relevant and is discussed below.

Cultural control

The following cultural practices are known to decrease the risk of gastropod damage: avoidance of crops favourable to gastropods in rotations, clearing of crop residues from fields after harvest, improvement of soil structure, seed-bed preparation and drilling in optimal circumstances and, finally, cultivation of less susceptible cultivars. Some of these practices are more desirable than others. Their advantages and disadvantages are discussed below.

Crop rotations have to be designed to fulfil multiple aims principally profitability but also maintenance of soil fertility, minimization of the risk of attack by diseases and pests, and to achieve efficient weed control (Jordan and Hutcheon, 1996). For this reason, farmers may well choose rotations with greater risks of gastropod damage than would be the case if gastropods were the only consideration (Glen *et al.*, 1996b). Oilseed-rape crops are normally grown in cereal-dominated crop rotations. Where cereal straw is incorporated, especially by non-inversion tillage, cereal-dominated rotations are favourable to gastropods (Glen *et al.*, 1996b). Because winter barley is harvested earlier than other cereals, it is often favoured as a crop to precede winter oilseed rape, which must be sown in late August or early September. Spring rape is generally at less risk than winter rape, but Jordan and Hutcheon (1996) reported that gastropod damage to spring oilseed rape was severe where cover crops had been grown over winter. As a result, they avoided growing cover crops before spring oilseed rape, despite the value of cover crops in restricting nitrogen losses from soil.

Removal of vegetation cover (weeds and volunteer seedlings germinating from the previous crop) and crop residues after harvest drastically decreases the food and shelter for gastropods. Moreover, a proportion of gastropods are destroyed when weeds and plant residues are physically removed from the soil surface by soil cultivations. As described by Glen and Moens (Chapter 12, this volume), incorporation of crop residues into soil provides a supplementary supply of food and shelter, which results in increased gastropod populations. While increased favourability for gastropods is a negative outcome from the perspective of pest management, the improvements in soil structure mean that incorporation is the preferred method of disposing of crop residues in integrated farming systems designed to combine profitable farming with environment protection (Jordan and Hutcheon, 1995). Improvements in soil structure also make it possible to prepare finer seed-beds. This is particularly important on heavy soils, where crevices and coarse aggregates furnish excellent shelter for gastropods. High-risk patches are located where there is a high clay content close to the surface – for example, on hillsides where the loam topsoil is thin, so that heavy clay is ploughed to the surface. In this context, non-inversion tillage methods may help to improve seed-bed structure and thus reduce the risk of poor establishment, even though such methods, as described earlier, may result in higher gastropod densities in the seed-bed compared with ploughing.

It follows from the above point that on heavy soils additional cultivations before drilling, and soil consolidation after drilling, are necessary to provide a finer and firmer seed-bed, which reduces gastropod activity and accelerates germination and growth. It should be borne in mind that consolidation is beneficial only when done on spongy, well-structured soils and in good weather conditions. Because of their small size, rape seeds are sown at shallow depth, just covered by soil or a little deeper in dry conditions (Pouzet, 1995). This contrasts with cereal seeds, which are normally sown at 30 mm depth (Wibberley, 1989). Because rape seeds are not vulnerable to gastropod damage, shallow sowing is not thought to increase the damage risk and may, by speeding germination, reduce the risk for rape. On coarse-textured seed-beds, a higher drilling rate is advisable, although this measure provides little protection during germination and emergence, as seedlings in these early stages are extremely vulnerable. In later stages, a dense vegetation cover is better able to resist gastropod attack because more leaves are produced and plants can thus compensate for leaf destruction. Optimal plant fitness in well-prepared seed-beds also maximizes the rate of leaf production and decreases crop vulnerability.

As discussed earlier, the palatability of oilseed-rape seedlings to gastropods is inversely related to the concentration of glucosinolates in seeds and seedlings. As modern cultivars contain low concentrations of glucosinolates, they are highly susceptible to gastropod damage. Selection of cultivars with rapid germination and growth, reducing the duration of the most critical stages, is the most appropriate response (Moens *et al.*, 1992a).

Observations on gastropod damage to oilseed rape (Speiser, 1999) suggested that fresh, anaerobically digested organic material from a biogas production plant was molluscicidal. Laboratory studies confirmed that fresh material killed *A. lusitanicus*, *A. distinctus* and *D. reticulatum* in the laboratory. However, molluscicidal properties were rapidly lost in storage and also after application in the field, and this is not considered to be a reliably effective, practical method of control.

Biological control

Natural enemies (predators, parasites and pathogens) of gastropods are described by Moens *et al.* (1992b) and in a related volume (Barker, 2002). Mass rearing of predators such as carabid beetles to control gastropod damage in oilseed rape would not be practical. Such predators can best be exploited by providing them with conditions under which they can have a maximum impact on gastropod populations. However, greater understanding is needed of the interactions between these beetles and their prey before the contribution of these generalist predators to integrated control of gastropods can be achieved. Studies in an oilseed rape crop in July–September, before and after harvest, have shown that gastropods are important prey for adults of *Pterostichus melanarius* (Illiger) (Symondson *et al.*, 1996). This species is probably the commonest and most widespread large carabid beetle in arable land in western Europe. Further information on the impact of carabid beetles on gastropod populations in arable fields is given by Glen and Moens (Chapter 12, this volume). It is important to note here that the main period of activity of adult *P. melanarius* is from June to September, i.e. for *c.* 3 months up to the time when winter oilseed rape crops are sown in late August/early September and for *c.* 1 further month during the critical period of crop establishment. For this reason, and based on the evidence reviewed by Glen and Moens (Chapter 12, this volume), conservation measures for *P. melanarius* could be particularly valuable in integrated control of gastropod damage to oilseed rape. Since carabid beetles are killed by methiocarb-based molluscicides, whereas metaldehyde-based molluscicides are harmless (Kennedy, 1990), it may be preferable to use the latter molluscicide at establishment of winter rape.

One possible selective means of controlling gastropod damage is the inundative application of the gastropod-parasitic rhabditid nematode *Phasmarhabditis hermaphrodita* (Schneider) to soil. The biology of this nematode and its use as a biological control agent are summarized by Glen *et al.* (1996a), Glen and Wilson (1997) and Morand *et al.* (2002). The nematode can significantly reduce gastropod damage to oilseed rape when it is applied to soil at the time of sowing or 1–2 weeks before sowing rape (Wilson *et al.*, 1995; Speiser and Andermatt, 1996). Because this nematode requires moisture for survival and because the soil surface often dries rapidly during and after seed-bed preparation for oilseed rape, there may be reduced efficacy of nematodes applied to the soil surface (Wilson *et al.*, 1995, 1996). However, this problem can be diminished by working the soil to a depth of 5–10 cm immediately after nematodes are applied, in order to incorporate them into the surface-soil layers (Wilson *et al.*, 1996). This mixing of nematodes with soil can be done during normal cultivations for seed-bed preparation and does not require any additional operations. Despite the technical feasibility of using this nematode for control of gastropod damage in oilseed rape, its high cost in comparison with chemical molluscicides will preclude its use for control of gastropod damage in arable crops until the economics of usage change significantly (Wilson *et al.*, 1995; Glen and Wilson, 1997). Application of *P. hermaphrodita* to soil when rape was sown did not affect the subsequent development of gastropod populations in rape crops (Wilson *et al.*, 2000). Thus, inoculative application of nematodes to prevent build-up of gastropod populations in rape crops does not seem to be feasible.

Chemical control

Chemical molluscicides should be used as part of an integrated-control approach to gastropod damage on oilseed rape in order to avoid or reduce the severity of gastropod attacks during the crucial vulnerable early growth stages (GS 02–25). In high-risk situations, molluscicidal baits should be broadcast immediately after drilling or just before susceptible growth stages appear in order to kill the gastropods active on the soil surface during this vulnerable phase of the crop. The importance of this is demonstrated by the results of a trial in France (Anon., 1995). Molluscicidal bait pellets applied at crop emergence increased the number of surviving plants from 3 m^{-2} to 15 m^{-2}. Pellets applied at drilling gave better results (33 plants m^{-2}), while pellets applied at drilling and crop emergence gave the best results (42 plants m^{-2}). Treatments before cultivation are likely to be less effective, as they are for wheat (see Glen and Moens, Chapter 12, this volume). In France (Anon., 1997), it is recommended that, in high-risk areas, treatment at sowing should be followed by surveillance at 5–6-day intervals until the crop four-leaf stage, with molluscicide applications repeated as necessary. Similar recommendations have been made by Voss *et al.* (1998).

Where there is a lower risk of gastropod damage, it is possible to wait until seedling emergence to assess the degree of gastropod damage before pellets are applied. However, the danger of waiting until emergence is that gastropods may eat off all growing points and cotyledons as soon as they appear, so that severe gastropod damage may be mistaken for slow emergence. Moreover, severe gastropod damage can be caused by relatively few gastropods, as described earlier. Some crops drilled at high densities in optimal conditions and covering the soil with a dense luxuriant canopy can resist moderate gastropod attacks without treatment, because seedling destruction and leaf consumption are compensated for by growth. In such cases, it will be important to assess regularly the severity of gastropod attack, with control decision depending on damage severity and plant fitness: experience indicates that treatment is not justified if more than 50 healthy seedlings m^{-2} are present.

As explained earlier, oilseed rape bordering wild-flower strips has been shown to be at particular risk from gastropod damage, especially by *A. lusitanicus*. Severe damage close to wild-flower strips can be effectively controlled by broadcasting molluscicide pellets (Frank, 1998c) and their use can be restricted to a band 50 cm wide adjacent to the strips (Friedli and Frank, 1998).

Methanol extracts of two plant species, *Saponaria officionalis* Linnaeus (Caryophyllaceae) and *Valerianella locusta* (Linnaeus) Laterrade (Valerianaceae), applied to the cotyledons of oilseed rape, have been shown to deter feeding by *A. lusitanicus* (Barone and Frank, 1999). This suggests that antifeedant plant extracts could make a useful contribution to integrated control of gastropod damage to oilseed rape, but problems of volatility, persistence and cost need to be solved (Barone and Frank, 1999). As mentioned previously, Frank and Barone (1999) found that the presence of either *C. bursa-pastoris* or *S. media* protected rape seedlings from attack by *D. reticulatum* at low density (10 m^{-2}), but not at 20 m^{-2}. In contrast, the efficacy of metaldehyde pellets was independent of *D. reticulatum* density over this range.

Conclusions

The severity of gastropod damage to oilseed rape at establishment in western Europe has increased since the 1970s for a number of reasons. Among the most important of these has been the European Community (EC) policy to favour cultivars with low concentrations of glucosinolates in the seeds, which has substantially diminished this crop's natural chemical defences and thus rendered the seedlings considerably more palatable to the main pest species, *D. reticulatum*. Over the same period, other agronomic changes have provided favourable conditions for gastropod populations. These practices include incorporation of residues of the previous crop into the soil (rather than burning *in situ* or removal), cover cropping (to reduce nitrate leaching) and the deployment of techniques of reduced or zero tillage. Improved weed control may also have contributed to increased damage, by removing weed seedlings, some of which are alternative foods for gastropods. Weather conditions in recent years (mild winters and wet summers) have also been favourable to high gastropod population densities.

Optimum control of gastropod damage to oilseed rape is achieved through integrated cultural and chemical control. Cultivation, especially by ploughing, decreases gastropod populations and therefore damage potential. Preparation of fine, firm seed-beds reduces gastropod population densities and promotes rapid germination of oilseed rape. Choice of faster-growing cultivars reduces the time period during which crops are at greatest risk. Molluscicidal treatments should be applied only when necessary and should be cost-effective. However, it is difficult for farmers and their advisers to take rational decisions on the need for control measures because, for best results, it is necessary to apply molluscicide before damage is visible. However, it is not possible at present to forecast accurately the severity of gastropod damage. Moreover, there are few data on the influence of gastropod damage on yield of oilseed rape. Predictions of gastropod damage risk and the need for control measures need to be considerably improved, with a system based on weather data, gastropod population densities, soil type, cultural practices and other important parameters.

In crops with a high risk of gastropod damage, molluscicidal bait pellets should be applied at drilling or, at latest, before germinating seedlings have started to emerge. The crop should then be monitored at 5–6-day intervals until the fourth-true-leaf stage, in order to assess whether further treatments may be beneficial. While inundative biological control of pests in oilseed rape is technically feasible, it is not economically viable at present. However, recent results suggest that natural biological control by predatory carabid beetles, especially *P. melanarius*, may restrict the growth of gastropod populations during the summer period (June to August) immediately before winter oilseed rape crops are established. Since *P. melanarius* are still active at the time of rape establishment, it may be beneficial to use metaldehyde-based molluscicides for control of gastropod damage to rape, as such molluscicides are known to be harmless to carabid beetles.

Acknowledgements

IACR receives grant-aided support from the Biotechnology and Biological Sciences Research Council of the UK.

References

Andersen, A. (1996) Naturally occurring and introduced slug and snail pests in Norway. In: Henderson, I. (ed.) *Slug and Snail Pests in Agriculture.* Symposium Proceedings No. 66, British Crop Protection Council, Farnham, pp. 333–334.

Anon. (1995) Limaces et oléagineux. *Oléoscope, Bulletin du CETIOM* 28, 11–18.

Anon. (1997) Les insecticides du colza. *Cultivar* 426, 36–39.

Barker, G.M. (ed.) (2002) *Natural Enemies of Terrestrial Molluscs.* CAB International, Wallingford, UK.

Barone, M. and Frank, T. (1999) Effects of plant extracts on the feeding behaviour of the slug *Arion lusitanicus. Annals of Applied Biology* 134, 341–345.

Bowman, J.G. (1989) Breeding for vegetable oils in Northern Europe. In: *Production and Protection of Oilseed Rape and Other Brassica Crops.* Aspects of Applied Biology 23, pp. 31–35.

Briner, T and Frank, T. (1998) The palatability of 78 wildflower strip plants to the slug *Arion lusitanicus. Annals of Applied Biology* 133, 123–133.

Bunting, E.S. (1984) Oilseed rape in perspective: with particular reference to crop expansion and distribution in England 1973–1983. In: *Agronomy, Physiology, Plant Breeding and Crop Protection of Oilseed Rape*. Aspects of Applied Biology 6, pp. 11–21.

Byrne, J. and Jones, P. (1996) Responses to glucosinolate content in oilseed rape varieties by crop pest (*Deroceras reticulatum*) and non-pest slug species (*Limax pseudoflavus*). In: *Tests of Agrochemicals and Cultivars 17, Annals of Applied Biology* 128 (suppl.), 78–79.

Chabert, A., Guinot, J. and Tisseur, M. (1997) Suivi des populations de limaces au champ. Vers la mise au point d'un système de prévision des risques. *Phytoma* 50, 16–20.

Davies, D.H.K., McRoberts, N., Foster, G.N., Evans, K.A., Whytock, G.P., McKinlay, R., Wale, S. and Hunter, A. (1997) The COIRE project: methods to investigate the interaction of weed distribution with geography, management and pests. *Aspects of Applied Biology* 50, 441–450.

Davis, R.P., Thomas, M.R., Garthwaite, D.G. and Bowen, H.M. (1993) *Arable Farm Crops in Great Britain 1992*. Pesticide Usage Survey Report 108, MAFF, London.

Evans, K.A. and Scarisbrick, D.H. (1994) Integrated insect pest management in oilseed rape crops in Europe. *Crop Protection* 13, 403–412.

Frank, T. (1998a) Slug damage and number of slugs (Gastropoda: Pulmonata) in winter wheat in fields with sown wildflower strips. *Journal of Molluscan Studies* 64, 319–328.

Frank, T. (1998b) Slug damage and numbers of slugs in oilseed rape bordering on grass strips. *Journal of Molluscan Studies* 64, 461–466.

Frank, T. (1998c) Slug damage and numbers of the slug pests, *Arion lusitanicus* and *Deroceras reticulatum*, in oilseed rape grown beside sown wildflower strips. *Agriculture, Ecosystem and Environment* 67, 67–78.

Frank, T. (1998d) The role of different slug species in damage to oilseed rape bordering on sown wildflower strips. *Annals of Applied Biology* 133, 483–493.

Frank, T. and Barone, M. (1999) Short-term field study on weeds reducing slug feeding on oilseed rape. *Zeitschrift für Pflanzenkrankheiten und Pflanzenschutz* 106, 534–538.

Frank, T. and Friedli, J. (1999) Laboratory food choice trials to explore the potential of common weeds to reduce slug feeding on oilseed rape. *Biological Agriculture and Horticulture* 17, 19–29.

Friedli, J. and Frank, T. (1998) Reduced applications of metaldehyde pellets for reliable control of the slug pests *Arion lusitanicus* and *Deroceras reticulatum* in oilseed rape adjacent to sown wildflower strips. *Journal of Applied Ecology* 35, 504–513.

Garthwaite, D.G., Thomas, M.R. and Hart, M. (1995) *Arable Farm Crops in Great Britain 1994*. Pesticide Usage Survey Report 127, MAFF, London.

Giamoustaris, A. and Mithen, R. (1995) The effect of modifying the glucosinolate content of oilseed rape (*Brassica napus* ssp. *oleifera*) on its interaction with specialist and generalist pests. *Annals of Applied Biology* 126, 347–363.

Glen, D.M. and Wilson, M.J. (1997) Slug-parasitic nematodes as biocontrol agents for slugs. *Agro-Food-Industry Hi-Tech* 8(2), 23–27.

Glen, D.M. and Wiltshire, C.W. (1988) Distribution of slug species. In: *Report of IACR-Long Ashton Research Station for 1987*, p. 28.

Glen, D.M., Milsom, N.F. and Wiltshire, C.W. (1989) Effects of seed bed conditions on slug numbers and damage to winter wheat in a clay soil. *Annals of Applied Biology* 115, 177–190.

Glen, D.M., Jones, H. and Fieldsend, J.K. (1990a) Damage to oilseed rape seedlings by the field slug *Deroceras reticulatum* in relation to glucosinolate concentrations. *Annals of Applied Biology* 117, 197–207.

Glen, D.M., Milsom, N.F. and Wiltshire, C.W. (1990b) Effect of seed depth on slug damage to winter wheat. *Annals of Applied Biology* 117, 693–701.

Glen, D.M., Wilson, M.J. and Hughes, L. (1996a) Exploring and exploiting the potential of the rhabditid nematode *Phasmarhabditis hermaphrodita* as a biocontrol agent for slugs. In: Henderson, I. (ed.) *Slug and Snail Pests in Agriculture*. Symposium Proceedings No. 66, British Crop Protection Council, Farnham, pp. 271–280.

Glen, D.M., Wiltshire, C.W., Walker, A.J., Wilson, M.J. and Shewry, P.R. (1996b) Slug problems and control strategies in relation to crop rotations. In: *Rotation and Cropping Systems, Aspects of Applied Biology* 47, pp. 153–160.

Grimm, B., Paill, W. and Kaiser, H. (2000) Daily activity of the pest slug, *Arion lusitanicus* Mabille. *Journal of Molluscan Studies* 66, 125–130.

Hommay, G. (1986) Les différentes espéces de limaces présentés en grandes cultures. *Phytoma,* February, 19–22.

Hommay, G. (1994) Contribution à la biologie et à l'écologie des limaces (Mollusques Gastéropodes Pulmonés) de grandes cultures. Thèse de Doctorat en Sciences Biologiques, Rennes.

Jordan, V.W.L. and Hutcheon, J.A. (1995) Less intensive farming and the environment: an integrated farming systems approach for UK arable crop production. In: Glen, D.M. Greaves, M.P. and Anderson, H.M. (eds) *Ecology and Integrated Farming Systems.* John Wiley & Sons, Chichester, pp. 307–318.

Jordan, V.W.L. and Hutcheon, J.A. (1996) Multifunctional crop rotation: the contributions and interactions for integrated crop protection and nutrient management in sustainable cropping systems. In: *Rotations and Cropping Systems. Aspects of Applied Biology* 47, pp. 301–308.

Kennedy, P.J. (1990) The effects of molluscicides on the abundance and distribution of ground beetles (Coleoptera, Carabidae) and other invertebrates. PhD thesis, University of Bristol.

Kimber, D.S. and McGregor, D.I. (1995) The species and their origin, cultivation and world production. In: Kimber, D.S. and McGregor, D.I. (eds) Brassica *Oilseeds, Production and Utilization.* CAB International, Wallingford, pp. 1–7.

Mallet, C. and Bougaran, H. (1970) Dégâts de limaces dans les colzas. In: *Journées Internationales sur le Colza*, pp. 245–249.

Moens, R. (1989) Factors affecting slug damage and control measure decisions. In: Henderson, I.F. (ed.) *Slugs and Snails in World Agriculture.* Monograph No. 41, British Crop Protection Council, Thornton Heath, pp. 227–236.

Moens, R., Couvreur, R. and Cors, F. (1992a) Influence de la teneur en glucosinolates des variétés de colza d'hiver sur les dégâts de limaces. *Bulletin Recherches Agronomiques Gembloux* 27, 289–307.

Moens, R., Latteur, G. and Fayt, E. (1992b) Contribution to an integrated slug control. *Parasitica* 48, 83–105.

Morand, S., Wilson, M.J. and Glen, D.M. (2002) Nematode parasites. In: Barker, G.M. (ed.) *Natural Enemies of Terrestrial Molluscs.* CAB International, Wallingford, UK.

Mouchart, A. (1984) Les attaques de limaces des grandes cultures. *Phytoma* November, 49–52.

Port, C.M. and Port, G.R. (1986) The biology and behaviour of slugs in relation to crop damage and control. *Agricultural Zoology Reviews* 1, 255–299.

Pouzet, A. (1995) Agronomy. In: Kimber, D.S. and McGregor, D.I. (eds) Brassica *Oil-seeds, Production and Utilization.* CAB International, Wallingford, pp. 65–92.

Ramans, M. (1989) Oilseed rape breeding for politics or varietal improvement? In: *Production and Protection of Oilseed Rape and Other Brassica Crops. Aspects of Applied Biology* 23, pp. 47–56.

Speiser, B. (1999) Molluscicidal and slug-repellent properties of anaerobically digested organic matter. *Annals of Applied Biology* 135, 449–455.

Speiser, B. and Andermatt, M. (1996) Field trials with *Phasmarhabditis hermaphrodita* in Switzerland. In: Henderson, I. (ed.) *Slug and Snail Pests in Agriculture.* Symposium Proceedings No. 66, British Crop Protection Council, Farnham, pp. 419–424.

Symondson, W.O.C., Glen, D.M., Wiltshire, C.W., Langdon, C.J. and Liddell, J.E. (1996) Effects of cultivation techniques and methods of straw disposal on predation by *Pterostichus melanarius* (Coleoptera: Carabidae) upon slugs (Gastropoda: Pulmonata) in an arable field. *Journal of Applied Ecology* 33, 741–753.

Tebrugge, F. and Wagner, M. (1996) Vorerntesaat von Raps in Getreidebestande. Erste Ergebnisse und Erfahrungen. *Landtechnik* 51, 192–193.

Thomas, M.R., Garthwaite, D.G. and Banham, A.R. (1997) *Arable Farm Crops in Great Britain 1996.* Pesticide Usage Survey Report 141, MAFF, London.

Voss, M.C., Ulber, B. and Hoppe, H.H. (1998) Impact of reduced and zero tillage on activity and abundance of slugs in winter oilseed rape. *Zeitschrift für Pflanzenkrankheiten und Pflanzenschutz* 105, 632–640.

Weiss, E.A. (1983) *Oilseed Crops.* Longman, Harlow, UK, 660 pp.

Wibberley, E.J. (1989) *Cereal Husbandry.* Farming Press, Ipswich, UK, 258 pp.

Wilson, M.J., Hughes, L.A. and Glen, D.M. (1995) Developing strategies for the nematode, *Phasmarhabditis hermaphrodita*, as a biological control agent for slugs in integrated crop management systems. In: *Integrated Crop Protection: Toward Sustainability?* Symposium Proceedings No. 63, British Crop Protection Council, Farnham, pp. 33–40.

Wilson, M.J., Hughes, L.A., Hamacher, G.M., Barahona, L.D. and Glen, D.M. (1996) Effects of soil incorporation on the efficacy of the rhabditid nematode, *Phasmarhabditis hermaphrodita*, as a biological control agent for slugs. *Annals of Applied Biology* 128, 117–126.

Wilson, M.J., Hughes, L.A., Hamacher, G.M. and Glen, D.M. (2000) Effects of *Phasmarhabditis hermaphrodita* on non-target molluscs. *Pest Management Science* 56, 711–716.

Index